GENETIC ENGINEERING OF PLANTS

An Agricultural Perspective

BASIC LIFE SCIENCES

Alexander Hollaender, General Editor

Associated Universities, Inc., Washington, D.C.

Recent volumes in the series:

Volume 16 • **DEVELOPMENT AND NEUROBIOLOGY OF DROSOPHILA**
Edited by O. Siddiqi, P. Babu, Linda M. Hall, and Jeffrey C. Hall

Volume 17 • **GENETIC ENGINEERING OF SYMBIOTIC NITROGEN FIXATION**
Edited by J. M. Lyons, R. C. Valentine, D. A. Phillips, D. W. Rains, and R. C. Huffaker

Volume 18 • **TRENDS IN THE BIOLOGY OF FERMENTATIONS FOR FUELS AND CHEMICALS**
Edited by Alexander Hollaender, Robert Rabson, Palmer Rogers, Anthony San Pietro, Raymond Valentine, and Ralph Wolfe

Volume 19 • **GENETIC ENGINEERING OF MICROORGANISMS FOR CHEMICALS**
Edited by Alexander Hollaender, Ralph D. DeMoss, Samuel Kaplan, Jordan Konisky, Dwayne Savage, and Ralph S. Wolfe

Volume 20 • **MOLECULAR AND CELLULAR MECHANISMS OF MUTAGENESIS**
Edited by J. F. Lemontt and W. M. Generoso

Volume 21 • **GENETIC TOXICOLOGY: An Agricultural Perspective**
Edited by Raymond A. Fleck and Alexander Hollaender

Volume 22 • **PLANT CELL CULTURE IN CROP IMPROVEMENT**
Edited by S. K. Sen and Kenneth L. Giles

Volume 23 • **INDUCED MUTAGENESIS: Molecular Mechanisms and Their Implications for Environmental Protection**
Edited by Christopher W. Lawrence

Volume 24 • **ORGAN AND SPECIES SPECIFICITY IN CHEMICAL CARCINOGENESIS**
Edited by Robert Langenbach, Stephen Nesnow, and Jerry M. Rice

Volume 25 • **BASIC BIOLOGY OF NEW DEVELOPMENTS IN BIOTECHNOLOGY**
Edited by Alexander Hollaender, Allen I. Laskin, and Palmer Rogers

Volume 26 • **GENETIC ENGINEERING OF PLANTS: An Agricultural Perspective**
Edited by Tsune Kosuge, Carole P. Meredith, and Alexander Hollaender

---

A Continuation Order Plan is available for this series. A continuation order will bring delivery of each new volume immediately upon publication. Volumes are billed only upon actual shipment. For further information please contact the publisher.

GENETIC ENGINEERING OF PLANTS

An Agricultural Perspective

Edited by

Tsune Kosuge and Carole P. Meredith

University of California
Davis, California

and

Alexander Hollaender

Associated Universities, Inc.
Washington, D.C.

Technical Editor
Claire M. Wilson

PLENUM PRESS • NEW YORK AND LONDON

Library of Congress Cataloging in Publication Data

Main entry under title:

Genetic engineering of plants.

(Basic life sciences; v. 26)
"Proceedings of a symposium held August 15-19, 1982, at the University of California, Davis, California"—T.p. verso.
Includes bibliographical references and index.
1. Plant genetic engineering—Congresses. 2. Plant-breeding—Congresses. I. Kosuge, Tsune. II. Meredith, Carole. III. Hollaender, Alexander.
SB123.G376 1983 631.5'2 83-4102
ISBN-13: 978-1-4684-4546-6 e-ISBN-13: 978-1-4684-4544-2
DOI: 10.1007/978-1-4684-4544-2

Proceedings of a symposium held August 15-19, 1982, at the University of California, Davis, California

Softcover reprint of the hardcover 1st edition 1983

A Division of Plenum Publishing Corporation
233 Spring Street, New York, N.Y. 10013

DEDICATION

This volume is dedicated to Dr. Elijah Romanoff, who retired from the National Science Foundation in the Fall of 1982 after thirteen years of service, most recently as Program Director for Metabolic Biology.

Dr. Romanoff is a scientist with unusually broad interests and vision. His feeling for and efforts in behalf of plant sciences have encouraged many of us to become involved in this important field. The rapid progress of the application of genetic engineering, as illustrated in this volume, gives us an opportunity to express our appreciation of Dr. Romanoff's encouragement - particularly when it was difficult to obtain support from other sources - and his deep interest in the basic aspects of applied science. We speak for many in this field of research when we wish Dr. Romanoff many years of continued productive involvement in basic biology.

The Editors

ACKNOWLEDGEMENT

The editors gratefully acknowledge the generous cooperative support of the following contributors: Agrigenetics Corporation; Allied Chemical; Arco Plant Cell Research Institute; Asgrow Seed Company; Calgene, Inc.; Cal Crop Improvement Association; Cal Planting Cotton Seed Distributors; Campbell Soup Company; CETUS Corporation; CIBA-GEIGY Corporation; DeKalb Ag Research, Inc.; DNA Plant Technology Corporation; Ford Foundation; Funk Seed International; Goldsmith Seeds, Inc.; International Plant Research Institute; Martin-Marietta Corporation; Monsanto Company; Moran Seeds, Inc.; National Science Foundation; North American Plant Breeders; Northrup King Company; Petoseeds Company, Inc.; Pfizer Central Research; Pioneer Hi-Bred International, Inc.; Plant Genetics, Inc.; Rockefeller Foundation; Standard Oil Company; Sungene Technologies Corporation; and W-L Research, with the U.C. College of Agricultural and Environmental Sciences and the Food Protection and Toxicology Center, who made our gathering of over 700 scientists from more than 30 countries possible at Davis.

It is our speakers, the authors of this volume, who present an exciting variety of topics in modern plant engineering.

We appreciate the editorial efforts of the session chairmen: Robert Allard, Albert Ellingboe, Kenneth Frey, C. Edward Green, Michael Holland, Eugene Nester, Calvin Qualset, D. William Rains, Donald Rasmusson, Charles Rick, Raymond Rodriguez, Robert Shepherd, William Taylor, and William Timberlake who assisted in the review of these chapters.

In order to prepare and coordinate the on-site details, we are thankful for the expertise of Ms. Carroll Miller, Ms. Carolyn Norlyn, and their assistants who tirelessly accommodated our symposium requirements.

The production crew - Larry West, Sarah Anders, Diane Fink, and Linda Pilkington - together with the staff of Plenum Press, have joined to rapidly publish these proceedings on Genetic Engineering of Plants: An Agricultural Perspective.

The attractive logo, prepared for our printed materials, was designed by Jan Conroy under the art direction of Rob Maddox, both of the Repro Graphics Department, U.C. Davis.

SCIENTIFIC ORGANIZING COMMITTEE: Carole Meredith and Tsune Kosuge with George Bruening, J. Eugene Fox, Alexander Hollaender, Roy Huffaker, Clarence Kado, Tom Orton, D. William Rains, S. R. Snow, and Raymond Valentine, and their colleagues.

CONTENTS

OPENING REMARKS

Charles E. Hess, Dean
College of Agricultural and Environmental Sciences
University of California
Davis, California 95616

It is a real pleasure to welcome you to the Davis Campus and to the Conference on "Genetic Engineering of Plants".

The Davis Campus was established in 1905 as the farm for the University of California, Berkeley agriculture students. Gradually the Campus evolved with the development of a four-year program in agriculture in 1926, a School of Veterinary Medicine in 1946, a College of Letters and Science in 1951, a College of Engineering in 1962, a School of Law in 1966, a School of Medicine in 1968, and finally, in 1981, a School of Administration. There are approximately 19,000 undergraduate and graduate students at Davis living and learning in an idyllic college town of 30,000 with at least 29,000 bicycles!

We feel it is very appropriate to hold a conference on genetic engineering of plants in California. First, we have a food and fiber producing system whose annual farm-gate value is between 13-14 billion dollars from some 250 different commodities.

Second, we are in a state in which the public is very concerned about the environment and the potential detrimental effects from the use of chemicals in agriculture. We are currently faced with the problem of nematodes attacking grape root stocks and for which there is no longer an effective chemical available. There are some 600,000 acres of grapes in California so the problem is a big one. One hope is that a gene or genes which can impart resistance will be found through tissue culture or transferred into grape root stocks by other techniques. This approach of course holds much promise for many other insect and disease problems in agriculture.

Another major problem in California is water availability and quality. High salt levels in irrigation water and soil limits potential productivity. So there is great interest in osmoregulation as an area of inquiry and in the potential to develop plants which will tolerate high salt and other conditions of stress.

As you are aware, we have been intensely interested in the area of biological nitrogen fixation because of the possibility of increasing the efficiency of existing systems as well as extending the range of crops which fix nitrogen. We wish to reduce our dependence upon synthetic nitrogen fertilizers as well as to reduce the potential for groundwater pollution. The point I am making is that there is a huge market for new ideas and new and improved plants coupled with an agriculture industry which is ready and able to apply those new ideas and plants.

Another reason I feel it is appropriate to hold a conference on genetic engineering of plants here is the fact that the Davis Campus has a research continuum from molecular biology to plant breeding and field evaluation. The Conference is designed to provide that blend and interaction. The special issue of California Agriculture (Univ. Cal., Berkeley. Vol. 36 #8, Aug., 1982) reflects that blend. Also, I have noticed that a number of the new genetic engineering firms have endorsed this concept by associating with seed companies not only to deliver new products, but also to provide the intellectual stimulation between the fields of basic and applied research.

I want to especially thank Carole Meredith and Tsune Kosuge, coordinators of the Conference, and Carroll Miller, Coordinator of Special Events, for the planning and organization of the Conference. We warmly thank the sponsors of the Conference listed in this volume for their generous support which has made the Conference fiscally sound. And, finally, I want to express a special tribute to Dr. Alexander Hollaender who is a continuing source of inspiration and encouragement for conferences such as this one. Without him, we would not be here today.

I hope that you enjoy your days in Davis and that you leave with many new ideas for future research and the application of the knowledge that the research generates.

AN OVERVIEW OF CROP IMPROVEMENT:

CHAIRMAN'S INTRODUCTION

Kenneth J. Frey

Department of Agronomy
Iowa State University
Ames, Iowa 50011

Plant breeding is really older than sedentary agriculture. This statement is based on the fact that non shattering plant types that produced quite well in single-crop stands must have been available to these early farmers. Such strains were selected from among the natural variants that nature provided. What is astounding, however, is the fact that the plant breeding technology used by those early farmers to select crop genotypes for sedentary agriculture 10,000 to 12,000 years ago was still in use with few refinements until 1900--i.e., just 80 years ago. A number of researchers such as Camerarius, Weissman, De Vries, etc. made many discoveries about sex and heredity in plants before 1900, but it was the rediscovery of Mendel's Laws of Heredity in 1901 that ushered in the idea that humankind could recombine genes from natural sources into combinations that were desired for agricultural or horticultural production, but could not be found in natural populations.

Thus, the two phases of plant breeding, i.e., creating genotypic variability and selection among genotypes, came into place eight decades ago. Since then, plant breeding has been honed and refined in three areas of activity: (a) selection techniques, procedures, and designs have been innovated and refined; (b) vast quantities of natural genetic variability have been collected and stored for past, present, and future use in developing better adapted and more productive genotypes; and (c) certain techniques such as colchicine treatment, embryo culture, induced mutation, etc. have expanded the breadth of germplasm that the plant breeder could use. Concurrently, geneticists, biochemists, and cell biologists over the past three decades have been using sophisticated instruments and viruses and microbes to learn about the genetic material, DNA, and how it functions to produce phenotypes. But to date, plant breeders have made little use of this knowledge about heredity learned from research on viruses and microbes. Several of these chapters, authored by biochemists, cell biologists, and geneticists, explain this newly gained knowledge about genetics and how it might be used in plant breeding for the future.

To set the stage for this volume on Genetic Engineering, Professor N.W. Simmonds has been requested to give us an assessment of the "Plant Breeding State of the Art."

PLANT BREEDING: THE STATE OF THE ART

N.W. Simmonds

Edinburgh School of Agriculture
West Mains Road
Edinburgh EH9 3JG
Scotland

INTRODUCTION

The readership of this book is a varied one. Some will know a great deal about plant breeding but little about genetic engineering; others will have reciprocal knowledge. Yet there is clearly an area of overlap of interests which it is the object of the book to explore. I am honored by having been invited to discuss plant breeding in an introductory way, to help to set the stage, so to speak, for the more detailed discussions to follow. In doing so, I am conscious that I must concentrate upon the wider issues of plant breeding, not the details, and that I cannot evade the responsibility of stating some kind of view as to the probable place of genetic engineering. My treatment is inevitably broad and I have taken the view that a little provocation here and there will not be amiss. With this approach in mind I shall first outline the general character of plant breeding (with apologies to the professionals in the audience who know it all already) and then enumerate what I take to be the principal problem areas of the subject, the areas in which we really need much more hard thinking, calculation and experiment.

Recent general works on plant breeding, including crop-by-crop outlines, include references 17, 18, 28, 35, 36, and 39.

GENERAL FEATURES OF PLANT BREEDING

Plant breeding has, I suggest, the following main features.

(1) It is a science-based technology directed towards economic objectives. It is nothing if not economically successful and it has long since passed the stage of being an "art".

(2) Genetics is basic to rational plant breeding but the technology appeals to numerous other sciences as well, with agricultural science, chemistry and plant pathology being perhaps especially prominent. Biometry is crucial to plant breeding because it lies at the heart of all selection and trials problems. Biometrical genetics, in the narrow sense, however, has had little practical effect. Often enough plant breeders do first what biometrical geneticists (and plant physiologists) later interpret. In only one crop I suspect (maize, 20) has formal biometrical genetics reached a level at which it influences breeding plans at the practical level.

(3) Plant breeding can also be thought of as applied evolutionary science. The breeder generates new genotypes of superior adaptation to environments which may be static or, perhaps more often, changing and he does so by manipulating gene frequencies in populations. The genetic variation is polygenic rather than oligogenic and the improved genotype requires a subtle balance (so far unanalyzable) of characters. All the components of neo-Darwinian micro-evolution are plainly identifiable: generation of variation, recombination, selection and isolation of the products. The plant breeder, it seems, is not only a technologist, but also an evolutionist.

(4) Plant breeding is but one component of crops research. The other, of course, is agronomy or crop husbandry research. The universal experience of crop yields in technologically-based agricultures of the past 50 years is that, with hesitations, they have gone steadily upwards. For this, both improved environments (E or husbandry effects) and improved genotypes (G or plant breeding effects) are generally agreed to have been responsible, in roughly equal measure. The interaction (GE) component has, I believe, been historically more important than is generally acknowledged (37) and I shall suggest later that it is deserving of more study. At all events, G, E and GE effects are the essential outcomes of crops research and plant breeders are responsible for something like half the achievement.

(5) Whatever the crop, the general pattern of plant breeding is a cyclical system of generationwise assortative mating (GAM): the best products of one cycle become the parents of the next (Fig. 1). The system is not closed, however, because there is (or should be) a steady trickle of new genetic material entering it from outside sources. There are, of course, endless complications of detail due to biological differences between crops and socio-economic requirements: wheat and sugarcane breeding programs, say, bear little superficial resemblance to each other. But the fundamental cyclical GAM pattern is common to both.

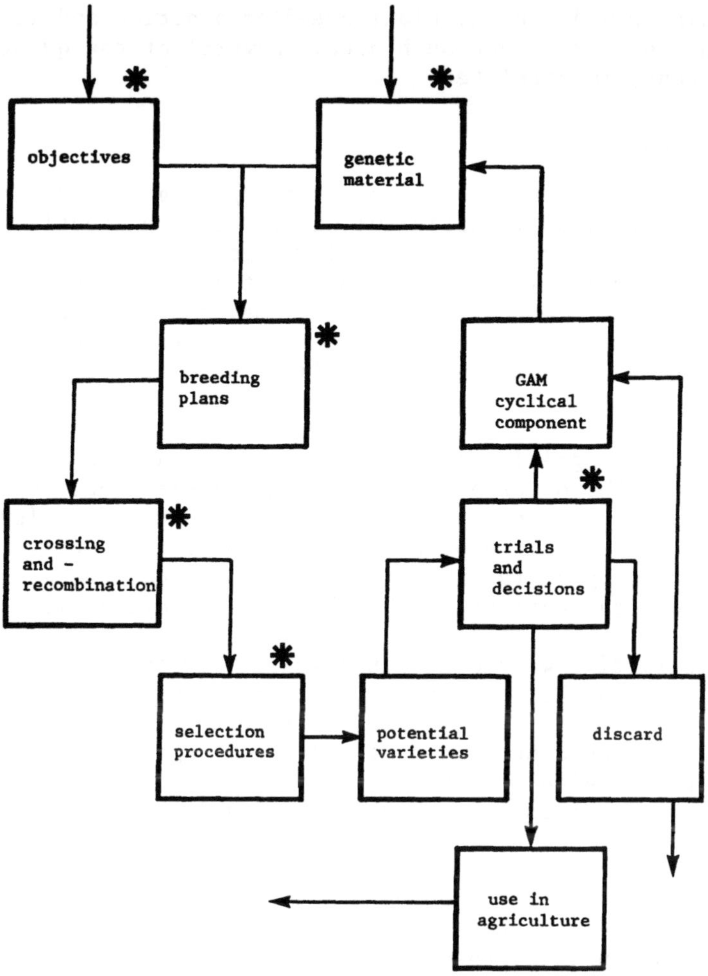

Fig. 1. The cyclical nature of plant breeding. Stars mark the problem areas identified in the text.

QUESTIONS ABOUT PLANT BREEDING

Despite the undoubted economic success of plant breeding as a major component of crops research and the existence of a vast body of experience and knowledge, there are, nevertheless, many areas of uncertainty, areas in which our understanding ought to be better. Different students would write different lists of these problem areas. My own list, a personal one of course, is given in this section. It covers the areas in which I think that improved understanding would enhance plant breeding practice and it refers to genetic engineering as one such area. Several of the questions are, it will appear, interrelated.

The Genetic Base

The cyclical pattern of plant breeding (Fig. 1) implies that, if there is no, or an insufficient, inflow of new genetic material into the system, improvement of a finite gene-pool will be limited. Any such program must, in time, run out of useful genetic variability. The recognition that this may happen is relatively recent but the importance of the genetic base for continued progress is now generally acknowledged (3,8,44). There are eight points to make:

(1) The recognition of a narrow genetic base is usually based upon the observation that progress seems poor despite vigorous breeding efforts or upon study of pedigrees which reveal small numbers of entries at the effective start of the program (as in potatoes and sugarcane generally, European barleys and U.S. maize and soybeans).

(2) Strictly, we should distinguish between a narrow genetic base in the breeding population and one in the producing populations of commercial varieties. The latter must be of concern because of the inherent agricultural and pathological risks but the former is fundamentally the more serious because therein lies future adaptation.

(3) Four sources of new genetic material may be recognized as follows. First, we have locally adapted varieties newly produced by other breeders in similar or homologous environments; these are universally used but suffer from the defect that they are likely to be closely related to the local base anyway. Would a U.S. corn breeder or a Scottish barley breeder get anything fundamentally unrelated from within the U.S.A. or within Europe respectively? Second, there are foreign unadapted varieties which may be valued for specific attributes by backcrossing but which are usually discouraging as parents in general. Third, there are land-race materials to which the preceding comments apply with even greater force. They have to be fairly fundamentally changed genetically if they are to provide anything more than a few specific genes (usually disease resistances). And finally, fourth, we have related species, often wild but sometimes cultivated, which can, of their own nature, do little or no more than provide specific genes for backcrossing.

(4) In practice, I think it is a fair generalization that unadapted genotypes have been used in backcrossing

programs following screening for specific attributes or not used at all. There have been many successes (examples in 26) but also many failures (the R-genes in potatoes, numerous cereal programs). However, such backcross programs do little for the genetic base.

(5) Most of the genetic variability in any crop, therefore, lies in the mass of unadapted varieties and land-races and is largely inaccessible to plant breeding in the short term.

(6) The practical question, then, is how to broaden the genetic base of a program, using unadapted materials, beyond the narrow limits imposed by backcrossing specific genes. The answer, I think, clearly lies in recognizing the need for back-up programs designed to develop new populations at a fairly high level of adaptation from primitive or otherwise unadapted materials (34,44). One recognizes that this will generally be a slow process that may require decades of hard work; there will be epistatic bottlenecks (day-length, maturity, stature) to be passed; methods will have to be rough rather than refined. Only when the best of the new materials approaches a commercial level of performance will it be usable.

(7) The California barley composite crosses (3) provide a beautiful model which show that primitive materials can indeed be vastly improved by simple means. Methods have to be modified for other crops, of course, but there is now a general appreciation of the fact that assisted natural selection in composites can be a powerful technique (21). Weak selection pressures and the deliberate maintenance of recombination and variability are essential features of any such program. Systematic base-broadening operations are now in hand in potatoes and sugarcane in several countries and must be imminent in many other crops (44). In coming years all major programs will have to undertake such operations if progress is to be sustained.

(8) The value of crop genetic resource conservation (GRC) work has become generally appreciated only in recent years (15,16,22). Its importance is now widely understood but achievement still falls far short of need. Despite the development of important "gene-banks" (e.g., at Fort Collins, U.S.A., at the International Rice Research Institute, Philippine Islands and all the excellent efforts of the International Board for Plant Genetic Resources (IBPGR) (23), many important crops are yet poorly collected, stored and documented (24) and most

minor ones have hardly been touched. But collections are only a start, though a fundamental one. They are useful only if used. Traditionally they have provided little more than major-gene disease resistances (24). I hope the discussion above will have shown that in collections lie the materials for a much more fundamental contribution to plant breeding.

Yield and Limits

Yield is the dominant objective of plant breeding. Even where quality characters are important, success goes to the high yielder, quality having been satisfied. There are, of course, economic trade-offs between the two and it is not unknown for a market's quality demand to be modified by technical change or availability of a primary product. Several important points arise, as follows.

1. Yield is maximized by enhancing biomass (total tons of dry matter per hectare, say), by improving partition ("harvest index") and by minimizing losses due to accidents and disease (36).

2. Biomass is essentially fixed by site and season and maximum potential biomass is effectively predetermined at a level which seems to be (somewhat mysteriously) much less than simple theory based on photosynthetic considerations would suggest. Maximum potential biomass for a north European summer season appears to be about 20-25 t/ha for diverse crops, and other crops in different environments have different maxima. A major question for the future is, therefore: is potential biomass such a fixed quantity as it seems to be or can plant breeding somehow enhance net photosynthesis and thus raise the limit? We don't know the answer, though empirical experience seems discouraging.

3. Improved partition has been an important feature of rising yields in many crops, with a reduced vegetative skeleton freeing photosynthate for incorporation in the desired product, whether seed, tuber, fiber, latex or other. For example, the partition ratio (dry grain/total above-ground dry matter) has risen in the European small grain cereals from about 35 to 50 percent in recent decades (4,30) and semi-dwarfness is, as is well known, a prominent feature of the wheats and rices of the "Green Revolution" (9). But dwarfness, however important in most other cereals, has not been a feature of hybrid maize improvement (10) (though it seems to have important potential in tropical maize). The question for the future, therefore, is: how far can partition go in reducing the plant skeleton to

enhance the product? Can we construct cereals with partition ratios exceeding 50 percent?

4. Losses of biomass (and therefore of yield) are caused by accidents (such as wind damage) and, probably more important, diseases (including pests). A few diseases, it is true, are primarily cosmetic in effect but all serious ones inhibit growth and yield. In principle, reasonable resistance-breeding strategies are becoming fairly well understood, even if the practice is often fraught with uncertainty (31,36). The questions that have to be raised about a strategy concern such matters as whether or not to use major genes, the use of mixtures or other heterogeneous populations, the deployment of varieties in agriculture and relations with other (e.g., chemical) control measures.

5. If we accept the notions that potential net photosynthesis (and hence biomass) is fixed by nature and that we may be pushing the limits of partition, then we might already be approaching the limits of yield. Certainly, many crops have already been grown at what we now conceive to be the limit (e.g., wheat, potatoes, and maize). This seems discouraging, but it is not: environments are nearly always imperfect and there is yet a long way to go in exploiting responsiveness of our crops to stresses (physical and biological). There may be room to enhance biomass and/or partition still further; many crops, especially tropical ones, are only just at the beginning of improvement. Real yields are nearly always far below what we know to be the potential and, as ever, the plant breeder will have a large share of any progress in narrowing the gap.

Are Hybrid Varieties Necessary?

The spectacular success of hybrid maize, followed closely by sorghum, beets, onions, and several vegetables, has generated a widespread belief (one might almost say myth) to the effect that "hybrids" are ultimately the best way of breeding seed-propagated crops. There is hardly such a crop, inbred or outbred, for which statements of intention or interest are wanting. With the virtual collapse of efforts to breed hybrid wheats and barleys, interest is now, I think, declining, but several important points and questions remain (36).

1. Hybrid varieties offer inbuilt economic protection to the breeder in the absence of plant breeders' rights. Given effective rights schemes (which are developing rapidly),

the economic incentive to breeding hybrid varieties declines.

2. Over-dominance as a component of performance in hybrids is now generally discredited and the evidence in maize points overwhelmingly to additive and dominance effects and little else as the basis for economic traits (20). Indeed, the evidence for over-dominance in any organism (with the exception of odd situations such as sickle-cell in man) is thin. If this be accepted, we have to ask whether hybrid varieties are ever necessary (even in maize!), especially if there are productive inter-genotypic interactions in heterogeneous populations (see below).

3. The evidence from maize is that excellent progress in economic characters can be made by diverse population improvement methods (20,28). But no one has ever had the time and money to push big populations thus for decades. Hybrid maize is successful but it took decades of work on a huge scale to succeed. What would happen if we put a similar effort into population improvement?

4. One of the undoubted attractions of the hybrid variety is its phenotypic uniformity. This attraction may, I suspect, sometimes be overrated, but it is certainly important in some hybrid vegetables: I have heard brassica breeders argue that the uniformity of habit and maturity of hybrid Brussels sprouts far outweighs any (problematic) yield gain. Hybrid uniformity, of course, has no attractions in the inbreeders such as wheat and barley, where hybrid breeding could only be justified by yield heterosis.

5. It may be that the principal lesson we should learn from hybrid breeding experience is to ask questions (36): about population structure in outbreeders (what is the role of heterogeneity?); about some of the curious genetic effects of inbreeding in outbreeders (why is stability often so difficult to attain?); and about our methods of handling inbreeders (intervarietal heterosis without overdominance implies that we ought to be able to exploit a cross better than we do).

The Neglect of Heterogeneity

The narrowing of the genetic base (in both breeding and producing populations), which has been a feature of technology-based agriculture during this century, has been accompanied by a trend towards homogeneity in the farmer's field: one clone, one pure

line, one hybrid. For this trend several factors have been responsible: scientific (the belief that the best genotype must be the best performer), technical (ease of husbandry), bureaucratic (seed certification rules), and aesthetic. The trend is now being (belatedly) questioned, for two reasons: first, that extreme uniformity presents biological hazards (especially from diseases) (5,8), and second, that there may be neglected positive productive interactions in mixtures (34). Several points arise as follows.

1. The idea that heterogeneity for disease resistance, at both the within- and between-field level, can be biologically advantageous is now well established in the context of multi-lines and varietal mixtures (42). We may expect to see developments in both (if they are not overtaken by wider development of race-non-specific resistance to airborne pathogens).

2. The idea that genotypes may interact favorably in mixed populations was widespread in the last century and only recently revived in this. An earlier review (34) made it clear that there is indeed an array of interactions, some negative, some positive, but with a mean slightly greater than zero (maybe + 3% in yield) and some outstanding specific combinations. Subsequent literature (which has yet to be fully reviewed) agrees with this conclusion. Doubtless, damping-out of diseases is one component of the interaction but there must be more to it than that (42).

3. The question arises as to whether the effects found in "constructed" mixtures occur also in outbreeding populations (such as open-pollinated maize) and heterogeneous but predominantly inbred land-races. For the latter there is indeed evidence from the Californian barley composites (2).

4. In addition to making some contribution to performance, there is also some evidence of enhanced stability of performance (i.e., lesser GE effects and errors) in heterogeneous populations. However, good data are hard to come by and the question deserves more study.

5. We have no secure knowledge of the physiology of interaction in mixtures. But we know enough, I think, to assert that heterogeneity in crops presents a problem of extraordinary biological interest and, probably, one of non-trivial agricultural importance too.

6. I add here, for convenience, a point about perennial crops, even though it hardly follows logically on the preceding. A common trend apparent, in varying degree, in

the tropical tree crops (such as rubber, tea, coffee, cacao, oil palm, coconut) is to displace seedling populations (essentially synthetics from selected parents or parent populations) by clones, if and when clones can be made (36). There seems to be a belief that clones are somehow automatically superior to seedlings. They are not, though the best clones often will be. Tropical tree crops are peculiarly at hazard from diseases, so that one hopes that a cautious strategy for the use of clones will prevail: one would like to see numerous rather than few clones being grown in a mosaic, perhaps even mixed within a field, and accompanied by large areas of good synthetic seedling populations. The same argument applies to the use of clones in forestry, temperate or tropical.

Recombination

In an inbreeder, nearly all the recombination that is going to occur in a cross between two lines takes place in the first three generations of selfing. The probability of generating the best possible line from a cross is correspondingly diminished; and with many loci segregating, this probability is small anyway, even in the absence of linkage (36). The same argument applies to inbreds isolated from outbred populations, as in maize. In outbreeders, by contrast, recombination will be unrestricted unless intense selection leads to a degree of inbreeding and homozygosity of unbroken linkage blocks. Two conclusions follow.

1. We probably never achieve the selection limit in isolating pure lines from a cross between inbreeders. We have no idea (I think) how far we fall short and whether it would be worth interjecting secondary crossing cycles between selections. Maybe it doesn't matter, but one recalls that tobacco has shown a remarkably good response to recurrent crossing (25). Perhaps the defects of one cycle are repaired by the next GAM cycle of crosses as the selections themselves become both varieties and parents, just as second-cycle corn inbreds improve over their first cycle progenitors.

2. In operations designed to widen the genetic base starting from primitive and/or ill-adapted materials, it is fair to assume that recombination will be of crucial importance. Hence selection should be weak and progress deliberately slow. The Californian barley composites must owe much to this principle.

Optimal Selection Procedures

All plant breeding screens tens of thousands, even millions, of entities for each really useful survivor. Selection is based on visual criteria and measurements of several to many characters, sometimes considered sequentially, sometimes jointly. Plant breeding costs time, effort, and money, and all breeders strive for, in some sense, an optimal allocation of resources to the components of their programs. However, the choices are almost infinitely numerous and decisions as to what is optimal are intuitive rather than explicit. I shall return to the economics later and touch here on two points that apply when resources are fixed.

1. All plant breeding depends upon heritability as a basis for response to selection. For the predominant polygenic characters, we select visually, quickly, and roughly at the start of the cycle when numbers are large and replication low, progressively more thoroughly and carefully later on when there are relatively large quantities of few survivors with better replication and information about them. We have some rudimentary knowledge of how to maximize progress (or how to optimize distribution of effort between phases) (38) but can generally only hope that we are not too badly wrong or that optima are broad enough to be non-critical. A more restricted question has, however, been answered: how to reduce a set of near-varieties to the best one most efficiently. The conclusion is to apply roughly constant resources to each cycle, reducing numbers of entries as the quantity of each survivor rises (14). This is an area which deserves more study than it has had: some programs are certainly inefficient, generally (I think) erring in the direction of over-confident selection in the early stages (38).

2. Plant breeding deals with several to many characters, often simultaneously. In principle, selection indices, as developed by the animal breeders, ought to have a place because they are (usually) more efficient than successive truncation. In practice, formal indices would probably be slow and wasteful but indices are in fact used by plant breeders, though in an intuitive form (36). The question is, then: is there a place for wider application of explicit selection indices, especially, perhaps, at the trials stage when entries are few, data good and economic components potentially calculable?

Trials

We enter here what is probably the most difficult and critical set of questions facing plant breeders (27,36). Trials are done to discriminate between potential new varieties, to choose this one and reject that one on grounds of likely performance in actual agricultural practice. That is, they are assumed to predict performance. If there were no GE interactions, this might be relatively simple: one large trial or several smaller ones in a single year would often suffice to decide with sufficient confidence. In practice, interactions are rife, decisions difficult and extrapolation to agriculture uncertain. Four points are worth making.

1. Trials techniques have advanced enormously in recent years, thanks to small-plot mechanization, and progress continues. Statistically, we now have computers and the generalized lattices which permit high precision in the individual trial (27).

2. Trials are done in sets, replicated over sites and seasons because of the universal experience of GE interactions. Sites are traditionally chosen on an arbitrary/convenience/historical/a priori basis; years are not choosable. Analyses of sets of trials generally reveal substantial GE interactions in the order of VYS > VY > VS (variance components: V varieties, Y years, S sites). Therefore, for good mean comparisons, trials should be replicated over years rather than sites, an unwelcome conclusion. Regression methods are available for the further definition of GE effects (11,13), but results tend to vary between years and they give little guidance for decision making. In practice, decisions are generally taken on a mean yield basis qualified by some reference to quality factors, the possibility of specific local adaptation and the idea of stability/responsiveness. Something like the intuitive selection index mentioned above re-enters here and formalization of it might be useful.

3. The situation just described is hardly satisfactory (though perhaps a rough justice is done). One can foresee useful developments along three possible lines. First, it should be possible to choose sets of sites on a rational basis to maximize GE effects between them and reduce redundancy of information (1). This would amount to defining a set of local environments identifiable over years to which local adaptation would be sought. Second, it is the trials officer's pride to achieve high yields and trials yields are, in fact, notoriously higher than agricultural ones. It might be possible deliberately to pick low-yielding sites or to apply treatments which would mimic the effect of seasons by generating similar GE effects. An economy of years would be very attractive. Third, and finally, if decisions are uncertain, it follows

that it is better to choose generously and put out many new varieties rather than few, leaving farmers to make their own final decisions; this is contrary to the current trend but it is agriculturally (and pathologically) sensible.

4. Trials are based on the assumption that they predict agricultural performance. The assumption is rarely testable because we rarely have satisfactory agricultural yield data by varieties. The few tests available are rather discouraging: agreement in sugarcane and potatoes seems to be poor, but in rubber quite good. Three points need to be made. First, some overestimation of yield gains due to new varieties is inevitable, due to statistical "attenuation" by selection. Second, there are some discrepancies far too large to be explained by attenuation and here we have to look to data biases and GE or other effects yet unrecognized; it may be that the choice of trial sites/years distorts results or that farmers sometimes select sites for varieties or that there are cryptic diseases to confuse the situation. And third, in manufacturing industry, the prudent engineer goes through one or more pilot plant stages before building a full-scale factory. In plant breeding, we usually jump from trial to recommendation without an intervening development phase. Exceptions are provided by some tropical plantation crops (e.g., sugarcane and rubber), and U.S. maize, in which at least one major breeding company runs paired strips of contrasting varieties on the farm scale; one looks forward with interest to seeing the analysis of these excellent data that is now in hand (10).

5. In conclusion, I believe that major problems lie, not in individual trials, but in the design and interpretation of sets of them and, above all, in their rational extrapolation to agriculture.

The Place of Economics

Plant breeding is a technology that has had a very large economic impact yet we know surprisingly little about the economics of it. There is no coherent body of theory or experience to guide us. I suggest that it is high time we took the matter seriously. Economic ideas impinge at two levels, as follows.

1. Internally to a plant breeding program, any consideration of alternative selection schedules or optional procedures must lead directly to economic questions. For example, given fixed resources adequate to grow N plants in all,

how to partition N into $(n_1 + n_2)$ over two generations in such a way that genetic progress is maximized? Alternatively, what resources are required for a given expectation of progress? These sorts of questions are at the heart of rational breeding plans and they will usually have both biometrical and economic components (38). This is obvious, yet the economic component hardly ever emerges explicitly in practice.

2. Externally, the question becomes far harder. One must first ask who benefits. Cui bono? The answer depends on who gives it. The commercial breeder's objective might well be enhanced company profits; the farmer might look for lower costs of production per ton or per hectare; and industry might wish for lower materials prices or processing costs; the government funding agency might look to "social benefits". All these are legitimate but different economic objectives (36). Irrespective of who writes the equations, however, excellent new varieties make only transient profits for breeders, farmers, and processors because the market mechanism ensures (or should ensure) that enhanced agricultural efficiency finally emerges as lower product prices to the benefit of consumers at large. All plant breeders, whatever their primary objectives, are in the social benefit business.

3. We have a number of examples to show that excellent new varieties do indeed generate large net benefits in social cost-benefit terms (i.e., $B-C \gg 0$, $B/C \gg 1$). Unfortunately, the examples are (inevitably) selected and we are far from being able to state any confident generalization about plant breeding as a whole. Probably, it is socially very beneficial, being remarkably cheap in relation to outcome, free of "externalities" (the economists' word for adverse side-effects) and acceptable from the welfare/distributional viewpoint; the last two points cannot be generally asserted for other kinds of crops research. With all reasonable certainty, plant breeding is socio-economically benign, but we cannot place a number on that judgement (12,32,33,36).

4. In its socio-economic effects, plant breeding (G) cannot easily be isolated from husbandry (E) research because of the presence of substantial GE interactions. Historically we know little about them, but there is a widespread belief and mounting evidence to show that the rising yields of crops in recent decades have had a large GE component (maybe about one-third in cereals) (37). It seems that rising husbandry inputs have been matched by responsive varieties adapted to them (at which point we

touch again upon the trials problems discussed above). It would be of surpassing interest to have many more "historical" trials of old and modern varieties split across low and high input treatments, even with attempts (never yet made, I think) to match the husbandry of decades past. It need not be assumed, however, that all modern varieties are of the "responsive" kind; some certainly are not (37). There is no evident reason to suppose that we could not deliberately set about breeding high yielding but "stable" varieties as well, with reasonable expectation of success.

5. If we wished to understand what crops research has done and might achieve in the future, we should need a kind of analysis which has, I think, never yet been attempted. We should need a historical breakdown of factor costs of production by G, E, and GE components. It would emerge, I think, that for the G component, the first or breeding cost was nearly all, whereas for the E component, implementation costs would be substantial (more fertilizers and chemicals, new machinery) and externalities present. If this were true, the G, E, and GE gains might be comparable, yet G economically the most attractive. This would be a difficult field of work, the GE component would be hard to allocate, and good data would be hard to find; but it would be an extraordinarily interesting field of study and one which would tell us something about what crops research does in the real world.

6. I must add, finally, that though I am in favor of economic thinking at the operational level in plant breeding and in _ex post_ analysis of history, I am skeptical of formal cost-benefit analysis in any _ex ante_ role in decision making. I should rather rely on experienced judgement than on uncertain economics (see Wise, ref. 41). But the intrusion of more explicit economic ideas into plant breeding could only be beneficial.

Genetic Engineering

In vitro culture techniques are well established in plant breeding. They take two main forms (surveys 6,29,36,40). First, embryo culture has, for decades, been a valuable adjunct to making difficult interspecific crosses. Second, more recent but also well established, is shoot-tip culture, which finds uses in rapid clonal multiplication, development of virus-free clones and genetic resource conservation work. The latter still presents some technical problems, but its applications multiply. Both techniques depend upon the retention of organizational integrity of the meristem. A step further takes us to callus, cell, and protoplast cultures in which organization is lost but can sometimes be recovered; many such cultures (indeed, probably most) are not regenerable because of (yet poorly understood) cytogenetic changes (7). In several crops (e.g., sugarcane, potatoes), regenerants from disorganized cultures have shown genetic variation of unknown nature but some potential utility and this is now being extensively explored as a new kind of mutation induction. A step further still takes us to in vitro hybridization, which has, after regeneration, yielded interspecific amphidiploids. The technique may provide desired amphidiploids which cannot be made by conventional means, and there may be possibilities for somatic recombination by some variant of it.

The foregoing techniques are already in use and there will no doubt be interesting and useful developments from them (7). They might be described, not disrespectfully, as in vitro horticulture, but they are not "genetic engineering" (see Fig. 2). Genetic engineering springs from the marvelous achievements of the microbial geneticists and biochemists and it underlies a whole new chemical technology (especially in pharmaceuticals) of great economic potential. In possible applications to plant breeding (19) the following points seem to me to be relevant.

1. The following steps are characteristic: (a) identification of the gene (DNA sequence) which is to be transferred to the economic species; (b) isolation and "cloning" of that DNA; (c) transfer of the DNA by various manipulations (transformation, transduction, etc.) to recipient crop host cells; (d) integration, transcription, and translation of the DNA in the recipient cells; and (f) multiplication and use of the modified crop plant. The modified crop might conceivably produce a new gene product or it might produce an enhanced gene product by way of tandem repeats.

2. The program just outlined is formidable but a beginning has been made in transferring the microbial technology to higher plants (43). Techniques will no doubt develop rapidly and we must expect that it will become possible, in a decade or two perhaps, to transfer a wide range of genes from higher plants, microbes, maybe even animals, into crop plants. The question for the plant breeder, however, is: will it be useful to do so? I personally think that the uses will be limited. There may be occasions when a specific gene can be identified and transferred to good effect to produce a new product and it may sometimes be useful to enhance enzymes already present (in rubber or drug biosynthesis maybe?). But when major

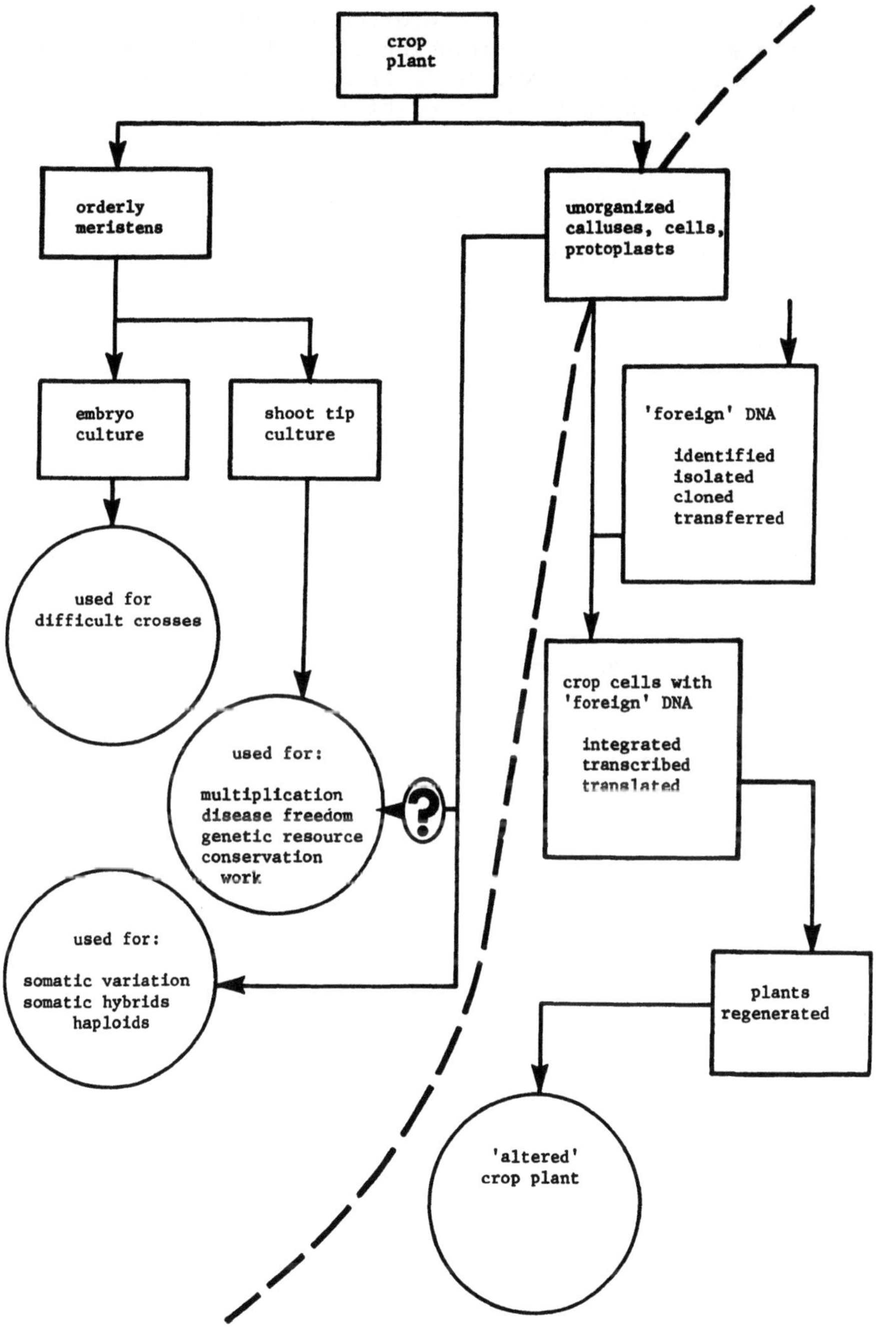

Fig. 2 Uses of in vitro methods in plant breeding. On the left of the heavy dashed line, established and developing in vitro techniques; on the right, prospective 'genetic engineering' developments.

genes are wanted, they are often available by conventional means and most plant breeding proceeds by the manipulation of polygenes; the wholesale transfer of (unidentifiable) polygenes seems beyond the skill of even the genetic engineers. The oft-quoted dream of making cereals fix their own nitrogen, of turning them into legumes, so to speak, would rest on the transfer of not just one gene, but of a multiplicity of them (19). A profound modification of the biochemical architecture of the cereal is implied.

3. I conclude that several in vitro techniques are well established and valuable in plant breeding, but that "genetic engineering" (in the sense of specific DNA transfer) will do beautiful things but is unlikely to be more than marginally useful. Plant breeding is not about to be revolutionized.

SUMMARY

Plant breeding is a science-based technology directed towards economic ends and it has been very successful. It appeals primarily to genetics as the underlying science, but with much reference to other scientific techniques, especially biometry. The general pattern is, with complications, one of generationwise assortative mating and most of the variation dealt with is polygenic in nature. Outstanding problem areas are mentioned, including: the genetic base, the limits of yield, the place of hybrid varieties, the cult of homogeneity, the neglect of recombination, optimal selection procedures, the interpretation of trials, the place of economics, and the potential of genetic engineering. On the last point it is argued that the potential is limited because of the predominance of polygenic characters; applications are likely only where specific genes can be identified, isolated, and transferred. It is concluded that plant breeding is a technology of great power and effectiveness that is likely to be further refined in coming years but not revolutionized.

REFERENCES

1. Abou-el-Fittouh, H.A., J.O. Rawlings, and P.A. Miller. 1969. Classification of environments to control genotype by environment interactions, with an application to cotton. Crop Sci. 9:135-140.
2. Allard, R.W., and J. Adams. 1969. Population studies in predominantly self-pollinating cereals XII. Amer. Nat. 103:620-645.
3. Allard, R.W., S.K. Jain, and P.L. Workman. 1968. The genetics of inbreeding populations. Adv. Genet. 14:55-131.

4. Austin, R.B., J. Bingham, R.D. Blackwell, L.T. Evans, M.A. Ford, C.L. Morgan, and M. Taylor. 1980. Genetic improvements in winter wheat yields since 1900 and associated physiological changes. J. Agric. Sci. Cambr. 94:675-689.
5. Barrett, J.A. 1981. The evolutionary consequences of monoculture. In Genetic Consequences of Man-Made Change. J.A. Bishop and L.M. Cook, eds. London and New York: Academic Press.
6. Barz, W., E. Reinhard, and M.H. Zenk, eds. 1977. Plant Tissue Culture and its Biotechnological Application. Berlin: Springer.
7. Chaleff, R.S. 1981. Genetics of Higher Plants, Applications of Cell Culture. Cambridge: Cambridge University Press.
8. Committee on Genetic Vulnerability of Major Crops. 1972. Genetic Vulnerability of Major Crops. Washington, D.C.: National Academy of Sciences.
9. Dalrymple, D.G. 1978. Development and Spread of High Yielding Varieties of Wheat and Rice in the Less Developed Nations. Washington: USDA.
10. Duvick, D.N. 1982. Personal communication.
11. Eberhart, S.A., and W.A. Russell. 1966. Stability parameters for comparing varieties. Crop Sci. 6:36-40.
12. Evenson, R.E., and Y. Kislev. 1975. Agricultural Research and Productivity. New York and London: Yale University Press.
13. Finlay, K.W., and G.N. Wilkinson. 1963. The analysis of adaptation in a plant breeding program. Aust. J. Agric. Res. 14:742-754.
14. Finney, D.J. 1958. Plant selection for yield improvement. Euphytica 7:83-106.
15. Frankel, O.H., and J.G. Hawkes, eds. 1975. Crop Genetic Resources for Today and Tomorrow. (IBP 2). Cambridge: Cambridge University Press.
16. Frankel, O.H., and M. Soulé. 1981. Conservation and Evolution. Cambridge: Cambridge University Press.
17. Frey, K.J., ed. 1966. Plant Breeding. Ames, Iowa: Iowa State University Press.
18. Frey, K.J., ed. 1981. Plant Breeding II. Ames, Iowa: Iowa State University Press.
19. De Groot, B., and K.J. Puite. 1982. Genetic Engineering: Methods for Plant Breeding: A review and prospects. In Induced Variability in Plant Breeding. Wageningen, The Netherlands: PUDOC.
20. Hallauer, A.R., and J.B. Miranda. 1981. Quantitative Genetics in Maize Breeding. Ames, Iowa: Iowa State University Press.
21. Harrison, C.H., T.H. Busbice, R.R. Hill, O.J. Hunt, and A.J. Oakes. 1972. Directed mass selection for developing multiple pest resistance and conserving germ plasm in alfalfa. J. Env. Qual. 1:106-111.
22. Hawkes, J.G., ed. 1978. Conservation and Agriculture. London: Duckworth.

23. International Board for Plant Genetic Resources (IBPGR). 1979. A Review of Policies and Activities 1974-78 and of the Prospects for the Future. Rome.
24. Marshall, R.D., and A.D.H. Brown. 1981. Wheat genetic resources. In Wheat Science Today and Tomorrow. L.T. Evans and W.J. Peacock, eds. Cambridge: Cambridge University Press.
25. Matzinger, D.F., and E.A. Wernsman. 1980. Population improvement in self-pollinated crops. In World Soybean Research Conference II. F.T. Corbin, ed. Boulder: Westview Press; London: Granada Press.
26. Nelson, R.R., ed. 1973. Breeding Plants for Disease Resistance. University Park, Penn.: Pennsylvania State University Press.
27. Patterson, H.D., and V. Silvey. 1980. Statutory and recommended list trials of crop varieties in the United Kingdom. J. Roy. Stat. Soc. A 143:219-252.
28. Pollak, E., O. Kempthorne, and T.B. Bailey, eds. 1977. Proceedings of International Conference on Quantitative Genetics. Ames, Iowa: Iowa State University Press.
29. Reinert, J., and Y.P.S. Bajaj, eds. 1977. Plant Cell, Tissue and Organ Culture. Berlin: Springer.
30. Riggs, T.J., P.R. Hanson, N.D. Start, D.M. Miles, C.L. Morgan, and M.A. Ford. 1981. Comparison of spring barley varieties grown in England and Wales between 1880 and 1980. J. Agric. Sci. Cambr. 97:599-610.
31. Russell, G.E. 1978. Plant Breeding for Pest and Disease Resistance. London: Butterworth.
32. Ruttan, V.W. 1982. Agricultural Research Policy. Minneapolis: University of Minnesota Press.
33. Schuh, G.E., and H. Tollini. 1979. Costs and Benefits of Agricultural Research. (Working Paper 360). Washington, D.C.: World Bank.
34. Simmonds, N.W. 1962. Variability in crop plants, its use and conservation. Biol. Rev. 37:422-465.
35. Simmonds, N.W., ed. 1976. Evolution of Crop Plants. London: Longman.
36. Simmonds, N.W. 1979. Principles of Crop Improvement. London: Longman.
37. Simmonds, N.W. 1981. Genotype (G) Environment (E) and GE components of crop yields. Expl. Agric. 17:355-362.
38. Simmonds, N.W. Two-stage selection strategy in plant breeding. In preparation.
39. Sneep, J., and A.J.T. Hendriksen, eds. 1979. Plant Breeding Perspectives. Wageningen, The Netherlands: Centre for Agricultural Publishing and Documentation.
40. Vasil, I.K., M.K. Ahuja, and V. Vasil. 1979. Plant tissue cultures in genetics and plant breeding. Adv. Genet. 20:127-215.
41. Wise, W.S. 1981. The theory of agricultural research benefits. J. Agric. Econ. 32:147-157.

42. Wolfe, M.S., J.A. Barrett, and J.E.E. Jenkins. 1982. The use of cultivar mixtures for disease control. In Strategies for the Control of Cereal Disease. J.F. Jenkyn and R.T. Plumb, eds. Oxford: Blackwell.
43. Wullems, G.J., L. Molendijk, G. Ooms, and R.A. Schilperoort. 1981. Retention of tumor markers in F_1 progeny plants from in vitro induced octopine and nopaline tumor tissues. Cell 24:719-727.
44. Zeven, A.C., and A.M. van Harten. 1979. Broadening the Genetic Base of Crops. Wageningen, The Netherlands: PUDOC.

GENES: CHAIRMAN'S INTRODUCTION

William C. Taylor

Department of Genetics
University of California
Berkeley, California 94720

It is evident by now that there is a great deal of interest in exploiting the new technologies to genetically engineer new forms of plants. A purpose of this meeting is to assess the possibilities. The papers that follow are concerned with the analysis of single genes or small gene families. We will read about genes found within the nucleus, plastids, and bacteria which are responsible for agriculturally important traits. Given that these genes can be isolated by recombinant DNA techniques, there are two possible strategies for plant engineering. One involves isolating a gene from a cultivated plant, changing it in a specific way and then inserting it back into the same plant where it produces an altered gene product. An example might be changing the amino acid composition of a seed protein so as to make the seed a more efficient food source. A second strategy is to isolate a gene from one species and transfer it to another species where it produces a desirable feature. An example might be the transfer of a gene which encodes a more efficient photosynthetic enzyme from a wild relative into a cultivated species.

There are three technical hurdles which must be overcome for either strategy to work. The gene of interest must be physically isolated. It must then be transferred into the recipient plant in such a way that it is stably integrated into the recipient's genome. Finally, the transferred gene must be properly expressed in the recipient plant such that the appropriate quantity of gene product is present in the right cells at the right time. Experience from gene transfer attempts in animals suggests that the latter may be the greatest technical hurdle to overcome. As the following papers will show, we are rapidly learning a great deal about the structure of plant genes. However, we know distressingly little about the regulation of gene expression during plant development. This lack of knowledge could be the greatest obstacle in the use of recombinant DNA to genetically engineer new plants.

It should be stressed that recombinant DNA technology does not provide us with possible methods for genetically engineering new forms of plants in the strictest sense. It rather provides a means of gene engineering. Recombinant DNA technology can transfer one, or at most, several genes at a time into a recipient plant. Creation of a new form of plant requires the coordinated expression of a great deal of new genetic information. We are considering instead techniques which might allow us to change one characteristic of a plant.

I am not saying that the new technologies have no role in attempts to modify cultivated plants. As will become apparent from the papers in this session, there are a number of intriguing possibilities. But we should be aware of what gene engineering, in its present state, can and cannot do.

Looking to the future, there are possible uses of recombinant DNA technology which could in fact give rise to new forms of plants. If it becomes possible to isolate and manipulate the genes which control the developmental processes of a plant, then we are faced with the possibility of making profound changes in that plant. It may be that the alteration of a single gene can significantly change a specific developmental program. It is clearly too early to speculate at length about the possibility and desirability of altering developmental programs. It is obvious that such changes would have major consequences for the plant.

I anticipate that one conclusion from this conference will be that we need to know much more about how genes of interest are regulated during plant development. It is only by understanding how the expression of a specific gene fits into the process of development as a whole that we can hope to manipulate either the specific gene or the process.

NUCLEAR GENES ENCODING THE SMALL SUBUNIT OF RIBULOSE-1,5-BISPHOSPHATE CARBOXYLASE

A.R. Cashmore

Department of Cell Biology
Rockefeller University
New York, New York 10021

INTRODUCTION

Ribulose-1,5-bisphosphate (RuBP) carboxylase is the major stromal protein in chloroplasts from C3 plants. It performs the first step in the Calvin cycle and is composed of 8 large subunits and 8 small subunits. The large subunits, of 53,000 daltons, are encoded by chloroplast DNA and the small subunits, of 14,000 daltons, are encoded by nuclear DNA (1,2). The small subunit is synthesized as a 20,000 dalton precursor, on free cytoplasmic ribosomes (3-6). This precursor functions in the post-translational transport of the small subunit from the site of synthesis into chloroplasts (5,6).

The expression of the genes encoding the small subunit are regulated in a variety of ways. Light induces the synthesis of the small subunit in both pea (7,8) and lemna (9). This regulation occurs at the level of mRNA, and at least in lemna the induction is mediated via phytochrome (10). In C4 plants, the small subunit is synthesized in bundle sheath cells but not in mesophyll cells (11,12). In both of these regulatory examples, the expression of the nuclear genes encoding the small subunit is coordinated with the expression of the chloroplast genes encoding the large subunit. The means by which gene expression is coordinated between these two distinct biosynthetic systems is not known. This coordinated expression is particularly interesting in view of the vastly different copy number for the genes encoding these two polypeptides. The circular DNA molecules encoding the large subunit are present in 15-30 copies per chloroplast (13) and leaf cells may contain several hundred chloroplasts (14). Consequently, the gene for the large subunit may be reiterated several thousand fold per cell. In contrast, RNA/DNA solution hybridization studies (15), showed that the nuclear genes encoding the small subunit are present as relatively few copies per cell.

cDNA Sequences Encoding the Small Subunit of RuBP Carboxylase

We have previously reported the isolation and characterization of a cloned cDNA sequence (pSS15) from pea which is complementary to mRNA encoding the precursor for the small subunit of RuBP carboxylase (16). The cDNA was characterized by translation of hybrid-selected mRNAs and the translation products were identified by SDS-polyacrylamide gel electrophoresis and *in vitro* chloroplast uptake and processing studies. This cDNA has been sequenced and shown to contain all of the codons encoding the mature small subunit plus 148 nucleotides corresponding to the 3' non-translated region of the small subunit mRNA (17). In addition, the cDNA contained sequences encoding 33 amino acids of the small subunit transit polypeptide.

A second small subunit cDNA sequence has been characterized by Bedbrook et al. (18). This sequence differs from the cDNA sequence discussed above in both coding and non-coding regions. A quantitative analysis of these differences is discussed below.

Nuclear DNA Sequences Encoding the Small Subunit of RuBP Carboxylase

The cloned small subunit cDNA has been used as a hybridization probe to isolate the corresponding nuclear genes. Partial Eco Rl digests of pea DNA were cloned in the lambda phage Charon 4 and recombinant phage were screened by plaque hybridization. Positively hybridizing phage were selected and one was studied in detail. This phage (ChPS163) contained three Eco Rl inserts, of 3.6, 4.7, and 8.0 kb, two of which (3.6 and 8.0 kb) hybridized to the small subunit cDNA (Fig.1). These two hybridizing fragments have been shown to correspond to duplicated small subunit genes and by restriction analyses they are seen to lie on either side of the 4.7 kb Eco Rl fragment. The structure of one of these genes is shown in Fig. 2 and the sequence is shown in Fig. 3. The genes contain two intervening sequences and the coding capacity of the first of the three exons is devoted almost exclusively to encoding the transit sequence of the small subunit. The small subunit genes therefore represent excellent examples of genes being composed of exons where the subdivisions reflect functional domains within the encoded proteins. The intervening sequences of the small subunit gene is rich in AT and at the 5' and 3' boundaries of these intervening sequences are the dinucleotides GT and AG, as are found for intervening sequences from other eukaryotes. The 5' end of the small subunit gene SS3.6 contains the "promoter" sequence TATATATA and this sequence lies at the 3' end of an AT rich region of 50

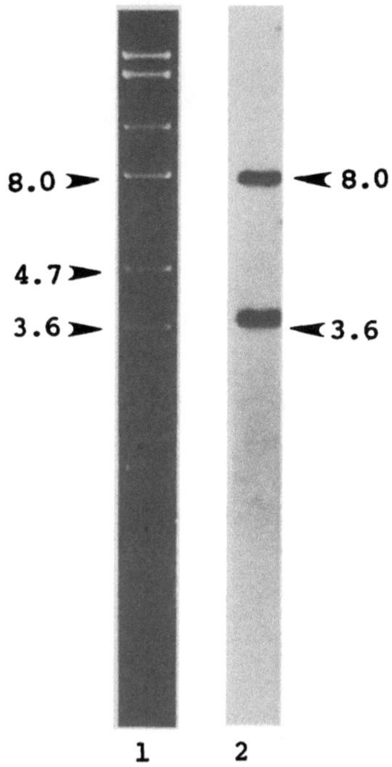

Fig. 1. Restriction of DNA from phage ChPS163. Phage DNA was restricted with Eco R1 and the restriction fragments were fractionated by gel electrophoresis in 0.8% agarose gels. The DNA was visualized by staining with ethidium bromide (1) and then transferred to nitrocellulose. Hybridization (2) was with ^{32}P-labeled small subunit cDNA (pSS15).

nucleotides which are 86% AT. The 3' non-coding end of the small subunit gene contains the sequence AATGAA. This sequence is also found in the small subunit cDNA (pSS15) (18) nucleotides from the poly A sequence (17) and thus these sequences appear to be analogous to the polyadenylation signal AATAAA (19).

From the DNA sequence of the small subunit gene we have obtained the complete amino acid sequence for the small subunit precursor (Fig. 4). It is observed that the transit sequence contains 57 amino acids, substantially more than the 44 amino acids found for the transit sequence for the small subunit from Chlamydomonas (20). Some homology is found at the N terminus of the pea and Chlamydomonas precursor with the sequence corresponding in four out of the first eight amino acids. A similar degree of homology is found at the N terminus of the mature polypeptides, with six out of the first twelve amino acids being identical. With the exception of the N terminus, the transit sequence from pea and

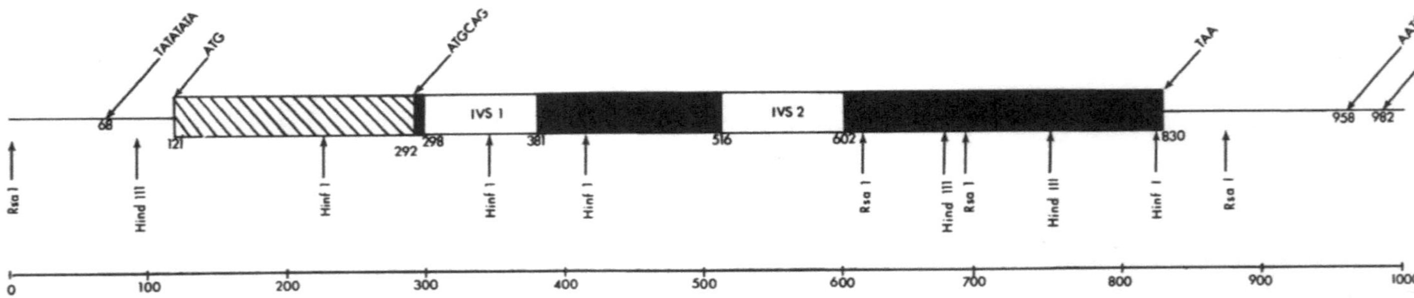

Fig. 2. Structure of the gene (SS3.6) for the small subunit of RuBP carboxylase. The small subunit gene SS3.6, from a pea genomic Eco R1 fragment of 3.6 kb, contains two introns (IVS 1 and IVS 2) and three exons. The first exon encodes primarily the small subunit transit sequence (▨).

CTAACAAGATTGGTACTAGGCAGTAGCTAATTACCACAATATTAAGACCATAATATTGGAAATAGATAAATAAAA 75

MetAlaSerMetIleSer
ACATTATATATAGCAAGTTTTAGCAGAAGCTTTGCAATTCATACAGAAGTGAGAAAAATGGCTTCTATGATATCC 150

-50 -40 -30
SerSerAlaValThrThrValSerArgAlaSerArgGlyGlnSerAlaAlaValAlaProPheGlyGlyLeuLys
TCTTCCGCTGTGACAACAGTCAGCCGTGCCTCTAGGGGGCAATCCGCCGCAGTGGCTCCATTCGGCGGCCTCAAA 225

-20 -10
SerMetThrGlyPheProValLysLysValAsnThrAspIleThrSerIleThrSerAsnGlyGlyArgValLys
TCCATGACTGGATTCCCAGTGAAGAAGGTCAACACTGACATTACTTCCATTACAAGCAATGGTGGAAGAGTAAAG 300

CysMetGln
TGCATGCAGGTGACAGAAACATATACATATATATATATAGTTGAATATCAGTAATGATTCAAGTTTGTTAACCGT 375

10 20
ValTrpProProIleGlyLysLysLysPheGluThrLeuSerTyrLeuProProLeu
TTATGTTGAATATTTAGGTGTGGCCTCCAATTGGAAAGAAGAAGTTTGAGACTCTTTCCTATTTGCCACCATTGA 450

30 40
ThrArgAspGlnLeuLeuLysGluValGluTyrLeuLeuArgLysGlyTrpValProCysLeuGluPheGluLeu
CGAGAGATCAATTGTTGAAAGAAGTTGAATACCTTCTGAGGAAGGGATGGGTTCCATGCTTGGAATTTGAGTTGG 525

Glu
AGGTTTCATATTCATTCCTTTTTTCAATGATTATATAAATACTTTTGTTTGAAACCGTAATGAGTTGATTTTGAC 600

50 60
LysGlyPheValTyrArgGluHisAsnLysSerProArgTyrTyrAspGlyArgTyrTrpThr
TGTTTGGTTGCAGAAAGGATTTGTGTACCGTGAGCACAACAAGTCACCAAGATACTATGATGGAAGATACTGGAC 675

70 80 90
MetTrpLysLeuProMetPheGlyThrThrAspAlaSerGlnValLeuLysGluLeuAspGluValValAlaAla
AATGTGGAAGCTTCCTATGTTTGGTACCACTGATGCTTCTCAAGTCTTGAAGGAGCTTGATGAAGTTGTTGCCGC 750

100 110
TyrProGlnAlaPheValArgIleIleGlyPheAspAsnValArgGlnValGlnCysIleSerPheIleAlaHis
TTACCCTCAAGCTTTCGTTCGTATCATCGGTTTCGACAACGTTCGTCAAGTTCAATGCATCAGTTTCATTGCACA 825

120
ThrProGluSerTyr
CACACCAGAATCCTACTAAGTTTGAGTATTATGGCATTGGAAAAGGTGTTTCTCTTGTACCATTTGTTGTGCTTG 900

TAATTTACTGTGTTTTTTTTTTCGGTTTTTGGTTTCGGACTGTAAAATGGAAATGGATGGAGAAGAGTTAATGAA 975

TGATATGGTCCTTTTGTTCATTCT

Fig. 3. Nucleotide sequence of the gene (SS3.6) for the small subunit of RuBP carboxylase. Putative transcriptional and polyadenylation signals are underlined. The site for polypeptide cleavage and the analogous site for polyadenylation (pSS15) are marked with arrowheads.

Chlamydomonas shows no obvious homology. This was not unexpected as it was known that the Chlamydomonas precursor is not processed by chloroplasts from higher plants (6). The pea small subunit transit sequence is quite basic, containing four lysines and three arginines compared with a single acidic amino acid (aspartic acid). It had been proposed for Chlamydomonas that the basic characteristic may reflect the binding of the small subunit precursor to the negatively charged chloroplast envelope (20).

The small subunit cDNA was used as a hybridization probe to estimate the number of genes encoding the pea small subunit. Pea DNA was restricted with Eco R1, fractionated by agarose gel electrophoresis, and then transferred to nitrocellulose.

										10						
SS 3.6	met	ala	ser	met	ile	ser	ser	ser	ala	val	thr	thr	val	ser	arg	ala
Chlamy	met	ala	val	ile ser	ala	lys	ser	ser	val	ser	ala	ala	val	ala	arg	pro
				20										30		
a	ser	arg	gly	gln	ser	ala	ala	val	ala	pro	phe	gly	gly	leu	lys	ser
b	ala	arg	ser	ser	val	arg	pro	met	ala	ala	leu	lys	pro	ala	val	lys
								40								
a	met	thr	gly	phe	pro	val	lys	lys	val	asn	thr	asp	ile	thr	ser	ile
b	ala	ala	pro	val	ala val	ala	pro	ala	glu	ala	asn	asp				
		50							▼			60				
a	thr	ser	asn	gly	gly	arg	val	lys	cys	met	gln	val	trp	pro	pro	ile
b										met	met	val	trp	thr	pro	val
a	gly	lys	lys	lys	phe											
b	asn	asn	lys	met	phe											

Fig. 4. The amino acid sequence of transit peptides. The seqtences have been derived from the pea small subunit sequence SS3.6 and by sequencing the small subunit polypeptide precursor from Chamydomonas (20).

Hybridization with ^{32}P-labeled small subunit cDNA showed the presence of five hybridizing Eco R1 bands of 11.0, 8.0, 5.4, 3.6, and 3.3 kb (Fig. 5). Two of these bands (3.6 and 8.0 kb) correspond to the small subunit genes that have been isolated and characterized. From these studies it appears that there are approximately five small subunit genes, two of which are tightly linked. I had previously estimated from solution hybridization studies that there were from one to five small subunit genes (15).

Diverged Genes Encoding the Small Subunit of RuBP Carboxylase

Four pea DNA sequences complementary to mRNA encoding the small subunit of RuBP carboxylase have now been characterized. No two of these sequences are identical. The two genomic sequences are distinct and neither of these are the structural genes for the two cDNA sequences that have been characterized. However, the two genomic sequences (SS3.6 and SS8.0), and the cDNA sequence that we have previously characterized (pSS15) (17) are similar and encode an identical mature small subunit. This polypeptide differs by eight amino acids from the mature small subunit polypeptide encoded by the cDNA sequence (pSSU1) characterized by Bedbrook et al. (18). These observations may reflect the expression of multiple genes encoding distinct small subunit polypeptides. It should be noted, however,

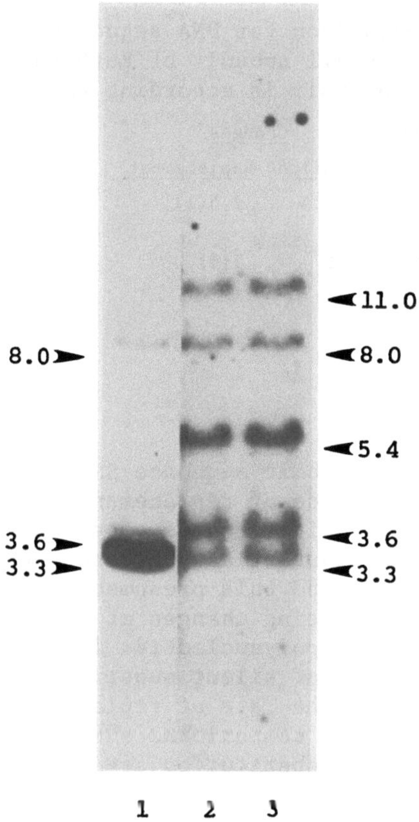

Fig. 5. Hybridization of the cDNA (pSS15) for the small subunit of RuBP carboxylase with Eco R1 restricted pea DNA. In track 1 are Eco R1 restricted pBR327 (3.3 kb) and the cloned small subunit genomic Eco R1 fragments of 3.6 and 8.0 kb. Tracks 2 and 3 contain Eco R1 digests of pea DNA. The restriction fragments were electrophoresed in 0.8% agarose gels and then transferred to nitrocellulose prior to hybridization.

that the two cDNA sequences have been obtained from two different pea varieties (P. sativum var. Progress #9 and P. sativum var. Feltham First) and that no heterogeneity for the small subunit has been observed, either at the cDNA level (18), or at the polypeptide level (21), for one of these varieties (var. Feltham First). These observations raise the interesting possibility that there are intervarietal differences, either at the level of small subunit gene structure or expression.

There is an additional interesting feature concerning the comparative analyses of the various small subunit DNA sequences. From a quantitative comparison of silent and replacement nucleotide substitutions (Tab. 1) it is observed, on comparing the cDNA

Table 1. Sequence comparison for DNA sequences SS3.6 and pSSU1 encoding the small subunit of RuBP carboxylase. The divergence analysis is according to Perler et al. (22).

	Nucleotides	Changes		% Corrected Divergence	
		Silent	Replacement	Silent	Replacement
Exon #2	135	0.5	2.5(1)	2.1	2.5
Exon #3	228	2	7(7)	4.5	4.0

sequence pSSU1 with the genomic sequence SS3.6, that these sequences show a relatively high degree of replacement substitution relative to silent substitution. Initial rates of silent substitution usually occur at a rate substantially greater than found for replacement substitution, and this presumably reflects the stronger selection pressures restricting changes at the polypeptide level relative to changes at the polynucleotide level. In globin genes, for example, initial rates of silent substitution occur at approximately seven times the rate of replacement substitutions (22). In contrast, for the comparisons shown in Tab. 1, the rates of silent and replacement substitution are essentially the same.

If we assume that the rate of silent nucleotide substitution for the small subunit DNA sequences is approximately constant, then we can interpret the above results in two opposing ways. It is possible that the two DNA sequences, pSSU1 and SS3.6, have diverged from one another under conditions where there were relatively few constraints for replacement substitutions. This interpretation would infer that whatever the role of the small subunit of RuBP carboxylase is, that in this particular example, this role is not affected by relatively unrestricted changes in the polypeptide sequence. It should be noted that this is certainly not a general characteristic of small subunit polypeptide sequences as indicated, for example, by the identity of the mature polypeptides encoded by the diverged DNA sequences SS3.6, SS8.0, and pSS15. An alternative interpretation of the above results is that used by Perler et al. (22) in their analysis of the pair of diverged rabbit β globin genes. According to this interpretation there has been a rapid divergence of the two genes, SS3.6 and pSSU1, and this divergence was due to positive selection forces favoring specific changes in the small subunit polypeptide. To distinguish between these two explanations for the rapid accumulation of amino acid changes will require a much greater level of understanding of the functional role of the small subunit of RuBP carboxylase than is presently available.

ACKNOWLEDGEMENTS

I thank Parkash Cashmore for valuable technical assistance. This work was supported by U.S. Department of Energy Research Contract DE-AC02-80ER10581 and National Institutes of Health Research Grant GM31137 01 and the U.S. Department of Agriculture Research Grant 59-2368-1-1-740-0.

REFERENCES

1. Ellis, R.J. 1979. The most abundant protein in the world. Trends in Biochem. Sci. 4:241-244.
2. Lorimer, G.H. 1981. The carboxylation and oxygenation of ribulose-1,5-bisphosphate: The primary events in photosynthesis and photorespiration. Ann. Rev. Plant Physiol. 32:349-383.
3. Dobberstein, B., G. Blobel, and N.-H. Chua. 1977. In vitro synthesis and processing of a putative precursor for the small subunit of ribulose-1,5-bisphosphate carboxylase of Chlamydomonas reinhardtii. Proc. Natl. Acad. Sci. USA 74:1082-1085.
4. Cashmore, A.R., M.K. Broadhurst, and R.E. Gray. 1978. Cell-free synthesis of leaf protein: Identification of an apparent precursor of the small subunit of ribulose-1,5-bisphosphate carboxylase. Proc. Natl. Acad. Sci. USA 75:655-659.
5. Highfield, P.E., and R.J. Ellis. 1978. Synthesis and transport of the small subunit of chloroplast ribulose-bisphosphate carboxylase. Nature 271:420-424.
6. Chua, N.-H., and G.W. Schmidt. 1978. Post-translational transport into intact chloroplasts of a precursor to the small subunit of ribulose-1,5-bisphosphate carboxylase. Proc. Natl. Acad. Sci. USA 75:6110-6114.
7. Bedbrook, J.R., S. Smith, and R.J. Ellis. 1980. Molecular cloning and sequencing of cDNA encoding the precursor to the small subunit of chloroplast ribulose-1,5-bisphosphate carboxylase. Nature 287:692-697.
8. Smith, S.M., and R.J. Ellis. 1981. Light-stimulated accumulation of transcripts of nuclear and chloroplast genes for subunits of the chloroplast enzyme ribulosebisphosphate carboxylase. J. Mol. and Applied Genetics 1:127-137.
9. Tobin, E.M., and J.L. Suttie. 1980. Light effects on the synthesis of ribulose-1,5-bisphosphate carboxylase in Lemna gibba L. Plant Physiol. 65:641-647.
10. Tobin, E.M. 1981. Phytochrome-mediated regulation of messenger RNAs for the small subunit of ribulose-1,5-bisphosphate carboxylase and the light-harvesting chlorophyll a/b protein in Lemna gibba. Plant Mol. Biol. 1:35-51.
11. Huber, S.C., T.C. Hall, and G.E. Edwards. 1976. Differential

localisation of Fraction I protein between chloroplast types. Plant Physiol. 57:730-733.

12. Kirchanski, S.J., and R.B. Park. 1976. Comparative studies of the thylakoid proteins of mesophyll and bundle sheath plastids of Zea mays. Plant Physiol. 58:345-349.
13. Whitfield, P.R., D. Spencer, and W. Bottomely. 1973. In The Biochemistry of Gene Expression in Higher Organisms pp. 504-552. J.K. Pollak and J.W. Lee Australia and New Zealand Book Co., Sydney.
14. Possingham, J.V., and W. Saurer. 1969. Changes in chloroplast number per cell during leaf development in spinach. Planta 86:186-194.
15. Cashmore, A.R. 1979. Reiteration frequency of the gene coding for the small subunit of ribulose-1,5-bisphosphate carboxylase. Cell 17:383-388.
16. Broglie, R., G. Bellemare, S.G. Bartlett, N.-H. Chua, and A.R. Cashmore. 1981. Cloned DNA sequences complementary to mRNAs encoding precursors to the small subunit of ribulose-1,5-bisphosphate carboxylase and a chlorophyll a/b binding polypeptide. Proc. Natl. Acad. Sci. USA 78:7304-7308.
17. Coruzzi, G., R. Broglie, A.R. Cashmore, and N.-H. Chua. 1982. DNA sequence of cDNA clones encoding two chloroplast proteins. J. Biol. Chem. (in press).
18. Bedbrook, J.R., S. Smith, and R.J. Ellis. 1980. Molecular cloning and sequencing of cDNA encoding the precursor to the small subunit of the chloroplast enzyme ribulose-1,5-bisphosphate carboxylase. Nature 287:692-697.
19. Proudfoot, N.J., and G.G. Brownlee. 1976. 3' non-coding region sequences in eukaryotic messenger RNA. Nature 263:211-214.
20. Schmidt, G.W., A. Devillers-Thiery, H. Desruisseaux, G. Blobel, and N.-H. Chua. 1979. NH_2-terminal amino acid sequences of precursor and mature forms of the ribulose-1,5-bisphosphate carboxlylase small subunit from Chlamydomonas reinhardtii. J. Cell Biol. 83:615-622.
21. Takruri, I.A.H., D. Boulter, and R.J. Ellis. 1981. Amino acid sequence of the small subunit of ribulose-1,5-bisphosphate carboxylase of Pisum sativum. Phytochemistry 20:413-415.
22. Perler, F., A. Efstratiadis, P. Lomedico, W. Gilbert, R. Kolodner, and J. Dodgson. 1980. The evolution of genes: The chicken pre-poinsulin gene. Cell 20:555-565.

OSMOREGULATORY (Osm) GENES AND OSMOPROTECTIVE COMPOUNDS

A.R. Strøm*, D. LeRudulier**, M.W. Jakowec, R.C. Bunnell, and R.C. Valentine

Plant Growth Laboratory
University of California
Davis, California 95616

ABSTRACT

A series of compounds, including glycine betaine and proline, known to accumulate in plants during osmotic stress, have been found to function as osmoprotective compounds for bacteria. In fulfilling "Koch's Postulates" for the biological activity of these compounds, they have been found to protect against osmotic stress when added to the growth medium in relatively low concentration, or when synthesized in the cell. Cells may accumulate very high intracellular levels corresponding to the osmolarity of the medium using uptake systems that appear to be osmotically modulated. A proline over-producing mutation conferring osmotic tolerance has been constructed. Molecular cloning of an osmotic tolerance gene has been achieved. A unified concept of osmoregulation in microorganisms, animals and plants is discussed with some possible applications being pointed out.

INTRODUCTION

The purpose of this paper is to discuss the general picture of osmoregulation, cellular adaptation to osmotic stress, that has emerged during the past few decades. Broadly speaking, the field

* Institute of Fisheries, University of Tromsø, N-9001 Tromsø, Norway; ** Laboratoire de Physiologie Végétale, Univerité de Rennes I, 35042 Rennes Cedex, France.

can be divided into three areas:

Classification of various osmoregulatory compounds which accumulate during stress in plants (e.g., seeds, pollen), animals and microorganisms.

Studies of the mechanism (physiology, biochemistry, genetics) of osmoregulation.

Applications with emphasis on drought and salinity tolerance in crop plants.

It is safe to say that the major incentive in this field comes from the age-old dream of plant scientists to create more hardy varieties of crop plants resistant to drought and other forms of environmental stress. In addition, there are major economic incentives most obvious in areas where water is becoming scarce or more costly to apply. However, it should be emphasized that crops depending on natural rainfall are subjected to a series of "mini-droughts" during the normal growing season, perhaps the most important aspect of water stress on crop production in the USA.

It is also becoming clear that osmoregulation represents a fertile area for fundamental science and it is easy to predict that many exciting discoveries are in store.

Given the dual incentive of major economic importance and exciting science, the National Science Foundation chose to sponsor a series of symposia covering various aspects of this area during the last few years (several symposia volumes have appeared or will be published shortly; e.g., see refs. 1-4). These symposia and others (5) have helped to focus attention on this important field and have helped to guide research goals and priorities. The emerging research program in this laboratory has been shaped by the high priority given by the administration at UC Davis and by inputs from several established laboratories at UC Davis. At least two other authors in this volume will be addressing questions of osmoregulation in plants; the reader is referred to their papers for additional information.

We have chosen to study the molecular biology of osmoregulation, research which combines aspects of the biochemistry and genetics of this process. The various sections below are organized to emphasize three major points: i) evidence that a series of organic compounds known to accumulate during stress in plants function as osmoprotective compounds; ii) elucidation of a gene governing osmotic tolerance via overproduction of proline as an osmoprotective compound; iii) speculations on a unified concept of cellular adaptation to osmotic stress with broad applications in the plant, animal and microbial worlds.

Osmoprotective Compounds

It has been known for many years that various organic compounds may accumulate in plants following periods of water stress (see Fig. 1). Likewise, seeds which represent a severely dehydrated state have been found to be a rich source of these compounds. This has led to the interesting hypothesis that these compounds work as osmoregulators with the biological role of balancing the osmotic strength of the cytoplasm with that of the environment. For a comprehensive review, see reference (5,6). Unfortunately, the correlative evidence currently available linking these compounds to osmoregulation while tempting, does not provide sufficient evidence for this point and, consequently, has been a "thorn in the side" of advocates for some time.

Indeed, the best evidence now comes from bacteria and dates back more than a quarter of a century (7,8). For example, in a pioneering experiment, Christian found that the growth of *Salmonella* spp. under conditions of inhibitory osmolarity was stimulated by supplementation of the medium with yeast extract or later with the amino acid L-proline. Shortly afterward, the Carnegie group demonstrated that proline levels in the cell reflected the osmotic strength of the medium with high osmotic strength resulting in extremely high proline levels (9). These early experiments provided an extremely valuable approach to the nagging question of the biological activity of suspected osmoregulatory compounds since they allowed the investigator to simply supplement a tester strain of bacteria with the compound. About a decade passed before Avi-Dohr and coworkers (10,11) demonstrated that glycine betaine, often found in plants, stimulated respiration at high osmolarities in a marine bacterium; however, the levels used in most experiments were very high and seemed to preclude a natural role for this compound as a cytoplasmic osmoticum in this bacterium. This objection has now been overcome by the discovery of Le Rudulier that enteric bacteria, such as *Escherichia coli*, respond to concentrations some 100 times lower than reported earlier (12). This finding provides the basis for the discussion in this section (see references 13-16 for work on proline in osmoregulation, experiments which will not be emphasized here).

Biological activity. For the purposes of this paper, a compound is considered to be osmoprotective if, at inhibitory osmolarities, it stimulates the growth of a tester strain of bacteria. The structures of a series of compounds (and their derivatives) found to protect microbial cells are summarized in Fig. 1.

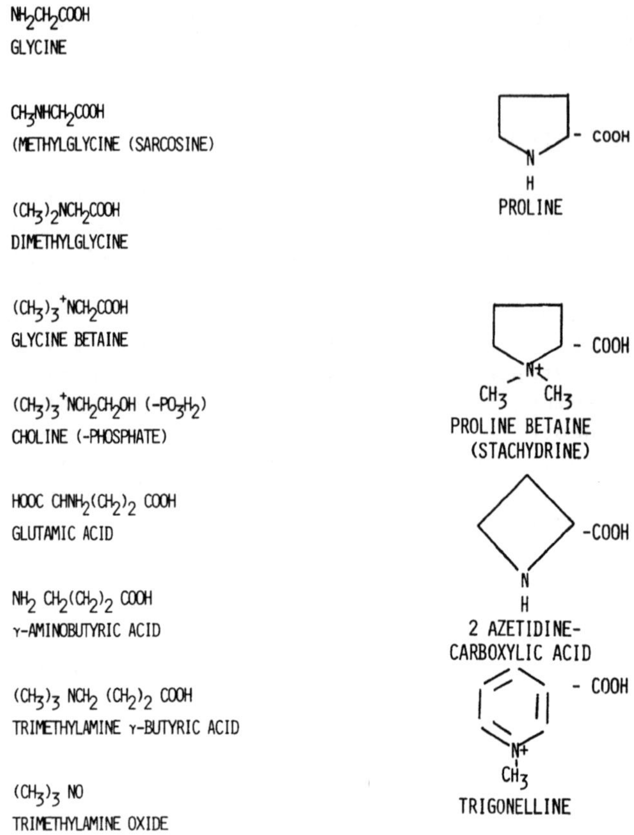

Fig. 1. Structures of several osmoprotective compounds and their derivatives used in this study. Note that many of these compounds have been found to accumulate in green plants during stress.

Enteric bacteria, including the much studied *E. coli*, *Salmonella* spp. and *Klebsiella* spp. are suitable tester organisms. This point is illustrated in Fig. 2 where *E. coli* was streaked on Petri dishes with glucose minimal medium containing 0.8 M NaCl and supplemented with 1 mM of various possible osmoprotective compounds. In this simple experiment, growth occurred on agar medium with betaine glycine (BET), choline (CHO) and dimethylglycine (DMG), the former compound giving the fastest growth response. No growth ensued on media with γ-amino-butyrate (GAB) and glutamate (not shown), found to accumulate in high levels in this organism during osmotic stress (17), as well as on media with glycine (GLY), sarcosine (SAR), choline phosphate, trigonelline, trimethylamine oxide and on control with no supplements (not shown). Although these results are also not shown, it should be noted that proline betaine (stachydrine) and trimethylamine γ-butyrate have been found to be among the most potent compounds.

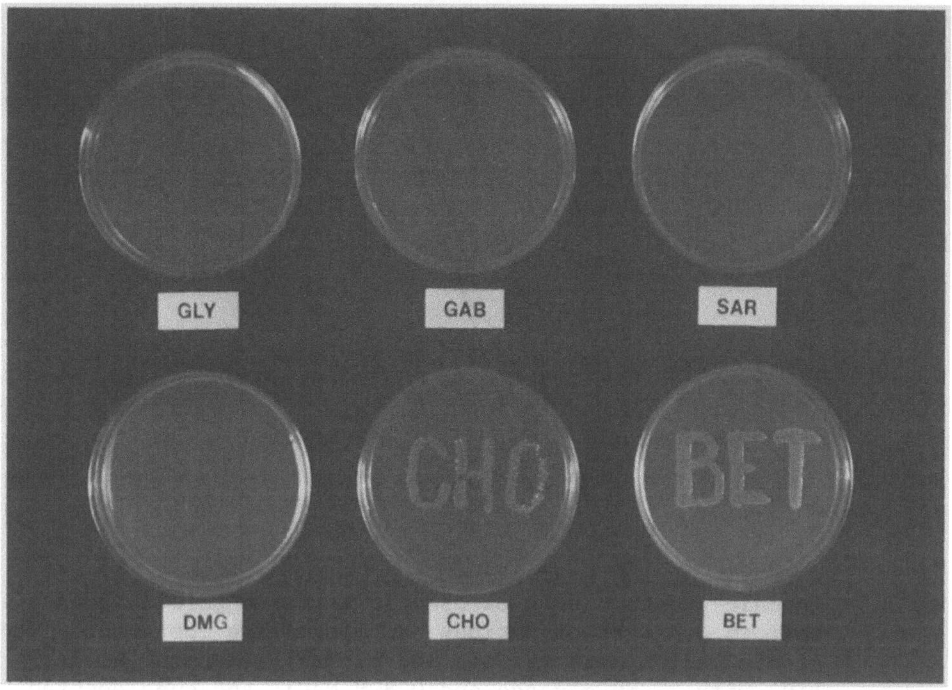

Fig. 2. Stimulation of growth of E. coli at inhibitory osmolarity (0.8 M NaCl) by various osmoprotective compounds. The agar medium used was M63 minimal medium of Miller (19) with 22 mM glucose and 0.8 M NaCl, and supplemented with 1 mM of glycine (GLY), γ-aminobutyrate (GAB), methylglycine (SAR), dimethylglycine (DMG), choline (CHO), and glycine betaine (BET). The bacteria were streaked onto the Petri dishes and incubated aerobically at 37°C. In all experiments with E. coli the k10 strain was used.

The growth response of E. coli to glycine betaine and some of its derivatives are further summarized in Fig. 3. Note the hierarchy of responses observed in liquid minimal media with glycine betaine being most potent followed by dimethylglycine and choline, with glycine being inactive. As is characteristic of this strain under these conditions (0.8 M NaCl), slow growth occurs after a prolonged lag period in media with no protective compounds. The nature of the intracellular osmotica produced by these cells following the adaptation is currently being investigated. The probable important role of the accumulation of K^+ in these organisms during osmotic stress is discussed by Epstein and coworkers and is not further described. (For latest papers and previous articles, see ref. 18).

In Fig. 4 we have plotted maximum growth yield of E. coli in liquid glucose media (1M NaCl) supplemented with various concentrations of glycine betaine. Up to about 0.2 mM betaine limited the

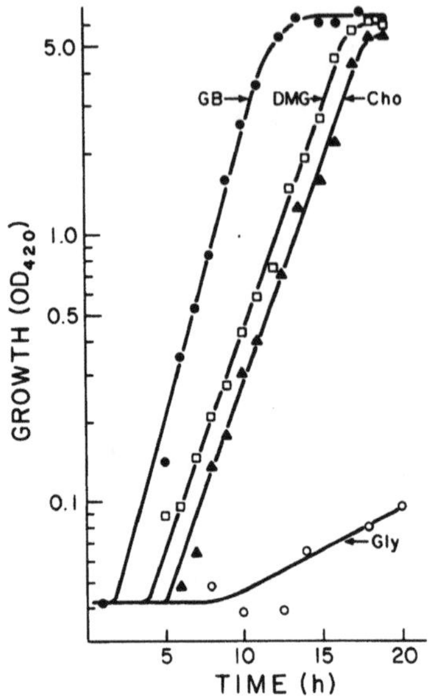

Fig. 3. Aerobic growth curves of osmotic stressed (0.8 M NaCl) E. coli in presence of various osmoprotective compounds. The growth medium used was the M63 minimal medium of Miller (19) with 22 mM glucose and 0.8 M NaCl, and supplemented with 1 mM of glycine betaine (●-●), dimethylglycine (□-□), choline (▲-▲), and glycine (○-○). The cultures were inoculated with an overnight culture of the organism and incubated with shaking at 37°C. Optical density was recorded at 420 nm (OD_{420}) using a Gilford, Stasar II, spectrophotometer.

maximum growth of the organism, and as little as 0.02 mM gave a clear stimulation of the growth yield. Together with the plate assays depicted in Fig. 5, this demonstrated that growth of E. coli at higher NaCl concentration may be developed into a bioassay for osmoprotective compounds. The upper panel of Fig. 5 illustrates that tiny colonies developed on Petri dishes with medium (0.8 M NaCl) containing approximately 0.02 mM glycine betaine (labeled 0.005 ml) and that the sizes of colonies increased as the concentration of glycine betaine was increased to 2.0 mM (labeled 0.5 ml). Similarly, the growth of the tester organism was stimulated by addition of increasing amounts of extract of sugar beet seeds, known to contain betaines (Fig. 5, lower panel).

We have reported previously that the process of N_2 fixation in Klebsiella pneumoniae is particularly sensitive to osmotic stress with the osmoprotective compounds displaying dramatic stimulation of this system (12). The potential of this reaction as a bioassay

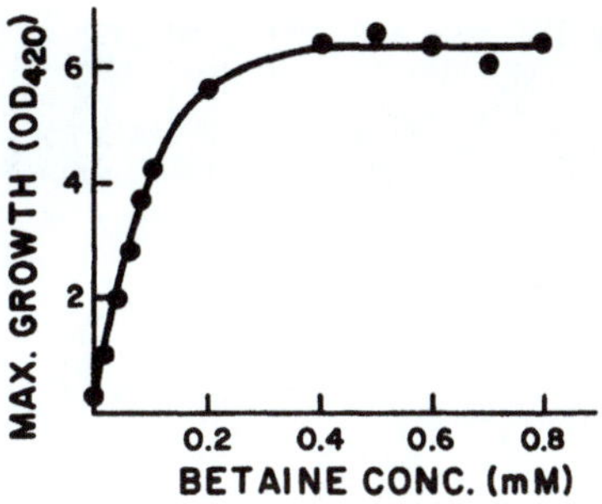

Fig. 4. Stimulation of aerobic growth yield of E. coli k10 grown under osmotic stress (1 M NaCl) in presence of various concentrations of glycine betaine. Basic medium and growth measurement are described in Fig. 3. The figure shows maximum optical density (OD_{420}) recorded for the cultures.

system for osmoprotective compounds is summarized in Fig. 6 where N_2 fixation activity is plotted as a function of the number of methyl moieties of the glycine betaine series (methylglycine, 1 methyl; dimethylglycine, 2 methyls; glycine betaine, 3 methyls).

In summarizing this section, in addition to inorganic potassium ions at least eight organic molecules have been implicated in osmotic tolerance in enteric bacteria. These are glycine betaine, choline, dimethylglycine, proline betaine, proline, trimethylamine γ-butyric acid, glutamate, and γ-aminobutyrate. Of the eight organic molecules, the six former behave as osmoprotective compounds when added exogenously. Further refinements of the bioassay procedures, along with construction of suitable tester organisms, might permit development of quantitative bioassays based on the osmoprotective properties of these and other compounds.

Metabolism. Studies on the metabolism and biochemistry of the various osmoprotective compounds are aimed at determining how the cell manufactures these molecules and how their biosynthesis is modulated. Recent reports have dealt with the metabolism of proline and glycine betaine with emphasis on higher plants (5,12). Emphasis in this section will be on biosynthetic pathways of proline and glycine betaine in enteric bacteria with the next section dealing with accumulation from the medium. A detailed literature review of proline and glycine betaine biosynthesis will not be given in favor of discussion of two specific points related to the production of these compounds during osmotic stress. As shown in the diagram below, the first enzyme in the proline biosynthetic pathway called γ-glutamylkinase (pro B gene product) catalyses the phosphorylation of glutamate in a reaction strongly feedback inhibited by proline.

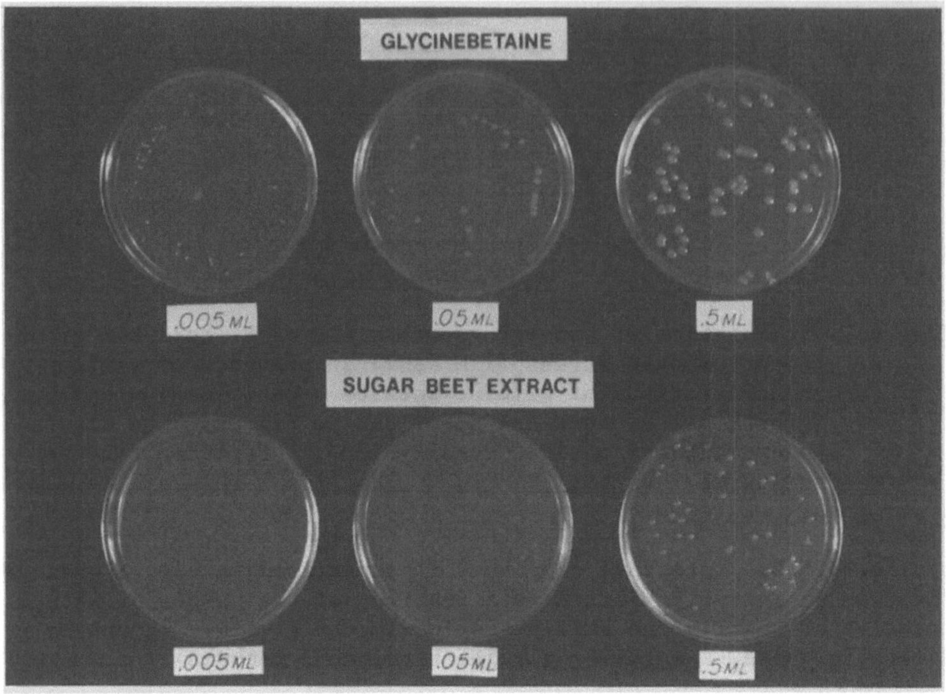

Fig. 5. Stimulation of growth of E. coli at inhibitory osmolarity (0.8 M NaCl) by glycine betaine versus extract of sugar beet seeds. Beet extract was prepared by treating 4 g (dry weight) of seed with ethanol, and redissolving the residue in 5 ml water after removal of ethanol on a rotary evaporator 20). Growth medium used is described in legend of Fig. 2. To approximately 25 ml of solidified agar medium were added indicated amounts of 1 M glycine betaine and beet extract. Petri dishes were inoculated with a proper dilution of overnight culture of bacteria, and incubated aerobically at 37°C.

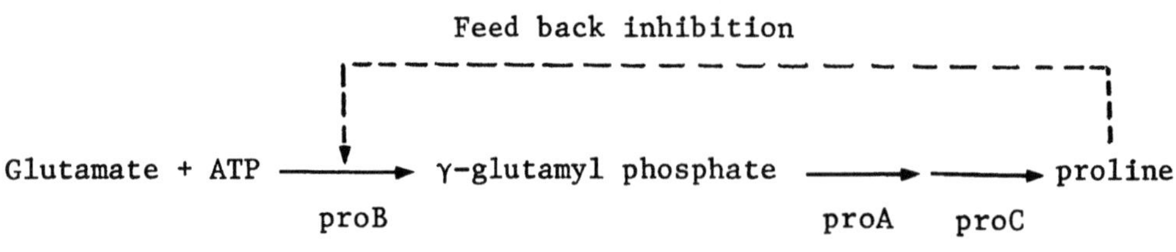

Indeed, initial characterization of this enzyme was based on an assay system dependent on "proline inhibitable" activity units for the enzyme (21). Regulatory mutants of the kinase less responsive to modulation by proline have been reported (22). It should be pointed out that, during stress, enteric bacteria do not normally

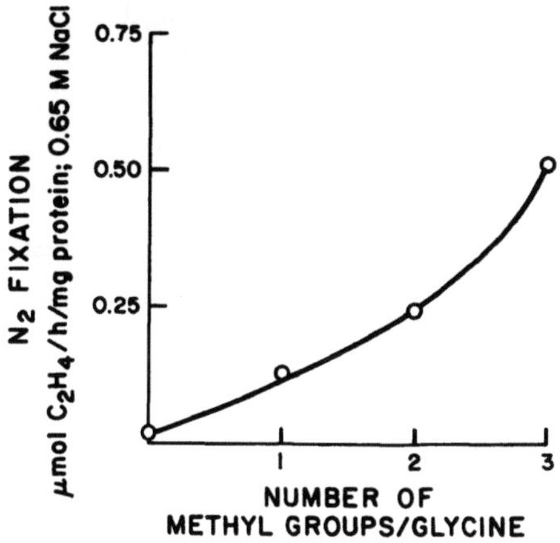

Fig. 6. Stimulation of N_2 fixation activity of Klebsiella pneumoniae as a function of the number of N-methyl groups of the glycine moiety (methylglycine, 1 methyl; dimethylglycine, 2 methyls; glycine betaine, 3 methyls). Conditions were as described previously except the inhibitory level of NACl was 0.65 M. The wild type strains of Klebsiella (M5A1) was used.

overproduce proline by this pathway, indicating that feedback inhibition of the kinase, as well as other control systems modulating proline levels in the cell operate more or less normally during stress; however, this is not the case for accumulation of proline from the medium, which is greatly accelerated (see below). The proline overproducing mutation constructed by Csonka in this laboratory (15) has all of the characteristics of a special class of feedback inhibition mutants of the kinase. It is interesting to speculate that the kinase of the overproducing strain has been mutationally altered such that the proline binding site is allosterically responsive to osmotic conditions of the cytoplasm, changes which are closely linked to perturbations in osmotic strength of the medium. The net effect of this mutation would be to place the proline pathway, probably rate-limited by the kinase step, in synchrony with the rising need for proline as osmoprotectant during stress. Characterization of the putative mutant kinase is currently underway and should help answer the interesting question of whether the mutant protein is, indeed, a sort of "osmo-sensing" protein whose activity is modulated as a function of the osmotic strength of the environment.

The next experiment involves the biosynthesis of glycine betaine from choline, a reaction which is postulated to occur by one of the two routes outlined below:

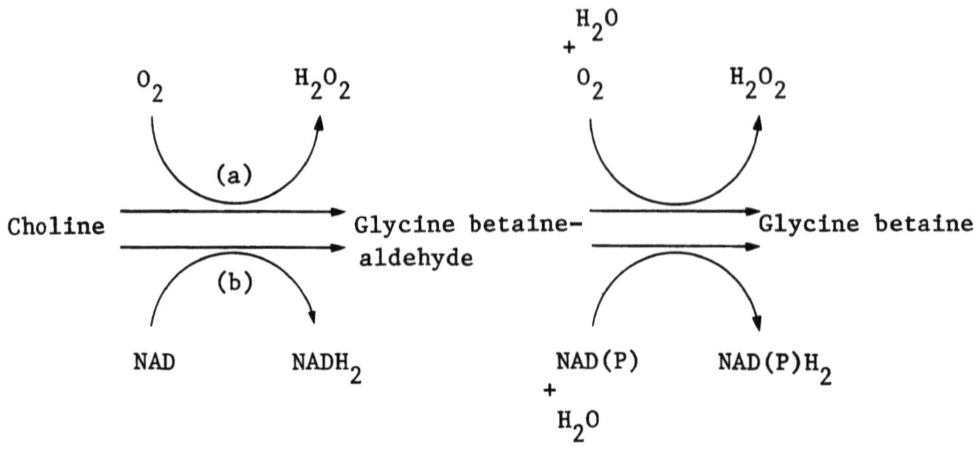

The top reaction (a) involving molecular oxygen (O_2) is referred to as the oxidase pathway reported to be present in some bacteria (23); the dehydrogenase route (b) is apparently active in other bacteria (24). The details of this pathway are not yet known in enteric bacteria; however, as summarized in Fig. 7, E. coli during osmotic stress rapidly converts radioactive choline to glycine betaine (preliminary identification) which accumulates internally. Also, in a marine bacterium, the effect of choline on respiration at high osmolarity has been attributed to its conversion to glycine betaine (25). It is also interesting to note that Galinski and Truper (26) recently reported the de novo synthesis of glycine betaine in a strongly salt tolerant, photosynthetic bacterium; the role of choline was not reported in this organism.

Uptake. Organisms may acquire their needed supply of osmoprotective compounds via synthesis or by uptake from the environment (or a combination of both). Experiments in this section are focused on the nature of the novel uptake systems which work to accumulate these molecules (Fig. 1) often in massive doses in the stressed cell. Generally speaking, there are two striking features of such uptake systems. The first point is that cells accumulate these compounds only during conditions of osmotic stress. This is an exciting feature of these systems and clearly deserves further study. Historically, the osmotic modulation of proline accumulation was first observed during the pioneering experiments of the Carnegie Group studying metabolism of E. coli. These workers observed a linear relationship between environmental osmolality and intracellular levels of proline (9). In a continued study of this proline uptake system, Csonka hypothesized on genetic evidence that a new, osmotically activated proline permease is responsible for the accumulation of high intracellular levels of proline during osmotic stress (L. Csonka, in press).

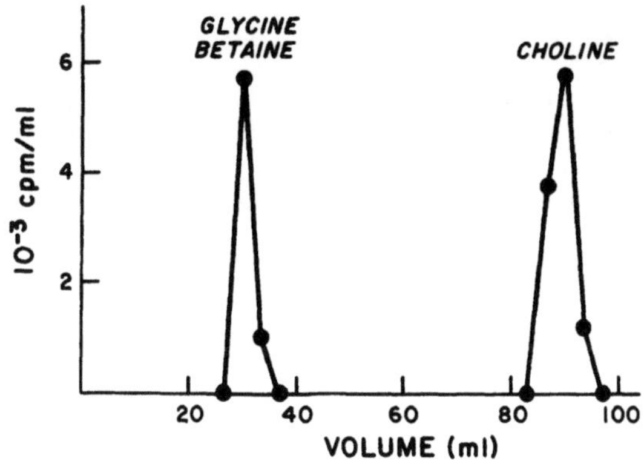

Fig. 7. Chromatographic evidene of the uptake and conversion of choline to glycine betaine during osmotic stress in E. coli. The organism was grown aerobically at 37°C in minimal glycose medium (see Fig. 2) containing 0.8 M NaCl and 1 mM [methyl-^{14}C] choline (sp. act. 0.011 mCi/mmol). The cells were harvested in late exponential growth phase and extracted with 0.6 M perchloric acid. Extract neutralized with KOH was applied to a 60 x 1 cm column of Amberlite CG-50 Type 1 (Fisher Scientific Company, N.J.). The resin was equilibrated with citrate-phosphate buffer, pH 7.3 and the sample was eluted with the same buffer, Ph 5, as described (27). All radioactive material in the cells was eluted in the first peak corresponding to the eultion volume of betaine; the second peak on the chromatogram represents the elution of a standard of choline.

Thus, it was not surprising to find, as summarized in Fig. 8, that the accumulation of other osmoprotective compounds, such as glycine betaine is closely linked to the osmolarity of the medium. Whereas this physiological effect is striking with preliminary evidence suggesting that virtually all of the osmoprotective compounds listed in Fig. 1 behave similarly, there is still little information regarding the molecular basis of this dramatic change in the cells' ability to accumulate these compounds. The study of this phenomena at the level of membrane vesicles (28) where the effects of osmotic strength on uptake can be directly tested is of considerable interest for future work.

The second striking feature of these osmotically modulated uptake systems in the massive levels of such molecules which may finally be concentrated by the cell during stress. For example, cells challenged with a 1 M solution of NaCl and supplemented with as little as 10^{-5} - 10^{-4} M glycine betaine have been found to concentrate these molecules to a level of approximately about 1 M in the cytoplasm (12).

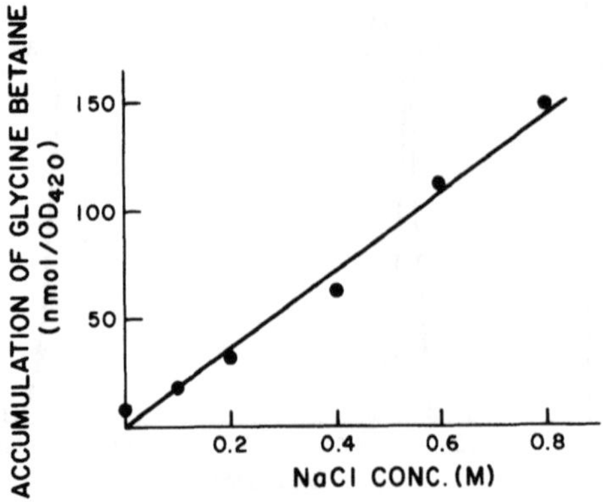

Fig. 8. Osmotically modulated accumulation of betaine glycine in E. coli. Inoculum was grown at 37°C in minimal glucose medium to OD_{420} equlas 1 (see Fig. 3). Cells were collected by centrifugation and resuspended in the same volume of medium containing various concentrations of NaCl and 1 mM [methyl-^{14}C] choline (sp. act. 0.011 mCi/mmol). Aerobic growth was continued at 37°C for 1.5 generation and the cells were removed by centrifugation. Accumulation of betaine in cells was measured by determining the disappearance of ^{14}C-labeled material from the cleared growth media using a liquid scintillation counter. All labeled material could be recovered by extractions of cells, and it chromatographed similarly to glycine betaine (Fig. 7).

In summarizing this section, it is clear that bacteria have evolved sophisticated systems for uptake of osmoprotective compounds from the environment. Uptake systems are somehow modulated by the osmotic strength of the medium. It should be pointed out that the logical role of these uptake systems is not only to concentrate molecules supplied intact from the environment, but also to insure that molecules once biosynthesized by the cell do not leak back into the environment. Plants and, to a lesser extent, animals and microbes would appear to represent the major sources of compounds such as choline and betaine in the biosphere with choline being an intermediate in synthesis of major classes of plant lipids.

Cloning Osm Genes

The identification of organic molecules which confer osmotic tolerance in bacteria and the study of osmotic tolerant mutants has led to the development of selection procedures for isolating Osm genes (Fig. 9). The first osmotic tolerance gene to be cloned was a

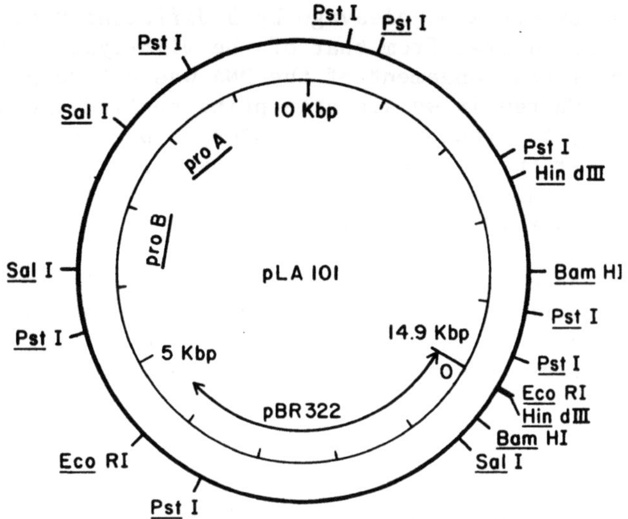

Fig. 9. Molecular cloning of proline overproducing gene conferring osmotic tolerance: Restriction endonuclease map of recombinant plasmid (see test for details).

proline over-producing mutation (gene) coding for osmotic tolerance (12), work to be described in detail elsewhere (M. Mahan and L. Csonka, in preparation). Since recombinant DNA techniques are in widespread use, the cloning will only be discussed briefly. First, DNA was isolated from the osmotolerant Salmonella strain and its wild-type parent (12). The DNA was cleaved at a large number of sites by the nucleotide-sequence specific edonuclease, ECOR1. The fragments generated were inserted into the cloning vector pBR322 and the hybrid plasmids were introduced into an E. coli mutant lacking proBA genes. The resultant cells were grown on a medium lacking proline, so that only those cells which received the proBA genes from Salmonella could grow. A number of such strains were obtained using DNA from both the osmotolerant mutant and the wild type parental strain. All the strains carrying the proBA genes of the wild type strain were sensitive to L-azetidine-2-carboxylate, whereas all the strains carrying the proBA genes of the mutant were resistant to the analogue, indicating that the mutation is very closely linked to proBA. In addition to being resistant to L-azetidine-2-carboxylate, strains carrying the mutant DNA were as osmotolerant as the original donor strain.

The results of a preliminary physical characterization of the recombinant plasmid carrying the proline over-producing mutation is summarized in Fig. 9. The mutation is present on a fragment which carries the proBA segment and whose size is 10.3×10^3 base pairs. It is interesting that the size of the insert and its physical map (as determined by sites of cleavage by 5 different restriction enzymes) does not differ from that of the wild-type region, suggesting that a major rearrangement of the DNA was not responsible for the mutation. Currently we are attempting to find the minimum length necessary to produce the osmotolerant phenotype, and to determine whether the proB or proA genes are required for its expression. Current results indicate that the necessary region is shorter than 4.35×10^3 base pairs.

Although the cloning vector pBR322 is a multicopy plasmid, the presence of the wild type proBA genes on it did not confer increased osmotolerance and pointed to the need for some kind of regulatory mutation to increase proline biosynthesis as mentioned above.

Unified Concept

The discussion in this section will be centered on Fig. 10 which is intended as a sort of summary of current concepts of the fundamentals of osmoregulation and how these ideas might be directed towards some potential applications. Some of these amount to "pet" ideas, and we will not attempt a comprehensive literature survey at this point.

The center of the figure is intended to represent osmoregulatory events at the cellular level with some possible areas of application shown at the points of the triangle. Researchers are now getting their teeth into the fundamentals of this phenomenon although much remains to be understood; for example, little information is available on the osmotic sensing systems sometimes referred to as "trigger mechanism" or sensing device which initiates the process, an important area for future research. The remaining topics listed represent some of the most interesting aspects of current research on the molecular basis of osmoregulation. In proceeding down this list it is now clear that many of the putative organic osmoregulatory compounds (osmotica) isolated from plant, microbial, and animal sources are indeed biologically active as osmoprotective compounds in bacteria, a subject covered in some detail above. Still, it is fair to say that the evidence for a direct biological role of these compounds in plants and animals is very thin. This is not to say that we believe the contrary; rather, it must be recognized, for the sake of remedying this situation, that some hard experimental evidence for the function of these compounds in plants is of the highest priority. Accepting the biological function of compounds such as glycine betaine as osmoprotecting compounds, at least for bacteria, the next question concerns the mechanisms of synthesis or acquisition of these compounds. For example, it was pointed out earlier that the overproduction of proline leading to osmotic tolerance may have its roots in a mutated proline feedback inhibition site on the surface of the enzyme, thousand or so different enzymes working at any one time in the E. coli cell are able to adapt to this shift in solute. Indeed, observations such as this have given rise to the use of the term "compatible solute" to describe such compounds. However, it now appears that many, and perhaps virtually all, proteins may be capable of mutational alteration such that their catalytic activity may become responsive to the osmotic strength in the cell. For example, several workers (29-31) have shown that cells harboring temperature sensitive (ts) enzymes which normally lead to growth inhibition at elevated temperatures are stabilized by increasing the osmotic strength of the medium. The point is that the conformation of these proteins is somehow restored by changing the osmotic strength of the medium which we now know results in a corresponding change in the cytoplasm. In a way, such a temperature sensitive protein behaves as a sort of osmotic probe able to monitor changes in osmotic strength of the cytoplasm. Obviously, natural selection might lead to similar osmo-sensing proteins as mentioned above. Next, a second type of mutant approach which points in a similar direction will be mentioned briefly. Unintentionally, during isolate of large numbers of glycine betaine osmoregulatory mutants (mutants unable to respond to glycine betaine during stress), we have constructed a variety of mutants with just the opposite behavior from the osmotically activated (ts) types discussed above. Glycine betaine appears to be "toxic" to these mutants since, at high osmotic strength, and in the presence of exogenous glycine betaine, these mutants display lesions in a variety of specific metabolic pathways (e.g., amino acids, purines, pyrimidines, etc.). When supplemented with these compounds these mutants are able to grow at high osmolarities in the presence of glycine betaine. It is interesting to speculate that the mutated enzymes somehow become inactivated when bathed in elevated glycine betaine concentrations which may arise during osmotic stress (Strom and Bunnell, unpublished). The study of the activation and inactivation of enzymes as a function of osmotic strength may provide additional information on the mode of action of these interesting osmoregulatory compounds and ultimately may lead to some important applications.

Little has been said here about the potential for osmotic modulation of gene expression. This is a very interesting possibility that has recently been raised by Epstein and coworkers who have hypotesized that expression of certain potassium transport genes in E. coli may be controlled by the osmotic strength (18).

From the above discussion, it is clear that much work is needed to develop the understanding of the cellular basis of osmoregulation. Nevertheless, certain trends are becoming evident. The first generalization is that osmoregulation is of cardinal importance for essentially all forms of life - plants, microbes, and animals. The unity that is emerging is most evident between the microbial and plant world. However, animals should not be neglected. A couple of examples might suffice to make this point.

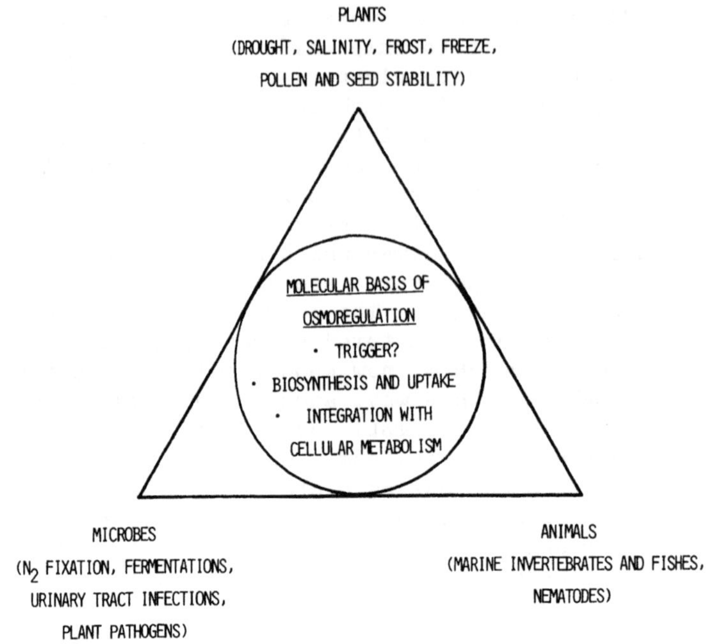

Fig. 10. A unified concept of osmoregulation and some potential applications (see text for details).

a mutation which, in essence, blocks feedback inhibition at high osmolarities. This might turn out to be an example of an "osmosensing" cytoplasmic enzyme. Indeed, the proteins which contribute to the osmotically modulated membrane uptake systems may represent natural examples of osmotically modulated membrane proteins. However, it should be kept in mind that the lipid portion of the membrane might undergo changes (i.e. both the protein part as well as the membrane component might be involved in osmotic modulation).

At first glance, the mode of action of osmoprotective compounds such as described above appears straightforward. As the name implies, they work to protect against dehydration of the cell by allowing the cell to balance its osmotic strength with that of the environment. Things never turn out to be so simple with recent studies showing us a new and more complex side of the general interaction between proteins and their solutes. The first point to recall is that during severe osmotic stress, the chemical composition of the cytoplasm which bathes all of the cellular machinery undergoes a major shift; for example, severely stressed cells of E. coli growing in 1 M NaCl adapt by accumulating an equivalent cytoplasmic level of a solute such as glycine betaine (i.e. ∿ 1 M cytoplasmic glycine betaine). In other words, the roughly one

All marine fishes and most marine invertebrates are known to accumulate high concentrations of trimethylamine oxide in their muscle tissue; the stability of the trimethylamine oxide pool in organisms not exposed to changes in salinity is shown to be extremely high (32,33). Trimethylamine oxide was postulated by Hoppe-Seyler more than 50 years ago to be an osmoprotective compound (for a review of early work, see 34). Various amounts of other betaines (i.e. glycine betaine, homarine, trigonelline and carnitine) are also found in marine invertebrates. The finding that the content of betaines are higher in the marine species than in their limnic counterparts (35,36) supports the notion that they play a role in cellular osmotic regulation. Finally, desiccant resistant cysts of nematodes, including forms of severe plant pests which survive for years in the soil, may be protected by high internal levels of osmoregulatory compounds (37).

Moving on to the microbial world, there is increasing evidence that microbes have evolved sophisticated mechanisms of osmoregulation as described above, mechanisms which, if understood, might lead to a more efficient harnessing of beneficial forms and more effective methods of combatting harmful pathogens. For example, *Rhizobium japonicum*, representing the most important form of seed inoculant for symbiotic N_2 fixation in the USA (soybeans), is prone to succumb quickly once inoculated onto the surface of dry seeds. This die-off may be reflected in the occurrence of a very weak set of Osm genes in these organisms, a project currently being researched in our laboratory. Similarly, more hardy varieties of fermentative organisms harboring more potent Osm genes might lead to increased yields of gasohol or pharmaceuticals. Pathogenic microorganisms for animals and plants might be easier to combat if there was better understanding of their mechanisms of osmoregulation. For example, *E. coli*'s ability to survive in a salty environment such as found in the urinary tract of humans, may help account for the virulence of the organism in the millions of cases of urinary tract infections reported each year, many of which are difficult to control. Similarly, the initiation and severity of many plant diseases are influenced by osmotic conditions of the host. Indeed, many obligate plant pathogens may have given up their own Osm genes in favor of those supplied by the host plant.

The last topic for discussion concerns osmoregulation in green plants, an area dominated by interest in drought and salinity tolerance (see papers in this volume by Wyn Jones and by Miflin and their reference lists for additional reading). In dispensing with one of the most often asked questions at this point it will clear the way for discussions below. Based on current concepts of osmoregulation (Fig. 10), would it be possible to breed crop plants more resistant to water stress? The answer is that this goal may be premature at this stage, a situation that could, however, change rapidly. A major request from plant breeders is the development of rapid assays for screening large numbers of samples of plant material for the presence or absence of various osmoprotective compounds. One approach is the development of tester strains of bacteria suitable for the rapid and quantitative bioassay of samples of plant tissues for osmoprotective compounds. Such strains are not available at this point; however, the state of the art of the microbiology is not that far away from this objective and, with effort, such strains could probably be developed.

In summary, we find it irresistible to accept the notion that osmoprotective compounds such as described in bacteria may play a similar critical role in plants. Where does this faith come from? The strongest points seem to be as follows:

> Molecules synthesized by plants during stress possess potent biological activity in protecting bacteria against osmotic stress.
>
> Seeds, pollen and other plant tissue subjected to dehydration often produce large amounts of these compounds.
>
> Chronologically, the levels of putative osmoprotective compounds increase (or decrease) in synchrony with stress.

Thus, it seems justified to go deeper into the nature of osmoregulation in plants with the knowledge that different plant parts and tissues may be genetically programmed to produce different patterns of osmoprotective compounds. For genetic engineers eyeing the isolation of Osm genes of plants, the tissue-specific response, particularly if it is of a high enough intensity, might be a blessing in disguise, permitting the fishing out of Osm genes already enriched in these tissues. If we have a takehome lesson for the plant breeder, it is not to give up, but to lower one's sights to specific targets where osmoregulation might represent a rate-limiting step at a particular time in crop development or in a particular tissue or seed. We cite as a possible example the case of what might be called "osmotic male sterility", which may be a particularly rewarding area (see discussion above on how tricks can be played on most enzymes to make them sensitive to, or dependent on, the osmoregulatory compounds synthesized by the stressed cell). Since pollen appears normally to produce high levels of these compounds, a simple mutation would be sufficient to sensitize any one of a multitude of critical enzymes potentially leading to male sterility.

ACKNOWLEDGEMENTS

This work was carried out under a grant from NSF (PFR 77-07301).

Any opinions, findings, and conclusions or recommendations expressed in this publication are those of the authors and do not necessarily reflect the views of NSF. M.J. was supported by a grant from the Kearney Research Foundation. One of us, A.R.S., was supported by a travel grant from the Norwegian Research Council of Fisheries. D.L.R. was supported by a grant from the C.N.R.S. in France where a part of this research was conducted.

REFERENCES

1. Raper, C.D., and P.J. Kramer, eds. 1982. Crop Reactions to Water and Temperature Stresses in Humid, Temperate Climates. Westview Press, Boulder, Colorado.
2. Rains, D.W., R.C. Valentine, and A. Hollaender, eds. 1980. Genetic Engineering of Osmoregulation: Impact on Plant Productivity for Food, Chemicals, and Energy. Plenum Press, New York.
3. Hollaender, A., J.C. Aller, E. Epstein, A. San Pietro, and O.R. Zaborsky, eds. 1979. The Biosaline Concept: An Approach to the Utilization of Underexploited Resources. Plenum Press, New York.
4. San Pietro, A., ed. 1982. Biosaline Research: A Look to the Future. Plenum Publishing Corporation, New York.
5. Paleg, L.G., and D. Aspinall, eds. 1981. The Physiology and Biochemistry of Drought Resistance in Plants. Academic Press, Sydney.
6. Wyn Jones, R.G., and R. Storey. 1981. Betaines. In The Physiology and Biochemistry of Drought Resistance in Plants. L.G. Paleg and D. Aspinell, eds. Academic Press, Sydney, pp. 171-204.
7. Christian, J.H.B. 1955. The influence of nutrition on the water relations of Salmonella orianenburg. Aust. J. Biol. Sci. 8:75-82.
8. Christian, J.H.B. 1955. The water relations of growth and respiration of Salmonella orianenburg at 30°C. Aust. J. Biol. Sci. 8:490-497.
9. Britten, R.J., and F.T. McClure. 1962. The amino acid pool in Escherichia coli. Bacteriol. Rev. 26:292-335.
10. Rafaeli-Eshkol, D., and Y. Avi-Dor. 1968. Studies on halotolerance in a moderately halophilic bacterium. Effect of betaine on salt resistance of the respiratory system. Biochem. 109:687-691.
11. Shkedy-Vinkler, C., and Y. Avi-Dor. 1975. Betaine-induced stimulation of respiration at high osmolarities in a halotolerant bacterium. Biochem. J. 150:219-226.
12. Le Rudulier, D., and R.C. Valentine. 1982. Genetic engineering in agriculture: Osmoregulation. Trends in Biochem. Sci. 427, (in press).
13. Csonka, L.N. 1980. The role of L-proline in response to osmotic stress in Salmonella typhimurium: Selection of mutants

with increased osmotolerance as strains which over-produce L-proline. In Genetic Engineering of Osmoregulation. D.W. Rains, R.L. Valentine, and A. Hollaender, eds. Plenum Press, New York, pp. 35-52.

14. Csonka, L.N. 1981. The Role of Proline in Osmoregulation in Salmonella Typhimurium and Escherichia Coli. In Trends in the Biology of Fermentations for Fuels and Chemicals. A. Hollaender, R. Rabson, P. Rogers, A. San Pietro, R. Valentine, and R. Wolfe, eds. Plenum Publishing Corporation, New York, pp. 533-542.
15. Csonka, L.N. 1981. Proline over-production results in enhanced osmotolerance in Salmonella typhimurium. Molec. Gen. Genet. 182:82-86.
16. Rains, D.W., L. Csonka, D. Le Rudulier, T. P. Croughan, S.S. Yang, S.J. Stavarek, and R.C. Valentine. 1982. Osmoregulation by organisms exposed to saline stress: physiological mechanisms and genetic manipulation. Biosaline Research: A Look to the Future. A.S. Pietro, ed. Plenum Publishing Corporation, New York, pp. 283-302.
17. Measures, J.C. 1975. Role of amino acids in osmoregulation in non-halophilic bacteria. Nature 257:398-400.
18. Laimins, L.A., D.B. Rhoads, and W. Epstein. 1981. Osmotic control of kpd operon expression in Escherichia coli. Proc. Natl. Acad. Sci. U.S.A. 78:464-468.
19. Miller, J.F. 1972. Experiments in Molecular Genetics. Cold Spring Harbor Laboratory, New York.
20. Le Rudulier, D., G. Goas, and F. Larher. 1982. Onium compounds, amides and amino acid levels in nodules and other organs of nitrogen fixing plants. Z. Planzenphysiol. 105:417-426.
21. Baich, A. 1969. Proline synthesis in Escherichia coli. A proline-inhibitable glutamic acid kinase. Biochim. Biophys. Acta 192:462-467.
22. Baich, A. and D.J. Pierson. 1965. Control of proline synthesis in Escherichia coli. Biochim. Biophys. Acta 104:397-404.
23. Ikuta, S., S. Imamura, H. Misaki, and Y. Horiuti. 1977. Purification and characterization of choline oxidase from Arthrobacter globiformis. J. Biochem. 82:1741-1749.
24. Nagasawa, T., Y. Kawabata, Y. Tani, and K. Ogata. 1975. Choline dehydrogenase of Pseudomonas aeruginosa A-16. Agric. Biol. Chem. 39:1513-1514.
25. Rafaeli-Eshkol, D., 1968. Studies on halotolerance in a moderately halophilic bacterium. Effect of growth conditions on salt resistance of the respiratory system. Biochem. J. 109:679-685.
26. Galinski, E.A. and H.G. Truper. 1982. Betaine, a compatible solute in the extremely halophilic phototrophic bacterium Ectothiorhodospira halochloris. FEMS Microbiol. Lett. 13:357-360.

27. Blau, K., 1961. Chromatographic methods for the study of amines from biological material. Biochem. J. 80:193-200.
28. Kaback, H.R., and T.G. Deuel. 1969. Proline uptake by disrupted membrane preparations from Escherichia coli. Arch. Biochem. Biophys. 132:118-129.
29. Kohno, T., and J.R, Roth. 1979. Electrolyte effects on the activity of mutant enzymes in vivo and in vitro. Biochemistry, 18:1386-1392.
30. Vinopal, R.T., S.A. Wartell, and K.S. Kolowsky. 1980. β-galactosidase from osmotic remedial lactose utilization mutants of E. coli. In Genetic Engineering of Osmoregulation. D.W. Rains, R.C. Valentine, and A. Hollaender, eds. Plenum Press, New York, pp. 59-72.
31. Fincham, J.R.S., and A.J. Baron. 1977. The molecular basis of an osmotically separable mutant of Neurospora crassa producing unstable glutamate dehydrogenase. Mol. Biol. 110:627-642.
32. Strøm, A.R. 1979. Biosynthesis of trimethylamine oxide in calanoid copepods. Seasonal changes in trimethylamine monoxygenase activity. Marine Biol. 51:33-40.
33. Agustsson, I., and A.R. Strom. 1981. Biosynthesis and turnover of trimethylamine oxide in the teleost cod, Gadus morhua. J. Biol. Chem. 256:8045-8049.
34. Shewan, J. M. 1951. The chemistry and metabolism of the nitrogenous extractives in fish. In The Biochemistry of Fish. R.T. Williams, cd. Cambridge: Biochemical Society Symposia 6, pp.28-48.
35. Schoffeniels, E., and R. Giles. 1970. Nitrogen constituents and nitrogen metabolisms in arthropods. In Chemical Zoology. M. Florkin, and B.T. Scheer, eds. Academic Press, New York. Vol. 5, part A, pp. 199-227.
36. Cilles, R. 1971. Mechanisms of ion and osmoregulation. In Marine Ecology. A Comprehensive, Integrated Treatise on Life in Oceans and Coastal Waters. O. Kinne, ed. John Wiley and Sons, Chichester. Vol. 2, part 1, pp. 257-347.
37. Wright, D.J. and D.R. Newall. 1981. Osmotic and ionic regulation in nematodes. In Nematodes as Biological Models. B.M. Zuckerman, ed. Academic Press, Inc., New York. Vol. 2, pp. 143-164.

DIFFERENTIAL REGULATION OF THE *Adh1* GENE IN MAIZE: FACTS AND THEORIES

Mitrick A. Johns, Mary Alleman, and Michael Freeling

Department of Genetics
University of California
Berkeley, California 94720

INTRODUCTION

All genes are differentially regulated in the sense that each gene is expressed at different levels in different tissues and times during development and senescence. A gene that was expressed to the same extent in every cell at all times would be highly unusual: even such basic housekeeping genes as those specifying the histones are differentially regulated over the cell cycle. Tissue and temporal specificity is thus a fundamental property of genes in higher organisms.

We have few clues as to how differential gene expression is controlled, although much has been learned in recent years about eukaryotic gene structure and expression through the application of molecular techniques, mainly in mammals, yeast, and *Drosophila*. Most of this knowledge seems to apply to plants; at least, no striking differences between plants and other eukaryotes have been found. These molecular studies are elucidating how a gene can be turned on or off, or have its level of expression modified, but rarely are the mechanisms of tissue specificity approached.

To what extent is tissue specificity of gene expression controlled by *cis* acting elements (such as promoters or RNA processing sequences) adjacent to or within coding sequences, and how much is expression controlled by events at other "regulatory" genes acting in *trans* on the structural gene? The alcohol dehydrogenase-1 (*Adh1*) gene in maize will be used as an example. If tissue specificity were completely determined elsewhere in the genome, the *Adh1* gene might be composed of a single promoter capable of being either on or off: a very simple gene, whose mutant alleles would affect every expression of *Adhl* in the same way. If tissue specific regulation of *Adhl* were largely controlled by events at the *Adhl* gene itself, the *cis* acting regulatory elements would probably be complex, perhaps composed of several interacting promoters, modifiers of gene "shape", and perhaps sites involving RNA processing. Different alleles of such a gene should be capable of very different patterns of expression during development. Of course, there is no *a priori* reason why both *cis*-acting sites and *trans*-acting regulatory genes cannot be informationally complex.

In the maize *Adhl* gene we have an example of differential, tissue specific gene expression in which subtle genetic changes have been detected and studied. When the amount of gene product produced by two standard, naturally occurring alleles of *Adhl* was compared in the scutellum and in the primary root, Drew Schwartz (24) found that in the scutellum, the *1F* allele makes as much alcohol dehydrogenase (ADH) as the *1S* allele, while in the root, the *1F* allele makes twice as much as ADH as the *1S* allele. Moreover, the genetic cause of this differential gene expression is a part of the *Adhl* complementation group. The comparison of alleles was done in plants heterozygous for the two alleles whose products differ in electrophoretic mobility, allowing them to be separated by starch gel electrophoresis. This technique allows even small differences between the activities of the two alleles or their products to be detected. By examining the effects of particular induced mutations and naturally occurring variations of *Adhl*, we are learning about how a housekeeping gene is quantitatively regulated, and in particular, how the difference between *Adhl* expression in the root and the scutellum is controlled.

The Maize Adhl Gene-Enzyme System

Information about the genetics and biology of alcohol dehydrogenase in maize, consisting of over 90 papers published since the original description of the system (23), has been reviewed recently (12). What follows is a brief description of the essentials of *Adhl* necessary to understand the arguments in the present paper; only the most important citations are given.

The *Adhl* gene is at map position 127 on the long arm of chromosome 1. It has been located physically to an area 80 to 90 percent of the distance from the centromere to the telomere on the long arm of chromosome 1 (3). *Adhl* codes for a polypeptide of about 39,000 dalton molecular weight. In tissues where the *Adh2* gene (see below) is not active, such as the scutellum, all of the ADH1 poplypeptides combine to form active dimers. Thus, when a plant is heterozygous for "fast" (*1F*) and "slow" (*1S*) electrophoretic alleles of *Adhl*, three bands of ADH activity appear on electrophoresis gels, corresponding to the F F, F S, and S S dimers (examples in Fig. 3).

The ratios of enzyme activity in the three bands is highly

characteristic of the tissue and genotype used. In the scutellum of a 1F/1S heterozygote, the ratios of F·F, F·S, and S·S enzyme activities is about 1:2:1, which is what would be expected from two alleles producing equal numbers of subunits which had equal specific activities and which combined at random to form active dimers. In primary roots, especially when subjected to anaerobiosis for 12 to 24 hours, the ratios of F·F, F·S, and S·S are about 4:4:1, or what would be expected if the 1F allele made twice as many subunits as the 1S allele (24). The 4:4:1 pattern is also seen in the basal, uninduced ADH from aerobic roots, and from roots treated with the synthetic auxin 2,4-D.

A gene on the fourth chromosome, Adh2, makes an alcohol dehydrogenase very similar to ADH1. However, the ADH2 homodimers are only about a tenth as active as ADH1 homodimers, as seen on starch gels stained for ADH under the standard conditions. ADH1 and ADH2 are so similar that they dimerize with one another to form heterodimers as readily as they form homodimers, and ADH1 and ADH2 share some antigenic sites. The heterodimers run to an intermediate position on electrophoresis gels, and their activity per molecule is equal to the average of ADH1 and ADH2 homodimers, or about 55% as active as ADH1 homodimers (9) (Fig. 1); Adh2 is found in many tissues, but always in association with Adh1.

"Null" alleles lack all traces of ADH1 enzyme activity or immunologically cross-reacting material (CRM). That such alleles have been isolated shows that ADH1 is not necessary for survival of maize under normal conditions. However, Adh1-null seeds and seedlings drown much more quickly than wild type seedlings (25). Thus, ADH1 appears to be necessary for survival during anaerobiosis.

When a seedling is subjected to anaerobiosis, either by submersion in water or by replacement of the air with argon gas, most protein synthesis is quickly stopped, and a small group of new proteins, the anaerobic proteins (ANPs) are made (22). The ANPs include both ADH1 and ADH2 (Fig. 2). The anaerobic response seems to be similar to the heat shock response seen in Drosophila, maize, and other organisms. However, the proteins made during heat shock do not appear to be the same proteins made during anaerobiosis (16). It has been hypothesized that the ANPs help the plant survive temporary flooding, but except for ADH1 and ADH2, no function has been positively ascribed to any of the ANPs.

Recently, work on Adh1 has been accelerated by the cloning of a cDNA copy of ADH1 messenger RNA (15). A copy of the genomic DNA encoding this message has also been cloned (Bennetzen and Swanson at the International Plant Research Institute; Taylor and Freeling at the University of California; unpublished results), and the basic structure of the gene is being rapidly determined. Molecular analyses of some of the interesting Adh1 mutants are summarized later in this paper (28).

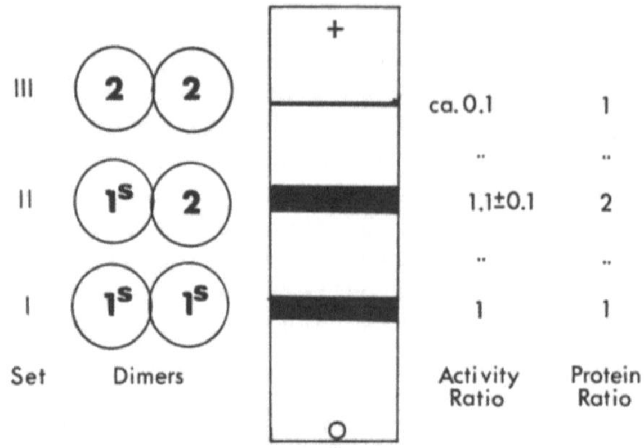

Fig. 1. The Adh gene system in maize. There are two anaerobically induced Adh genes specifying polypeptides that dimerize randomly, yielding three sets of enzymes. The electrophoretogram exemplifies the electrophoretic separation of the three types of isozyme dimers, 1·1, 1·2, and 2·2 in the standard activity ratio of 1:1.1:∿0.1 characteristic of the Adh1-1S, Adh2-2N standard line. This activity ratio actually represents a 1:2:1 ratio at the protein level. "O" denotes the origin. 1^S denotes ADH1-S subunits, etc. From Freeling and Birchler, ref. 12.

Mutants Induced with EMS

Schwartz was the first to attempt to isolate regulatory mutants at Adh1. He and his co-workers have isolated over 100 mutant Adh1 alleles induced by ethyl methanesulfonate (EMS); many of these mutants are listed, along with their properties, in Freeling and Birchler (12). Schwartz's methods involved soaking dry seeds in an EMS solution, then growing them into plants and crossing their ears with pollen from tester plants carrying a different Adh1 electrophoretic mobility allele. The resulting ears were then screened for clones of mutant heterozygotes displaying alterations of the normal 1:2:1 allozyme ratio.

Some of the resulting mutants, such as S5657 and F207, make no product detectable by either enzyme activity or antibody titrations. Other mutants make products detectable by antibodies; however, they are inactive; have low activity; are faulty or unusual in dimerization properties; are sensitive to heat, electrophoresis, or dialysis; are altered in electrophoretic mobility; or have some combination of these properties. This group of mutants clearly consists of alterations in the protein coding sequence. The mutants in the former group, which are CRM^-, are harder to classify: the mutational defects could be either in coding sequences or in regulatory sequences.

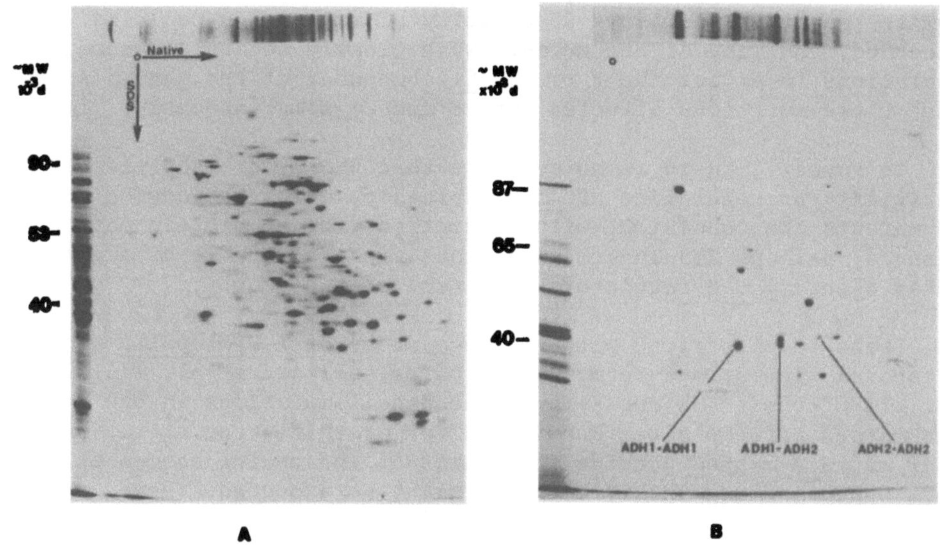

Fig. 2. Fluorographs of native/SDS two-dimensional polyacrylamide gels of labelled maize root proteins. (A) Aerobic-labelled for 5 hr with ^{35}S-methionine; (B) 17 hr anaerobic treatment labelled for the last 5 hr with ^{35}S-methionine. From Okimoto et al. (18).

Recent molecular studies have shown that of four mutants tested, three, *S5657* (CRM$^-$) (M.M. Sachs, pers. comm.), *S1015*, and *S719* (CRM$^+$), have approximately normal amounts of normal-sized ADH1 messenger RNA, while *S664* (CRM$^-$) appears to have a normal amount of mRNA divided into normal sized and slightly larger than normal species (Hake, this laboratory, pers. comm.). Having normal amounts of RNA suggests that transcriptional or RNA processing control, i.e., what is commonly considered "gene regulation", is not involved in these mutants. Several CRM$^-$ mutants are still uninvestigated, except that they are known to recombine as if they were point mutations and not insertions or deletions (11).

None of the EMS-induced mutants displays any sign of tissue specific behavior: all mutants act the same in scutellum, pollen, and anaerobic root. Some mutants change the surface charge of the subunits; these mutants continue to act like their progenitors with respect to their behavior in anaerobic roots. Mutants in the *1S* allele that now have the "F" electrophoretic mobility make only half as much ADH in anaerobic roots as the *bona fide 1F* allele. Thus it is demonstrated that the tissue specific action of *Adhl* does not depend on the electrophoretic mobility of the proteins, but rather on some other property of the *Adhl* gene.

A similar study of EMS-induced mutants in the naturally occurring duplication *Adhl-FCm* by Birchler and Schwartz (2) led to similar results. No tissue specificity of expression could be discerned, and many of the mutants were clearly coding sequence alterations in either the F or the Cm component of the gene. Also, none of the mutations affected both F and Cm simultaneously.

In summary, there is no evidence that EMS alters the tissue specificity or regulation of Adh1 expression. The phenotypes of all the mutants are consistent with alterations in the protein coding sequence; perhaps EMS-induced mutations are not capable of significantly altering the regulatory component of the gene.

These results agree with findings at several Drosophila genes. In careful fine structure mapping studies, Gelbart et al. (14) found that all of their EMS and X-ray induced null mutations in the ry gene mapped between sites known to alter protein structure. None of their mutants mapped outside the limits of the coding sequence, where noncoding regulatory elements might be expected. However, Chovnick et al. (7) mapped two naturally occurring variants that altered the amount of gene product to sites outside the limits of the coding sequence, demonstrating that such regulatory elements do exist. Paton and Sullivan (20) screened EMS treated Drosophila for mutants in the cinnibar gene which affected either eye color or Malpighian tubule color, but not both. They isolated many new alleles, but all of them affected both tissues equally. Less thorough studies of EMS induced mutants by Schwartz and Sofer (27) on the Drosophila Adh gene and by Bell and MacIntyre (1) on the Drosophila Acph-1 gene have identified many protein structure alterations, but no cases of tissue specific changes in activity. We believe that all of these data taken together indicate that EMS rarely if ever makes mutations which affect the differential regulation of the mutant gene.

Naturally Occurring Variants and the Reciprocal Effect

If EMS is incapable of making mutations affecting differential gene regulation, how can such mutants be generated? One way is to examine naturally occurring variants of Adh1, available from the many races of maize grown in Central and South America. Native peoples of these areas have cultivated maize for thousands of years under a wide variety of moisture, temperature, and altitude regimes. Thus, excellent prospects exist for the selection and fixation of different Adh1 alleles with interesting properties.

The problem with studying naturally occurring variants is that one can never be sure that a given alteration in DNA sequence is the cause of the observed regulatory change. Adh1 genes from different races may differ in many ways extraneous to the effect of interest. For this reason, geneticists prefer to use induced mutations, where the mutant can be compared to its progenitor allele before random variations can accrue. Nevertheless, important conclusions can be drawn from the study of naturally occurring variants. In any case they are much more numerous than induced regulatory mutations.

Woodman and Freeling (30) examined the enzyme activity ratios of F/S heterozygotes in 14 lines of maize having the F mobility and 6 lines having the S mobility. All the F lines were crossed to the standard S allele, 1S, and all the S lines were crossed to 1F. The resulting F_1 individuals were tested for their allozyme ratios in the scutellum and the anaerobic root by electrophoresis. The gels were scanned to give quantitative data, and the binomial distribution was used to calculate the fraction of the total ADH activity attributable to each allele.

A large amount of variation between the lines was found. The percentage of enzyme in the S•S peak varied from 20.3% to 41.2% (a theoretical 1:2:1 ratio would give 25% activity in the S•S peak). Similarly, in anaerobic roots, the percentage of S•S enzyme varied from 9.7% to 22.5%, where 1S/1F gave 11.3%, close to the 11.1% expected from the 4:4:1 ratio found when the F allele makes twice as many subunits as the S allele. All of the ratios closely fit the binomial distribution. It is clear that a high degree of quantitative variation does occur at Adhl, although no examples of variants that completely eliminate ADH from one or the other organ were found.

A striking phenomenon was discovered: when the percentage of activity due to the S allele in the scutellum was plotted against S activity in the root for the various combinations of alleles, it was found that all of the data points lay near a line (Fig. 3). This line represented a reciprocal relationship between enzyme activity in the two organs: if S allele activity was high in the root it would be low in the scutellum, and vice versa. This relationship is termed the reciprocal effect, and is an unusual example of tissue specific gene regulation.

Several useful control experiments were done to constrain possible explanations for the reciprocal effect. Backcrosses, self crosses, and crosses to a strain with a third electrophoretic mobility (Ct) were done to show that the cause of the various ratios found in combinations of the alleles 1S, 1F, 54S, and 33F was tightly linked to and acted in cis upon Adhl. It was estimated that the site of quantitative variation was no more than 0.45 map units from the site of electrophoretic variation in Adhl. The reciprocal effect can thus be considered an intrinsic part of the Adhl gene.

Two dimensional immunoelectrophoresis on heterozygotes of the four alleles just mentioned showed that the four ADH allozymes had identical specific activities. Also, measurement of ^{3}H-leucine incorporation into ADH allozymes in the anaerobic root led to the conclusion that the rates of synthesis of the allozymes accurately reflect the enzyme activities seen on starch gels. These results indicate that the difference in enzyme ratios seen in the reciprocal

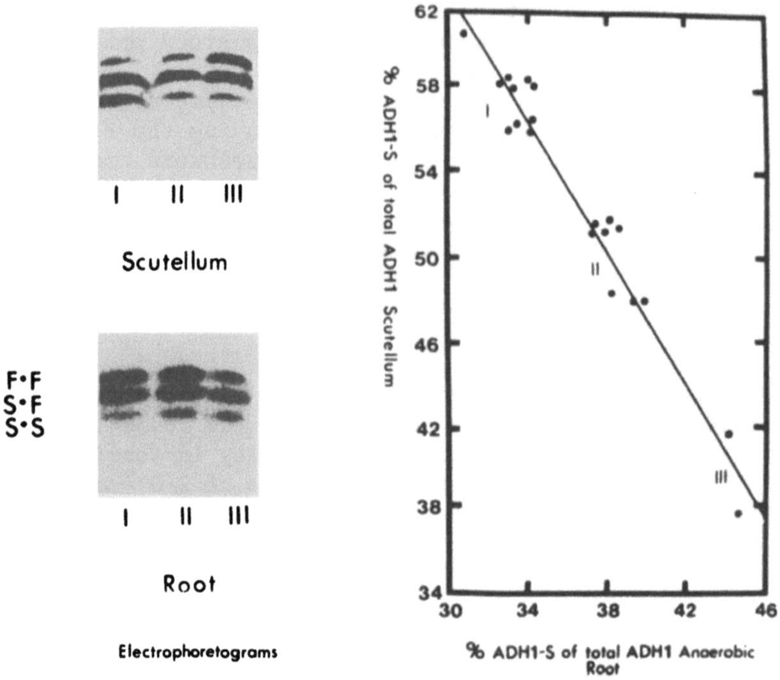

Fig. 3. The organ-specific reciprocal effect for the quantitative expression of naturally occurring Adh1 alleles. The graph plots the percent ADH1-S of the total ADH1 subunits for 10 Adh1-"F"/Adh1-"S" heterozygotes of diverse origin. The electrophoretograms are examples of each of the three groups (I, II, and III). Examples of allelic pairs for each group are as follows: Group I: 1F/1S; Group II: 1S/33F and 54S/1F; Group III: 54S/33F. From Freeling and Brichler (12).

effect variants are probably due to changes in gene or transcript activity and not to the more trivial changes in enzyme activity per molecule.

The reciprocal effect also appears to apply to variation within a given genotype. If an individual seed had a particularly high allozyme ratio in one organ, i.e., a ratio more than one standard deviation from the mean for that genotype, that individual would almost inevitably have an unusually low ratio for the other organ. This effect on individuals seems to imply that the level of gene expression in both tissues is set using a single program for the whole organism. Perhaps it is set before the precursor cells to the two tissues have become separate. If different organs had their Adh1 genes set separately, one would expect that the random occurrence of a high ratio in one tissue would not necessarily lead to an unusually low ratio in the other tissue. The fact that a reciprocal effect exists even within individuals of the same genotype argues for close collaboration between the setting events in the two tissues.

The reciprocal effect is an example of tissue specific gene regulation involving cis-acting genetic variation at the Adh1 gene itself. It is important because it shows that organ specificity can occur at the structural gene. Schwartz's (24) description of differences in the expression of 1S and 1F has thus been shown to be a part of the organ specific reciprocal effect. Other examples of allelic differences in tissue specific expression have been described in *Drosophila* and mice (8,19). The notion that gene activity is apparently preset coordinately in the root and the scutellum argues that even if there are separate promoters for Adh1 expression in these tissues, one must invoke other sequences for a reasonable explanation of the reciprocal effect.

These conclusions shed some light on a model Schwartz (26) proposed to explain the difference between the allozyme ratios in the root and the scutellum. Schwartz's model includes two promoters of equal strength for both the 1S and 1F alleles. In the 1S allele, one of the promoters is active only in the scutellum, while for 1F, both promoters are active in both root and scutellum. Thus, in the scutellum, both 1F and 1S have two active promoters and so the alleles make equal numbers of subunits. In the root, 1F has two active promoters, while 1S has only one active promoter, so twice as many F subunits are made as S subunits. Essentially, Schwartz has proposed that the level of expression of the Adh1 gene can be varied by changing the number of cis-acting regulatory elements controlling the gene rather than by changing the structure (sequence) of a single regulatory element. Similar models have been proposed by Brink et al. (6) to explain paramutation at the maize R locus, and by Wallace and Kass (29) to explain the positive heterotic effects of X-rays in *Drosophila*.

The reciprocal effect data make a strict form of this simple model untenable, for several reasons. It now appears that the 1:2:1 and 4:4:1 allozyme ratios, seen for 1S/1F in scutellum and root respectively, are due to a fortuitous choice of alleles, rather than being the product of a strict doubling of the activity of one allele in one tissue compared to its activity in another tissue. Also, both F and S alleles can affect the allozyme balance in both tissues; in the simple model, only the S allele (1S) has any tissue specificity. Finally, the existence of a reciprocal relationship between allozyme ratios in root and scutellum seems to imply a closer relationship between the activities of the Adh1 genes in the two tissues than would be expected from two independent promoters.

Schwartz's idea of two promoters for Adh1 is still a useful way of explaining the difference between allozyme ratios in root and scutellum, although perhaps not in its original form. No combination of alleles gives exactly equal ratios in the two organs; in the root, F is always expressed more than S, while in the scutellum, S is usually expressed more the F. The finding that the mouse alpha-amylase gene is transcribed from separate promoters in the salivary gland and the liver (3) makes the idea of multiple promoters for Adhl, one for the root and one for the scutellum, seem reasonable. The reciprocal effect implies that, however, if two tissue specific promoters do exist they probably interact with each other. Perhaps there is a single regulatory site which can exist in two alternative states, one for the root and one for the scutellum. Alterations in this element lead to simultaneous changes in Adhl expression in both tissues; if there were separate promoters for the two tissues, variants might be expected to alter one but not the other, some of the time.

A potentially important tool in the study of the reciprocal effect is the isolation of two organ specific mutants (13). The allele S1951a was derived from the standard ls2p line after bombardment of the immature tassel with accelerated neon ions and selection of the pollen for underexpression of Adhl using allyl alcohol (method described in ref. 12). This mutant underexpresses ADH in the scutellum but overexpresses ADH in the anaerobic root: it acts like a naturally occurring reciprocal effect variant. Since we know the progenitor allele it should be possible to locate the nucleotide sequence alteration responsible for the organ specific changes that have occurred.

A second allele, FkF3037, is also of interest. This allele spontaneously arose in our Funk G4343 tester line. The phenotype of FkF3037 is overexpression in the scutellum but normal expression in the anaerobic root, thus violating the reciprocal effect relationship. This allele is the only case known that violates the reciprocal effect, and we hope that it will shed light on why all other mutants and variants do follow the reciprocal effect.

The Gene Competition Model

The first example of Adhl regulation was observed by Schwartz (24), in the roots of 1F/1S heterozygotes. On starch gels, there is a 4:4:1 ratio of F•F, F•S, and S•S allozymes, which implies that the 1F allele makes twice as many subunits as the 1S allele. This in turn implies that, if the genetic background is held constant, an 1F/1F homozygote should have twice as much total ADH as an 1S/1S homozygote. However, Schwartz found that 1F/1F, 1F/1S, and 1S/1S seedlings taken from a single ear (to randomize genetic background) had approximately equal ADH activities. In other words, the 1F alleles in 1F/1F individuals were making as much ADH as the 1S alleles in 1S/1S individuals, while in 1F/1S seedlings the 1F allele was making twice as much ADH as the 1S allele. This is a clear case of regulatory behavior.

Schwartz proposed a gene competition model to explain this phenomenon. In this model, a factor necessary for the expression of Adhl is present in limiting quantity, and the two Adhl alleles compete for it. The 1F allele has twice the competitive ability of the 1S allele, so 1F is expressed twice as much as 1S. In homozygous individuals, the amount of factor is the same as in heterozygotes, so the amount of ADH made is the same. In other words, the quantity of limiting factor, encoded by a separate gene, determines the amount of ADH that is made, while the Adhl alleles themselves merely determine how the production of ADH is partitioned. Freeling (10) subsequently argued that the Adh2 alleles also compete for the factor, with a competitive ability equal to the Adhl-1S allele. This interpretation was supported by data showing that the amount of ADH2 made in Adhl-1S/Adhl-1S individuals was greater than the amount of ADH2 made in Adhl-1F/Adhl-1F individuals where the two strains shared a common randomized genetic background.

The gene competition model is unusual because it implies that the dosage of Adhl is irrelevant to the amount of ADH produced. In all other known structural genes, the amount of gene product made is directly proportional to the number of copies of the gene present. However, as Schwartz points out, this general result of gene dosage effect is expected when several genes compete for the same limiting factor; total gene dosage compensation is approached as the number of gene competitors approaches one. Schwartz tested the effect of varying gene dosage using a translocation of most of the chromosome arm (1L) on which Adhl resides. He indeed found that the amount of ADH in the scutellum was the same whether there were two doses of Adhl or three doses. Subsequently it was shown that compensation in the scutellum held for 1, 2, 3, and 4 doses of the long arm of chromosome 1 (4). Birchler also retested the gene competition model using translocations that led to the duplication of a smaller, 18 centimorgan region containing Adhl, and found that the amount of ADH increased in direct proportion to the number of doses of Adhl present: three doses of Adhl gave 150% of the wild type ADH level, and four doses gave 200%, as one would expect for a simple gene with no compensatory mechanism. Birchler further showed that a region included in Schwartz's original translocation but not included in the smaller duplication had a negative regulatory effect on Adhl. The more doses of this region that were present, the less ADH was made. In Schwartz's original experiments, the increase in ADH due to an increased dosage of Adhl was balanced out by a decreased production of ADH caused by the negative regulatory locus, causing no net change in ADH levels. This finding makes the original version of the gene competition model untenable: Birchler showed that the dosage of Adhl, rather than the amount of "limiting factor", limits the amount of ADH made.

One criticism about the relevance of Birchler's work to the gene competition model is that Schwartz studied ADH in the root and scutellum and Birchler studied ADH in the scutellum only. It is possible that gene competition occurs in the root but not in the scutellum because his Adh1 dosage series showed no effect on scutellar ADH. If Schwartz had had Birchler's smaller duplication he would have concluded that gene competition did not occur in the scutellum, with no implication for the root at all.

One of us (M.A.) recently tested the effect of 2 versus 4 doses of the Adh1 region on ADH1 levels in the anaerobic root by comparing a segmental tetrasomic line with its euploid progenitor (5). The expression of Adh1 during anaerobiosis was measured by scanning autoradiographs of roots labelled with ^{3}H-leucine under argon. The results (Table 1) show that ADH1 increases from 6.1% to 13.8% when the dosage of Adh1 is doubled from 2 to 4. The amount of ADH2, whose structural gene is on chromosome 4, does not change; none of the other anaerobic proteins (ANPs) increased by more than 20% when the Adh1 dosage was increased. The total amount of label incorporated into ANPs also did not change much. The fact that ADH1 level approximately doubled when Adh1 dosage was doubled indicates that the dosage of Adh1 does indeed limit its own expression in the root. This finding makes the gene competition model as originally proposed untenable.

The discovery of the reciprocal effect also makes gene competition in the root unlikely, because the reciprocal effect predicts that Adh1 in the root and scutellum are regulated by a common mechanism. If gene competition occurs in the root, those alleles which are highly expressed in the root should be very good at competing for the limiting factor. Then, if no competition occurred in the scutellum, this competitive ability should have no effect on gene expression, so Adh1 expression in the scutellum should be independent of expression in the root, rather than reciprocally related.

Although the gene competition model as originally proposed appears to be incorrect, Schwartz's original observation that 1F/1F roots do not make twice as much ADH as 1S/1S roots still needs to be explained. There is probably some kind of compensatory mechanism regulating the amount of ADH made, possibly involving the competition of several genes. At present no convincing explanation for the data exists.

Birchler's identification of a negative compensatory locus on the first chromosome probably represents a trans-acting regulatory gene controlling Adh1. It seems to affect other, unrelated genes in the same way (Birchler, pers. comm.). The negative compensatory locus has not been precisely localized, and it may not even be a single discrete gene; at present it is merely a phenomenon associated with a large region of chromosome 1L between positions 0.20 and 0.72 on the physical map.

Another trans-acting regulatory gene affecting Adh1 has been described: the Adr1 locus (17). Normally, the amount of ADH1 in the

Table 1. Gene dosage effects in anaerobic roots.

genotype	Adh1 doses	%ADH1	%ADH2	total densitometer units
S/F	2	6.1	3.1	190.6
SF/SF	4	13.8	3.1	141.2

The strains T 1-3 (5267)/T 1-3 (5242) and (S/F) Dp 1-3 (5267):1-3 (5242)/Dp 1-3 (5267):1-3 (5242) (SF/SF) were derived as described in Birchler, Alleman and Freeling (1981). Primary roots 8 cm long were subjected to 20 hr of anaerobiosis in an argon atmosphere, then labelled for 6 hr with ^{3}H-leucine under argon. The roots were then subjected to extraction and 2-dimensional gel electrophoresis and fluorography as described in Sachs, Freeling and Okimoto (1980). All protein spots named in Sachs et al. (1980) were scanned with a densitometer, and the area under each peak was determined. The areas under the ADH1-F•F, -F•S and -S•S peaks were summed and divided by the total area under the ANP peaks to give the %ADH1. The results shown are from a typical experiment.

scutellum decreases 2-3 fold during the first 10 days after germination. A variant line was found in which ADH1 levels remained high during this period. The "wild type" pattern (a decline in ADH1) appears to be dominant, and the variant trait behaves as a single locus when followed in a series of crosses. Lai and Scandalios propose that Adr1 regulates the degradation of ADH1 rather than its synthesis. The discovery of the Adr1 locus demonstrates that the control of Adh1 is complex and may involve control at levels beyond the nucleus. (It should be noted that our nomenclature differs from that of Scandalios: our Adh1 is his Adh2.)

We have performed a small experiment looking for trans-acting mutants (as opposed to variants or chromosomal abberations) affecting Adh1 based on the following rationale. Every year we screen for new mutants by treating pollen from mutagenized plants with allyl alcohol and crossing it onto tester ears. Usually about 1,000 seeds develop from about 10,000 ears pollinated. Of these seeds, perhaps 30 will look unusual when their allozyme ratios are examined by starch gel electrophoresis. The rest of the seeds have normal allozyme ratios. Why did the pollen grains that fertilized these seeds survive the allyl alcohol treatment? Possibilities include wild type "escapers", mutations in other genes conferring allyl alcohol resistance, mutations at Adhl conferring allyl alcohol resistance without affecting the ability to metabolize ethanol and, most importantly, mutations in other genes causing the recessive loss of ADH1 activity. Mutations in the latter category would be complemented by the wild type allele from the female parent, so would not affect the scutellar allozyme ratio. Only if the mutant was made homozygous or if it was expressed in the haploid pollen grain would the mutant phenotype, no ADH1 activity, be expressed. Thus, recessive, trans-acting mutants affecting Adhl would have no visible phenotype in the seed, but would show a 1:1 segregation of wild type and ADH1-null pollen grains when the seeds were grown into mature plants. The null grains carry the mutant allele and the normal grains carry the wild type allele from the female parent.

Following this rationale, we grew 934 seeds whose male parents carried either Robertson's Mutator (Mu) or Ac-Ds (both transposable element systems capable of making mutations at Adhl) and which had been treated with allyl alcohol. Pollen from each of these plants was stained for ADH. None of the plants showed null or very low activity for ADH, although several had 1:1 segregations of normal and aborted pollen grains, and several were sterile or had abnormal morphology. We conclude that, if trans-acting genes necessary for the expression of Adhl exist, they must be rare, hard to mutate with Mu or Ac-Ds, or incapable of causing a major loss of ADH1 activity in pollen. Perhaps a more efficient mutagen, such as EMS, would have been a better choice, but the necessity of applying mutagens premeiotically creates technical complications.

Mutations Induced by Robertson's Mutator (Mu)

Robertson (28) described a line of maize that gave rise to recessive mutations at a frequency fifty times background; these mutants were often unstable. Freeling, Cheng, and Alleman (13) crossed the Mu genotype into the standard line carrying Adhl-1S and the marker Adh2-P alleles (1s2p) and recovered a mutant allele following allyl alcohol selection. This mutant, S3034, has about 40% of the wild type level of ADH in the scutellum. The mutant acts in cis on Adhl, and no recombinants separating the genetic site responsible for the 40% activity and the electrophoretic site in Adhl have been recovered. The protein product is apparently completely normal, in contrast to EMS induced mutants, which lower ADH activity by altering the protein product.

In addition, S3034 is usually unstable: at a frequency of about 10^{-4}, stained pollen grains show increased or decreased amounts of ADH. This instability was exploited to select derivatives with even lower amounts of ADH, using the standard allyl alcohol selection system with increased amounts of allyl alcohol. The first two derivatives recovered were S3034a and S3034b, with 0% and 13% of wild type activity respectively. Both of these derivatives are still unstable.

None of these three alleles displays any tissue specificity or unusual reciprocal effect: in all tissues the amount of S gene product is about 40%, 13%, or 0% of the wild type level. However, these mutants might be considered regulatory mutants because they lower the amount of gene product without apparently altering its structure. Also, the unstable nature of these alleles might lead to the generation of new alleles which are tissue specific.

Quantitative Northern blots were performed by Strommer, Hake, Bennetzen, Taylor, and Freeling (28) to compare ADH1 messenger RNA levels of 1S (the progenitor), S3034, S3034a, and S3034b in primary roots after 12 hours of anaerobiosis. A cDNA probe to the 3' half of the message, which apparently hybridizes to a negligible extent to Adh2, was used (15). The mRNA levels correlated with the enzyme activity levels characteristic of the alleles. This result contrasts with results from several EMS induced mutants, all of which specify approximately normal amounts of ADH1 mRNA. S3034a is the only allele known so far which does not have any ADH1 mRNA. This correlation between enzyme activity and mRNA levels allows these three alleles to be operationally defined as regulatory mutants. Probing the Northern blots with cDNAs for other anaerobic proteins shows equal amounts of RNA for anaerobic genes other than ADH1; therefore, the mutant phenotype is not due to a general decrease in RNA levels.

Possession of the cDNA clone allowed Strommer et al. (28) to compare the region around the Adh1 gene of the progenitor and the mutant alleles using restriction maps. The summarized results are shown in Fig. 4. S3034 apparently has an insert of about 1.5 kilobases in the 5' side of the gene. This insert contains a Bst EII site. The insert lies between an Xba I site and a Hind III site, and all DNA fragments spanning this region are increased by 1.5 kb, while all other DNA fragments are unaltered.

Interestingly, the derivatives S3034a and S3034b map to the same location as S3034, and both of them have a Bst EII site in the same place as S3034. No difference between S3034 and its derivatives has yet been detected in the DNA. It is clearly necessary to clone and sequence these alleles to determine what causes the difference in their enzyme activity phenotypes; this work is now in progress. The regulatory differences may result from internal rearrangements or replacements of the inserts, or by small changes at the ends. In any case, we have no evidence for transposition.

More recently, we have isolated two new Adh1 mutants caused by Mu, S4477, and S4478. These mutants have the same phenotype as S3034: they decreased the level of ADH to 40% of the wild type

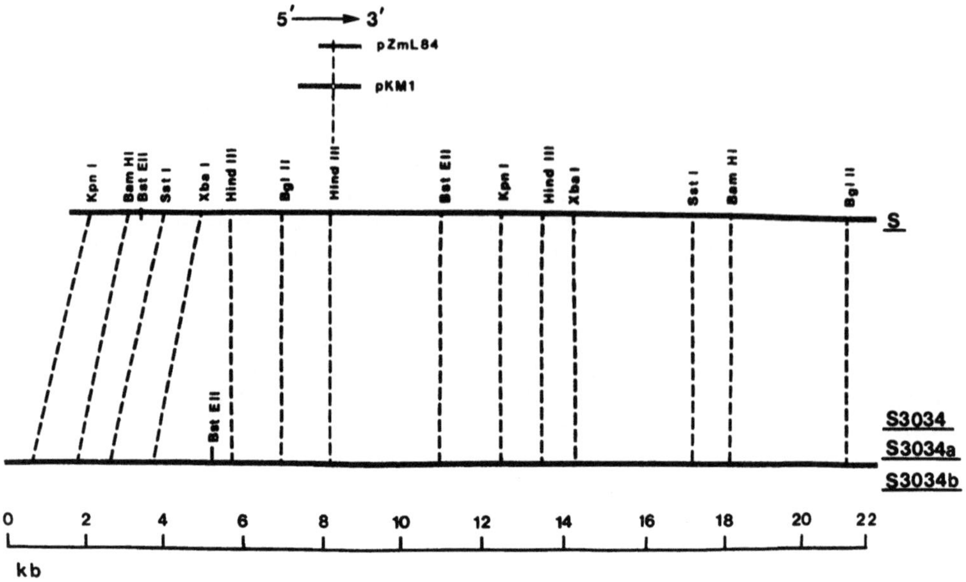

Fig. 4. Endonuclease site maps of genomic DNA containing 1S (top) and the mutant alleles S3034, S3034a, and S3034b (bottom). The two cDNA probes pZmL84 and pKm1 are aligned by their homologous Hind III sites. Placement of each endonuclease digestion site was confirmed with at least two sets of two-enzyme digests. Maps of the different alleles were generated independently. It should be noted that the Bst EII site seen 5' to hybridizing sequences in 1S is presumed present in S3034 and its derivatives but cannot be detected due to the appearance of a new Bst EII site closer to the region of probe homology. Plasmid pZmL84 was the generous gift of Gerlach et al. (see ref. 15). Redrawn from Strommer et al. (28).

level, are unstable, and have no tissue specificity to their action. When restriction maps were prepared for S4477 and S4478, it was found that they contain inserts which map to identical sites (± 50 bp) slightly 3' to the site of insertion in S3034 (data to be published elsewhere). This position is clearly within the Adh1 transcription unit. S4477 and S4478 have 1.5 kb insertions with a Bst E2 site in the same relative position as S3034. It is important to note that S4477 and S4478 had different parents, both male and female, and they both had many wild type siblings; S4477 and S4478 are independent insertions, perhaps into a mutational "hot spot" for Mu.

The fact that the three independent Mu induced mutations have the same regulatory phenotype (40% of normal ADH activity) and that over 100 EMS induced mutants show no sign of regulatory behavior

demonstrates that different mutagens can lead to different kinds of mutants. This idea leads to the conclusion that a full understanding of gene regulation will require the selection of many mutants using several different mutagens.

The most important conclusion that can be drawn from our studies of Mu induced mutations is that regulatory mutants which do not alter the tissue specific expression of Adhl can exist. We now have two classes of regulatory mutants of Adhl: those which alter the allozyme balance between the anaerobic root and the scutellum (reciprocal effect mutants) and those which alter ADH levels in all tissues uniformly. As the number of mutants isolated and analyzed increases, so should our understanding of Adhl gene regulation.

REFERENCES

1. Bell, J.B., and R.J. MacIntyre. 1973. Characterization of acid phosphatase-1 null activity mutants of Drosophila melanogaster. Biochem. Genet. 10:39-55.
2. Birchler, J.A., and D. Schwartz. 1979. Mutational study of the alcohol dehydrogenase-1 FCm duplication in maize. Biochem. Genet. 17:1173-1180.
3. Birchler, J.A. 1980. The cytogenetic localization of the alcohol dehydrogenase-1 locus in maize. Genetics 94:687-700.
4. Birchler, J.A. 1981. The genetic basis of dosage compensation of alcohol dehydrogenase in maize. Genetics 97:625-637.
5. Birchler, J.A., M. Alleman, and M. Freeling. 1981. The construction of a segmental tetrasomic line of maize. Maydica 26:3-10.
6. Brink, R.A., E.D. Styles, and J.D. Axtell. 1968. Paramutation: Directed genetic change. Science 159:161-170.
7. Chovnick, A., W. Gelbart, M. McCarron, B. Osmond, E.P.M. Candido, and D.L. Baille. 1976. Organization of the rosy locus in Drosophila melanogaster: Evidence for a control element adjacent to the xanthine dehydrogenase structural element. Genetics 84:233-255.
8. Dickinson, W.J., and H.L. Carson. 1979. Regulation of the tissue specificity of an enzyme by a cis-acting genetic element. Proc. Natl. Acad. Sci. USA 76:4559-4562.
9. Freeling, M. 1974. Dimerization of multiple maize ADHs studied in vivo and in vitro. Biochem. Genet. 12:407-417.
10. Freeling, M. 1975. Further studies on the balance between Adhl and Adh2 in maize: Gene competitive programs. Genetics 82:641-654.
11. Freeling, M. 1978. Allelic variation at the level of intragenic recombination. Genetics 89:211-224.
12. Freeling, M., and J.A. Birchler. 1981. Mutants and variants of the alcohol dehydrogenase-1 gene in maize. In Genetic Engineering: Principles and Methods. J.K. Setlow and A.

Hollaender, eds. Plenum Press, New York.
13. Freeling, M., D.S.-K. Cheng, and M. Alleman. 1982. Mutant alleles that are altered in quantitative, organ-specific behavior. Devel. Genet. (in press).
14. Gelbart, W., M. McCarron, and A. Chovnick. 1976. Extension of the limits of the XDH structural element in Drosophila melanogaster. Genetics 84:211-232.
15. Gerlach, W.L., A.J. Pryor, E.S. Dennis, R.J. Ferl, M.M. Sachs, and W.J. Peacock. 1982. cDNA cloning and induction of the alcohol dehydrogenase gene (Adhl) of maize. Proc. Natl. Acad. Sci. USA 79:2981-2985.
16. Kelley, P., and M. Freeling. 1982. A preliminary comparison of maize anaerobic and heat-shock proteins. In Induction of Heat Shock Protein, M.J. Schlesinger, A. Tissiers, and M. Ashburner, eds. Cold Spring Harbor, New York (in press).
17. Lai, Y.K., and J.G. Scandalios. 1980. Genetic determination of the developmental program for maize scutellar alcohol dehydrogenase: Involvement of a recessive, trans-acting temporal-regulatory gene. Devel. Genet. 1:311-324.
18. Okimoto, R., M.M. Sachs, E.K. Porter, and M. Freeling. 1980. Patterns of polypeptide synthesis in various maize organs under anaerobiosis. Planta 150:89-94.
19. Paigen, K. 1979. Genetic factors in developmental regulation. In Physiological Genetics. J.G. Scandalios, ed. Academic Press, New York.
20. Paton, D., and D. Sullivan. 1978. Mutagenesis at the cinnibar locus in Drosophila melanogaster. Biochem. Genet. 16:855-865.
21. Robertson, D, 1978. Characterization of a mutator system in maize. Mut. Res. 51:21-28.
22. Sachs, M. M., M. Freeling, and R. Okimoto. 1980. The anaerobic proteins of maize. Cell 20:761-768.
23. Schwartz, D., and T. Endo. 1966. Alcohol dehydrogenase polymorphisms in maize-simple and compound loci. Genetics 53:709-715.
24. Schwartz, D. 1971. Genetic control of alcohol dehydrogenase - A competition model for regulation of gene action. Genetics 67:411-425.
25. Schwartz, D. 1969. An example of gene fixation resulting from selective advantage in suboptimal conditions. Am. Nat. 103: 479-481.
26. Schwartz, D. 1976. Regulation of the alcohol dehydrogenase genes in maize. Stadler Genetics Symp. 8:9-15.
27. Schwartz, M., and W. Sofer. 1976. Alcohol dehydrogenase-negative mutants in Drosophila: Defects at the structural locus. Genetics 83:125-136.
28. Strommer, J.N., S. Hake, J. Bennetzen, W.C. Taylor, and M. Freeling. 1982. Regulatory mutants of the maize Adhl gene caused by DNA insertions. Nature (in press).
29. Wallace, B., and T.L. Kass. 1974. On the structure of gene control regions. Genetics 77:541-558.

30. Woodman, J.C., and M. Freeling. 1981. Identification of a genetic element that controls the organ-specific expression of Adhl in maize. Genetics 98:354-378.

31. Young, R.A., O. Hagenbuchle, and U. Schlibler. 1981. A single mouse alpha-amylase gene specifies two different tissue-specific mRNAs. Cell 23:451-458.

CYTOPLASMIC MALE STERILITY

C.S. Levings, III

Department of Genetics
North Carolina State University
Raleigh, North Carolina 27650

INTRODUCTION

The trait cytoplasmic male sterility (cms) is common in higher plants. Edwardson (1) reported that the cms trait had been observed in at least eighty unique plant species. Although the trait may manifest itself in different fashions among the various species, cms plants have in common the inability to produce viable pollen. In maize, plants are terminated by an inflorescence called the tassel. The tassel is composed of many male spikelets (flowers) which, at maturity, exert their anthers from which pollen is shed. The young male gametophyte is borne within the pollen grain. In its most severe manifestation (e.g., cms-T), the tassel of a cytoplasmic male-sterile plant does not exert anthers and no pollen is shed. Frequently, as in the case of the S sterile cytoplasm (cms-S), deformed anthers, termed "sticks", are exerted. However, they contain only aborted pollen grains. The cms trait normally does not affect female fertility.

The cytoplasmic male sterility trait is inherited in an extra-chromosomal or non-Mendelian fashion. It has been assumed that mutations of cytogenes which control male fertility are responsible for the cms trait. In many species, especially maize, there is a growing body of evidence which suggests that the cms trait is encoded on the mitochondrial genome. In a few species the chloroplast genome or a viral basis has been implicated with the sterility trait. In this treatment we will primarily consider cytoplasmic male sterility in maize because it has been more intensely studied and is better understood. The cms trait has been extensively used in the commercial production of hybrids because it circumvents the costly procedure of hand emasculation. Certain genes, called "restorers", which countermand the cms trait and restore fertility, are utilized in seed production so that male fertility is again achieved in the farmer's field.

THE PLANT MITOCHONDRIAL GENOME

Plant mitochondrial genomes are very large when compared with those of other organisms. Animal mitochondrial DNAs (mtDNA) range from 9 to 12 Mdal in size, while those of fungi and protozoa range from 18 to 49 Mdal (2). In contrast, plant mtDNAs which range from 220 to 1600 Mdal have been reported (3) among members of the cucurbit family (e.g., 1600 Mdal for muskmelon, 1000 Mdal for cucumber, 560 Mdal for zucchini squash, 220 Mdal for watermelon, 240 Mdal for pea, and 320 Mdal for corn). These size values were obtained by reassociation kinetic studies, and in some cases were verified by sizing DNA restriction fragments. These examples serve to illustrate the large and variable nature of plant mitochondrial genomes.

The large size of plant mitochondrial genomes, as contrasted with other organisms, predicts that they could encode additional genes or regulatory sequences. Although there are certainly nowhere near the number of extra genes anticipated from the genome size, it now seems evident that these mitochondrial genomes do carry a few genes unique to higher plants. The paucity of plant mitochondrial genes has led to the suggestion that most of the mtDNA may serve no function or a sequence independent function (3).

Many have assumed that the information encoded on the mitochondrial genomes of fungi and animals would also be found in plant mtDNAs. For the most part, the informational content of the mitochondrial genomes of yeast and several animals (humans and mice) is well established. These genomes encode polypeptides involved with ATPase, cytochrome b, cytochrome oxidase, maturase, and the ribosomes. To date, only a single gene, coding for the cytochrome oxidase subunit II, mox 1, has been reported in plants (4). This gene was identified and sequenced from the maize mitochondrial genome.

In addition, it is now clear that plant mitochondrial genomes code for their own ribosomal RNAs (rRNA). It has been shown in wheat that the mtDNA codes for 26S, 18S, and 5S rRNA species (5). Interestingly, the genes for the 18S and 5S DNA are in close proximity to each other, while the 26S RNA is more distantly located. The relative locations of the 5S, 18S, and 26S rRNA genes has also recently been reported in maize (6). Mitochondrial rRNA have been identified in several other plant species (7,8,9,10,11). These rRNAs were distinct from chloroplast and cytoplasmic rRNA, and therefore, they were presumed to be mtDNA gene products. Species of tRNAs, which differ from those observed in the cytosol, have been found in the mitochondria of higher plants (12,13,14). Species of tRNA are reported to be scattered throughout the mitochondrial genome of wheat (5).

CYTOPLASMIC MALE STERILITY

As pointed out previously, there is substantial evidence suggesting that the factors responsible for the cytoplasmic male sterility trait are encoded on the mitochondrial genome. In the following sections, experimental results which are related to the location and function of these factors will be discussed.

Restriction endonucleases are very important for investigating DNAs. This has been especially true for plant mtDNAs because their small complexity has permitted digestion and electrophoretic fractionation of the resulting fragments into patterns which could be visibly compared. When mtDNAs from maize with normal (fertile) T, C, and S cytoplasms were digested and fractionated by electrophoresis, the resulting fragment patterns from the four cytoplasms were easily distinguished (15,16,17). These results clearly demonstrated that the four distinctive cytoplasms, N, T, C, and S, contain unique mtDNAs. When similar studies were performed on chloroplast DNAs (ctDNAs) from N, T, C, and S cytoplasms, the four ctDNAs were almost identical (17). Subsequently, it was observed (18) that a variety of restriction endonucleases, including Bam HI, Eco RI, Hind III, Sal I, Sma I, and Xho I, easily distinguished the mtDNAs of N, T, C, and S cytoplasms and this technique has now become an acceptable means of identifying the different sterile cytoplasms of maize. These analyses have also pointed out that substantial variation exists among the mtDNAs of N, T, C, and S cytoplasms because the various restriction endonucleases recognize different cleavage sites. In a similar study, it was shown (16) by restriction enzyme analysis that small but repeatable distinctions exist in the mtDNA from the various normal (fertile) cytoplasms. These results suggest that heterogeneity is present in normal maize which does not involve those sequences responsible for male-fertility factors. In several other sterile cytoplasms of maize, E, RB, BB, and ES, mtDNA variation was found (19). These cytoplasms appear to be variants of the cms-C type because they are restored by the same inbred lines which restore the C cytoplasm.

Leaver's group has carried out interesting studies on protein synthesis (in vitro) with isolated maize mitochondria. Twenty electrophoretically discrete polypeptides were elaborated by this technique (20). Although functional identifications were not attempted, several distinctive polypeptides were found associated with male-sterile cytoplasms (21). Additional variant polypeptides of 13,000 Mr (apparent molecular weight derived from gel electrophoresis) and 17,500 Mr were observed in cms-T and cms-C, respectively, which were not seen in the N cytoplasm. In the N cytoplasm, polypeptides of 21,000 Mr and 17,500 Mr were found which were absent in T and C, respectively. Eight new minor polypeptides of high molecular weight were detected in cms-S that ranged from 42,000 to 88,000 Mr. They also observed that fertility restoration of the T-cytoplasm with nuclear restorer genes, Rf and Rf-2, appeared to drastically suppress the 13,000 Mr polypeptide. Other polypeptide distinctions were not identified between restored and non-restored cytoplasms of C, T, or S. Even though the function of these polypeptide variants remains unknown, these data have demonstrated polypeptide differences which could be linked to factors responsible for the cms traits.

Substantial differences between the mtDNAs of normal and sterile cytoplasms have also been shown by electron microscopy and electrophoretic studies. Electron microscopy investigations of mtDNA of normal, T, and S maize cytoplasms have distinguished each type because they contain unique distributions of DNA molecules (22). For example, the most abundant classes of mtDNA circular molecules were 21, 17, and 25 μm for N, S, and T cytoplasms, respectively. Small mtDNA molecules (minicircles), which are less than 5 μm in length, also differ among the various cytoplasms. The different cytoplasms can be identified by the size of their minicircles when fractionated by gel electrophoresis (23,24).

Maize carrying the T cytoplasm is preferentially attacked by two fungal pathogens, *Phyllostica maydis* and race T of *Helminthosporium maydis* (25). Cms-T was widely used in hybrid maize production before severe disease outbreaks in the late sixties forced the industry to abandon its use. Numerous investigations have suggested that the mitochondrion is the target site involved in the maternally-inherited susceptibility of cms-T maize to southern corn leaf blight caused by *H. maydis* (18). An inseparable association between disease susceptibility and male sterility appears to exist in the T cytoplasm. It is not yet clear if this association is due to a close linkage or to a pleiotropic phenomenon. Several have speculated (ref. 26, for example) that intermembrane proteins may be responsible for the disease susceptibility-cms trait because several of these proteins are likely candidates of mitochondrial gene products.

Particular nuclear genes that are called restorers of fertility (Rf) can countermand the cytoplasmic male sterility trait in maize and restore it to full pollen fertility. The various cms types are distinguished on the basis of the specific nuclear genes that restore pollen fertility (27). Sets of inbred lines containing combinations of Rf genes are most often used for the identification of the different cms types (28,29). The dominant genes, Rf and Rf-2, located on chromosomes 3 and 9, respectively, are needed to restore full pollen fertility to cms-T. These genes behave in a complementary manner in restoring fertility in that plants carrying cms-T and Rf and Rf-2 are restored fertiles. Rf and Rf-2 loci do not restore fertility to cms-C or cms-S. Cms-T restoration is at the sporophytic level. A single gene, Rf-3, restores pollen fertility to cms-S. The Rf-3 locus is located on chromosome 2 and pollen restoration is gametophytic in nature (30). Rf-3 does not restore cms-T or cms-C. Cms-C is restored by the Rf-4 gene (31); its chromosomal location is unknown.

Differences in the degree of pollen restoration have been observed within cms types (27). This could be due to minute diversity within each cms type or to variations in the nuclear restorer genes. Genes which modify the degree of pollen fertility are also known in maize. Finally, environmental conditions can appreciably influence pollen fertiliay.

The effect of restorer genes on the various cms types illustrates that the male-fertility trait is controlled by both nuclear and cytoplasmic genes. The nature of this interaction is unknown, but undoubtedly will be the subject of future investigations.

Very interesting results bearing on cytoplasmic male sterility have come from selection and regeneration studies with tissue cultures of cms-T maize (32,33). Resistant cells were obtained by selection in T cytoplasm callus grown in the presence of the T toxin from H. maydis. Plants differentiated from the newly arisen resistant callus were resistant to both the toxin and the fungus and were male fertile. Furthermore, it was established by genetic analysis that these changes were cytoplasmic and not nuclear. Restriction fragment analysis with Xho I of mtDNA of four lines derived from resistant, male-fertile regenerated plants (34) indicated that they contained 1 to 3 additional bands not found in the control cms-T mtDNA. Clearly, variations in the mtDNA were demonstrated in plants regenerated from tissue culture; however, it is not known if the mtDNA changes were related to the phenotypic change from susceptible, male-sterile to resistant, male-fertile plants. Finally, it should be pointed out that selection in the presence of the T toxin is not needed to obtain resistant and fertile plants. Brettell et al. (35) reported finding thirty-one resistant and male-fertile plants among sixty plants regenerated from unselected T cytoplasm callus cultures.

Cytological investigation of microsporogenesis in maize (36,37) has implicated the mitochondria in the expression of male sterility. During the tetrad stage, mitochondrial degeneration was seen in the tapetum and middle layer of anthers from cms-T. Similar mitochondrial degeneration was not observed in normal maize cytoplasms. In contrast, plastids and other organelles did not change in structure until very late in anther development.

PLASMID-LIKE DNA MOLECULES

The S cytoplasm, cms-S, is unusual in that associated with its mitochondrial genome are two unique plasmid-like DNAs, designated

S-1 and S-2, which are 6.2 and 5.2 kilobases in size, respectively (38). Both molecules have linear configurations with unique ends as observed by electron microscopy. Since linear configurations could have arisen from breakage of native circular molecules, linearity should be regarded as a tentative judgement.

Plasmid-like DNAs were thought to be solely associated with the S cytoplasm of maize until recently. Similar elements, which differ a little from those found in cms-S, have now been observed in normal cytoplasms (39). The plasmid-like DNAs, S-1 and S-2, have been detected only in mitochondrial preparations from cms-S; attempts to find these molecules in nuclear or chloroplast preparations were unsuccessful. Additional studies have demonstrated that the plasmid-like DNAs are transmitted strictly through the maternal parent.

Electron microscopic studies have shown that the S-1 and S-2 are terminated by inverted repeats, which are approximately 200 base pairs long (18). Furthermore, these investigations have verified that the linear molecules are terminated by non-heterogeneous ends. Heteroduplex analysis has revealed sequence homology between the S-1 and S-2 molecules (40). Homology was detected between the inverted repeats and 1428 base pairs immediately adjacent to one of the inverted repeats. These results suggest the possibilities of similar origins and/or functions. Sequence homology has also been demonstrated between the S-1 and S-2 DNAs and the mitochondrial chromosomal DNAs of normal and male-sterile cytoplasms of maize (41,42,43). Thus, it is clear that some of the sequences found in S-1 and S-2 plasmid-like DNAs are also present in the mitochondrial genome itself. The origin of the plasmid-like DNAs and the cms trait may be related to this sequence homology.

REVERSION TO MALE FERTILITY

The T and C cytoplasms are notoriously stable, a fact which has made them useful in hybrid seed production. On the other hand, the S cytoplasm is unstable and frequently reverts to male fertility. Many instances have been observed (30,44) where cms-S changes from the male-sterile to male-fertile phenotype. Two types of changes have been established by genetic analysis: (1) cytoplasmic mutation from the male-sterile to the male-fertile, and (2) mutations giving rise to new nuclear restorer genes. In most cases the newly arisen male-fertiles were due to cytoplasmic changes which were stable through subsequent generations of propagation. Interestingly, cytoplasmic reversion occurs much more frequently in certain inbred lines than in others (45). For example, 10.9% reversion to male fertility was found in the line M825, while none was seen in N6. These results indicate an interaction between nuclear and cytoplasmic genes. For interested readers, the Laughnans have thoroughly reviewed their studies on the effect of nuclear background on reversion in the S group of cytoplasms (45).

A second class of male-fertile revertants that manifested a behavior anticipated from newly arisen nuclear restorer genes was discovered by Laughnan's group (30,44). These revertants resembled the naturally occurring restorer of cms-S, Rf-3, in that they exhibited the gametophytic mode of pollen restoration. However, thorough studies of these newly arisen restorers have shown that they differ from Rf-3 in several characteristics. Unlike Rf-3, they display reduced transmission through the female gametophyte, a reduced kernel size, and, in most cases, a lethality of the restorer homozygote. More importantly, studies indicate that the newly arisen restorers generally map at unique chromosomal locations and are not allelic to Rf-3. Based upon these results, the investigators postulated a male-fertility element with episome-like characteristics. They proposed that the new restorer genes occurred by the integration of the fertility element into chromosomal sites. Alternatively, they suggested that when the male fertility element is fixed in the cytoplasm, a cytoplasmic change results which causes the reversion to male fertility.

Additional studies with the cytoplasmic type of male-fertile revertants from cms-S have revealed interesting changes in their mitochondrial genomes (46). MtDNAs of newly arisen revertants, and for comparison, male-sterile members of the same families in which the revertants occurred, were characterized by gel electrophoresis and hybridization techniques. Normally, cms-S is made up of equimolar quantities of S-1 and S-2 DNAs. When the revertant and nonrevertant types were compared, it was observed that they differed in the amounts of plasmid-like DNAs present. Even though nonrevertant male-steriles contained both the S-1 and S-2 molecules, they were no longer found in equimolar amounts, but instead, the S-2 DNA occurred in lesser quantities. These results suggested a possible relationship between the unstable nature of the cms-S and the composition of plasmid-like DNAs.

Analysis of seven cytoplasmic revertant strains showed that in each case the S-1 and S-2 had disappeared or virtually disappeared. This observation has been extended to 16 other cytoplasmically reverted male-fertile strains, and in every case the S-1 and S-2 DNAs were absent (45). Consequently, the disappearance of S-1 and S-2 plasmid-like DNAs seems to be a general event associated with the cytoplasmic reversion phenomenon.

A contrast of total mtDNA from nonrevertant (steriles) and revertant (fertile) types (46) indicated that changes had taken place which involved the main mtDNA. When the mtDNAs were digested with restriction endonucleases, for instance Xho I, and fractionated by gel electrophoresis, one to three new bands were discovered in the revertants. These changes indicated that alterations had occurred in the large mitochondrial chromosomal DNAs of the revertants. Hybridization by the Southern method (47) with S-1 and S-2 labelled probes showed that the new bands in the revertants most frequently were composed of sequences homologous with the plasmid-like DNAs. Equally important was the observation that the new banding and hybridization patterns differed among the various revertants. These events led to the proposal that the plasmid-like DNAs were acting like transposable elements and that S-1 and/or S-2 sequences have been inserted into the large mitochondrial chromosomal DNA. Correlated with this happening, free forms of the S-1 and S-2 DNAs have disappeared and the plant's phenotype has reverted to male-fertility. Finally, it is worth noting, in considering this phenomenon, that controlling elements which move and bring about changes in genetic expression have long been recognized in maize (see ref. 48 for review).

Additional plasmid-like DNAs have now been discovered in maize and teosinte, a close relative of maize. A survey of South American maize accessions revealed twelve strains that carried plasmid-like DNA which differed from those found in *cms*-S (39). These new species have been designated R-1 and R-2. The R-2 and S-2 molecules appear identical when compared by restriction fragment analysis and heteroduplexing studies. The R-1 and S-1 molecules differ in that R-1 is approximately 1000 bp larger. However, R-1 and S-1 do share about 4900 bp of common sequence, including their terminal inverted repeats. The interesting distinction between strains carrying the R and S plasmid-like DNAs is in regard to male-fertility. Unlike the S cytoplasm, all strains carrying the R plasmid-like molecules appear to be male-fertile. Perhaps a comparison of the R and S cytoplasmic types will help us understand the molecular basis of male sterility. In *Zea diploperennis*, which is a perennial diploid teosinte, two new plasmid-like DNAs have been found (49) and designated D-1 and D-2. As yet, these new elements have not been thoroughly characterized; however, they do seem very similar to the R-1 and R-2 molecules. Interestingly, it appears that the presence of D-1 and D-2 DNAs is also not associated with male sterility. These results suggest that plasmid-like DNAs are more widely spread than previously believed and that they may carry different genetic determinants.

EPILOGUE

We have reviewed the evidence that implicates mitochondrial genes with cytoplasmic male sterility in maize. Recent molecular studies, especially with the S cytoplasm, have moved us close to identifying those genes responsible for the *cms* trait. Once these factors are identified, it should be possible to manipulate these genes to achieve new and useful forms of male sterility. Perhaps the maize genes responsible for pollen fertility can be used as molecular probes for identifying fertility genes in other species. In this event, it may be possible, through genetic engineering, to create male sterility in species where it did not previously exist. In some species this would be particularly beneficial because it would simplify crossing procedures for plant breeders and perhaps permit commercial hybrid seed production for the exploitation of hybrid vigor.

Studies of cms-S have identified plasmid-like DNAs which behave like transposable elements. These DNAs may have application as transfer vectors for introducing genes into plant cells. In maize these elements appear to insert DNA sequences into the mitochondrial genome which result in heritable, phenotypic changes. It may be practical to use these DNA species, or parts of them, for the construction of transfer vectors for the modification of plant genomes.

Several important questions have not been addressed in this treatment. It is not clear why genes effecting male fertility should be encoded on the mitochondrial genome or what function these genes serve in the development of pollen. Although cytoplasmic male-sterility studies have clearly demonstrated interactions between genes in different organelles, the nucleus, and mitochondrion, they have provided little insight into the mechanism by which these interactions are brought about. These questions and others remain to be answered.

REFERENCES

1. Edwardson, J.R. 1970. Cytoplasmic male sterility. Bot. Rev. 36:341.
2. Borst, P. 1976. Structure and function of mitochondrial DNA. In International Cell Biology, p. 237, B.P. Brinkley and K.P. Porter, eds. New York: Rockefeller Univ. Press.
3. Ward, B.L., R.S. Anderson, and A.J. Bendich. 1981. The size of the mitochondrial genome is large and variable in a family of plants (Cucurbitaceae). Cell 25:793.
4. Fox, T.D., and C.J. Leaver. 1981. The Zea mays mitochondrial gene coding cytochrome oxidase subunit II has an intervening sequence and does not contain TGA codons. Cell 26:315.
5. Bonen, L., and M.W. Gray. 1980. Organization and expression of the mitochondrial genome of plants 1. The genes for wheat mitochondrial ribosomal and transfer RNA: Evidence for an unusual arrangement. Nucleic Acid Res. 8:319.
6. Stern, D.B., T.A. Dyer, and D.M. Longsdale. 1982. Organization of the mitochondrial ribosomal RNA gene in maize. Nucleic Acid Res. 10:3333.
7. Leaver, C.J., and M.A. Harmey. 1973. Plant mitochondrial nucleic acids. Biochem. Soc. Symp. 38:175.
8. Leaver, C.J., and M.A. Harmey. 1976. Higher plant mitochondrial ribosomes contain a 5S rRNA component. Biochem. J. 157:275.

9. Pring, D.R. 1974. Maize mitochondria: Purification and characterization of ribosomes and ribosomal ribonucleic acid. Plant Physiol. 53:677.
10. Cunningham, R.S., and M.W. Gray. 1977. Isolation and characterization of ^{32}P-labeled mitochondrial and cytosol ribosomal RNA from germinating wheat embryos. Biochim. Biophys. Acta 475:476.
11. Bonen, L., R.S. Cunningham, M.W. Gray, and W.F. Doolittle. 1977. Wheat embryo mitochondrial 18S ribosomal RNA: Evidence for its prokaryotic nature. Nucleic Acid Res. 4:663.
12. Meng, R.L., and L.N. Vanderloef. 1972. Mitochondrial tyrosyl transfer ribonucleic acid in soybean seedlings. Plant Physiol. 50:298.
13. Guderian, R.H., R.L. Pulliam, and M.P. Gordon. 1972. Characterization and fractionation of tobacco leaf transfer RNA. Biochim. Biophys. Acta 262:50.
14. Guillemant, P., A. Steinmetz, G. Burkard, and J.H. Weil. 1975. Aminoacylation of tRNA species from *Escherichia coli* and from the cytoplasm chloroplasts and mitochondria of *Phaseolus vulgaris* by homologous and heterologous enzymes. Biochim. Biophys. Acta 378:64.
15. Levings, C.S., III, and D.R. Pring. 1976. Restriction endonuclease analysis of mitochondrial DNA from normal and Texas cytoplasmic male-sterile maize. Science 193:158.
16. Levings, C.S., III, and D.R. Pring. 1977. Diversity of mitochondrial genomes among normal cytoplasms of maize. J. Hered. 68:350.
17. Pring, D.R., and C.S. Levings, III. 1978. Heterogeneity of maize cytoplasmic genomes among male-sterile cytoplasms. Genetics 89:121.
18. Levings, C.S., III, and D.R. Pring. 1979. Molecular basis of cytoplasmic male sterility of maize. In Physiological Genetics, vol. 5, p. 171, J.G. Scandalios, ed. New York: Academic Press.
19. Pring, D.R., M.F. Conde, and C.S. Levings, III. 1980. DNA heterogeneity within the C group of maize male-sterile cytoplasms. Crop Sci. 20:159.
20. Forde, B.G., and C.J. Leaver. 1980. Nuclear and cytoplasmic genes controlling synthesis of variant polypeptides in male-sterile maize. Proc. Natl. Acad. Sci. U.S.A. 77:418.
21. Forde, B.G., R.J.C. Oliver, and C.J. Leaver. 1978. Variations in mitochondrial translation products associated with male-sterility in maize. Proc. Natl. Acad. Sci. U.S.A. 75:3841.
22. Levings, C.S., III, D.M. Shah, W.W.L. Hu, D.R. Pring, and D.H. Timothy. 1979. Molecular heterogeneity and mitochondrial DNAs from different maize cytoplasms. In Extrachromosomal DNA. ICN-UCLA Symposia on Molecular and Cellular Biology, vol. XV, p. 63, D.J. Cummings, P. Borst, I.G. Dawid, and S.M. Weissman, eds. New York: Academic Press.

23. Kemble, R.J., and J.R. Bedbrook. 1980. Low molecular weight circular and linear DNA molecules in mitochondria from normal and male-sterile cytoplasms of Zea mays. Nature 284:565.
24. Kemble, R.J., R.E. Gunn, and R.B. Flavell. 1980. Classification of normal and male-sterile cytoplasms in maize. II. Electrophoretic analysis of DNA species in mitochondria. Genetics 95:451.
25. Ullstrup, A.J. 1972. The impacts of the Southern corn leaf blight epidemics of 1970-71. Annu. Rev. Phytopathol. 10:37.
26. Peterson, P.A., R.B. Flavell, and D.H.P. Barratt. 1975. Altered mitochondrial membrane activities associated with cytoplasmically-inherited disease sensitivity in maize. Theor. Appl. Genet. 45:309.
27. Duvick, D.N. 1965. Cytoplasmic pollen sterility in corn. Adv. Genet. 13:1.
28. Beckett, J.B. 1971. Classification of male-sterile cytoplasms in maize (Zea mays L.). Crop Sci. 11:724.
29. Gracen, V.E., and C.O. Grogan. 1974. Diversity and suitability for hybrid production of different sources of cytoplasmic male sterility in maize. Agron. J. 65:654.
30. Laughnan, J.R., and S.J. Gabay. 1975. An episomal basis for instability of S male sterility in maize and some implications for plant breeding. In Genetics and Biogenesis of Mitochondria and Chloroplasts, p. 330, C.W. Birkey, P.S. Perlman, and T.J. Beyers, eds. Columbus: Ohio State University Press.
31. Kheyr-Pour, A., V.E. Gracen, and H.L. Everett. 1981. Genetics of fertility restoration in the C-group of cytoplasmic male sterility in maize. Genetics 98:379.
32. Gengenbach, B.G., and C.E. Green. 1975. Selection of T-cytoplasm maize callus cultures resistant to Helminthoporium maydis race T pathotoxin. Crop Sci. 15:645.
33. Gengenbach, B.G., C.E. Green, and C.M. Donovan. 1977. Inheritance of selected pathotoxin resistance in maize plants regenerated from cell cultures. Proc. Natl. Acad. Sci. U.S.A. 74:5113.
34. Pring, D.R., M.F. Conde, and B.G. Gengenbach. 1981. Cytoplasmic genome variability in tissue culture derived plants. Environ. Exp. Bot. 21:369.
35. Brettell, R.I.S., B.V.D. Goddard, and D.S. Ingram. 1979. Selection of Tms-cytoplasm maize tissue culture resistant to Drechslera maydis T-toxin. Maydica 24:203.
36. Warmke, H.E., and S.L.J. Lee. 1977. Mitochondrial degeneration in T cytoplasmic male-sterile corn anthers. J. Hered. 68:213.
37. Warmke, H.E., and S.L.J. Lee. 1978. Pollen abortion in T cytoplasmic male-sterile corn: A suggested mechanism. Science 200:561.
38. Pring, D.R., C.S. Levings, III, W.W.L. Hu, and D.H. Timothy. 1977. Unique DNA associated with mitochondria in the "S" type cytoplasm of male-sterile maize. Proc. Natl. Acad. Sci. U.S.A. 74:2904.

39. Weissinger, A.K., D.H. Timothy, C.S. Levings, III, W.W.L. Hu, and M.M. Goodman. 1982. Unique plasmid-like mitochondrial DNAs from indigenous maize races of Latin America. Proc. Natl. Acad. Sci. U.S.A. 79:1.
40. Kim, B.D., R.J. Mans, M.F. Conde, D.R. Pring, and C.S. Levings, III. 1982. Physical mapping of homologous segments of mitochondrial episomes from S male-sterile maize. Plasmid 7:1.
41. Thompson, R.D., R.J. Kemble, and R.B. Flavell. 1980. Variations in mitochondrial DNA organization between normal and male-sterile cytoplasms of maize. Nucleic Acid Res. 8:1999.
42. Spruill, W.M., Jr., C.S. Levings, III, and R.R. Sederoff. 1980. Recombinant DNA analysis indicates that the multiple chromosomes of maize mitochondria contain different sequences. Dev. Gen. 1:363.
43. Spruill, W.M., Jr., C.S. Levings, III, and R.R. Sederoff. 1981. Organization of mitochondrial DNA in normal and Texas male sterile cytoplasms of maize. Dev. Gen. 2:319.
44. Laughnan, J.R., and S.J. Gabay. 1975. Nuclear and cytoplasmic mutations to fertility in S male-sterile maize. In International Maize Symposium: Genetics and Breeding, p. 427, D.B. Walden, ed. New York: Wiley.
45. Laughnan, J.R., and S. Gabay-Laughnan. 1981. Characteristics of cms-S reversion to male fertility in maize. Stadler Symp. 13:93.
46. Levings, C.S., III, B.D. Kim, D.R. Pring, M.F. Conde, R.J. Mans, J.R. Laughnan, and S.J. Gabay-Laughnan. 1980. Cytoplasmic reversion of cms-S in maize: Association with a transpositional event. Science 209:1021.
47. Southern, E.M. 1975. Detection of specific sequences among DNA fragments separated by gel electrophoresis. J. Mol, Biol. 98:503.
48. Fincham, J.R.S., and G.R.K. Sastry. 1974. Controlling elements in maize. Annu. Rev. Genet. 8:15.
49. Timothy, D.H., C.S. Levings, III, W.W.L. Hu, and M.M. Goodman. 1982. Zea diploperennis may have plasmid-like mitochondria DNAs. Maize Genet. Coop. News Letter 56:133.

GENETIC ENGINEERING OF SEED STORAGE PROTEINS

Brian A. Larkins

Department of Botany and Plant Pathology
Purdue University
West Lafayette, Indiana 47907

INTRODUCTION

A list of traits that would be valuable to genetically engineer in plants would include characteristics of disease and herbicide resistance, enhanced photosynthetic capacity, the ability to fix nitrogen, drought tolerance, and increased protein nutritional quality. While it would be a tremendous achievement to incorporate all of these into a major crop species, the ability to isolate a gene or genes responsible for even one of them and transfer it from one plant to another by a non-sexual process would be a highly significant accomplishment. Unfortunately, we presently know so little about the biochemical and physiological processes involved in these responses or processes that it is difficult or impossible to identify such genes. In view of the progress being made in gene isolation and characterization, we may be able to genetically engineer plants before we have identified which genes are responsible for the traits we wish to modify.

One notable exception to this generalization is the genes encoding seed storage proteins. During the past five years there has been significant progress in the characterization of these proteins and the biochemical mechanisms involved in their biosynthesis and deposition in developing seeds. Furthermore, the genes for several different types of storage proteins have been isolated and characterized. As a result, seed storage protein genes present one of the first systems for exploring the potential of genetic engineering in plants.

Application of Genetic Engineering to Seed Storage Proteins

Before considering specific storage proteins for which genetic engineering techniques would be appropriate, I would like to review some of the characteristics that are unique to seed storage proteins. A more detailed review of the biochemistry of these proteins can be obtained from several review articles (7,22).

In addition to starch and lipid, storage proteins serve as one of the major food reserves for the developing seed. The amount of protein varies with genotype or cultivar; generally, cereals contain 10% of the dry weight as protein, while in legumes, protein content varies between 20 and 30% of the dry weight. In many seeds, the storage proteins account for 50% or more of the total protein.

The storage proteins have no apparent enzymatic activity. Their primary function is to store nitrogen and amino acids which can be utilized by the developing seedling. It has been suggested that the amino acid composition of the storage proteins reflects what is nutritionally balanced for the young seedling, but while this may be true, it is difficult to prove or disprove.

T.B. Osborn (32) was one of the first scientists to carefully characterize the chemical nature of seed proteins. He used a classification system based on the solubility of proteins in different solvents. He called proteins soluble in water "albumins", proteins soluble in 5% saline "globulins", and proteins soluble in aqueous alcohol (70% ethanol) "prolamines". The proteins that remained following these extractions were extracted with dilute acid or alkali, and these he called "glutelins".

After comparing proteins isolated from a variety of monocot and dicot seeds, Osborn found that most cereals contained primarily prolamine type proteins, while legumes contained globulin proteins.

The polypeptides composing the prolamine fractions have been difficult to characterize because of their insolubility in aqueous buffers. Many of the early analyses were based solely on amino acid composition, and showed them to be rich in glutamine and proline, hence the name "prolamine". These analyses also showed that prolamines contained very little lysine or tryptophan. With the advent of SDS-polyacrylamide gel electrophoresis it became apparent that most cereals contained prolamines of several distinct molecular weights. Each of these can be resolved into a number of differently charged forms by 2-dimensional gel electrophoresis, thus indicating the existence of families of proteins for the major molecular weight species.

Danielsson (6) described the basic physical properties of the storage globulins of the legumes. He showed the globulin fraction contained two major types of proteins having sedimentation coefficients of 7S and 11S. These two are found in nearly equal proportions in some legumes, while one or the other predominates in others. As is true of the cereal prolamines, amides constitute a significant proportion of the total amino acids of the globulins. The sulfur-containing amino acids, methionine and cysteine, are generally found in very low amounts in both of these proteins, but they are particularly low in the 7S globulins.

Both the 7S and 11S proteins are composed of multiple subunits. In most of the species that have been examined in detail, the 7S protein is composed of three polypeptides of similar size (7). The 11S protein is more complex; it is composed of a mixture of large acidic, 40,000 Mr (apparent molecular weight derived from gel electrophoresis) and small basic (20,000 Mr) polypeptides. There is some debate regarding the organization of these polypeptides within the 11S molecule, but there is general agreement that each acidic polypeptide is associated with a basic one, and there appear to be six acidic and six basic polypeptides per 11S complex. Although some structural and chemical differences exist between legume and cereal storage proteins, that they have similar biological functions implies that they share some common characteristics. Among these are their high content of amide amino acids, which allows for nitrogen storage, and the fact that both types of proteins are deposited in an insoluble form in the developing seed. Cereal prolamines form insoluble aggregates called protein bodies inside the lumen of the rough endoplasmic reticulum (RER). In legumes, and presumably other seeds storing globulin-type proteins, the proteins are transported into the vacuole where they form protein bodies. In some seeds, other constituents, such as phytic acid, are also included in the protein body (26).

There are a number of reasons why it is important that the storage proteins be deposited in an insoluble form. Since the seed must eventually become desiccated, storage proteins cannot be osmotically active. These proteins must be stable as there is often a long period between the time of seed maturation and seed germination. Finally, it is necessary that the structure of these proteins be compatible with the proteolytic enzymes produced by the plant during germination, so peptides and amino acids can be rapidly released and transported to the developing seedling.

The low levels of lysine and tryptophan in cereal prolamines and methionine and cysteine in legume globulins are not beneficial to animals dependent on these seeds for essential amino acids. There has been a great deal of effort to try and overcome these amino acid limitations by breeding and selecting for more nutritionally balanced varieties. Attempts have been made to mutagenize plants in hopes of recovering individuals with more nutritious storage proteins. Neither of these approaches has been entirely successful, although some naturally occurring and artificially produced mutants of cereals were shown to contain more nutritionally balanced amino acid compositions (29). These mutations cause a significant reduction in the amount of storage protein synthesized and thereby result in a higher percentage of lysine in the seed. The reduction in storage protein also causes the seed to become more brittle; as a result, these seeds shatter more easily during storage. The lower levels of prolamine also result in flours with unfavorable functional properties which causes brittleness in the baked products. Thus, no satisfactory solution has been found for improving the amino acid composition of storage proteins.

One direct approach to this problem would be to modify the nucleotide sequence of genes encoding storage proteins so that they contain higher levels of essential amino acids. This is conceptually a simple thing to do; however, answers to a number of questions must be known before this approach is practical. In theory, it should be relatively straightforward to isolate storage protein genes and change their coding sequences. However, changes must not be in regions of the protein that perturb the normal protein structure; otherwise the proteins might be unstable or could prevent the seed from undergoing its normal desiccation process. It is also necessary to know how many genes encode the storage proteins. If there are more than one or two, it is important to know the extent to which they are transcribed. Modification of a more actively transcribed gene should have a greater effect on the overall amino acid composition than modification of a less active gene. One also needs to know the structure and organization of these genes in the genome, as well as the nature of their developmental regulation. Once the gene or genes have been isolated and modified, they must be reinserted into the genome so that they are developmentally regulated and efficiently expressed during seed development.

It is possible that we will not be able to make these structural modifications and stably maintain them in the storage proteins. Obviously these genes have been around for several million years, and if they could tolerate substitution of these amino acids, it might have appeared through normal mutation processes. On the other hand, if these changes can only be tolerated at specific sites in the proteins, the random frequency with which this might occur would be considerably reduced. It appears that the seedling does not require these amino acids in the protein, so there would be no selection pressure to maintain them.

Should it not be possible to change the normal storage proteins by site directed mutagenesis, there are alternative approaches that might be practical. One of these is to transfer heterologous storage protein genes that encode storage proteins with higher levels of the desired amino acids. For example, the soybean 11S storage globulin contains adequate levels of lysine. If these genes were transferred to maize and could be efficiently expressed, the resulting seed would contain a well-balanced amino acid composition. As yet, it is unknown whether the storage cells in a maize seed can transcribe soybean genes, translate the mRNAs, and perform the post-translational modifications necessary to assemble these proteins into protein bodies. However, this can be tested.

Another possibility for improving the nutritional quality of seed proteins is to identify proteins which are normally found in seeds and contain high levels of the requisite amino acids. Although these proteins generally occur in low amounts, it might be possible to select for seeds with higher levels or place these genes under transcriptional control mechanisms that would enhance their expression. It is first necessary to identify these genes and know the basis of their transcriptional regulation.

To assess what is known about seed storage proteins and the genes directing their synthesis, I will describe several different types and consider problems related to their potential modification.

Maize Zein Proteins

The storage proteins of maize seed consist of a group of prolamine proteins called zeins. Zein proteins are synthesized by membrane-bound polyribosomes in the developing endosperm and are deposited as aggregates called protein bodies within the RER. Because of this, one observes RER membranes surrounding protein bodies in electron micrographs of maize endosperm (Fig. 1). It is often difficult to observe continuity between membranes surrounding the protein bodies and the RER, but previous studies have shown the existence of similar populations of polyribosomes on the surface of both membranes (20). Perhaps the most convincing evidence that zein protein bodies form simply by protein aggregation within the RER is the observation that structures with the same physical characteristics as protein bodies can be isolated from Xenopus laevis oocytes previously injected with zein mRNAs (16).

A 2-dimensional gel analysis of the alcohol-soluble proteins contained within the protein bodies reveals a mixture of polypeptides (Fig. 2). the most abundant of these have apparent molecular weights of 22,000 and 19,000, but there are also some smaller polypeptides of 15,000 and 10,000 daltons. As can be seen from the 2-dimensional gel separation shown in Fig. 2, there is significantly more charge heterogeneity among the 22,000 Mr and 19,000 Mr zeins than the smaller molecular weight zein proteins.

We have constructed cDNA clones of zein mRNAs and determined the DNA sequence for representative 22,000 Mr, 19,000 Mr, and 15,000 Mr zein proteins (27,34). The sequence for a 19,000 Mr zein has also been reported by Geraghty et al. (10). From knowledge of the DNA sequence we were able to determine the complete primary amino acid sequence of the polypeptides and compare them for structural similarities.

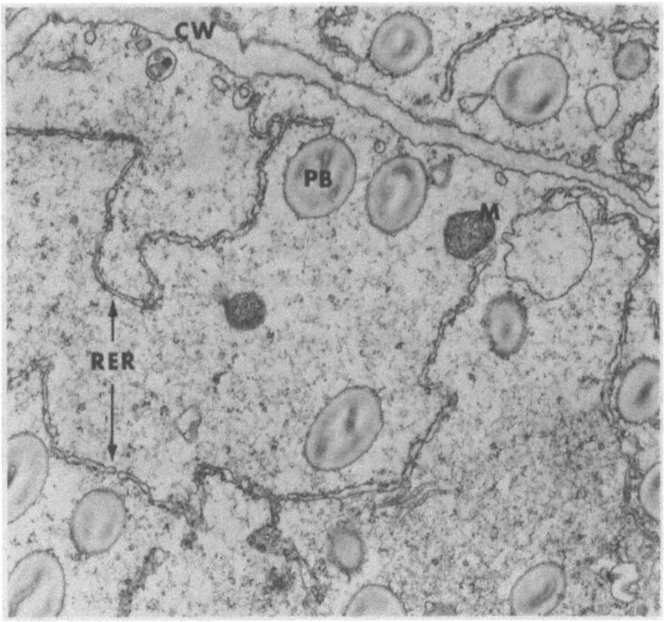

Fig. 1. Electron micrograph of developing maize endosperm. CW = cell wall, PB = protein body, M = mitochondria, RER = rough endoplasmic reticulum. (X17,800) (From ref. 20 with permission.)

This analysis revealed that the zeins were significantly larger than expected, based on their mobility on SDS polyacrylamide gels. Zein proteins that had been estimated to have molecular weights of 22,000 and 19,000 were found to be closer to 27,000 and 23,000, respectively (Table 1). This analysis also comfirmed the presence of signal peptides on the zein proteins (Fig. 3). These signal sequences were previously demonstrated to be removed when the protein is transported into the lumen of the RER (21).

The amino acid composition predicted from the polypeptide sequence is similar to that previously found for mixtures of zein proteins (23). Glutamine, leucine, proline, alanine, and glycine account for the majority of the amino acids, and lysine and tryptophan are absent from all of them (Table 1). It is interesting to note that methionine, which is deficient in most legume storage proteins, accounts for a significant percentage of the 15,000 Mr zein. In fact, cysteine and methionine account for 11% of the total amino acids in this polypeptide.

A particularly interesting feature of the protein sequence is the occurrence of a conserved, tandemly repeated peptide in both of

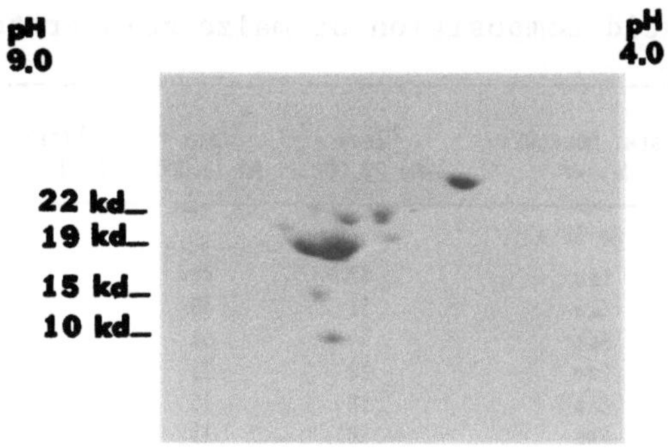

Fig. 2. Two-dimensional polyacrylamide gel separation of zein proteins. Isoelectric focusing is from left to right, SDS polacrylamide gel electrophoresis is from top to bottom. Mole weights of polypeptides are indicated. (From ref. 16 with permission.)

the 22,000 Mr and 19,000 Mr zeins (Fig. 3). The first of these repeat sequences begins 35-36 amino acids after the NH_2-terminus, and is repeated nine times in each polypeptide. The COOH-terminal sequence following the repeats is slightly longer in the 22,000 Mr zeins; this accounts for the size difference between the two polypeptides. Most of the amino acids in these repeats are nonpolar, while the repeated peptide is sequentially polar, nonpolar, polar, nonpolar, polar.

Circular dichroism measurements of mixtures of zein proteins indicate from 45-55% α-helical structure (1), and this percentage correlates well with the proportion of amino acids in these repeated peptides. To determine if the repeats have the potential to form α-helices, we compared their amino acid sequences with those found in proteins having α-helical structure (1). Although a comparison to soluble proteins shows little propensity for these repetitive sequences to be α-helical, they do have α-helical properties when compared with sequences found in some hydrophobic proteins. In view of the hydrophobic nature of zein proteins, it seems reasonable to predict an α-helical structure for them.

If these are α-helices, how might these nine repeats be organized into a 3-dimensional structure? Fig. 4A shows that when the consensus repeat is placed in an α-helical wheel, the polar amino acids are distributed at three symmetrical sites. If the repeats are tandem, and if they fold back upon one another in an antiparallel arrangement, two polar groups in each repeat can hydrogen bond with each of two adjacent repeats (Fig. 4B). The nine helices would then interact to form a roughly cylindrical, rod-shaped molecule. The cylinder would collapse in the center to accommodate the nonpolar tails of the amino acids (Fig. 3B). As

Table 1. Amino acid composition of maize zein proteins.

Apparent Molecular Weight	[A]Zein α Mr 22,000	[B]Zein β Mr 19,000	[C]Zein γ Mr 15,000
Amino Acid			
Leu	42	44	15
Gln	31	39	28
Ala	34	31	18
Pro	22	21	13
Ser	18	15	11
Phe	8	13	0
Asn	13	9	2
Ile	11	9	1
Tyr	6	8	16
Val	17	6	4
Gly	2	4	12
Thr	7	4	5
Arg	4	3	7
His	3	3	4
Cys	1	2	6
Glu	1	1	5
Met	5	1	11
Asp	0	1	4
Lys	0	0	0
Trp	0	0	1
Total	225	214	163

[A] Marks and Larkins (1982) J. Biol. Chem. (in press).
[B] Pedersen et al. (1982) Cell (in press).
[C] Pedersen (unpublished results).

these protein molecules associate within the endoplasmic reticulum, the third polar group, which is on the surface of the helix, would hydrogen bond to a different zein molecule. This arrangement also allows the glutamine residues, which lie at the ends of the helices, to hydrogen bond with neighboring protein molecules in the protein body (Fig. 4C).

Although we do not know if zein proteins conform to this structure, the model does explain many of the physical properties of the proteins (1). It will be necessary to analyze x-ray diffraction patterns of protein crystals to confirm the accuracy of our model. If the interaction of these α-helices is important in structuring the polypeptide and aggregating it into a protein body, altering the amino acid sequence of these repeated regions could deleteriously affect the protein's structure. It would seem more advantageous to change the NH_2-terminal or COOH-terminal turn sequences which lie outside the repeat structures.

Another important question that comes to bear on the problem of genetically engineering zein proteins is how many genes encode these

Zein protein	Signal Sequence	Sequence positions
Z22	M A T K I L S L L A L L A L F A S A T N A	1-21
Z19	M A A K I F C L I M L L G L S A S A A T A	1-21
	N-Terminal Turn	
Z22	S I I P Q C S L A P . S S I I P Q F L P P V T S M A F E H P A V Q A Y R	22-56
Z19	S I F P Q C S Q A P I A S L L P P Y L S P A M S S V C E N P I L L P Y R	22-57
	Repeat Sequences	
Z22 & Z19 9 repeats	Q Q F/L L P A/F N Q L A/L A/V A N S P A Y L Q Q	57-237 58-225
	C-Terminal Turn	
Z22	Q Q Q L L P Y N R F S L M N P V L S R Q Q P I V G G A I F	238-266
Z19	. . . . Q Q P I I G G A L F	226-235

Fig. 3. Structural domains of zein polypeptides. Amino acid sequences were derived from the DNA sequence of zein clones (see refs. 27 and 34).

polypeptides? This question can be experimentally approached by one of several procedures, although each has its limitations. By determining the rate at which zein cDNAs or cDNA clones hybridize to maize DNA, it is possible to determine the average reiteration frequency for these genes in the maize genome. Data from such experiments yield estimates of 4-10 copies of a given gene sequence (33,42). But this number is simply an estimate of the average reiteration of all zein genes, or of a specific sequence when a zein cDNA clone is used as a probe. It does not indicate the total number of genes, nor does it say which of these are expressed.

An alternative approach to estimate the number of zein genes is to use a cDNA clone as a probe in a Southern blot hybridization of maize DNA. If maize DNA is digested with a restriction enzyme that does not cut within the zein genes, it is possible to estimate the number of genes hybridizing to that particular sequence by the number and intensity of bands in the autoradiograph. A reconstruction of the gene number can be made with the cDNA clone to show the intensity of hybridization of one, two, or five gene copies. In order to make the estimation reasonably quantitative, one must also know the sequence homology among different cDNA clones to establish the complexity of the mRNA population and the extent to which different cDNA clones cross-hybridize.

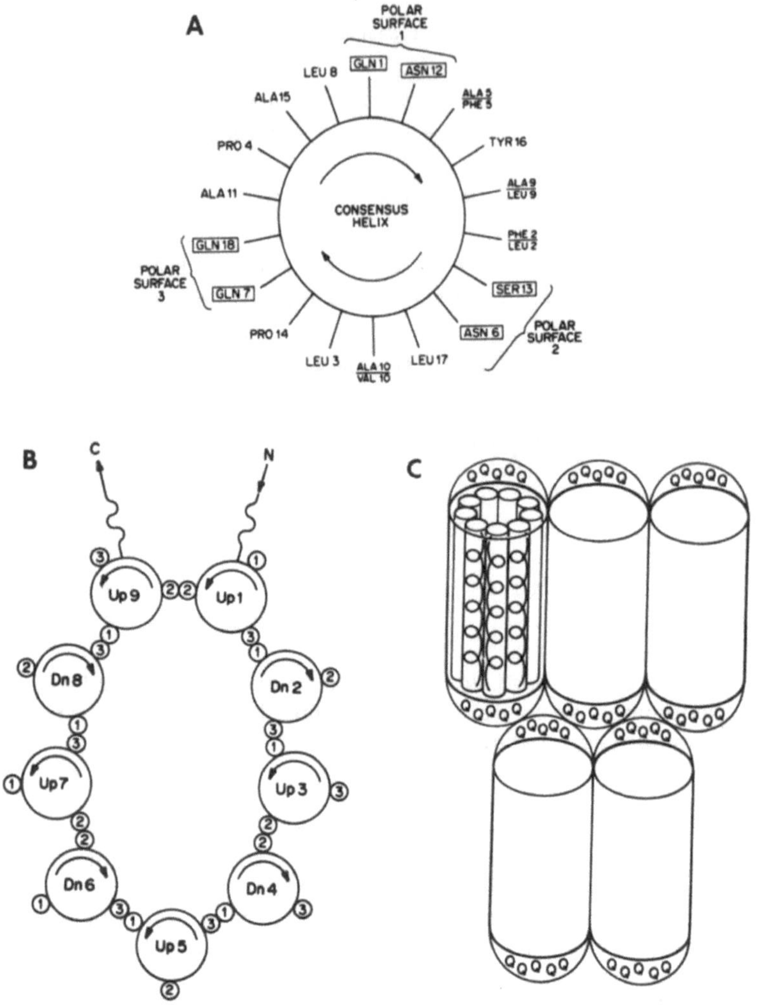

Fig. 4. Structural model for maize zein proteins. (A) Analysis of consensus repeat on an α-helical wheel. (B) Organization of nine repeated α-helices in flattened cylindrical protein structure. The numbers 1, 2, and 3 on each repeat indicate the positions of the three polar regions. "Up" and "Down" indicate the antiparallel orientation of the helices. (C) Possible model for arrangement of zein proteins within protein body. "Q"s indicate the positions of glutamine residues at the ends of the repeated peptides (1).

The Southern blot shown in Fig. 5 shows an analysis of a cDNA clone corresponding to one of the 19,000 Mr zein proteins. Based upon the reconstruction analysis, the four smaller and fainter bands have intensities of single genes, while the three upper bands

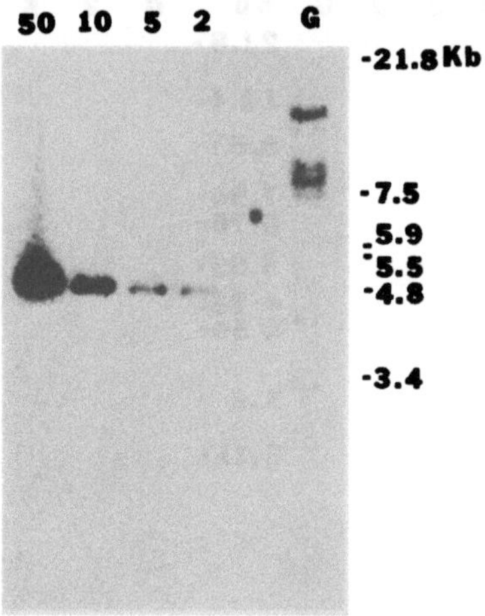

Fig. 5. Southern hybridization of a zein cDNA clone (pZ19.1) to restriction endonuclease digested maize DNA. The hybridization and wash criteria were Tm-40°C and Tm-24°C, respectively. The DNA was restricted with the enzyme Eco RI (lane G) which does not cut in the coding region of the gene. The reconstruction lanes show the hybridization intensity of 50, 10, 5, and 2 gene copies. (From 34 with permission.)

represent sequences present in three to five copies. In this case, the DNA was cut with the restriction enzyme Eco RI, which does not cut in this gene sequence (34). Thus, this zein sequence appears to be a member of a small gene family of approximately 10 members.

We detected two major types of 19,000 Mr and 22,000 Mr cDNA sequences in the genotype from which this mRNA was cloned. At Tm-30° the two different 19,000 Mr sequences do not cross-hybridize, whereas all the 22,000 Mr sequences do. Therefore, by hybridizing with a 22,000 Mr cDNA clone at this criterion, we can estimate the total number of the 22,000 Mr genes rather than those specific for a given 22,000 Mr sequence. Fig. 6 shows a Southern blot that was hybridized at two different criteria. At Tm-30° we can presumably detect all sequences homologous to the 22,000 Mr genes, while at Tm-20° we detect only those that are highly homologous to the probe sequence. From the gene copy reconstruction we estimate that there are a total of 20-25 genes corresponding to the 22,000 Mr genes in this inbred, with 10-12 being highly homologous to the probe sequence. In experiments using the 15,000 Mr sequence as a probe and similar hybridization stringencies, we find only two to three copies of this gene in the genome.

When we compile the total number of genes estimated from these experiments, it appears that there are about 40-50 zein genes in

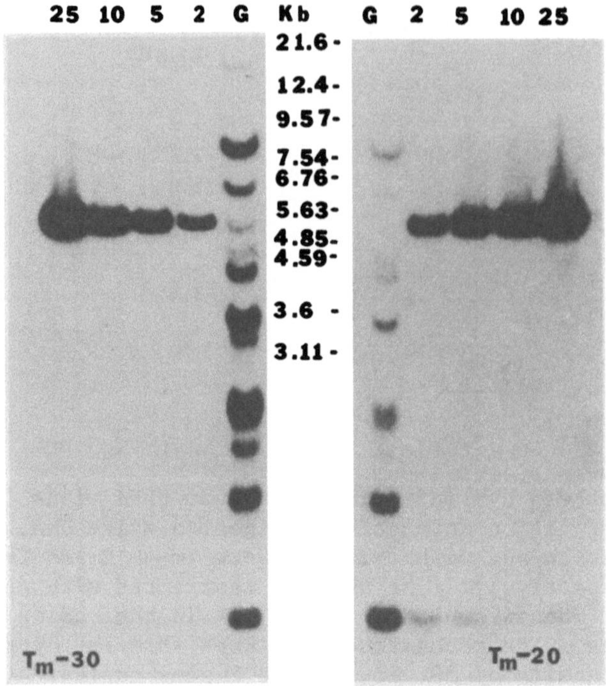

Fig. 6. Southern hybridization of a zein cDNA clone (pZ22.1) to restriction endonuclease digested maize DNA. The hybridization and wash criteria are given in the figure. Duplicate sample of maize DNA were digested with the enzyme Eco RI (lanes G) which does not cut in the coding region of the gene. The reconstruction lanes show the hybridization intensity of 25, 10, 5, and 2 gene copies at the given Tms.

this maize inbred. This number is less than others have estimated by similar techniques (13,42), and the variation may be a reflection of differences in experimental techniques. Since we do not know whether each of the bands in the Southern analysis represents a gene or genes that are actually expressed, this number must be considered a maximum estimate. The best way to determine how many of these genes are expressed is to determine the number of different zein cDNA clones that can be derived from the polysomal mRNA population. The accuracy of this method is dependent on extensive screening of zein cDNAs to detect mRNAs present in low abundance. In our analysis of the mRNA population in the inbred W64A, we isolated 5 different cDNA clones for 22,000 Mr zeins, 4 different clones for 19,000 Mr zeins, and 1 clone for a 15,000 Mr zein (27). These data show clearly that zein genes occur in small multigene families and that several members of these families are expressed. We do not know if some members are expressed more than others, nor how tightly their expression is coordinated. Nevertheless, one cannot make a significant impact on the amino acid composition of the zein fraction by engineering one or two genes; instead, it will be necessary to introduce a number of genes under conditions that will favor their expression. It might also prove useful to take advantage of known regulatory mutants, such as opaque-2 and floury-2, which suppress the synthesis of zein proteins. These mutations could be used to reduce the level of endogenous zeins, while the genetically engineered genes could be placed under alternative transcriptional regulation. This would allow for higher levels of products from the engineered genes.

It should soon be possible to obtain answers to some of the questions regarding the modification of zein genes. Several laboratories have isolated clones of zein genes (24,43,45), and the nucleotide sequences of these genes are being reported (34). To date it appears that zein genes do not contain intervening sequences. The R-loops of several different zein clones indicate no intervening sequences (34,35,45) and none were found in a completely sequenced zein gene (34). The flanking regions of the gene were found to contain consensus sequences typical of other eukaryotic genes, but it is premature to speculate on their function.

The availability of cloned genes will allow us to begin experiments to answer questions regarding the stability of structurally modified zein proteins and the factors affecting the expression of these genes in the genome. Based on experiments that have been done to date, the synthesis and processing of these storage proteins is a process common to quite divergent eukaryotic cells (16,21). Since these proteins self-associate into protein bodies in cells as divergent as Xenopus oocytes, it seems likely that this process could be duplicated in other plant cells.

Soybean Storage Globulins

Most of the protein in soybean seeds can be extracted with a buffered saline solution. When this protein is separated by density gradient centrifugation, it resolves into three major fractions with sedimentation coefficients of 2S, 7S, and 11S. The 2S band contains a mixture of proteins, including several enzymes and protease inhibitors. The 7S and 11S fractions correspond to the major storage proteins, conglycinin and glycinin, respectively. The 11S fraction is usually the most abundant, and it accounts for 50-60% of the total protein. Neither the 7S nor 11S fractions are homogeneous proteins. The 7S fraction contains at least two proteins which are called β- and γ-conglycinin. The 11S protein appears to be more homogeneous, although there is substantial microheterogeneity among the polypeptides composing this molecule (28).

Amino acid analysis of the 7S and 11S proteins (Table 2) reveals that they, like the maize zein proteins, contain high levels

Table 2. Amino acid composition of legume storage globulins.

	Glycine	max.	Pisum	sativum
Amino Acid	[A]11S	[B,C]7S	[D]11S	[D]7S
Glx	19	24	20	18
Asx	12	12	12	13
Leu	7	9	8	9
Gly	8	5	7	6
Ala	7	5	6	5
Ser	7	7	7	7
Pro	6	5	5	5
Arg	6	7	9	6
Cys	4	6	5	8
Ile	5	5	4	5
Val	6	4	5	5
Thr	4	2	4	3
Phe	4	6	4	5
Tyr	2	3	2	2
His	2	2	3	2
Cys	1	0	1	1
Met	1	0	0	2

[A] Kitamura and Shibasaki, 1975.

[B] Average of the six isomers.

[C] Than and Shibasaki, 1978.

[D] Grant and Lawrence, 1964.

of amides with glutamine and asparagine making up 30% of the total amino acids. Unlike zeins, these proteins comtain 4-6% lysine, but very low levels of the sulfur-containing amino acids methionine and cysteine. Each of these represents about 1% of the amino acids in the 11S protein, but they are barely detectable in the 7S protein (Table 2). Since 7S and 11S proteins share many structural features with similar protein fractions in other legumes, including pea and broad bean, the deficiency in sulfur-containing amino acids is a problem common to most legumes.

SDS polyacrylamide gel electrophoresis shows that both the 7S and 11S proteins are made up of multiple subunits. β-conglycinin, which is the major 7S form, contains four polypeptides with molecular weights of 84,000, 75,000, and 54,000. An individual 7S molecule is composed of three of these polypeptides, although the combination is variable (40). The 11S protein separates into two distinct sets of polypeptides. The larger of these has an apparent molecular weight of about 40,000 and the smaller group is about 20,000. There is significant charge heterogeneity within these two molecular weight groups. The 40,000 dalton polypeptides, which have acidic pIs, can be separated into at least six differently charged forms. The 20,000 dalton polypeptides, which have basic pIs, can be separated into five major charged forms (28,30). There is some

Table 3. Amino acid composition of the glycinin subunits.

Amino acid	Subunit											
	A_{1a}	A_{1b}	A_2	A_3	A_4	A_5	A_6	B_{1a}	B_{1b}	B_2	B_3	B_4
Asx	30.1	28.5	34.4	37.2	41.5	9.0	29.2	25.5	23.5	24.3	19.2	20.7
Thr	9.8	10.1	10.1	12.7	9.6	3.9	9.3	8.1	7.8	9.1	6.2	5.4
Ser	15.0	15.5	13.4	22.2	19.2	7.7	16.2	13.5	14.1	12.4	12.1	12.4
Glx	69.7	69.7	70.6	74.9	75.7	14.9	56.3	22.5	23.9	22.7	24.8	21.0
Pro	19.6	20.4	17.4	27.7	22.7	8.1	21.0	10.5	10.4	10.8	10.2	9.1
Gly	25.3	22.3	24.4	24.4	18.3	7.9	18.6	11.1	11.3	10.4	13.4	16.1
Ala	11.8	13.0	14.8	8.9	5.1	5.0	4.4	15.6	14.7	14.3	12.4	11.2
Val	9.7	10.1	12.5	14.2	9.9	3.8	8.7	11.4	12.2	10.8	17.0	19.2
Met	2.9	3.4	4.7	2.0	1.1	1.1	1.8	2.3	1.5	2.7	0	1.3
Ile	14.4	13.6	12.5	10.0	8.5	5.1	6.6	9.2	8.0	9.8	7.0	7.3
Leu	16.4	15.3	16.4	17.8	11.4	10.7	11.2	17.9	18.0	17.4	18.1	18.1
Tyr	6.0	7.1	5.4	4.6	3.6	2.0	5.7	2.8	4.5	2.5	5.8	8.4
Phe	10.0	14.6	10.1	9.8	6.3	0.9	5.7	8.6	10.4	9.1	6.0	5.7
His	4.9	3.9	2.1	11.5	7.8	2.8	7.8	2.1	3.0	2.7	4.8	4.2
Lys	17.3	13.0	12.2	12.1	15.4	3.9	8.0	5.9	6.0	5.9	7.0	6.5
Arg	14.8	17.3	18.6	18.2	23.2	3.1	13.3	8.9	12.5	9.9	10.9	12.5
Cys	3.7	N.D.	3.5	2.9	0.6	N.D.	N.D.	1.7	N.D.	1.5	0.2	1.5

variability in the amino acid composition of the different acidic and basic polypeptides (Table 3) although the members in each of the two families of polypeptides are clearly related.

If the 11S protein is separated by SDS polyacrylamide gel electrophoresis in the absence of reducing agents, a complex made up of one acidic and one basic polypeptide is isolated (38). These complexes have molecular weights of approximately 56,000 and are formed by disulfide linkage between specific 40,000 Mr and 20,000 Mr polypeptides (Table 3). It is important to note that certain of these complexes have a higher methionine content than others. For example, the A2B1a complex has 7-8 methionine residues, while the A4B3 and A5B3 complexes have only one. This could have important implications for the genetic improvement of the methionine content of the 11S protein.

The 7S and 11S proteins are synthesized in the cotyledons of the developing soybean embryo and are deposited in an insoluble form in protein bodies. Although only some of the details of this process have been established in soybean, a fairly extensive study has been done on the 7S and 11S proteins in pea (3,4).

Messenger RNAs directing the systhesis of both 7S and 11S proteins are translated on RER. As the protein is secreted into the lumen of the membrane, the 7S proteins undergo glycosylation, although the 11S polypeptides apparently do not. The mRNAs for the 11S polypeptides do not direct the synthesis of 40,000 Mr and 20,000 Mr polypeptides, but rather they synthesize a 60,000 Mr precursor (41). The precursor contains a short signal peptide which directs transport of the protein into the lumen of the RER. The remainder

Table 4. Covalent linkage of complexes.

SUBUNIT			ACIDIC COMPONENT		BASIC COMPONENT	
COMPOSITION	M_R	METHIONINE	M_R	METHIONINE	M_R	METHIONINE
$A_{1A}B_2$	≃ 56,000	6-7	≃ 37,000	3-4	≃ 19,000	3
$A_{1B}B_{1B}$	≃ 56,000	5-6	≃ 37,000	3-4	≃ 19,000	2
A_2B_{1A}	≃ 56,000[1]	7-8	≃ 37,000[2]	5	≃ 19,000[3]	2-3
A_3B_4	≃ 59,000	3	≃ 40,000	2	≃ 19,000	1
A_4B_3	≃ 57,000	1	≃ 37,000	1	≃ 20,000	0
A_5B_3	≃ 30,000	1	≃ 10,000	1	≃ 20,000	0
A_6B_{UNK}	–	–	≃ 30,000	UNK	UNK	UNK

[1]51,500 DALTONS [2]31,600 DALTONS [3]19,900 DALTONS BY CHEMICAL DETERMINATION

of the molecule is composed of a covalently linked acidic polypeptide region, a short linker, and then the basic polypeptide. This covalent linkage is undoubtedly responsible for the specific pairing of the complexes (Table 4).

After entering the RER, both proteins are eventually transported to vacuolar membranes where they aggregate and form the dense masses called protein bodies. The precise mechanism by which they are transported to the vacuole is not clear, but it appears to primarily involve transport through ER vesicles (4,14). It is not until the 11S protein reaches the vacuole that cleavage of the 11S acidic-basic polypeptide complexes takes place (4), and presumably the association of these polypeptides into an 11S molecule. On the other hand, association of the 7S polypeptides into a holoprotein appears to occur immediately within the RER (4). Eventually, both the 7S and 11S proteins end up in the same protein body (5).

Although there is little known about the 3-dimensional structure of 7S and 11S proteins from x-ray crystallographic studies, distinctive structures have been seen in electron micrographs (2). The 11S protein appears to be made up of two monomers stacked one atop the other. Each monomer is made up of three acidic and three basic polypeptides which alternate and lie in the same plane. The 7S protein has a simpler structure. In this case, the molecule is seen as an aggregate of three spherical polypeptides (Beachy, personal communication).

Genomic clones for both the 7S and 11S protein genes have been isolated and partially characterized (8, N.C. Nielsen, personal communication; R.N. Beachy, personal communication). Although none of these genes has been completely sequenced, many of the relevant structural details are beginning to emerge. Dr. R.B. Goldberg and his associates have studied a number of these genes by R-looping. The data presented in Fig. 7 show their analyses of several genomic clones. In contrast to the maize zein genes, which contain no

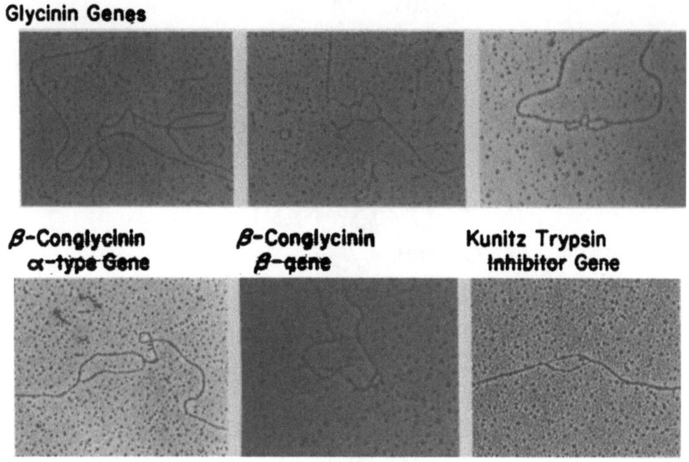

Fig. 7. Intragenic structure of seed protein genes. R-loop analysis of soybean glycinin and conglycinin genes. The R-loop analysis was performed as described by Fisher and Goldberg (8).

introns, several members of the 11S family do. There is one large intron as well as one or more small introns in each of the 11S genes. Preliminary sequence analysis of several different glycinin genomic clones reveals that the large intron occurs in the basic polypeptide coding sequence near the linker region connecting the acidic and basic components (N.C. Nielsen, personal communication). There also appear to be some small introns in the conglycinin genes (R.N. Beachy, personal communication). This observation is consistent with a report for the 7S genes of another legume, Phaseolus vulgaris (39). Although a complete structural characterization of these genes will require determination of their DNA sequences, these studies illustrate that the structure of the storage globulin genes is more complex than the maize zein genes, albeit simpler than the storage protein genes of some animal eggs (15,44).

The 7S and 11S genes also exist in small multigene families. Using a glycinin cDNA as a probe, Fisher and Goldberg (8) detected three soybean DNA fragments that hybridized with the intensity of single copy genes. This number agrees closely with that obtained by solution hybridization (11), and suggests that there are relatively few copies of these genes in the genome. However, in view of the protein sequence data of Moreira et al. (28) and the characterization of the acidic-basic complexes (38,41), there must be at least seven different genes encoding glycinin precursor polypeptides. A plausible explanation for the discrepancy between the DNA hybridization results and the protein sequence data is that the gene families are sufficiently divergent such that they do not

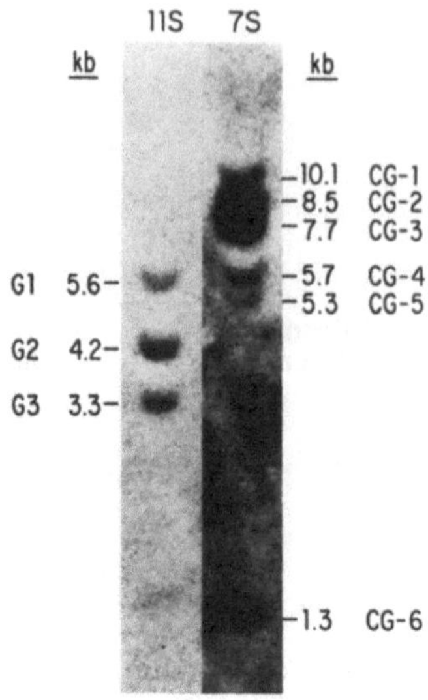

Fig. 8. Genome blots. Southern blot analysis of soybean DNA with 11S and 7S cDNA clones as probes. The DNA was digested with Eco R1; conditions for hybridization were as described by Fisher and Goldberg (8).

cross-hybridize at the criteria used in these experiments. Similar divergence was shown to have occurred with some of the maize zein genes (27).

Southern blots of the 7S genes reveal a more complex pattern than is obtained for the 11S genes. The simplest interpretation of these results is that there are perhaps 7-10 members of each of these gene families (Fig. 8). Understanding the relationships of the 7S genes is further complicated by the fact that the 7S clones hybridize to two different sizes of mRNAs of approximately 2500 and 1700 nucleotides in length (11). The relationships among members of this complex family are not yet clear and will require additional study.

The nature of most of the problems of engineering amino acid changes in legume storage proteins is similar to that described for the cereal prolamines. Again we are faced with the presence of multiple genes, although there are potentially fewer globulin genes than prolamine genes. The ease with which we can manipulate expression of genetically engineered genes relative to the background of endogenous genes will depend on how much we learn about the regulation of transcription and developmental regulation of gene expression.

Unfortunately, we still know little about the structure of storage globulins, so we can only guess the impact of adding more sulfur-containing amino acids to the proteins. In this regard it is interesting to note that certain of the acidic-basic complexes, and hence the genes encoding them, contain more methionine (Table 3). Knowledge of the positions of these amino acids relative to the shape of the protein would be valuable in predicting how to modify these genes. Of course, the existence of such genes implies that it might eventually be possible to increase the methionine content by breeding for favorable combinations of acidic-basic subunits.

The observation that the 7S proteins appear to have a simpler structure than the 11S proteins suggests that it might be more feasible to modify the genes encoding these polypeptides. On the other hand, the fact that they normally contain little or no methionine and cysteine may be significant.

When contemplating the structural alterations that might be made to these proteins, it is worth remembering that there are several post-translational modifications that occur between the time the protein is synthesized and when it is deposited in a protein body. Not all of these reactions are known, but they at least involve glycosylation of the 7S polypeptides and proteolytic cleavage of the 11S acidic-basic complexes and their association into an 11S molecule. Clearly, whatever structural changes are made must not alter these reactions. Consideration must also be given to the organization of introns and exons in the coding regions of these genes. It may be that genetic engineering of the storage globulin genes might be more complex than the maize prolamine genes.

Soybean Protease Inhibitor Genes

An alternative approach for improving seed protein quality is to increase levels of endogenous proteins that have more favorable amino acid compositions. This is illustrated by the trypsin inhibitors of the soybean seed.

Inhibitors of proteolytic enzymes occur widely among plants, and they are very often found in high concentrations in storage tissues such as seeds (37). At least three different protease inhibitors have been identified in soybean seed. These include the Kunitz inhibitor (Mr = 21,500), which inhibits trypsin activity, the Bowman-Birk inhibitor (Mr = 8000), which inhibits both trypsin and chymotrypsin activity (see Fig. 9), and its related family of isoinhibitors PI I-IV (Mr = 7000-8000), which also inhibit trypsin activity (17). Of these three, the Bowman-Birk and the related family of isoinhibitors PI I-IV are the most prevalent. In the soybean cultivar Tracy, the Bowman-Birk inhibitor accounts for 4% and PI I-IV account for 2% of the total seed protein (18).

These small proteins contain only 70-75 amino acid residues and are composed of about 17% aspartic acid-asparagine, 14% serine, and

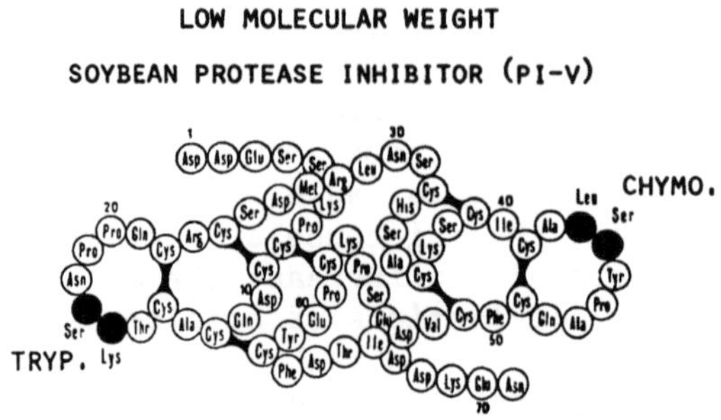

Fig. 9. Complete covalent structure of the Bowman-Birk protease inhibitor. (From Odani and Ikenaka, ref. 31, with permission).

20% cysteine (Table 5). The high content of cysteine in these proteins is remarkable. Even though the protease inhibitors account for only 6% of the total seed protein, if one calculates the absolute amount of sulfur-containing amino acids they contribute relative to the total seed protein, they account for over half.

The high content of cysteine in these proteins is directly related to their structure. Their small size facilitated determination of their complete amino acid sequence (17,31), and from this Odani and Ikenaka (31) developed a structural model for the Bowman-Birk inhibitor (Fig. 9). This protein is essentially pretzel shaped as a result of the cross-linkage of the half-cysteine residues. Because of this planar shape, the trypsin-reactive (Ser-Lys) and the chymotrypsin-reactive (Leu-Ser) sites are readily exposed, and this presumably facilitates the activity of the molecule.

Although it seems obvious that one could significantly increase the content of sulfur-containing amino acids in the soybean seed by increasing the amounts of the protease inhibitors, it also seems reasonable to ask what impact this would have on the nutritional value of the soybean seed. It is widely assumed that because these proteins inhibit bovine and porcine trypsins in vitro, they also inhibit human trypsin and chymotrypsin in vivo. However, this is probably not the case. Based on the very weak inhibition of trypsin 1, Liener (25) concluded that soybean protease inhibitors do not have a significant effect in human nutrition. Furthermore, PI I-IV are generally only weak inhibitors of proteolytic activity (17). It is therefore unlikely that larger amounts of these proteins will affect the digestability of the soybean seed. It should also be remembered that soybean seed must be cooked to destroy lectins and other antinutritional factors prior to being used as a food source. So it is unlikely that higher levels of the trypsin inhibitors would create a problem.

In considering the biological impact of increasing these proteins, some attention must be given to their normal function in

Table 5. Amino acid composition of the Bowman-Birk protease inhibitor and the related family of isoinhibitors (PI I-IV).

Amino Acid	Bowman-Birk[a]	Inhibitor I	II	III	IV
Lys	5	4	4	4	4
His	1	1	1	1	1
Arg	2	5	5	5	5
Asx	11	11	11	12	13
Thr	2	3	3	3	3
Ser	9	8	8	9	12
Glu	7	5	5	5	6
Pro	6	5	5	5	5
Gly	0	1	1	1	1
Ala	4	0	0	0	0
Half-cys	14	14	14	14	14
Val	1	0	0	0	0
Met	1	3	3	3	3
Ile	2	1	1	1	1
Leu	2	3	3	3	3
Tyr	2	2	2	2	3
Phe	2	1	1	1	1
Trp	0	0	0	0	0
Total	17	67	67	70	75

[a]Frattali,V. (1969) J. Biol. Chem. 244,274-280.

the seed. Although Richardson (37) makes several suggestions as to their possible function, it is not really known. One possibility would be the inhibition of endogenous proteases, but thus far no one has identified a protease whose activity is inhibited by these proteins. It was suggested that protease inhibitors aid as a seed dispersal mechanism by inhibiting the digestion of the seed in the guts of birds. While this is possible, it has never been demonstrated. It is known that these proteins are rapidly leached from the seed when it is placed in an aqueous environment (18), and this has led to the suggestion that the proteins may inhibit proteases of fungi or other soil microorganisms and thus help establish the young seedling. Richardson (37) points out that these proteins might simply serve as a sulfur depot for the seed, although this is difficult to reconcile in view of the protein's structure. It is also possible that protease inhibitors fulfill all these functions. Whatever their function, to date no one has found a mutant soybean in which they are entirely lacking, so they appear to have some relevant function.

Information regarding the molecular biology of these genes is just beginning to emerge. Goldberg et al. (12) identified a cDNA clone for the Kunitz trypsin inhibitor in a cDNA library of soybean embryo mRNAs. The mRNA for this protein showed a similar pattern of developmental regulation to the mRNAs for the major seed proteins glycinin and conglycinin. This cDNA clone was subsequently used to isolate a genomic clone containing the Kunitz trypsin inhibitor gene. R-loop analyses of this gene indicated no intervening sequences (R.G. Goldberg, personal communication).

We have partially purified mRNAs directing the synthesis of the Bowman-Birk inhibitor and one of the isoinhibitors, PI IV (9). This mRNA fraction has been used to construct cDNA clones, and from these we have identified clones for both these protease inhibitors (R.W. Spencer, unpublished results). Since the isoinhibitors PI I-IV are immunologically cross-reactive and share the same amino acid sequence except for the first nine amino acids, we expect that some of these clones correspond to PIs I-III.

We have used these cDNA clones to isolate the corresponding nuclear genes, and experiments are in progress to characterize these genes and determine their interrelationships as well as the factors modulating their transcriptional activity.

The trypsin inhibitor proteins are but one example of a protein that is normally present at low levels in the seed, but could have significant impact on seed protein quality if its level could be increased. Another example that is being actively investigated is the enzyme urease (36).

CONCLUSION

The protein quality of seeds is primarily determined by the storage proteins. With respect to human nutrition, most seeds do not provide a balanced source of protein because of deficiencies in one or more essential amino acids. Genes encoding storage proteins with a more favorable amino acid balance do not exist in the genomes of major crop plants, so one way of overcoming this problem would be to change the coding capacity of the genes and so genetically engineer more nutritionally balanced seed storage proteins.

We have learned much in recent years about the synthesis of the major types of seed storage proteins; however, there is still a great deal to learn about their structures and the mechanisms by which they associate into protein bodies. Rapid progress in gene isolation and DNA sequencing techniques has resulted in the structural characterization of several of these genes. It should now be possible to begin experiments directed at genetic engineering. Should it not be possible to change the endogenous storage proteins so they contain high levels of certain desired amino acids, an alternative would be to introduce other seed protein genes which would correct the amino acid deficiency. It may also be possible to increase the level of expression of endogenous genes encoding proteins that have a more complete amino acid balance.

There are many important unknowns that must be dealt with before any type of genetic engineering will be possible. Ways must be found to introduce genes into plant cells so that they become stably integrated and developmentally regulated. We also must learn a great deal more about transcriptional regulation so that it is possible to control the level of expression of desired genes. These are areas of research that are currently the subject of intense investigation, and hopefully answers will be forthcoming. Providing support for this research can be increased, it is not unreasonable to expect dramatic changes in the genetic improvement of crop plants in the future.

ACKNOWLEDGEMENTS

Part of the research described in this chapter was supported by NSF grant PCM-8003757 and DOE contract DE-AC02-80ER10715.A000. This is journal paper number 9143 of the Purdue Agricultural Experiment Station.

REFERENCES

1. Argos, P., K. Pedersen, M.D. Marks, and B.A. Larkins. 1982. A structural model for maize zein proteins. J. Biol. Chem. 257:9984-9990.
2. Badley, R.A., D. Atkinson, H. Hauser, D. Oldani, J.P. Green, and J.M. Stubbs. 1975. The structure, physical and chemical properties of the soybean protein glycinin. Biochem. Biophys. Acta 412:214-228.
3. Chrispeels, M.J., T.J.V. Higgins, S. Craig, and D. Spencer. 1982. Role of the endoplasmic reticulum in the synthesis of reserve proteins and the kinetics of their transport to protein bodies in developing pea cotyledons. J. Cell Biol. 93:5-14.
4. Chrispeels, M.J., T.J.V. Higgins, and D. Spencer. 1982. Assembly of storage protein oligomers in the endoplasmic reticulum and processing of the polypeptides in the protein bodies of developing pea cotyledons. J. Cell Biol. 93:306-313.
5. Craig, S., and A. Millerd. 1981. Pea seed storage proteins--Immunocytochemical localization with protein A-gold by electron microscopy. Protoplasma 105:333-339.
6. Danielsson, C.E. 1949. Seed globulins of the Gramineae and Leguminosae. Biochem. J. 44:387-400.
7. Derbyshire, E., D.J. Wright, and D. Boulter. 1976. Legumin and vicilin, storage proteins of legume seeds. Phytochemistry 15:3-24.

8. Fischer, R.L., and R.B. Goldberg. 1982. Structure and flanking regions of soybean seed protein genes. Cell 29:651-660.
9. Foard, D.E., P.A. Gutay, B. Ladin, R.N. Beachy, and B.A. Larkins. 1982. In vitro synthesis of the Bowman-Birk and related soybean protease inhibitors. Plant Molec. Biol. (in press).
10. Geraghty, D., M.A. Peifer, I. Rubenstein, and J. Messing. 1981. The primary structure of a plant storage protein: Zein. Nucleic Acids Res. 9:5163-5174.
11. Goldberg, R.B., G. Hoschek, S.H. Tam, G.S. Ditta, and R.W. Breidenbach. 1981. Abundance, diversity, and regulation of mRNA sequence sets in soybean embryogenesis. 1981. Dev. Biol. 83:201-217.
12. Goldberg, R.B., G. Hoschek, G.S. Ditta, and R.W. Breidenbach. 1981. Developmental regulation of cloned superabundant embryo mRNAs in soybean. Dev. Biol. 83:218-231.
13. Hagen, G., and I. Rubenstein. 1981. Complex organization of zein genes in maize. Gene 13:239-249.
14. Harris, N., and M.J. Chrispeels. 1980. The endoplasmic reticulum of mung-bean cotyledons: Quantitative morphology of cisternal and tubular ER during seedling growth. Planta 148:293-303.
15. Heilig, R., F. Perrin, F. Gannon, J.L. Mandel, and P. Chambon. 1980. The ovalbumin gene family: Structure of the X gene and evolution of duplicated split genes. Cell 20:625-637.
16. Hurkman, W.J., L.D. Smith, J. Richter, and B.A. Larkins. 1981. Subcellular compartmentalization of maize storage proteins in Xenopus oocytes injected with zein messenger RNAs. J. Cell Biol. 89:292-299.
17. Hwang, D.L.-R., K.-T. Davis Lin, W.-K. Yang, and D.E. Foard. 1977. Purification, partial characterization, and immunological relationships of multiple low molecular weight protease inhibitors of soybean. Biochem. Biophys. Acta 495:369-382.
18. Hwang, D.L., W.-K. Yang, D.E. Foard, and K.-T.D. Lin. 1978. Rapid release of protease inhibitors from soybeans. Immunochemical quantitation and parallels with lectins. Plant Physiol. 61:30-34.
19. Kitamura, K., and K. Shibasaki. 1975. Isolation and some physico-chemical properties of the acidic subunits of soybean 11S globulin. Agric. Biol. Chem. 38:1083-1085.
20. Larkins, B.A., and W.J. Hurkman. 1978. Synthesis and deposition of zein proteins in maize endosperm. Plant Physiol. 62:256-263.
21. Larkins, B.A., K. Pedersen, A.K. Handa, W.J. Hurkman, and L.D. Smith. 1979. Synthesis and processing of maize storage proteins in Xenopus laevis oocytes. Proc. Natl. Acad. Sci. U.S.A. 76:6448-6452.

22. Larkins, B.A. 1981. Seed storage proteins: Characterization and biosynthesis. In The Biochemistry of Plants: A Comprehensive Treatise, Vol. 6, P.K. Stumpf and E.E. Conn, eds. New York: Academic Press.
23. Lee, K.H., R.A. Jones, A. Dalby, and C.Y. Tsai. 1976. Genetic regulation of storage protein synthesis in maize endosperm. Biochem. Genet. 14:641-650.
24. Lewis, E.D., G. Hagen, J.I. Mullens, P.N. Mascia, W.D. Park, and I. Rubenstein. 1981. Cloned genomic segments of Zea mays homologous to zein mRNAs. Gene 14:205-215.
25. Liener, I.E. 1979. Protease inhibitors and lectins. In International Review of Biochemistry, Biochemistry of Nutrition IA, A. Neuberger and T.H. Jukes, eds. Baltimore: University Park Press.
26. Lott, J.N.A. 1980. Protein bodies. In The Biochemistry of Plants: A Comprehensive Treatise, Vol. 1, P.K. Stumpf and E.E. Conn, eds. New York: Academic Press.
27. Marks, M.D., and B.A. Larkins. 1982. Analysis of sequence microheterogeneity among zein messenger RNAs. J. Biol. Chem. 257:9976-9983.
28. Moreira, M.A., M.A. Hermodson, B.A. Larkins, and N.C. Nielsen. 1981. Comparison of the primary structure of the acidic polypeptides of glycinin. Arch. Biochem. Biophys. 210:633-642.
29. Nelson, O.E. 1969. Genetic modification of protein quality in plants. Advances in Agronomy 21:171-194.
30. Nielsen, N.C., M. Moreira, P. Staswick, M.A. Hermodson, N. Tumer, and V.H. Thanh. 1981. The structure of glycinin from soybeans. Abhdlg. Akad. Wiss. DDR, Abt. Math Naturwiss., Techn. (in press)
31. Odani, S., and T. Ikenaka. 1973. Studies on soybean trypsin inhibitors VIII. Disulfide bridges in soybean Bowman-Birk protease inhibitor. J. Biochem. 74:697-715.
32. Osborn, T.B. 1909. The Vegetable Proteins. London: Longmans, Green and Co.
33. Pedersen, K., K.S Bloom, J.N. Anderson, D.V. Glover, and B.A. Larkins. 1980. Analysis of the complexity and frequency of zein genes in the maize genome. Biochemistry 19:1644-1650.
34. Pedersen, K., J. Devereux, D.R. Wilson, E. Sheldon, and B.A. Larkins. 1982. Cloning and sequence analysis reveals structural variation among related zein genes in maize. Cell 29:1015-1026.
35. Pintor-Toro, J.A., P. Langridge, and G. Feix. 1982. Isolation and characterization of maize genes coding for zein proteins of 21,000 dalton size class. Nucleic Acids Res. (in press).
36. Palacco, J.C., and E.A. Havir. 1979. Comparison of soybean urease isolated from seed and tissue culture. J. Biol. Chem. 254:1707-1715.
37. Richardson, M. 1977. The proteinase inhibitors of plants and microorganisms. Phytochemistry 16:159-169.

38. Staswick, P.E., M.A. Hermodson, and N.C. Nielson. 1981. Identification of the acidic and basic subunit complexes of glycinin. J. Biol. Chem. 256:8752-8755.
39. Sun, S.M., J.L. Slighton, and T.C. Hall. 1981. Intervening sequences in a plant gene--comparison of the partial sequence of cDNA and genomic DNA of french bean phaseolin. Nature 289:37-41.
40. Thanh, V.H., and K. Shibasaki. 1978. Major proteins of soybean seeds. Subunit structure of β-conglycinin. J. Agric. Food. Chem. 26:695-698.
41. Tumer, N.E., J.D. Richter, and N.C. Nielsen. 1982. Structural characterization of glycinin precursors. J. Biol Chem. 257:4016-4018.
42. Viotti, A., E. Sala, R. Marotta, P. Alberi, C. Balducci, and C. Soave. 1979. Genes and mRNAs coding for zein polypeptides in Zea mays. Eur. J. Biochem. 102:213-222.
43. Viotti, A., D. Abildsten, N. Pogna, E. Sala, and V. Pirrotta. 1982. Multiplicity and diversity of cloned zein cDNA sequences and their chromosomal localisation. The EMBO. J. 1:53-58.
44. Wahli, W., I.B. David, G.U. Ryffel, and R. Weber. 1981. Vitellogenesis and the vitellogenin gene family. Science 212:298-304.
45. Wienand, U., P. Langridge, and G. Feix. 1981. Isolation and characterization of a genomic sequence of maize coding for a zein gene. Mol. Gen. Genet. 182:440-444.

VECTORS: CHAIRMAN'S INTRODUCTION

Eugene W. Nester

Department of Microbiology and Immunology, SC-42
University of Washington
Seattle, Washington 98195

In this section, the authors will discuss various aspects of introducing DNA into plants. In my view, this section represents the heart of the theme of this volume. Dr. Anne DePicker, who is currently involved in active research in one of the most highly regarded laboratories of crown gall research, discusses the overall aspects of crown gall, a natural genetic engineering system in plants. In this system, a defined piece of bacterial plasmid DNA (T-region) is transferred from the inciting strain of *Agrobacterium* to the plant cell where it is integrated into plant nuclear DNA. By integrating foreign DNA into this defined piece of plasmid DNA, it is possible to transfer and integrate the foreign DNA into plant cells. An important feature of this system is that cells into which T-DNA is integrated can be selected on the basis that callus tissue is able to grow in the absence of exogenously added auxin and cytokinin, two phytohormones required by the untransformed tissue. *Agrobacterium* also has a remarkably broad host range infecting most dicotyledonous plants. At this time, the crown gall system seems to be the best vehicle for the stable incorporation of genes into plants.

Another potential vector is represented by cauliflower mosaic virus (CAMV), a double stranded DNA virus, which infects many members of the cruciferae. This virus multiplies in the nucleus, but its DNA is not integrated into the plant DNA. It results in a systemic infection of the plant. One drawback to its use as a vector is the fact that only very small pieces of DNA, less than the size of an average gene can be stably integrated into the viral genome. Dr. Richard Gardner, one of the pioneers in the study of this system, will evaluate the prospects of this system in the genetic engineering of higher plants in light of the properties of the virus.

Although tumors can be induced in whole plants by inoculating whole cells of *Agrobacterium* into a wound site, new insights might be gained if it were possible to infect plant protoplasts with isolated plasmid DNA. In addition, the use of plant protoplasts might extend the range of plants that *Agrobacterium* strains can infect to monocots. However, to protect the DNA from nucleases, it is necessary to protect the DNA at the time it is added to the plant protoplasts. One such way is to encapsulate the DNA in artificial lipid vesicles, liposomes, and have the plant protoplasts take up this encapsulated form of DNA. A comparison of the various techniques by which naked DNA and RNA can be introduced into plant cells will be reviewed by Dr. Rob Fraley. Dr. Fraley has been a pioneer in the use of liposomes for introducing nucleic acids into plants.

The chapter by Dr. Russell Malmberg, discusses his studies on the isolation and uptake of chromosomes by plant protoplasts. The transfer of whole chromosomes opens up the possibility of transferring a large number of linked genes simultaneously. Therefore, the techniques of isolating and transferring a group of genes as a single unit is of great interest.

Each of the authors of chapters in this section is actively engaged in research at the cutting edge of plant cell transformation. The insight that they bring coupled with a report on some of their most recent data makes this very stimulating and informative reading.

PLANT VIRAL VECTORS: CaMV AS AN EXPERIMENTAL TOOL

Richard C. Gardner

Calgene Inc.
1910 Fifth Street
Davis, California 95616

USES OF PLANT VIRAL VECTORS

Components of a Plant Vector

Viruses provide natural examples of genetic engineering, since viral infection of a cell results in the addition of new genetic material which is expressed in the host. In both microbial and mammalian systems, viruses have played important roles in vector development, so it is not unexpected that plant viruses are considered as candidates for plant gene vectors. Additional genetic material incorporated into the genome of a plant virus might be replicated and expressed in the plant cell along with the other viral genes.

In fact, a simple but economically significant example of plant genetic engineering using a viral vector already exists. One phenotype that is usually conferred onto a plant by viral infection is called cross-protection--a plant infected by one virus usually cannot be superinfected by a second strain of a related virus. This phenomenon was utilized in tomato greenhouses, where persistent tobacco mosaic virus (TMV) infections can be a major problem. Before the introduction of resistance genes in the late 1970s, it was common practice to inoculate tomato seedlings with a symptomless TMV strain produced by nitrous acid mutagenesis. The mild strain provided cross-protection against infection by the more severe isolates endemic to greenhouses (1).

I would like to use this simple example to illustrate three essential components of a viral vector system. The first component is a gene of agronomic importance, which is to be introduced into a crop plant. In this particular case, the gene is inherently part of the vector, rather than representing a separate gene which might have been obtained or constructed using recombinant DNA techniques. Other articles in this volume focus on this component of genetic engineering, and it will not be considered further here. A second component of the vector system is the virus itself. The virus should replicate in the host plant, and its symptoms must be attenuated so that yield of the host plant is essentially unaffected. The third component is the method of introducing the vector bearing its agronomic gene onto the crop in the field. Fortunately with many viruses, including TMV in the example mentioned above, the problem is solved by simply rubbing the virus onto the leaves of the plant. From a few infected leaf cells, the virus spreads by cell-to-cell movement to systemically infect the whole plant. This ability of viruses to move systemically bypasses the need for regeneration of crop plants from tissue culture which is required for the introduction of most other types of vector.

Changing a Plant Virus into a Vector

A number of modifications of the biological properties of most viruses must be made before they can become useful plant vectors. Most obviously, the virus should be an attenuated strain that does not affect plant yield. Symptomless strains of many viruses occur naturally, or have been obtained by mutagenesis. However, there have been few yield studies with such strains. The symptomless TMV mutant discussed above has been reported to reduce yields by up to 5%, depending on the conditions (2). Trials with cryptic (symptomless) viruses of *Cacao* and *Vicia* showed no detectable effects on yield (1,3). In practice, some residual yield losses caused by infection with a viral vector may be acceptable, provided that they are more than offset by the gains resulting from introduction of the gene(s) carried on the vector.

The attenuation of the virus must be a stable characteristic, so that reversion to a severe strain does not occur. Choice of a virus that is known to have a low mutability in the field should help in this regard. It might also be important to use a virus which has been debilitated in its natural host range, or in its ability to be transmitted naturally (by insects, for example). Since viruses that are mild on one host can still be severe on others, these modifications would reduce potential damage resulting from the maintenance of a reservoir of a virus in a crop plant. Another potential objection to the use of viral vectors comes from the observation that some viruses in combination are more severe than either alone. Hence, systemic infection by a symptomless viral vector might increase other viral problems in a crop. Again, a careful choice of virus/host plant combination should overcome this objection. Finally, it may be advantageous to block the phenomenon of cross-protection in certain circumstances. For example, if it is desired to sequentially introduce genes into a crop, it would be necessary to develop a new vector for each gene unless this step was performed.

Once the biological characteristics of a virus have been suitably selected and manipulated, new genes must be incorporated into the virus in such a way that they are expressed in the plant. These molecular aspects of converting a plant virus into a vector are discussed in more detail in later sections of this article.

Viral Versus Integrating Vectors

In order to outline the reasoning behind the study of viral vectors, I would like to emphasize the differences between viral vectors and integration-type vectors, such as the Ti-plasmid. Imagine a time in the future when both types of vector will be available to a genetic engineer. Imagine that a suitable gene is available, and that it can be incorporated either into an integration vector that delivers it to the plant chromosome, or into a viral vector that systemically infects the host plant without any adverse affects. We than ask the question: Which of these vectors is most suitable for a particular crop plant?

For crops planted annually, viral vectors are generally at a disadvantage, primarily because of the cost of mechanically inoculating large numbers of plants. However, there are some examples where this disadvantage may be overcome:

(a) There are a few examples where the rate of seed transmission of a virus is high [e.g., tobacco ringspot virus in soybeans (4), cryptic viruses in beet (5) and *Vicia* (6)]. In these cases, a viral vector may have an advantage, since regeneration of the cultivar from tissue culture is not required.

(b) Some annual crops are planted from vegetative "seed pieces" (e.g., potato, cassava, sugar cane, pineapple, etc). Hence, inoculation of the parental stocks may be economically feasible.

(c) For some high cash value, intensively-farmed crops, such as the greenhouse tomato, the cost of mechanical inoculation is not prohibitive.

However, if we consider crops that are perennial, rather than annual, viral vectors frequently offer considerable advantages over an integration-type vector. This is particularly true for crops with a long lag time between planting and production. For example, imagine the choice offered to an apple orchardist: inoculate each tree in the orchard and get systemic resistance to a pathogen without any loss of production, or replace the entire orchard with a new resistant variety which will take five seasons to bring to full production. Even for crops without a significant lag time, a viral vector might have advantages where inexpensive mechanical inoculation could be devised (e.g., inoculation of perennial forage crops during harvesting).

The choice between the two types of vectors discussed above is also influenced by the choice of gene to be delivered. For example, since viruses are usually present at a high copy number per cell, they may be useful for engineering overproduction of a particular protein. Hence, they may have an advantage when it is desired to alter the nutritional value of a particular food crop, or to over-produce a secondary metabolite either in tissue culture or in the intact plant (for example, alkaloid production in leaves of *Solanum laciniatum*).

One final factor that could be mentioned in this context is that viral vectors inherently carry viral resistance via cross-protection, as discussed in the opening example. Viral diseases are very significant economically, particularly in perennial and vegetatively-propagated crops, where virus-based vectors may be most suitable. Indeed, many varieties developed in regions where viruses are a problem have been found to contain mild strains that provide cross-protection [e.g., citrus tristeza virus in Israel and Australia, cacao swollen shoot virus in West Africa (1)].

The discussion above suggests that there is an economic justification for developing vectors based on plant viruses, primarily as a result of their ability to move systemically through existing plants.

Viral Vectors as an Experimental Tool

In addition to this economic consideration, study of viral vectors appears certain to provide valuable experimental tools for the whole field of plant vector development, particularly in the early stages. The best example of this comes from the SV40 system, which has made a vital contribution to vector development in mammalian cells in spite of offering little opportunity to become a practical vector *per se*. In particular, SV40 vectors have provided a rapid assay system for gene expression, since the virus replicates to give up to 100,000 copies per cell. Thus, gene products can be identified within individual cells two days after infection, compared to a wait of a month or more for vectors with lower copy number. In addition, mammalian viruses (including SV40) have provided a source of promoter sequences which direct high rates of transcription, which operate constitutively, and which may be thought of as promiscuous in that they often function in a wide range of different hosts. Some plant viruses also possess these characteristics, so that it is reasonable to suppose that plant viruses will provide equally important experimental tools in the development of plant vectors.

Another possibility is that a plant viral replicon may ultimately be used to develop multicopy plasmids. Such a vector would need to be engineered to ensure stable maintenance in the plant through both mitotic and meiotic divisions.

Choice of Virus

Over three hundred plant viruses have been identified and classified into 25 different groups (7). Biological characteristics which affect the choice of a potential viral vector include host range, virulence, ease of mechanical transmission, and rate of seed transmission, as discussed above. The potential to carry additional genetic information is also an important consideration since strict packaging limitations are found for some bacterial and mammalian viruses. Thus, viruses whose capsid is filamentous or rod-shaped, or viruses which possess a multipartite genome or a helper or satellite component, offer the potential for carrying extra nucleic acid. An additional alternative is offered by TMV and tobacco rattle virus, where the RNA does not require packaging for its infectivity (8).

The most important characteristic of a virus affecting its suitability as a vector is that the genetic material must be able to be manipulated and reintroduced into the plant in a biologically active form. In practice, this means that DNA clones of the virus should be infectious. As far as I am aware, the only group of viruses for which this step has been established is the caulimoviruses, whose genome consists of dsDNA. Progress with this group will be discussed in the second part of this article.

The geminiviruses, whose genome is ssDNA, will probably be the next group to get through this practical barrier. Goodman and his coworkers have shown that infected cells contain a double-stranded replicative form of the virus, and that this dsDNA is infectious (9). Thus, it seems likely that clones of this virus will also be infectious. Cloning and sequencing of members of the geminivirus group is underway in at least two laboratories. The properties of the geminivirus group (host range, possession of a divided genome) suggest that some members have the potential to make useful vectors (see reviews 10 and 11).

The huge majority of plant viruses have genomes composed of ssRNA. Within this group there are a number of very well-characterized viruses which have properties suited to their development into vectors (e.g., TMV, tobacco rattle virus, tobacco ringspot virus, alfalfa mosaic virus). However, the practical barrier to using RNA viruses remains. At least two potential ways of circumventing this barrier are available. Bacteriophage Qβ and poliovirus are RNA viruses which have no known DNA stage in their life cycle. However, full-length cDNA clones of both of these viruses are infectious when inoculated to a suitable host (12,13). Thus, the possibility exists that this result will hold for plant RNA viruses. An alternative procedure might be to engineer a cDNA copy of TMV, for example, in such a way that RNA copies could be produced in an *in vitro* system, and these copies used for infection (for example, putting cloned TMV downstream from a bacterial promoter and producing TMV RNA in an extract of *E. coli*). These options will undoubtedly be experimentally tested in the near future.

DEVELOPMENT OF CaMV AS A VECTOR

Biological Properties

The caulimovirus group consists of 6-10 viruses, each of which has a relatively limited host range (14,15). The best known member, cauliflower mosaic virus (CaMV), infects many members of the Cruciferae (cabbage, cauliflower, turnips, brussel sprouts, rapeseed, *Arabidopsis*, etc.) and *Datura stramonium*. Virulence of different CaMV strains varies from very mild (essentially latent) to lethal. The virus is readily transmitted mechanically and by aphids, but there is no evidence for seed transmission.

The viral particle of most caulimoviruses is icosahedral, and contains circular dsDNA of about 8 kb. An unusual feature of the structure of the DNA is the presence of three (two or four in some caulimoviruses) site-specific single-stranded discontinuities in the DNA. Sequence analysis of the DNA at these discontinuities in CaMV have shown that they in fact consist of short overlaps of 6-18 nucleotides so that there is a terminal redundancy in one strand of the DNA at the site of the discontinuity (16). Presumably as a result of these overlaps, the DNA can be isolated in a variety of twisted circular forms resembling supercoils and also as a linear form due to breakage. The function of the discontinuities in the DNA is not clear, although they are not required for infectivity (see below).

In infected cells, CaMV particles accumulate in cytoplasmic inclusion bodies. Early labelling studies showed that radioactivity accumulates rapidly in these inclusion bodies, suggesting that viral replication occurs there. However, recent evidence from several labs shows that replication and transcription of CaMV probably occur in the nucleus. Guilfoyle (17) showed that transcription of CaMV in isolated nuclei is α-amanitin sensitive, suggesting that RNA polymerase II transcribes the genome. Ansa et al. (18) have shown that isolated nuclei from infected plants can incorporate radioactive label into viral DNA. Recently, two laboratories (19,20) have reported that a covalently closed form of CaMV without the gaps may be isolated from infected cells, and that this molecule is probably packaged into nucleosomes (20).

The host range and other biological properties of CaMV discussed above suggest that it is not likely to become an economically important vector in the future. However, as a DNA virus whose genome is known to be packaged in nucleosomes and transcribed by RNA polymerase II, it is presently more suited than any other plant virus to exploitation as an experimental tool, in the same way as SV40 and polyoma have been used in mammalian systems. The knowledge gained from a CaMV model system should contribute to the development of other types of plant vectors.

Molecular Organization of CaMV

The complete nucleotide sequence has been determined for two strains of CaMV (16,21), with two more expected to be forthcoming soon (22). Six open reading frames occur on one strand of the DNA, essentially back-to-back and containing no apparent introns. Recently, in a review, Hohn et al. (22) suggested that two more smaller reading frames, conserved between the DNA sequences, may also code for proteins (see Fig. 1).

Of these six to eight possible gene products, only two have been identified and assigned to a reading frame. Region IV was suggested to code for the viral capsid protein on the basis of its amino acid content (16). This suggestion has been experimentally confirmed by the demonstration that clones containing region IV direct the expression in _E. coli_ of a protein which reacts with antibody raised against the viral particle (23). Another protein of 61,000 MW has been identified in CaMV-infected tissues, and shown to be a component of the viral inclusion bodies (24,25,26). An RNA species which can direct expression of the 61,000 MW protein _in vitro_ (26,27,28) hybridizes to region VI on the CaMV genome (28,29). Therefore, region VI probably codes for an inclusion body protein of unknown function. No protein or function has been identified for any of the other reading frames.

Details of the transcription of CaMV are also not well understood. Transcription is confined to the strand that contains the open reading frames (30). Two prominent polyadenylated

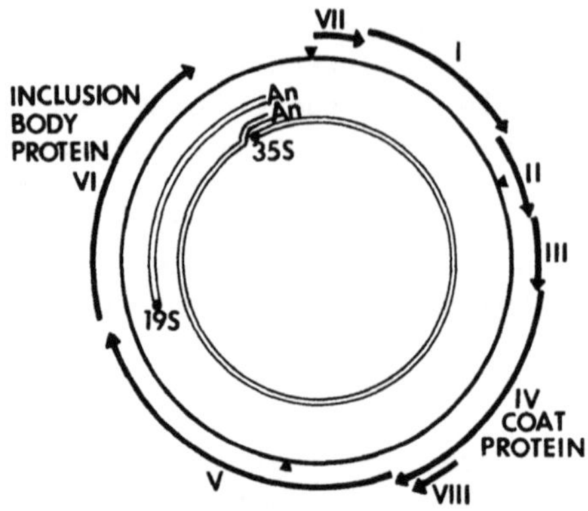

Fig. 1. Genomic organization of CaMV. The 8 kb genome of CaMV is indicated by the circle, with the three gaps marked by ▼. The eight open reading frames in the CaMV DNA sequence are identified by arrows outside the circle. The two principal polyadenylated transcripts (19S and 35S) are shown inside.

transcripts are found in infected cells. The larger one (35S) appears to be a greater-than-full-length transcript that originates at nucleotide position 7435 and terminates near position 7615 after 1.02 circuits of the molecule (31,32,33). A second transcript (19S) begins at nucleotide position 5765 and terminates at the same site as the full-length transcript (31,32). The nucleotide sequence shows typical eukaryotic promoters (including TATA boxes) just upstream from the initiation of these two transcripts, and the usual polyadenylation signal (AATAAA) near their common terminus. Several smaller, minor transcripts are found in infected cells, but their origin is unclear; one possibility is that they are spliced out of the larger message. There do not appear to be eukaryote transcription initiation signals upstream from the other reading frames, with the possible exception of region IV (21).

At present there is no information available concerning the number or location of replication origins on CaMV, or whether DNA replication occurs via a host- or viral-coded polymerase. The unusual overlap structures in CaMV DNA suggest that a viral-coded enzyme may be involved in their formation.

Infectivity of Cloned DNA

Shepherd et al. (34) originally demonstrated that CaMV DNA, in the double-stranded circular form, was infectious when mechanically inoculated onto turnip leaves using an abrasive such as carborundum. The key step in the development of CaMV as a vector was the

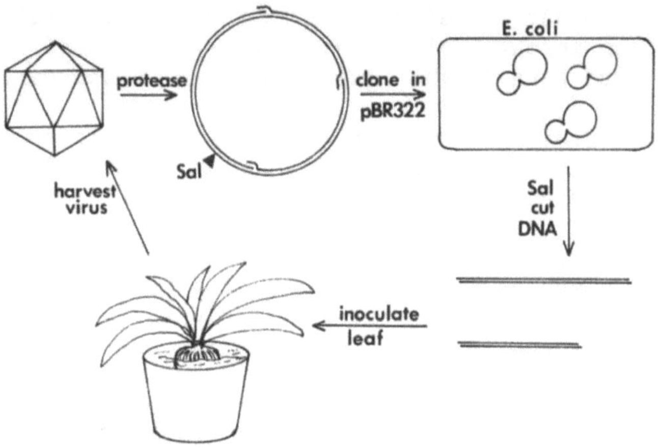

Fig. 2. Cloned DNA of CaMV is infectious. Circular viral DNA isolated from CaMV particles may be cloned in E. coli vector plasmids. Preparations of DNA from E. coli that have been cleaved with restriction enzymes to yield full length linear CaMV molecules cause typical symptoms when inoculated to turnips. Preparations of viral DNA from plants infected with the clone again contain the three gaps typical of CaMV.

demonstration by Howell and colleagues (35) that clones of the viral DNA were also infectious on turnips, provided that the bacterial plasmid used to propagate CaMV in E. coli was excised. This procedure is outlined in Fig. 2. Since then, a number of different strains of CaMV have been cloned in different plasmid vectors at different unique sites and subsequently shown to be infectious (22).

I would like to emphasize several points about this infection process:

(a) Very little DNA is required; 0.1-0.5 μg DNA is sufficient to inoculate one plant with a high probability of success.

(b) The results can be analyzed quite rapidly. Symptoms are seen within two to three weeks, and within three to four weeks kilogram quantities of leaf tissue, completely infected with CaMV at a high titer, can be obtained. Viral DNA can be rapidly reisolated from infected tissues for analysis (36).

(c) The gaps or overlaps present in the DNA are not required for infectivity, since the E. coli DNA used to inoculate the plant contains no nicks. However, viral DNA isolated from infected plants has the gaps as usual. The role of these gaps in the infection cycle is unclear.

(d) The CaMV DNA used for infection can be circular or linear, and the linear ends can be of any sort: 3' extension, 5' extension, or blunt (22).

These experiments are one of three confirmed cases of transformation of plant cells by naked DNA; the others are infection of tobacco protoplasts by purified Ti plasmid (37,38) and infection of bean protoplasts and leaves by gemini virus DNA (39,40). The ability to transform plants with cloned CaMV DNA provides the opportunity to manipulate the genome in *E. coli* using recombinant DNA techniques, and test the effects of the manipulations in the host plant. In the remainder of this article, I will describe some of the manipulations that have been performed with the aim of developing CaMV as a vector.

"In Vitro" Recombination between CaMV Strains

The availability of clones of different CaMV strains with different properties has been utilized to construct hybrids between the strains. CM1841 is a relatively mild strain of CaMV that lacks the ability to be transmitted by aphids. In contrast, Cabb B-Davis is a severe strain (often lethal to *Brassica campestris* cv "Just Right", the usual glasshouse host), and is aphid transmissible. The restriction maps of these two strains are suited for *in vitro* recombination: they are sufficiently similar so that they possess certain common sites, but sufficiently different so that hybrids between them can be identified from the parental types. Both strains have been cloned, and recombinants between the two have been constructed using common unique cleavage sites for *Sac*I, *Bst*EII and *Xho*I (Li, Daubert, Gardner and Shepherd, unpublished results).

The results of these recombination experiments are shown in Fig. 3. All of the hybrid CaMV molecules were infectious on turnip plants. The DNA responsible for aphid transmission has been mapped between the *Bst*EII and *Xho*I sites on the genome. Thus, open reading frames I, II, or VII must code for the "aphid transmission factor" first described by Lung and Pirone (40). In addition, most of the difference in severity of symptoms between the strains is accounted for by the top half of the genome; changing the coat protein gene, for example, has little influence on symptoms. The insertion mutagenesis work discussed below suggests that the inclusion body protein (region VI) is an important genetic determinant of symptom severity.

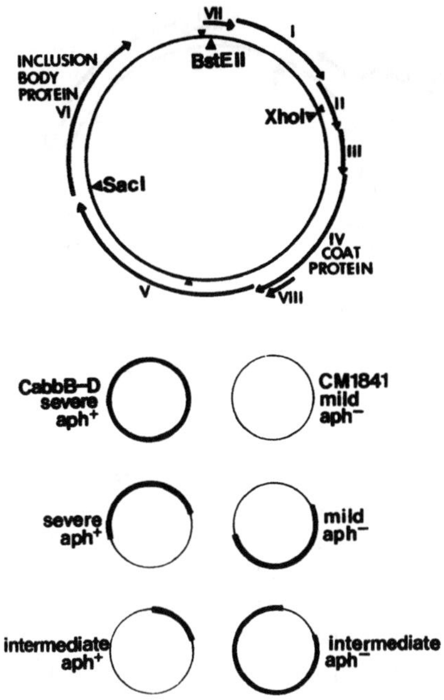

Fig. 3. *In vitro* recombination between CaMV clones. Segments of the genome of Cabb B-Davis (clone pCaMV12) were substituted into the genome of CM1841 (pCaMV10) using unique sites for *SacI*, *BstEII*, and *XhoI*. Recombinants were identified based on differences between the strains in their *EcoRI* and *SalI* restriction patterns. DNA from each recombinant was inoculated to turnips, and symptoms estimated for their severity (severe, mild, and intermediate) in terms of plant stunting, leaf size, and amount of leaf curling. Aphid transmission was performed as described by Lung and Pirone (41). The aph^+ = aphid transmissible, aph^- = not aphid transmissible.

While the generality of this method is limited, it provides one way to map phenotypes on the CaMV genome. Defining the genetic determinants that affect symptoms and host range is critical for vector development.

Generalized Insertion Mutagenesis

In this section I would like to discuss two experiments in which insertions were made at multiple sites around the genome of CaMV. Howell et al. (42) inserted *EcoRI* linkers (8 bp) into *AluI* sites on the CaMV genome. All of the insertions abolished infection, except for one in region VII, which is discussed further

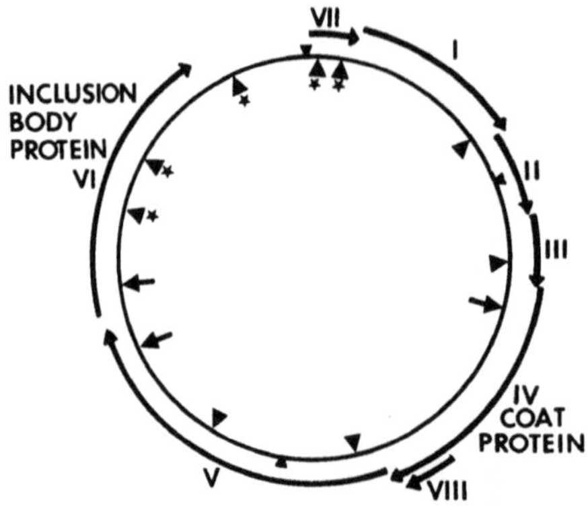

Fig. 4. Generalized insertion mutagenesis of CaMV. Insertions of 12 bp (▼) or 30 bp (arrows) were made at random Sau3A and EcoRI sites, respectively, of a clone of Cabb B-Davis (pCaMV12). In each case the insertion contained a unique SalI site (pCaMV12 lacks a site for SalI). Infectivity was assayed by examining symptoms (marked with a star), the result was confirmed by isolating viral DNA and assaying for the presence of the SalI site.

below. I would like to present in more detail some work carried out primarily by Steve Daubert in R.J Shepherd's laboratory (Daubert, Gardner, and Shepherd, unpublished) in which insertions of 12 bp or 30 bp containing a new SalI site were made at a number of different sites in CaMV. Insertions of 12 bp cause an insertion of four amino acids in the protein, and do not result in a frame shift. Hence, this experiment provided us with a potential method of obtaining attenuation mutants in the various proteins. The experiment and the results are summarized in Fig. 4.

The results show that insertions made in open coding regions I, III, IV (the coat protein gene), and V were lethal. One insertion early in region VI (the gene for inclusion body protein) was lethal, but two other insertions in VI reduced the severity of symptoms. The addition of a lac operator fragment (65 bp) to make an out-of-phase insertion in one of these two mutants abolished infectivity. Thus, region VI is probably essential to the life cycle of CaMV, and the two insertions modified the activity of the protein, but did not abolish it completely. Insertions in the large intergenic region or in region VII had no effect on the appearance of symptoms in turnips.

These experiments have identified a region of the CaMV genome where foreign DNA may be inserted, and have provided a series of lethal mutations in different parts of the genome with which complementation studies have been attempted (see below). Once a protoplast transformation system is established, they may also provide a way of identifying the function of the gene products from the different coding regions (e.g., by detection of the exact stage of CaMV infection that is blocked during synchronous infection). In addition, the non-lethal mutations in region VI have implicated the inclusion body protein in the severity of symptoms; the nature of these mutations, particularly their effect on viral titer, needs to be investigated further.

Insertions in Region II

Gronenborn et al. (43) constructed a series of insertions in the unique *XhoI* site present in region II. This site was chosen because a naturally occurring deletion mutant of CaMV, strain CM4-184, has been shown by restriction and sequence analysis to have lost 421 bp of DNA in this region (44). The deletion is entirely within region II, and includes the *XhoI* site as well as one of the three gaps on the CaMV genome. Therefore, this region was presumed to be non-essential.

Fragments of bacterial DNA of different sizes were inserted at this unique *XhoI* site in a full-length strain CM1841 (the progenitor strain of the deletion strain CM4-184), and the DNA inoculated to turnips. Insertions of the *lac* promoter-operator DNA in either orientation (one 65 and one 66 bp), and of a fragment of bacteriophage λ (256 bp), were maintained in the CaMV genome after passage through turnips. In contrast, two larger insertions, one of 531 bp of λ DNA and one of a 1200 bp kanamycin resistance gene from Tn903, gave rise to deletions of most or all of the insertion (see Table 1).

This experiment confirmed that region II was not essential for CaMV propagation, although a slight change in symptoms was observed in the inserted derivatives. More importantly, the experiment demonstrated that CaMV could be used to carry foreign (bacterial) DNA through cycles of replication in the plant, and therefore function as a vector. The inability to propagate large fragments may be accounted for by a size limitation in the packaging capacity of the CaMV virion, although additional insertions are needed to establish this point clearly. Some instability of the *lac* promoter-operator insertion was observed during several serial transfers. The instability may be a property of the particular inserted DNA or of the CaMV vector.

Insertions in Region VII

I would next like to describe a recent series of experiments in which insertions have been made in putative coding region VII of the

Table 1. The effect of inserting bacterial DNA into the unique XhoI site of Region II in strain CM1841 (data from ref. 43).

Inserted DNA	Size	Result
→ lacpo	65 bp	maintained
← lacpo	66 bp	maintained
λ fragment	256 bp	maintained
λ fragment	531 bp	deletions
Kanamycin resistance gene	1200 bp	deletions

CaMV genome (Gardner, Daubert, and Shepherd, unpublished). The results of these experiments (shown in Fig. 5) can be summarized as follows:

(a) Small insertions (up to 16 bp) at four closely spaced sites of two different CaMV strains were successfully propagated through the plant and had no effect on viral symptoms.

(b) Two larger insertions (65 and 270 bp) of bacterial DNA were successfully propagated through the plant in CaMV strain Cabb B-Davis.

(c) Four larger insertions (120-550 bp) abolished infectivity in CaMV strain CM4-184, which contains the 421 bp deletion in region II.

The basis of the difference in the ability of these two strains to tolerate insertions is not clear. One hypothesis is that deletion of the region II discontinuity in CM4-184 limits its ability to package additional DNA. However, the different bacterial DNAs used for insertion in each strain, or some other factor, could be responsible for the difference.

For each of four lethal insertions in CM4-184, infectious derivatives were obtained that had either neatly deleted the inserted DNA, or deleted the inserted DNA together with about 130 bp of surrounding CaMV sequences (see Fig. 5). Nucleotide sequencing of one of the larger deletions revealed that 129 bp of CaMV sequences (position 92-221) had been removed. This deletion, like that which generated strain CM4-184, occurred at short direct repeats (see Fig. 6). It is likely that the other class of deletions that were observed, where only the inserted DNA was excised, occurred at short direct repeats formed during restriction

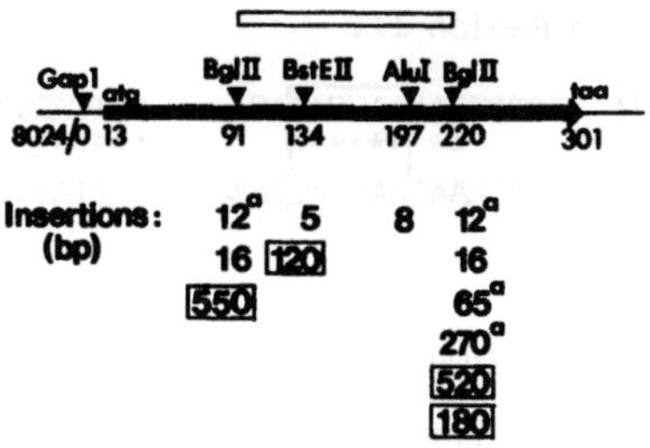

Fig. 5. Insertion mutagenesis in region VII. The figure summarizes insertions made in the region VII area of the CaMV genome in two strains, CM4-184 and Cabb B-Davis (superscript a). Open reading frame VII, together with the nucleotide positions of relevant restriction sites, is depicted above. The size in bp of insertions made at each site are also shown. The small insertions contained new restriction sites for EcoRI (8 bp, 42), SalI (12 bp, see Fig. 4), or PvuI (16 bp, Gardner unpublished), or were obtained by filling in the unique BstEII site (5 bp, Gardner unpublished). The larger insertions consisted of: a lac operator fragment (65 bp, 43), the same insertion plus a random EcoRI* fragment of λ(270 bp, Gardner and Daubert, unpublished), a BstEII fragment of λ(120 bp, Gardner, unpublished), a part of the Tn903 kanamycin resistance gene (550 bp, Gardner and Shepherd, unpublished), a fusion product between the lac promoter and the DHFR gene from R388 (520 bp, inserted in both orientations, Gardner and Shepherd, unpublished), and a deletion formed from this last insertion that retained only the DHFR coding region and had lost CaMV sequences from 134-220 (180 bp, Gardner, unpublished). Insertions that were not infectious are indicated by boxes. A deletion that arose *in vivo* from three of these non-infectious insertions (120, 520, 18 bp) is shown by the box above (see text, Fig. 6).

enzyme manipulations used to create the insertion. Similar repeats have been implicated in deletion formation in both bacterial and mammalian systems (45).

The insertions and deletions described here clearly indicate that region VII is not essential for CaMV infectivity. The results also confirm the earlier finding that CaMV can stably maintain some insertions of foreign DNA in its genome. The inability to maintain larger, gene-sized insertions is tentatively ascribed to a packaging limitation, which may be affected by the number of single-stranded gap structures in the viral DNA.

Deletion in Region II:

```
1374  TTTATAAAAAGGATAC       TATTATTAGA  139
             ** ******       ******** *
1795  GATCAAAGAAGGATTAAAGAATATTATTGGC 1826
```

Deletion in Region VII:

```
92    GATCTTTCAAGATCAAAAC  110
      *****   ********* *
221   GATCTAAAAAGATCAAGAA  239
```

Fig. 6. Direct repeats involved in deletion formation in CaMV. This figure shows the nucleotide sequence of the short direct repeats that occur at the borders of both the naturally occurring deletion in region II of CM4-184 (from ref. 44) and one of the deletions obtained from insertions in region VII of CM4-184 (Gardner and Heidecker, unpublished). The homologies are indicated by asterisks, with the deletion occurring somewhere within the boxed sequences. Numbers refer to the first and last nucleotide positions of each portion of CaMV sequences shown (21).

Complementation and Helper Virus Systems

One possible means of overcoming packaging limitations is to use a helper virus system. In this approach, one or more genes of CaMV are replaced by a foreign gene, and the function of the deleted gene is supplied *in trans* by a helper virus. Such a system works well for SV40 (46).

Different laboratories have tested the ability of CaMV mutations to be complemented by a helper virus (42,47, Daubert and Shepherd, unpublished). In this system, two lethal insertions in different parts of the genome are coinoculated to turnip leaves. If the mutations complement, infection should result, with both mutants present in infected plants. However, in all cases examined, infectivity was the result of recombination between the two mutant viruses to give a wild-type CaMV.

Considerable effort has been expended to analyze the nature of the recombination events that occur. A number of different pair-wise combinations of CaMV mutants and subclones have been used, in an effort to reduce recombination to a level where complementation might be observed. In all cases, recombination was observed (see Fig. 7). The only requirement for recombination seems to be the presence of homology between the pairs. Neither different free ends of the DNA nor supercoiled inoculating molecules could block the recombination. Once precise data becomes available about

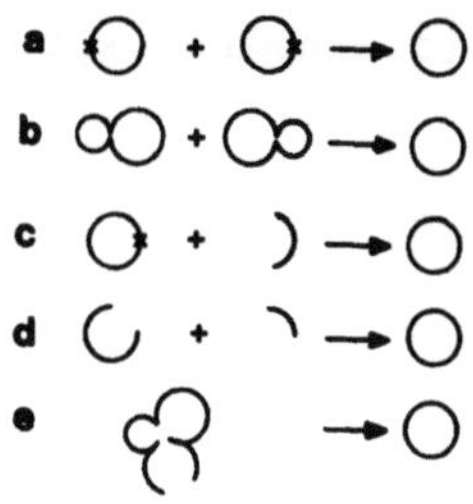

Fig. 7. In vivo recombination of CaMV. Inoculation with various combinations of CaMV clones and subclones gives rise by recombination to wild-type CaMV DNA. The inoculant can consist of pairs of lethal insertion mutants where the insertion is a small linker (a, 42) or the plasmid cloning vector (b, 47). A subclone from wild-type CaMV that covers an insertion can be used (c, 47), or two subfragments that together comprise one genome (d, 47). Finally an incomplete tandem clone of CaMV is infectious when inoculated without excising the plasmid vector (e, 47).

the location of genetic elements required in cis for the CaMV life cycle (e.g., replication origins, promoters and splice sites, cos-type packaging sites, etc.), it may be possible to devise a set of non-overlapping mutants that lack homology, and thus block recombination.

The biological assay used for these complementation studies (the appearance of symptoms in turnip plants) may be affecting the frequency of recombination observed. In order for CaMV symptoms to be observed, the initial infection of individual leaf epidermal cells must spread cell-to-cell throughout the remainder of the plant. Thus, a large number of independent cell infections must occur. In order for complementation to occur, both mutant strains have to coinfect each new cell. Although simultaneous coinfection does occur for viruses with multipartite genomes, it may be that the concentration of CaMV particles that move between cells is lower, so that the probability of coinfection by two independent particles is correspondingly lower. If this were the case, the selection pressure exerted for a recombination event could be very high.

Conclusions and Prospects for CaMV

The results described herein have identified non-essential regions of the CaMV genome and shown that CaMV can be used to propagate bacterial DNA in the plant. At present, the packaging limitation and the potential of CaMV to recombine severely limit the amount of foreign DNA which can be carried by the virus, so that it has not yet been possible to examine expression of entire genes cloned into CaMV.

Use of alternative experimental systems may overcome these limitations. Melcher and his colleagues have developed a leaf hybridization assay in which the leaf inoculated with CaMV DNA is essentially treated as a nitrocellulose filter (48). Hybridization with labelled probes identifies CaMV present in local areas of the leaf, so that only limited cell-to-cell movement is required before the results can be analyzed. Using this assay system, they have shown that the host range for CaMV multiplication is wider than the host range in which infection becomes systemic. They also suggest that large inserts of bacterial DNA can be propagated in the plant (49). These potentially very significant results need to be confirmed using other systems.

In the longer term, a more promising system appears to be infection of plant protoplasts with CaMV. Since there is no cell-to-cell movement in a protoplast system, there should be no requirement for packaging the DNA, and no selection for recombination over complementation. Several workers (30,49, Otsuki and Shepherd, unpublished results) have shown that CaMV particl7les can be used to obtain high level infection of Brassica protoplasts (up to 90%), and that the virus can be immunologically detected after forty-eight hours. At present there are no reports of infection using viral or cloned DNA of CaMV. However, Haber et al. reported infection of bean protoplasts using DNA of bean golden mosaic virus (40). Thus, a rapid assay system for gene expression using CaMV vectors should soon be a reality.

The availability of DNA sequence data, together with a partial transcription map, allows two potential CaMV promoters to be identified quite precisely. Since the two transcripts from CaMV are quite prominent in infected plants, these promoters may prove valuable for the construction of selectable markers or for obtaining high level expression of genes of interest.

ACKNOWLEDGEMENTS

The unpublished results reported here were supported by Calgene, Inc., and by research grants to R.J. Shepherd from the National Science Foundation (PCM-7904960) and the United States Department of Agriculture (SEA 5901-0723). I would like to thank John Caton and Jack Kiser for excellent technical assistance, Bob Shepherd and Bob Goodman for helpful discussions and comments, and Nelinia Henry for typing the manuscript.

REFERENCES

1. Fletcher, T.J. 1978. The use of avirulent virus strains to protect plants against the effects of virulent strains. Ann. Appl. Biol. 89:110.
2. Channon, A.G., N.J. Cheffins, G.M. Hitchon, and J. Barker. 1978. The effect of inoculation with an attenuated mutant strain of tobacco mosaic virus on the growth and yield of early glasshouse tomato crops. Ann. Appl. Biol. 88:121.
3. Rothamsted Experimental Station report for 1980, part 1, p. 189.
4. Athow, K.L., and J.B. Bancroft. 1959. Development and transmission of tobacco ringspot virus in soybean. Phytopathology 49:67.
5. Kassanis, B., G.E. Russell, and R.F. White. 1978. Seed and pollen transmission of beet cryptic virus in sugar beet plants. Phytopathol. Z. 91:76.
6. Rothamsted Experimental Station report for 1979, part 1, p. . 176.
7. Mathews, R.E.F. 1981. Plant Virology. New York: Academic Press.
8. Mathews, R.E.F. 1981. Plant Virology. New York: Academic Press. P. 464 and pp. 230-232.
9. Ikegami, M., S. Haber, and R.M. Goodman. 1981. Isolation and characterization of virus-specific double-stranded DNA from tissues infected by bean golden mosaic virus. Proc. Natl. Acad. Sci. U.S.A. 78:4102.
10. Howarth, A.J., and R.M. Goodman. 1982. Plant viruses with genomes of single-stranded DNA. Trends in Biochem. Sci. 7:180.
11. Howell, S.H. 1982. Plant molecular vehicles: Potential vehicles for introducing foreign DNA into plants. Ann. Rev. Plant Physiol. 33:609.
12. Taniguchi, T., M. Palmieri, and C. Weissman. 1978. Qβ DNA-containing hybrid plasmids giving rise to Qβ phage formation in the bacterial host. Nature 274:223.
13. Racaniello, V.R., and D. Baltimore. 1981. Cloned polivirus complementary DNA is infectious in mammalian cells. Science 214:916.
14. Shepherd, R.J. 1979. DNA plant viruses. Ann. Rev. Plant Physiol. 30:405.
15. Shepherd, R.J., and R.H. Lawson. 1981. Caulimoviruses. In Handbook of Plant Virus Infections, E. Kurstak, ed. Amsterdam: Elsevier.
16. Franck, A., H. Guilley, G. Jonard, K. Richards, and L. Hirth. 1980. Nucleotide sequence of cauliflower mosaic virus DNA. Cell 21:285.
17. Guilfoyle, T.J. 1980. Transcription of the cauliflower mosaic virus genome in isolated nuclei from turnip leaves. Virology 107:71.

18. Ansa, O.A., J.W. Bowyer, and R.J. Shepherd. 1982. Evidence for replication of cauliflower mosaic virus DNA in plant nuclei. Virology 121:147.
19. Menissier, J., G. Lebeurier, and L. Hirth. 1982. Free cauliflower mosaic virus supercoiled DNA in infected plants. Virology 117:322.
20. Olszewski, N., G. Hagen, and T.J. Guilfoyle. 1982. A transcriptionally active, covalently closed minichromosome of cauliflower mosaic virus DNA isolated from infected turnip leaves. Cell 29:395.
21. Gardner, R.C., A.J. Howarth, P. Hahn, M. Brown-Luedi, R.J. Shepherd, and J. Messing. 1981. The complete nucleotide sequence of an infectious clone of cauliflower mosaic virus by M13mp7 shotgun sequencing. Nucleic Acids Res. 9:2871.
22. Hohn, T., K. Richards, and G. Lebeurier. 1982. Cauliflower mosaic virus on its way to becoming a useful plant vector. Curr. Topics Microbiol. Immun. 96:193.
23. Daubert, S., R. Richins, R.J. Shepherd, and R.C. Gardner. Mapping of the coat protein gene of cauliflower mosaic virus by its expression in a procaryotic system. Virology (in press).
24. Shepherd, R.J., R. Richins, and T.A. Shalla. 1980. Isolation and properties of the inclusion bodies of cauliflower mosaic virus. Virology 102:389.
25. Shockey, M.W., C.O. Gardner, Jr., U. Melcher, and R.C. Essenberg. 1980. Polypeptides associated with inclusion bodies from leaves of turnips infected with cauliflower mosaic virus. Virology 105:575.
26. Al Ani, R., P. Pfeiffer, D. Whitechurch, A. Lesot, G. Lebeurier, and L. Hirth. 1980. A virus specified protein produced upon infection by cauliflower mosaic virus (CaMV). Ann. Virol. (Inst. Pasteur) 131E:33.
27. Odell, J.T., and S.H. Howell. 1980. The identification, mapping, and characterization of mRNA for P66 a cauliflower mosaic virus-coded protein. Virology 102:349.
28. Covey, S.N., and R. Hull. 1981. Transcription of cauliflower mosaic virus DNA. Detection of transcripts, properties, and location of the gene encoding the virus inclusion body protein. Virology 111:463.
29. Odell, J.T., R.K. Dudley, and S.H. Howell. 1981. Structure of the 19S RNA transcript encoded by the cauliflower mosaic virus genome. Virology 111:377.
30. Howell, S.H., and R. Hull. 1978. Replication of cauliflower mosaic virus and transcription of its genome in turnip leaf protoplasts. Virology 86:468.
31. Covey, S.N., G.P. Lomonossof, and R. Hull. 1981. Characterisation of cauliflower mosaic virus DNA sequences which encode major polyadenylated transcripts. Nucleic Acid Res. 9:6735.
32. Dudley, R.K., J.T. Odell, and S.H. Howell. 1982. Structure and 5'-termini of the large and 19S RNA transcripts encoded by the cauliflower mosaic virus genome. Virology 117:19.

33. Dudley, R.K. 1982. Presentation at the Gordon Conference on Plant Molecular Biology.
34. Shepherd, R.J., G.E. Bruening, and R.J. Wakeman. 1970. Double-stranded DNA from cauliflower mosaic virus. Virology 41:339.
35. Howell, S.H., L.L. Walker, and R.K. Dudley. 1980. Cloned cauliflower mosaic virus DNA infects turnips (Brassica rapa). Science 208:1265.
36. Gardner, R.C., and R.J. Shepherd. 1980. A procedure for rapid isolation and analysis of cauliflower mosaic virus DNA. Virology 106:159.
37. Davey, M.R., E.C. Cocking, J. Freeman, N. Pearce, and I. Tudor. 1980. Transformation of Petunia protoplasts by isolated Agrobacterium plasmid. Plant Sci. Lett. 18:307.
38. Krens, F.A., L. Melendijk, G.J. Wullems, and R.A. Schilperoort. 1982. In vitro transformation of plant protoplasts with Ti-plasmid DNA. Nature 296:72.
39. Goodman, R.M. 1977. Infectious DNA from a white-fly transmitted virus of Phaseolus vulgaris. Nature 266:54.
40. Haber, S., M. Ikegami, N.B. Bajet, and R.M. Goodman. 1980. Evidence for a divided genome in bean golden mosaic virus, a geminivirus. Nature 289:324.
41. Lung, M.C.Y., and T.P. Pirone. 1974. Acquisition factor required for aphid transmission of purified cauliflower mosaic virus. Virology 60:260.
42. Howell, S.H., L.L. Walker, and R.M. Walden. 1981. The rescue of in vitro generated mutants of the cloned cauliflower mosaic virus genome in infected plants. Nature 293:483.
43. Gronenborn, B., R.C. Gardner, S. Schaeffer, and R.J. Shepherd. 1981. Propagation of foreign DNA in plants using cauliflower mosaic virus as vector. Nature 294:773.
44. Howarth, A.J., R.C. Gardner, J. Messing, and R.J. Shepherd. 1981. Nucleotide sequence of naturally occurring deletion mutants of cauliflower mosaic virus. Virology 112:678.
45. Albertini, A.M., M. Hofer, M.P. Calos, and J.H. Miller. 1982. On the formation of spontaneous deletions: The importance of short sequence homologies in the generation of large deletions. Cell 29:319.
46. Elder, J.T., R.A. Spritz, and S.M. Weissman. 1981. Simian virus 40 as a eukaryotic cloning vehicle. Ann. Rev. Genetics 15:295.
47. Lebeurier, G., L. Hirth, B. Hohn, and T. Hohn. 1982. In vivo recombination of cauliflower mosaic virus DNA. Proc. Nat. Acad. Sci. U.S.A. 79:2932.
48. Melcher, U., C.O. Gardner, Jr., and R.C. Essenberg. 1982. Clones of cauliflower mosaic virus identified by molecular hybridization in turnip leaves. Plant Mol. Biol. 1:63.
49. Melcher, U., C. Brannan, C.O. Gardner, Jr., and R.C. Essenberg. Leaf skeleton hybridization assesses cauliflower mosaic virus host range. Presentation at Beltsville Symposium VII on Genetic Engineering, May 1982.

50. Furusawa, I., N. Yamaoka, T. Okuno, M. Yamamoto, M. Kohno, and H. Kunoh. 1980. Infection of turnip Brassica-rapa cultivar perviribis protoplasts with cauliflower mosaic virus. J. Gen. Virol. 48:431.

PLANT CELL TRANSFORMATION BY AGROBACTERIUM PLASMIDS

A. Depicker, M. Van Montagu, and J. Schell

Laboratorium voor Genetica
Rijksuniversiteit Gent
B-9000 Gent, Belgium

AGROBACTERIUM TAXONOMY

Infection of many dicotyledonous plants by the genus *Agrobacterium* may lead to the plant diseases crown gall, cane gall, and hairy-root disease.

Crown gall and cane gall, characterized by tumorous overgrowths on the root crown and stem, are induced by *A. tumefaciens* and *A. rubi*, respectively. The hairy-root disease which involves abundant root proliferation at the infected site is attributed to *A. rhizogenes*. The non-oncogenic strains are grouped as *A. radiobacter*. However, this phytopathogenic subdivision does not correspond to a taxonomic classification since it can be explained by the finding that a large plasmid (Ti) is responsible for the oncogenic properties of *Agrobacterium* (83,116,138). The transfer between agrobacteria of such a Ti plasmid passes onto the acceptor strain pathogenic properties identical to those of the donor strain. Therefore, the designation of *Agrobacterium radiobacter*, *tumefaciens*, *rubi*, and *rhizogenes* mainly indicates the presence and type of Ti plasmid(s) the agrobacteria contain and, hence, their pathogenic properties.

The taxonomic classification of agrobacteria is based on a number of physiological and biochemical characteristics (56,61) and DNA hybridization studies (62,63). Agrobacteria are subdivided in two major biotype groups [1 and group 2] and a minor intermediate one [3], based on properties such as 3-ketolactase production (biotype 1 and 3), growth on erythritol as carbon source (biotype 2), and growth on 2% NaCl (biotype 3). Agrobacteria and the fast-growing rhizobia are closely related, while the slow-growing rhizobia are more distantly related; all are included in the family *Rhizobiaceae* (63). Oncogenic *Agrobacterium* strains isolated in nature from a wide variety of dicotyledonous plants belong to biotype 1 and 2, but agrobacteria isolated from grape vine tumors usually belong to biotype 3 whose Ti plasmids display a very narrow host range.

CROWN GALLS

As already postulated by Braun (5), a tumor-inducing principle (TIP) is transferred from the inciting bacteria to the host plant cell resulting in their stable transformation to crown gall tumor cells. It was found that the molecular basis for this TIP is a segment of the Ti plasmid (T-region) which is transferred to and maintained as T-DNA in the transformed plant cell by insertion into the plant nuclear genome (10,98). The term "transformation" can mean the introduction of foreign DNA into a cell (without necessarily resulting in a tumorous phenotype), while transformation *sensu stricto* limits the notion of transformation to the conversion of a normal cell to a tumorous one. Thus, the *Agrobacterium* Ti plasmid transforms the plant cell by introducing DNA; the T-DNA transforms the plant cell *sensu stricto* to a tumorous phenotype.

Crown gall, cane gall, and hairy root disease are characterized by an autonomous neoplastic growth and the synthesis of metabolites (opines) that are not present in normal plant cells. The transformed plant cells are not subject to normal morphogenetic controls. The tumor cells continue division in the absence of the inciting bacteria and they proliferate *in vitro* in axenic cultures without the addition of phytohormones, which are required for the growth of normal cells. It is assumed that the capacity for phytohormone-independent growth of crown gall cells is due to an alteration of their auxin and cytokinin production or release (6).

Phytohormone-independent growth and suppression of differentiation are observed in *Agrobacterium*-induced tumors on a very wide range of dicotyledonous plants (18) which might suggest that the introduced T-DNA alters common basic plant control mechanisms. Morphologically altered tumors induced by mutant Ti plasmid support the hypothesis that the Ti plasmid mainly influences the hormone-level in the plant cell (8a), either during the induction and/or the maintenance phase (see "THE TRANSFERRED DNA--Functional Organization of the T-DNA").

AGROBACTERIUM PLASMIDS

Introduction

Most strains of the *Agrobacterium* genus, both pathogenic and nonpathogenic, contain one or more large plasmids (16,81,121,138).

The different types of Ti plasmids which are responsible for the tumorigenic properties of Agrobacterium have a molecular weight in the range of 120 to 160 x 10^6 dalton. A subgroup of Ti plasmids inducing the hairy-root tumors are often referred to as Ri plasmids. Ti plasmids are most easily identified by their transfer to non-virulent strains (116,119). The transfer of virulence is correlated with the transfer of a Ti plasmid. Mixed infections in planta provided the first indication for the transmissibility of the pathogenic properties of Agrobacterium (58,59). Concurrent with the characterization of the Ti plasmid, the transfer of virulence was further demonstrated by R-plasmid mobilization of the Ti plasmid (3,53,117), by induction of the conjugative properties of Ti plasmids (41,60,93), and by transformation with purified Ti plasmids (50). These plasmid transfers also served to characterize other plasmid-linked phenotypes, such as plant host range, induced tumor morphology, opine synthesis and opine catabolism.

The Opine Concept

As a general rule, T-DNA-transformed plant cells synthesize opines, i.e., plant metabolites synthesized under the control of T-DNA and which can be specifically catabolized by the inciting bacteria, and thus used as carbon and nitrogen source. The presence of these opines in plant tissue is therefore the most conclusive criterium to indicate Agrobacterium as a causative agent for plant cell transformation, since opines discriminate crown galls from other plant cell proliferations (102). The Ti plasmids allow the host bacteria to utilize those opines they induce as specific energy, nitrogen, and carbon source (3), thus providing a competitive advantage in the crown gall rhizosphere to the inciting agrobacteria over other soil organisms. This phenomenon has been called "genetic colonization" (98). By now, many different opines have been described. No matter which combination of opines is considered (see Table 1), the genes coding for synthesis and catabolism of these opines are located on the same Ti plasmid, thus providing strong evidence for this concept. Several non-oncogenic plasmids also carry genes for opine catabolism, demonstrating the ecological importance of opines for bacteria (108). The fact that the opine synthase genes are genetically linked to "onc" genes on the T-DNA is probably significant since this results in a proliferation of the rare plant cells into which the opine synthase genes were initially transferred. Moreover, the opines released from the transformed plant cells induce the conjugative transfer of the inciting Ti plasmid, thus spreading the Ti plasmid in opportune circumstances (64,93).

Ti Plasmid Groups

Comparisons between different Ti plasmids lead to the following two conclusions (30,34,114,122):

(i) The different Ti plasmids are phylogenetically related; however, total homology does not exceed 10 to 20% of the total Ti plasmid and is localized in DNA segments which are mainly involved in virulence (plant/bacterium interaction and DNA transfer) and in oncogenicity (plant tumorous growth).

(ii) Despite the large segments without homology, the Ti plasmids show a remarkable structural and functional convergent organization, which is centered around opine synthesis and catabolism, the apparent driving forces of Ti plasmid evolution. Thus, Ti plasmids are subdivided according to the opines they specify (45). The different groups are listed in Table 1.

The classic octopine type plasmids (type I) direct the synthesis of both octopine (79,92) and agropine (36) in the transformed plant cells; these are respectively an arginine-pyruvic acid condensation product and a carbohydrate derivative. A third opine, mannopine, is found together with agropine and is presumed to be the open chain precursor in the agropine synthesis pathway (107). Octopine and agropine catabolism is also genetically linked to the Ti plasmid (3,45). All natural isolates are closely related and show nearly 100% DNA homology (101). On Kalanchoe these plasmids induce rough tumors with adventitious roots.

A second group of octopine Ti plasmids (type II) has been found to direct the synthesis of octopine only; these Ti plasmids are commonly referred to as narrow-host range Ti plasmids since they induce tumors efficiently only on grape vine (91). These plasmids share a low percentage of DNA homology with the octopine-type I Ti plasmids (114). However, it is remarkable that the genetically and physically completely distinct genes, coding for octopine synthesis in the tumor cells and for octopine catabolism in the bacteria are both very well conserved between these divergent Ti plasmids.

The nopaline Ti plasmids specify the synthesis of nopaline (80,92) in crown gall cells, nopaline being the condensation product between arginine and α-ketoglutaric acid. Other opines found in nopaline tumors are agrocinopine A and B, both sugar phosphate derivatives (31). Nopaline and agrocinopine are catabolized by Ti plasmid-linked operons (3,31). The nopaline-type plasmids are a more heterogenous group (101), though heteroduplex mapping shows that heterogenicity is mainly restricted to one contiguous part of the Ti plasmid with large insertion and/or deletion loops (Engler et al., in preparation). On Kalanchoe these nopaline Ti plasmids induce smooth tumors which frequently develop into teratomas with leaf-like structures.

The agropine Ti plasmids (formerly referred to as null-type plasmids) have the opine agropine and mannopine in common with the

Table 1. Opines, specified by the different groups of Ti plasmids.

Ti plasmids	Octopine	Agropine mannopine	Nopaline	Agrocinopine A + B	Agrocinopine C + D
Octopine-type 1 Ti plasmids	+ (tra)	+	-	-	-
Octopine-type 2 Ti plasmids	+	-	-	-	-
Nopaline Ti plasmids	-	-	+	+ (tra)	-
Agropine Ti plasmids	-	+ (tra)	-	-	+
Rhizogenes Ri plasmids	-	+ (tra)	-	+ (?)	?

The different groups of Ti plasmids with the opines they specify are discussed in the text. (tra) indicates that the opine induces conjugative transfer between the bacteria.

octopine-type I Ti plasmids (45). They also specify the synthesis of agrocinopines C and D, which are structurally related to agrocinopines A and B (31). Besides the common agropine catabolism and synthesis genes, the agropine Ti plasmids do not show any more homology with the octopine than with the nopaline Ti plasmids (30). The agropine Ti plasmids induce rough tumors without adventitious root proliferation on *Kalanchoe*.

The rhizogenes Ri plasmids have been considered as a different class of *Agrobacterium* plasmid, mainly because of their hairy root induction, which seemingly differs from the undifferentiated neoplasms or teratomata (14). However, there are indications that the Ri plasmids are not only functionally but also phylogenetically related to the other types of Ti plasmids (15,122). *Agrobacterium rhizogenes* has a more restricted host range than A. *tumefaciens* (19). The axenic root tissue cultures induced on carrot and potato (13,110,128) and the tumors induced on *Nicotiana glauca* (123) by the rhizogenes plasmid ATCC15834 contain both agropine and mannopine, the same related opines found in octopine-type I and in agropine tumors. Moreover, agrocinopines similar to the ones found in nopaline tumors have also been demonstrated in the root cultures (J. Ellis and J. Tempé, pers. comm.; 127 or 128).

Physical and Functional Comparisons of Octopine and Nopaline Ti Plasmids

The functional maps of octopine and nopaline Ti plasmids were established mainly be transposon insertion mutagenesis and by deletion mapping (20,25,37 or 38, 48,51,66,87). The location of insertions or deletions can be determined by the changes in restriction pattern or by heteroduplex analysis. This allows the assignment of altered phenotypes to particular regions of the Ti plasmid. Concurrent with the genetic mapping, restriction maps of octopine- and nopaline-type Ti plasmids have been constructed (24,28).

Octopine and nopaline are degraded via arginine and ornithine, whose Ti plasmid-determined utilization is under regulatory control of the opines (32,64) and physically linked (20,51). Most remarkably, the arginine-catabolizing functions in both types of Ti plasmids do not show DNA sequence homology (34).

In a detailed study of the homology regions between octopine and nopaline Ti plasmids, four major regions of homology were recognized. These are often interrupted by asymmetric substitutions (Figure 1) (34). Two of those homology stretches are involved in oncogenicity (A and D), one corresponds to the replication control region (B) and another encodes conjugative functions (C) (20,51).

The so-called "common" DNA (region A), is a 9 kb DNA sequence, internal to the T-DNA stretch found in octopine as well as in nopaline tumors. This "common" DNA is responsible for the tumorous phenotype. Mutations in this region result in attenuated Ti plasmids (reduced plant host range and tumor formation and/or slow tumor growth) or in induction of morphologically different tumors forming shoots or roots (see "THE TRANSFERRED DNA--Functional Organization of the T-DNA").

Many non-oncogenic Ti plasmid mutants map in the D homology region (also called the vir region) (20,37,51). Some mutants in the vir region can also results in an altered tumor phenotype (20,54) although DNA is not present in the transformed plant cell. This observation reflects the importance of the vir region for the early events. The influence of the induction phase on the tumor phenotype is also demonstrated by the observation that a cytokinin treatment during the induction phase of transformation with a root-inducing plasmid results in undifferentiated tumor growth (2). Auxins are required during the induction phase. Many onc$^-$ mutations can be complemented by the external addition of auxins. Mutations in a gene, located in this vir region and involved in auxin synthesis, are non-oncogenic in auxin-deficient *Agrobacterium* host strains (72).

Non-oncogenic mutants in the vir region can be complemented by broad-host range cosmids containing the corresponding wild-type Ti plasmid region (65). Only two physically linked insertion mutants could not be complemented; the authors suggest that this *cis*-dominant DNA sequence could correspond to the origin of transfer. As mentioned above, the octopine type II plasmids show a Ti plasmid-linked narrow-host range for plant tumor induction. The low degree of homology between type I and II octopine plasmids indicates a divergence in tumor induction, in T-DNA transfer, or in maintenance. The narrow-host range of octopine type II plasmids might be related to an alteration of the early oncogenicity functions (induction, transfer, and stabilization) rather than to the oncogenicity maintenance functions. Alternatively, the "onc" genes in the T-DNA might be altered in a way that the suppression of differentiation and activation of cell division is weak, thereby allowing recognizable tumor growth only on a limited number of plant species.

The relation between Ri plasmids and the Ti plasmids is not only functional (presence of opines) as mentioned above, but is also suggested by DNA sequence homology between the Ri plasmids and the octopine type I plasmids (122). The D-region in particular is conserved between all Ti and Ri plasmids, which may indicate that similar bacterium/plant cell interactions must occur in both types of infection. Support to this assumption is given by the fact that avirulent octopine Ti plasmid mutants can be complemented by the compatible Ri plasmids (54).

A third homology region (C) between octopine and nopaline Ti plasmids can be assigned to at least some of the tra genes (Fig. 1).

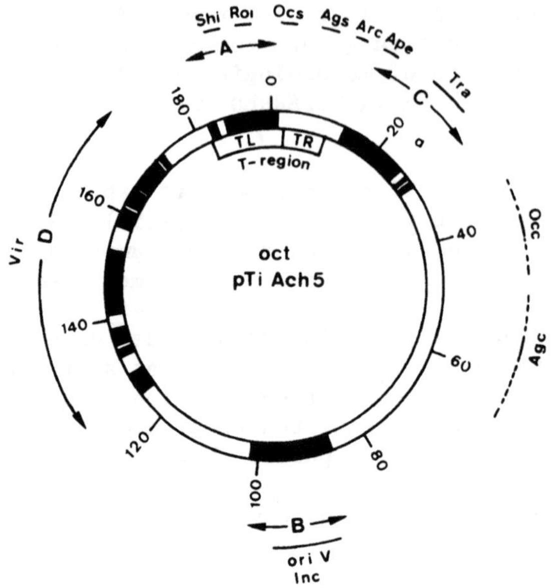

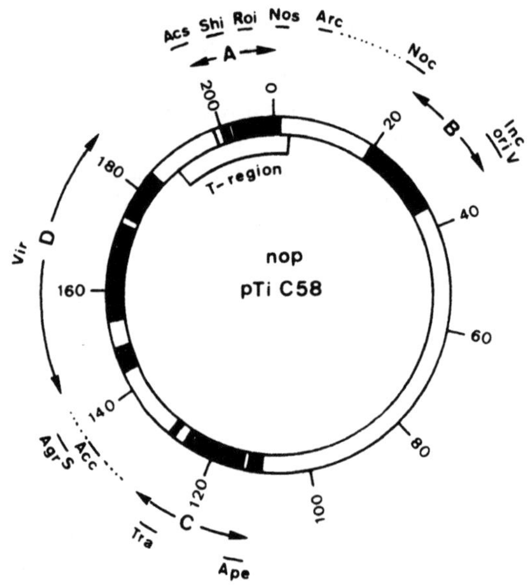

This homology region is located adjacent to the octopine cataboliz-ing genetic unit in octopine type I Ti plasmids and adjacent to the agrocinopine A and B catabolism region in nopaline-type Ti plasmids (31). Thus, the simultaneous induction of the tra genes and the opine catabolism genes by the opine is reflected in a similar functional organization. Agrocin 84 sensitivity (106) also maps in this region and is a genetic marker for nopaline Ti plasmids (33). The basis of this sensitivity is a high affinity uptake system (86) which belongs to the agrocinopine catabolizing operon.

Fig. 1. Functional organization of the octopine-type I and the nopaline Ti plasmids. The functional maps are basically as described by De Greve et al. (20) and by Holsters et al. (51). The numbers on the maps indicate the distances in kb. The black bars indicate the homology between the two types of Ti plasmids as determined by Engler et al. (34). The letters A, B, C, and D designate the four major blocks of homology between those two types of Ti plasmids. Especially B and D are conserved with other types of Ti or Ri plasmids (see text for discussion). Homology block A lays internal to the region transferred to the plant DNA. Shi and Roi, respectively shoot and root inhibition (previously defined as shoot and root induction), indicate the tumor phenotypes determined by those loci (69). The four indicated phenotypes Ocs (octopine synthesis), Ags (agropine synthesis), Acs (agrocinopine synthesis), and Nos (nopaline synthesis) direct the synthesis of opines in the transformed plant cells. The opine catabolism regions on the Ti plasmid are designated as Occ (octopine catabolism); Arc (arginine-ornithine catabolism), Agc (agropine catabolism), Noc (nopaline catabolism) and Acc (agrocinopine catabolism). The Acc operon corresponds to the mapped phenotype of AgrS (agrocine 84 sensitivity); the active uptake system for both compounds is mediated by the same enzymes. Homology region C contains at least part of the genes for the conjugative transfer (tra) between bacteria. Ape (phage AP1 exclusion) has been localized in this region. Homology region B is mainly involved in replication controls of the Ti plasmids (oriV). The incompatibility properties of octopine and nopaline Ti plasmids (Inc) could be mapped in the same region (see refs. 66 and 67). Homology region D contains many non-asymmetric substitution loops and is involved in the Agrobacterium/plant cell interaction during tumorigenesis (Vir). Transposon insertions in this region result in a reduction of tumor induction and plant host range (see text for discussion, "Physical and Functional Comparisons of Octopine and Nopaline Ti Plasmids").

The transfer of Ti plasmid genes to the plant cell can not involve the same mechanism as the one used for bacterial conjugation, although a similar thermosensitive step is involved (5,94, 109). Several non-conjugative Ti plasmid mutants were isolated that were still oncogenic; the reverse was found: many onc$^-$ Ti plasmid mutants were still conjugative. Two different mutants with onc$^-$ T-region deletions still induced the synthesis of octopine or agropine at the wounded site (69) suggesting that no T-region functions were required for the transfer of Ti plasmid DNA to the plant cell.

A fourth homology region (B) can be correlated with functions involved in Ti plasmid replication and incompatibility (Fig. 1) (34,67).

In conclusion, comparison of homology regions in different types of Ti plasmids, together with their genetic analysis, provides general insights into the mechanisms by which Agrobacterium interacts with host plant cells to cause plant cell transformation.

PHYSIOLOGY OF PLANT CELL TRANSFORMATION

Agrobacterium plant cell transformation requires wounding. The wound is supposed to be essential not only for removing a physical barrier, but also for conditioning the plant cell. Indeed, wounding stimulates the plant cell to switch from a resting stage to one of rapid cell division. Plant cell transformation is optimal after the wound response leads to dedifferentiation and just before the first mitotic cell division (71).

As already mentioned above, the Ti plasmid determines the host range of plant species that can be transformed to a tumorous growth 73,113). The physiological requirements for plant cell transformation are expected to vary in a similar way between different plant species as they do between different plant organs and tissues. The Kalanchoe tumors have a different morphology dependent on their position on the plant. Tumorigenesis by some Ti plasmid mutants in Kalanchoe leaves may prove to be very attenuated (no tumor induction), while normal tumors are induced by these same Ti plasmids on Kalanchoe stems (69). Certain plant factors also have an influence on the host range. Different species of tobacco and Kalanchoe can show a remarkably different response to the same Agrobacterium tumefaciens Ti plasmid (4,43). The sensitivity of the plant for crown gall formation and also the degree of tumor tissue organization can often be correlated with the regenerative capacity of the plant genus. Kalanchoe and tobacco species show a high sensitivity for a wide range of Ti plasmids. The nopaline plasmid T37 induces shoot teratoma on the upper half of stem or on leaves, while inducing only unorganized tumors on the lower half of a tobacco plant (65).

THE TRANSFERRED DNA

The first evidence indicating that Ti-plasmids were involved in a gene transfer system resulted from a genetic analysis of Ti-plasmid functions (3,96,97).

Direct evidence for such a transfer was obtained first by renaturation kinetics analysis (10). A more precise picture resulted from DNA/DNA hybridizations using the Southern gel blotting technique.

Physical Structure of the T-DNA

Southern blotting analysis of octopine and nopaline tumor lines demonstrated that a well-defined part of both Ti plasmids is transferred and stably maintained in plant cells. The octopine and nopaline T-region contain a central "core" of about 9 kb homology (Figure 2), often referred to as the "common DNA" (11,23,34). DNA sequence data (M. De Beuckeleer; J. Gielen; M. Lemmers; J. Seurinck; pers. comm.) confirm that this homology is indeed close to 90%. It was postulated that this common core might contain genes essential for oncogenicity. Data on the expression of the T-DNA and the study of the effects of insertion/deletion mutations of the common DNA have led to a confirmation and modification of this hypothesis involving two transforming systems which work cooperatively.

Four nopaline tumor lines have been studies by genomic hybridization (70). The most important conclusion is the colinearity of the T-DNA and the T-region. The internal nopaline T-region *Hin*dIII T-DNA fragment. Moreover, the same T-DNA segment is present in tumors on different plants. Also when T-DNAs occur in tandem copies, the colinearity is maintained to form direct repeats. Only the outermost T-region restriction fragments (*Hin*dIII-10 and *Hin*dIII-23) hybridize to fragments with different molecular weights, indicating that each has been processed to produce respectively left and right T-region borders. Outside this region, no Ti plasmid *Hin*dIII fragments have been found to hybridize to the transformed plant cell DNA.

These data indicated that the nopaline T-DNA is a genetic unit (of roughly 23 kb in size) with defined, or at least preferred, ends. Moreover, insertion of the 15 kb transposon Tn*7* in the T-region results in cotransfer and integration of the transposon with the T-DNA. Thus, the size of the T-DNA does not appear discriminatory (49).

The study of octopine tumor lines shows that the overall transfer and integration pattern of the T-DNA is similar (17,111, 112). All tumor lines contain the TL-DNA (13 kb) that includes the 9 kb "common" T-DNA region. The integration pattern is, however, complicated by the occurrence of a second T-DNA (TR). This TR-region, adjacent to the TL-region in the octopine Ti plasmids, seems to act as an independent T-DNA. The TR-DNA is not present in all tumor lines; moreover, if present, it can occur adjacent to foreign DNA rather than to the TL-DNA (111). In addition, the TR-DNA can be present in a different copy number than the TL-DNA (82,111). Only one octopine tumor line has been shown to contain a TL and TR contiguous T-DNA (84). In some tumor lines, the TL-DNA can occur in

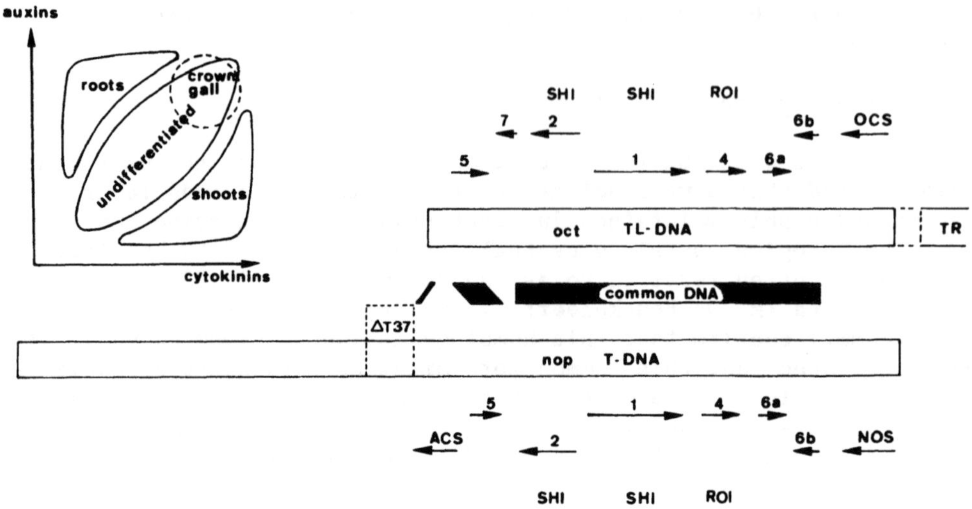

Fig. 2. The octopine and nopaline T-DNA map with their localized transcripts. Their homology (34) is indicated as a bar between both T-DNAs with the gene they specify. The 8 transcripts are mapped on the octopine T-DNA according to Willmitzer et al. (127). Only the 6 nopaline T-DNA transcripts, homologous with the octopine T-DNA are indicated; the other nopaline T-DNA transcripts will be presented elsewhere (129). Shi (shoot inhibition), Roi (root inhibition), Ocs (octopine synthase), and Nos (nopaline synthase) are fenotypes corresponding to the mapped loci on the T-DNA. The auxin-cytokinin diagram is a schematic representation of how the hormone levels can influence the callus morphology (104). Variations of the auxin-cytokinin ratio cause the callus to give rise to buds, shoots, or roots. A similar control could be exerted by the T-DNA in a direct or indirect way. Inactivation of T-DNA transcripts can cause a different crown gall morphology possibly as an alteration in the auxin-cytokinin ratio.

tandem arrays, as was already observed for the nopaline T-DNA (M. Holsters; R. Villarroel; pers. comm.).

Tumors or roots induced by Agrobacterium rhizogenes also contain defined T-DNA(s) (13,123,128). The same subset of Ri plasmid fragments have been demonstrated in independent tumor lines. By preparing DNA from purified nuclei, chloroplasts, and mitochondria isolated from crown gall tissues, it was demonstrated that the T-DNA was located in the nucleus (12,124).

T-Region and T-DNA Border Sequences

In order to analyze in detail the T-DNA integration pattern, T-DNA/plant junctions have been isolated by molecular cloning from transformed plant cell lines, either by screening with a T-region probe (112,132,139) or by direct selection for a T-DNA-containing selectable marker (52).

Analysis of these clones has shown that the T-DNA is covalently linked to plant DNA. The T-DNA/plant junction fragments hybridize to either highly or moderately repeated, or unique plant DNA sequences, indicating multiple insertion sites in the plant DNA. Furthermore, the DNA sequence of right and left ends of the nopaline and octopine T-region compared with the T-DNA/plant junctions, indicates that the "ends" of the T-region are actively involved in T-DNA integration. Comparison of the nucleotide sequence of the left and right T-region ends of octopine, as well as nopaline, Ti plasmids reveals one apparently common characteristic: direct repeats in a 25 bp box occur at either end of the T-region. The DNA sequence for the nopaline boxes is as follows:

nopaline left TGTGGCAGGATATATTGTGGTGTAAACAA

nopaline right TTTGACAGGATATATTGGCGGGTAAACCT

consensus nopaline --t$^{\text{G_CAGGATATAT}}_{-}$tg--g-^{GTA}aac--

The points of divergence between the T-regions and T-DNAs are indicated by arrows. These 25 base repeat regions both left and right lay in nonconserved heterologous DNA of the octopine as well as the nopaline T-regions (34,103,133,140). An especially striking feature is that two right nopaline T-DNA/plant border clones from different tumors diverge in DNA sequence from the T-region at exactly the same base pair just before the direct 25 bp repeat box (140).

Simpson et al (103) have described the sequence of an octopine left T-DNA border in which the homology with the T-region stops exactly after this box. Moreover, another left octopine T-DNA border, from an independent tumor line, was cloned, which ends at exactly the same base pair (M. Holsters; R. Villarroel, pers. comm.). Therefore, it appears that at least the right nopaline T-region and the left octopine T-region borders must be recognized by a specific mechanism.

The left nopaline T-DNA border seems to be less specific and subject to more rearrangements during or after integration, though involvement of the 25 bp box cannot be excluded, since a left T-DNA/plant junction (132,133) ended in the same repeat box. Zambryski et al. (140) characterized 3 left nopaline T-DNA borders which diverged from the T-region at approximately 100 bp from the repeat box with a variation of only 6 base pairs. These authors called this a "secondary" T-DNA left border, since evidence was obtained for rearrangements prior to stabilization of this left border. Two T-DNA border clones, containing the junction of two T-DNAs in tandem array, show respectively 135 and 22 bp as junctional sequences between both diverging points. The most characteristic feature of these junctional sequences is their construction of direct and indirect repeats of the right and left ends of the T-region. One of those repeats was not included in the stabilized T-DNA and was derived from the left T-region between the box and the secondary border (140), strongly suggesting that those T-DNAs evolved from longer T-DNA or T-region copies. A similar structure has been found in a junctional clone of a tandem array of the octopine TL-DNA (M. Holsters, R. Vliiarroel, and H. De Greve, pers. comm.). Also, T-DNA/plant junctions have been described to contain T-region-derived rearrangements (103). Several hundred base pairs joined to the octopine left T-DNA border consist of scrambled T-DNA sequences; these consist of a series of direct and indirect repeats varying in length between 10 and 228 bp derived from the T-region ends (inside the 25 bp boxes), thus creating extensive regions of dyad symmetry. All the above rearrangements are derived from the last 300 bp internal to the T-region; one exception has been described by Thomashow et al. (111,112) and Simpson et al. (103). They found an inverted repeat of 520 bp which was derived from the middle of the octopine T-DNA, in between the right octopine T-DNA and the plant DNA junction. The previous data provide sufficient arguments to postulate the involvement of the 25 bp repeat box as part of the recognition signals for the end of the T-region.

However, several Ti plasmid deletions, spanning the T-region borders, have been obtained without noticeable effect. Deletion of the right octopine TL-border regions (pGV2208) does not result in reduced tumor induction (69). Also, the left border of the nopaline T-region can be deleted without any noticeable effect on tumor induction frequency (H. Joos and A. Caplan, unpub. results). However, when the right border of the nopaline T-region was deleted, the Ti plasmid mutant was very attenuated (51). This means that either a 25 bp box at one border is sufficient to direct the interaction between Ti plasmid DNA and plant DNA or that recognition of one of the border sequences may reside in DNA sequences distant form the observed border sequences.

These data do not support the idea that the T-DNA behaves like a transposon in which both ends are required; however, the T-DNA has one property in common with bacterial and eukaryotic transposons as it moves as a discrete unit of DNA with the capacity to integrate in nonhomologous DNA. It is unlikely that the T-region ends recombine before integration, as one would expect to find circular permutations of the T-DNA inserts, except if the recombined left and right

T-region ends would show site specificity (like an λ "att" site) for integration. How the T-region is inserted into the nuclear plant genome is still obscure. It is not known whether integration is the result of plant or Ti plasmid-specified functions, but it is likely that both are involved. The fact that plant cells are only conditioned for transformation in the S phase before cell division might indicate the easier transfer of DNA to the nuclear DNA in the absence of the nuclear membrane (9). Moreover, the gradual transformation and temperature sensitivity (7) during the first 24-28 hours after infection might be related to the transfer (amplification) and subsequent stabilization of the T-DNA by integration.

The mechanism responsible for the creation of repeated sequences between the tandem T-DNA arrays or at the T-DNA plant junctions is still unknown. It could be the result of amplification before integration or of recombination between either two T-DNAs or T-DNA and plant DNA accompanied by (repair?) replication at the junctions. A similar integration pattern has been found for adenovirus. Although the DNA termini of those viruses are well conserved, they are lost after integration (about 50 bp); the host adenovirus junctions consist mainly of repeats of the ends of adenovirus DNA over several hundred base pairs (27).

Integration of the T-region in the plant DNA is not always accompanied by T-region end rearrangements, replication in scattered repeats or by tandem array organization; T-DNA/plant junctions have been found without rearrangements of the ends, such as the right nopaline T-DNA borders (140). The manner of integration with or without replication of the ends (or of the whole T-DNA), might reflect the different stages of cell division of competent plant cells.

In vitro transformation of plant protoplasts (74) results in segregation of all T-DNA-linked phenotypes. In some instances, the phytohormone-independent growth is lost after some passages while the opine synthesis remains stable (131). It is not yet clear whether this is related to an alteration of the stabilization of the transferred DNA. Similarly, selection for phytohormone-independent growth of tobacco protoplasts treated with the nopaline strains T37 or C58 resulted in calli that all expressed nopaline synthase. Only 40 to 60% of these calli also expressed agrocinopine synthesis (M. Van Lijsebettens, pers. comm.). Genomic blotting analysis indicated many rearrangements and proved that the lack of agrocinopine synthesis (acs) was due to the absence of the acs gene of the T-region (M. Van Lijsebettens, pers. comm.).

Expression of the T-DNA in Plant Cells

The first positive evidence that the T-DNA is transcribed within the plant cell was reported by Drummond et al. (29), who demonstrated that total RNA from crown gall cells hybridizes to a specific fragment of the Ti plasmid. Gurley et al. (44) found that the right part of the octopine T-DNA is transcribed most actively. Similarly, the nopaline T-DNA in teratoma was found to be transcribed most extensively at the left and right ends (134). A more detailed study was performed by Willmitzer et al. 124) who compared nuclear RNA with the poly(A)$^+$ polysomal RNA fraction. Labeled RNA from both sources was hybridized to Southern blots of the T-region fragments. It was shown that the octopine T-DNA is transcribed over its entire length without major differences between nuclear and polysomal RNA distribution. The opine synthesis genes at the right end are transcribed most actively. The transcription is host RNA polymerase II directed since the T-DNA transcription is inhibited by low concentrations of α-amanitin (0.7 μg/ml) (125).

Eight polyadenylated transcripts on the octopine TL-DNA have been characterized in octopine crown gall tissue by Northern blots (40;127;134; Willmitzer et al., in press). The transcripts were mapped by hybridization with different T-region probes and the 5' → 3' polarity was identified by hybridization to T-DNA-separated single- strand DNA. The octopine T-DNA transcripts are shown in Fig. 2. The transcripts occur in different orientations and concentrations, indicating one promoter site per transcript. Transcripts 3 and 7, being the most abundant, do not hybridize with the nopaline T-region (as expected from T-region homology studies, 34). Transcript 3 codes for octopine synthase, while the function of transcript 7 remains unknown. The other octopine TL transcripts 5, 2, 1, 4, 6a, and 6b are also found in nopaline tumors (Willmitzer et al., in prep.). Transcript 4 (inactivation results in root induction), and 6a and/or 6b are less abundant, while transcripts 1 and 2 (inactivation leads to shoot induction) are present in very low concentrations. Compared to other cellular transcripts, the T-DNA transcripts contribute to no more than 10^{-5} - 10^{-6} of the total poly(A)$^+$ RNA population; all the TL transcripts belong to a very low abundance class. At this time, there is no evidence for a transcriptional regulation of the T-DNA genes. The only differential expression of transcription that has been found is an altered relative abundance of transcripts 6a and 6b in the polysomal RNA from logarithmic versus stationary growing tumor suspension cells (126).

The opine synthase gene is expressed in callus or suspension cells and in differentiated tissues of stem, roots, and leaves, and appears to have a constitutive promoter (21,90). It will be important to know if the transcription of the "onc" genes, i.e., genes involved in suppression of shoot and root differentiation, is altered in crown gall tissues from which shoots and/or roots regenerate.

Translation of T-DNA-derived mRNA

T-DNA-specific transcripts, isolated by selective hybridization, are translationally active in wheat germ extracts as well as in rabbit reticulocyte lysates (76,99). The majority of T-DNA mRNAs are capped as suggested by the suppression of translational activity by 0.5 mM pm^{7}G (100). Four different proteins could be demonstrated above background, one of which corresponds with octopine synthase as the *in vitro* translation product is specifically immunoprecipitated by octopine synthase specific antibodies (85,99,100).

Functional Organization of the T-DNA

To study the detailed functional organization of both the octopine (38) and the nopaline T-region (Inzé et al., in prep.), transposons Tn*3* and Tn*5* were introduced into isolated DNA fragments derived from the T-region and substituted into the Ti plasmid by *in vivo* recombination. Specific deletions, removing one or more T-DNA transcripts, have been constructed for both octopine and nopaline T-DNA (69;55). These specific mutations and deletions can therefore be used to study the functions of some transcripts (Fig. 2). Several mutant Ti plasmids are apparently avirulent on some plants, but on Kalanchoe and tobacco, the mutations mainly change tumor morphology. Other mutations result in the loss of opine synthesis. Some mutations resulting in the inactivation of the mapped transcripts have no obvious phenotype.

Remarkably, all T-DNA functions shown to affect the tumor phenotype (and also host range) were located in the "common" T-DNA segment. Instead of undifferentiated octopine tumors, the mutant octopine tumors show root or shoot proliferation from the callus. Six different well-defined transcripts are derived from this common region of octopine and nopaline T-DNAs (i.e., 5, 2, 1, 4, 6a, and 6b). Inactivation of transcript 4 results in root formation from the tumor on *Kalanchoe* stems, but it renders the plasmid avirulent on *Kalanchoe* leaves. The product of this gene must therefore prevent root formation (*roi*, see Fig. 2) and can therefore be compared to the effect obtained with normal plant tissue by increasing cytokinin concentrations or lowering the auxin content (104).

The two transcripts 1 and 2 (*shi*, see Fig. 2) prevent shoot formation by both normal and transformed plant cells. The effect of those genes might be analogous to raising auxin or lowering cytokinin levels. Transcript 5 might inhibit the organization of transformed cells in teratoma (69). Transcript 6a and 6b mainly affect the tumor size (38).

Agropine-type tumors do not contain the homology to transcript 6a and 6b, but based on the presence of homologous DNA (30), should have the transcripts 5, 2, 1, and 4 in common with the octopine and nopaline tumors. The rhizogenes root cultures contain only homology between the gene for transcript 1 and the octopine TL-DNA (128). Octopine-type II plasmids have very little homology with these common genes, although homologous DNA was detected at low stringency in the region of transcript 1 (122).

It has been inferred that none of the T-DNA transcripts are essential for the transfer of T-DNA segments into the plant genome. Opines are synthesized at the wounds infected with any of the Ti plasmid mutants tested, including those unable to induce a tumorous phenotype (69; H. Joos, P. Zambryski, and J.-P. Hernalsteens, pers. comm.). This indicates that all the T-region mutants can still transfer the T-DNA genes. No single gene is absolutely necessary for tumor formation, perhaps because the common genes work cooperatively and synergistically. This may be illustrated best by the observation that mutants containing deletions of the genes for either transcripts 5, 7, 2, and 1 or 4, are able to induce tumors while deletions of all transcripts are not (69). The T-DNA may play a role in the production of phytohormones by the transformed plant cells, either directly, i.e., through the synthesis of auxins and/or cytokinins, or indirectly, i.e., through interaction with the plant biosynthetic pathways. There is evidence that phytohormones can determine some tumor phenotypes. Amasino and Miller (1) determined the endogenous levels of auxins and cytokinins in teratoma versus unorganized growths and found them to be similar to concentrations necessary to induce the same growth patterns in callus of normal plant cells. Skoog and Miller (104) described the development of roots or shoots from normal callus as the relative ratio of auxin to cytokinin as changed in the culture medium. T-DNA mutations resulting in shoots are analogous to decreases in auxin content of or to an increase in cytokinin content. Conversely, root-forming calli could have a higher auxin or a lower cytokinin content compared to undifferentiated tumors. Relatively high concentrations of both phytohormones could be sufficient to explain the suppression of morphogenesis in proliferating tumor cells (88).

Root-forming tumors, brought about by inactivation of transcript 4 in both octopine and nopaline Ti plasmids no longer produce roots when complemented by the external addition of cytokinins. This could be explained if the cytokinin content was lower in these tumors as compared to wild type tumors; moreover, extra cytokinins stimulate cell growth in tissue culture (75; H. Joos, D. Inzé, personal communications). Those mutant Ti plasmids resemble in many ways the Ri plasmids, which either induce unorganized tumors or proliferating roots, depending on the host plant. Hairy-root tissue contains agropine (13,128), indicating at stable transformation of the differentiated tissue.

Notwithstanding the remarkable correlations between the effects of genes 1 and 2 and the similar effects of exogenous auxins on the one hand, and of gene 4 and the similar effect of exogenous cytokinins on the other, it should be pointed out that there is as yet no formal proof that these genes are directly involved in the formation of these plant growth factors.

Opine Synthases

The transcripts of two T-DNA-linked genes, octopine synthase (ocs) and nopaline synthase (nos) have been studied extensively by RNA mapping and DNA sequence analysis. Both genes have been mapped by deletion and insertion mutations (20,37,38,49,51,89) and are located adjacent to the right of the common T-DNA core. Although both genes code for enzymes with very similar functions, they are nevertheless very different and must have a distinct genetic origin. Indeed both DNA/DNA hybridization and DNA sequence comparisons (22,26) demonstrate that these genes are not related. The opine synthases therefore illustrate a remarkable case of biological convergence.

In octopine tumors, a single monomeric enzyme octopine synthase catalyses the reductive condensation of pyruvate and arginine, lysine, histidine, or ornithine to yield, respectively, octopine, lysopine, histopine, or octopinic acid (46,90). The molecular weight of octopine synthase polypeptide is 38,000 daltons as determined by gel filtration and 39,000 as determined by SDS gel electrophoresis (47). This is in close agreement with the molecular weight of 38,700 (358 AA) derived from the DNA sequence (22). The ocs transcript has been characterized by S1 mapping and contains 1,294 nucleotides of which 26 nucleotides from the 5' leader sequence and 194 nucleotides constitute the 3' trailer sequence. The length of the mapped transcript correlates well with the estimated length of 1,400-1,500 nucleotides by Northern blot analysis (40,127), indicating that the poly(A) tail may range between 100 and 200 adenosine residues. Some of the control signals for transcription and translation are similar to those found in sequenced animal genes. First of all, no Pribnow box at -10, but a "Goldberg-Hogness" box 5'-TATTTAAA-3' is found 32 nucleotides upstream from the 5' mRNA start. Translation is initiated at the first AUG codon encountered after the 5' end mRNA; there is no Shine Dalgarno sequence for this gene (22).

Nopaline synthase exists as a homotetramer and catalyzes the condensation between α-ketoglutaric acid and arginine or ornithine (57). The genetically mapped T-DNA segment (49,51) corresponding with the nos gene codes for a transcript of 1,600 bp as determined by Northern analysis (Willmitzer et al., in press); S1 mapping indicates that the transcript has approximately 1,440 nucleotides as a coding sequence (26). Translation of this transcript reveals a polypeptide of 413 amino acids in an open reading frame from an initial AUG codon. Both octopine and nopaline synthase do not contain intervening sequences. The expression signals are similar to other characterized plant genes (Messing, this volume). Notwithstanding the fact that the transcription of both the octopine and nopaline synthase genes start near (± 340 bp) the junction with plant DNA, it could be shown experimentally that no more than ± 260 bp are required for expression, thus demonstrating that the whole of the opine synthase promoter consists of T-region sequences (C. Koncz and H. De Greve, pers. comm.).

As detailed knowledge of these opine genes is available, and the genes are constitutively expressed (21,90), the promoter regions can be adapted for the expression of other genes. The nos promoter region can give prokaryotic transcription activity (L. Herrera-Estrella, pers. comm.). It is not yet clear whether identical signals are used both in Agrobacterium and in the plant cell or whether the nos promoter region contains overlapping eukaryotic and prokaryotic signals. Also, it is not known whether this prokaryotic nos promoter has any biological function(s). It can not be excluded that some of the Agrobacterium T-region transcripts (39; A. Janssens, pers. comm.) might be involved, providing a selective advantage to the Ti plasmid. Moreover, the fact that both the genes coding for the synthesis of specific opines (active in plants after the transfer of the T-DNA) and the genes coding for the catabolism of the same opines by agrobacteria, are located on the same Ti plasmids but are not closely physically linked, reinforces the notion that these genes are reciprocally dependent on one another. Furthermore, the fact that the nos promoter would appear to allow both prokaryotic and eukaryotic (plant cell) expression distinguishes it as a versatile genetic signal.

Suppression of and Reversal from the Tumorous State

As described above, the T-DNA-containing tumor lines have different morphogenic phenotypes, either producing defective shoots (teratomas) or roots (hairy-root) or undifferentiated tumors, depending on the type of inciting Ti plasmid and host plant cell. Braun and Wood (8) were first to describe suppression of the tumorous phenotype by grafting normal looking shoots derived from a cloned BT37 nopaline tumor line, on host plants. The shoot-derived tissues synthesize nopaline and contain the complete T-DNA (though some alterations were later observed by genomic hybridization; 70,135). Moreover, those grafts were not susceptible to superinfection and reverted to the tumorous phenotype by plant wounding and cultivation on hormone-free medium (130). The reversible state of the morphogenetic controls suggests that the expression of T-DNA functions can be controlled either at the gene transcription level or epigenetically controlled (77,78).

De Greve et al. (21) have made an extensive study on regeneration of plants from transformed plant cells. A Tn7 insertion mutant in transcript 2 of the octopine T-region (pGV2100) was isolated that induces tumors only after prolonged incubation. These tumors produce shoots from the greenish tissue (69). Analysis of the shoots from tumors induced on *Nicotiana tabacum* cv Wisconsin 38 shows that most shoots do not contain octopine, but some independent tumor lines give a few shoots that were positive (1/250). One such shoot (rGV1) developed into a fully normal plant (21,90). Leaves, stem, and roots contain octopine synthase activity demonstrating the expression of this T-DNA-linked gene in all differentiated tissues. Moreover, seeds obtained by self-fertilization produce plants to which 75% contain octopine. T-DNA-linked genes go through meiosis as a Mendelian genetic trait (90). It was found by genomic hybridizations that the shoot retained only the T-DNA segment corresponding to transcript, i.e., octopine synthase, while the tumors - at least the majority of the cells - contained the whole T-DNA stretch with the inserted Tn*7* (21). It still has to be demonstrated whether those T-DNA deletions occur during the induction phase or subsequent to T-DNA integration in the plant genome. Either case may have implications for plant genetic engineering. If the deletion occurs during the induction phase of transformation, it means that the octopine synthase gene can be transferred by itself and be stably incorporated into the plant genome. Normally, this general gene transformation capacity would not be observed because of the absence of a selectable marker and because of the rapid growth of cells transformed by the wild-type T-DNA. It should be noted that the tumor induction *in vivo* or the phytohormone-independent growth *in vitro* is as yet the only available selectable marker, provided by the T-DNA, to characterize transformed cells. If the deletion can occur int he plant cell at this frequency, it might suggest that the T-DNA can be a hot spot for deletion formation in its neighborhood.

Similarly, shoot formation was induced in a BT37 line by a kinetin treatment (42). Some of the resulting shoots formed roots and genomic blots demonstrated that some of these plants retained 4 T-DNA fragments. These fragments corresponded to the left and right ends of the nopaline T-DNA, while most of the internal T-DNA with the common DNA was lost (136,137). It will be most interesting to determine the molecular processes which result in deletions and losses of most (and probably also all) of the T-DNA.

A second type of regenerate from transformed plant cells has been obtained by Wöstemayer et al. (130a). Shoots obtained from teratoma tumors, lacking the T-DNA genes 5, 2, and 1 were repeatedly grafted, resulting in a "normalization" and finally in fertile flowers. After selfing of the fertile grafts, two types of germinating plantlets were obtained: normally rooting plants without octopine synthesizing activity, and rootless plantlets with high levels of octopine synthase. When these rootless plantlets were cultured *in vitro* on hormone-free medium, they formed callus at their base and developed in teratoma-like tissue. These observations indicated that the octopine synthase activity and the root inhibition and growth on hormone-free medium were 100% linked and that also the "onc" gene 4 can be sexually inherited in a Mendelian fashion. Whether these regenerates originated as the result of transcription regulation or any other epigenetic effect is not known.

In conclusion, the neoplastic condition can be reversibly suppressed with regard to the capacity of differentiation, or can revert by loss of all or part of the T-DNA. Revertant plants still

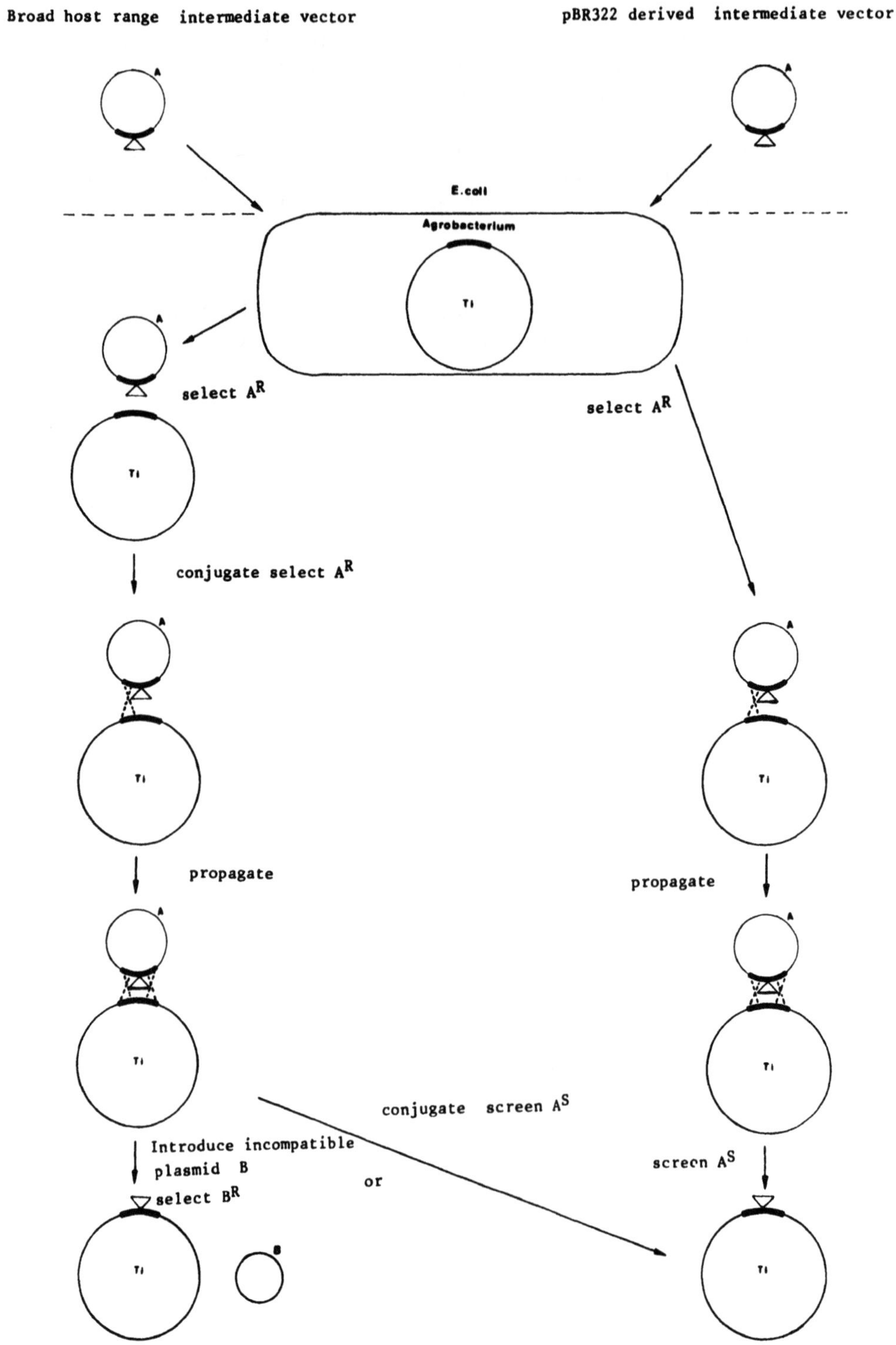
Broad host range intermediate vector
pBR322 derived intermediate vector
A
A
E.coli
Agrobacterium
Ti
A
select A^R
Ti
select A^R
conjugate select A^R
A
A
Ti
Ti
propagate
propagate
A
A
Ti
Ti
conjugate screen A^S
Introduce incompatible
plasmid B
select B^R
or
screen A^S
Ti
B
Ti

containing opine genes are normal in every respect, demonstrating the possibility to transfer genes to plants.

Ti Plasmids as Gene Vectors

Because of their natural ability to insert DNA into plant cells, the Ti plasmids have become the subject of extensive study for their use in genetic engineering of plants.

The study of signals involved and required for the transfer and stabilization of the T-DNA reveals how the Ti plasmid can be used as a vector for insertion of foreign genes into the plant genome. First, the transposon *Tn7* inserted in the T-region is cotransferred to and maintained in the plant cell without major rearrangements (49,52). The cotransfer of insertions at different positions in both the octopine and the nopaline T-regions demonstrates the involvement of "ends" for the stabilization of the T-region (68,75). Therefore, one can expect that any segment between those "ends" will be co-transferred and stabilized in the plant nucleus. Moreover, the study of the functional organization of the T-DNA reveals that none of the genes involved in oncogenesis or suppression of differentiation are required for T-DNA transfer. However, the "ends", presumably including the DNA sequences around the 25 bp repeat box, have to occur in *cis* of the Ti plasmid. A plasmid containing the T-region in *trans* of the Ti plasmid can not induce tumors (C. Koncz, pers. comm.). Therefore, genes have to be engineered into the T-region of the Ti plasmids.

Fig. 3. Schematic representation of the two different methods used to shuttle genes into the Ti plasmid. The intermediate vector contains homology (indicated as a bar) with the Ti plasmid and can be either a broad host range replicon, a P-type plasmid derivative, or a narrow-range replicon (pBR322-derivatives). The engineered segment is represented by a triangle. In the first method, the vector is mobilized to *Agrobacterium*, where it stably replicates. A cointegrate can be selected for by a second conjugation. During propagation of those cointegrates, a second cross-over can occur, resulting in an exchanged segment. This event can be selected for by incompatibility curing or can be screened for after a third conjugation. The second method selects immediately for the first cross-over event since pBR-derivatives cannot replicate in *Agrobacterium*. After propagation, the second cross-over can then be screened for by the loss of the vector marker (see text, "Ti Plasmids as Gene Vectors".)

Due to the large size of the Ti plasmids and the requirement of many genes for virulence, direct cloning into the Ti plasmid is impossible. Therefore, a technique to shuttle genes into the Ti plasmid via an intermediate vector has been worked out (68). The intermediate vector contains a region of homology with the T-region of the Ti plasmid into which genes can be inserted (95). The engineered segment of the intermediate vector can be transferred to the T-region of the Ti plasmid by homologous recombination. Two different types of intermediate vectors have been worked out. One is derived from a broad-host range vector (W- and P-type plasmids), capable of replicating stably in Agrobacterium. Recombinants between vector and Ti plasmid can be isolated by selecting for cotransfer of the cointegrate plasmid during conjugation (68). This method requires conjugation in order to select for the first recombination, and subsequent screening for the second recombination or immediate selection by incompatibility curing for the substituted recombinants (38,75). The second type of intermediate vector is a pBR322-derivative, which can be mobilized efficiently by F- or I-type plasmids when the mob-protein is complemented in trans (by a compatible ColE plasmid) (118). Since the pBR322-derivatives can not replicate in Agrobacterium, they are maintained only through homologous cointegration with the Ti plasmid. The second recombinant (or substituted T-region) is obtained after propagation of the cointegrates by screening for loss of the pBR322 vector markers (Van Haute et al., in prep.). The three possibilities are outline in Figure 3.

As the final goal in the breeding of normally engineered plants, the gene transfer by the T-DNA has to be separated from its tumorigenic activities. This has been shown to be possible, since none of the T-DNA genes are essential for T-region transfer. Regenerates are fertile and transmit T-DNA-linked genes in a Mendelian way to the progeny (90). The T-DNA can also be a source of expression signals for plant genes. Currently, T-DNA expression signals are being adapted for the insertion of foreign genes in an intermediate expression vector. The plant host range is limited to those plants on which tumor formation can take place and comprises most of the dicotyledonous plants. However, DNA transformation sensu largo, without the concomitant induction of tumors, would not be recognized. Therefore, the possibility still exists that Agrobacterium can transmit DNA even to monocotyledonous plants. In conclusion, our knowledge about the Ti plasmids allows us to transfer any DNA sequence into plant cells, from which normal plants can be regenerated.

ACKNOWLEDGEMENTS

We acknowledge A. Caplan, L. Herrera-Estrella, P. Tenning, and J.-P. Hernalsteens for helpful discussions and critical reading of the manuscript. We thank Ms. M. De Cock for typing the manuscript, and Mr. A. Verstraete and Ms. C. Morita for drawing the figures.

REFERENCES

1. Amasino, R.M., and C.O. Miller (1982) Hormonal control of tobacco crown gall tumor morphology. Plant Physiol. 69:389-392.
2. Beiderbeck, R. (1973) Wurzelinduktion an Blättern von Kalanchoë daigremontiana durch Agrobacterium rhizogenes und der Einfluss von Kinetin auf diesen Prozess. Z. Pflanzenphysiol. 68:460-467.
3. Bomhoff, G., P.M. Klapwijk, H.C.M. Kester, R.A. Schilperoort, J.P. Hernalsteens, and J. Schell (1976) Octopine and nopaline synthesis and breakdown genetically controlled by a plasmid of Agrobacterium tumefaciens. Mol. Gen. Genet. 145:177-181.
4. Bopp, M., and F. Resende (1966) Crown gall tumoren bei verschiedenen arten und bastarden der Kalanchoidae. Portugaliae Acta Biologica 9:327-366.
5. Braun, A.C. (1947) Thermal studies on the factors responsible for tumor initiation in crown gall. Amer. J. Bot. 34:234-240.
6. Braun, A.C. (1956) The activation of a growth-substance system accompanying the conversion of normal to tumor cells in crown gall. Cancer Res. 16:53-56.
7. Braun, A.C. (1958) A physiological basis for autonomous growth of the crown gall tumor cell. Proc. Natl. Acad. Sci. USA 44:344-349.
8. Braun, A.C., and H.N. Wood (1976) Suppression of the neoplastic state with the acquisition of specialized functions of cells, tissues, and organs of crown gall teratomas of tobacco. Proc. Natl. Acad. Sci. USA 73:496-500.

8a. Braun, A.C. (1978) Plant tumours. Biochem. Biophys. Acta 516:167-191.

9. Capecchi, M.R. (1980) High efficiency transformation by direct microinjection of DNA into cultured mammalian cells. Cell 22:479-488.
10. Chilton, M.-D., H.J. Drummond, D.J. Merlo, D. Sciaky, A.L. Montoya, M.P. Gordon, and E.W. Nester (1977) Stable incorporation of plasmid DNA into higher plant cells: The molecular basis of crown gall tumorigenesis. Cell 11:263-271.
11. Chilton, M.D., M.H. Drummond, D.J. Merlo, and D. Sciaky (1978) Highly conserved DNA of Ti-plasmids overlaps T-DNA, maintained in plant tumors. Nature 275:147-149.
12. Chilton, M.-D., R.K. Saiki, N. Yadav, M.P. Gordon, and F. Quetier (1980) T-DNA from Agrobacterium Ti plasmid is in the nuclear DNA fraction of crown gall tumor cells. Proc. Natl. Acad. Sci. USA 77:4060-4064.
13. Chilton, M.-D., D.A. Tepfer, A. Petit, C. David, F. Casse-Delbart, and J. Tempé (1982) Agrobacterium rhizogenes inserts T-DNA into the genomes of the host plant root cells. Nature, London 295:432-434.
14. Costantino, P., P.J.J. Hooykaas, H. den Dulk-Ras, and R.A. Schilperoort (1980) Tumor formation and rhizogenicity of Agrobacterium rhizogenes carrying Ti-plasmids. Gene 11:79-87.
15. Costantino, P., M.L. Mauro, G. Micheli, G. Risuelo, P.J.J.

Hooykaas, and R.A. Schilperoort (1981) Fingerprinting and sequence homology of plasmids from different virulent strains of Agrobacterium rhizogenes. Plasmid 5:170-182.

16. Currier, T.C., and E.W. Nester (1976) Evidence for diverse types of large plasmids in tumor inducing strains of Agrobacterium. J. Bacteriol. 126:157-165.
17. De Beuckeleer, M., M. Lemmers, G. De Vos, L. Willmitzer, M. Van Montagu, and J. Schell (1981) Further insight on the transferred-DNA of octopine crown gall. Mol. Gen. Genet. 183:283-288.
18. De Cleene, M., and J. De Ley (1976) The host range of crown-gall. Botan. Rev. 42:389-466.
19. De Cleene, M., and J. De Ley (1981) The host range of infectious hairy-root. Bot. Rev. 47:147-194.
20. De Greve, H., H. Decraemer, J. Seurinck, M. Van Montagu, and J. Schell (1981) The functional organization of the octopine Agrobacterium tumefaciens plasmid pTiB6S3. Plasmid 6:235-248.
21. De Greve, H., J. Leemans, J.P. Hernalsteens, L. Thia-Toong, M. De Beuckeleer, L. Willmitzer, L. Otten, M. Van Montagu, and J. Schell (1982a) Regeneration of normal and fertile plants that express octopine synthase, from tobacco crown galls after deletion of tumour-controlling functions. Nature, London (in press).
22. De Greve, H., P. Dhaese, J. Seurinck, M. Van Montagu, and J. Schell (1982b) Nucleotide sequence and transcript map of the Agrobacterium tumefaciens Ti plasmid-encoded octopine synthase gene. J. Mol. Appl. Genet. 1:499-512.
23. Depicker, A., M. Van Montagu, J. Schell (1978) Homologous DNA sequences in different Ti-plasmids are essential for oncogenicity. Nature, London 275:150-153.
24. Depicker, A., M. De Wilde, G. De Vos, R. De Vos, M. Van Montague, and J. Schell (1980a) Molecular cloning of overlapping segments of the nopaline Ti-plasmid pTiC58 as a means to restriction endonuclease mapping. Plasmid 3:193-211.
25. Depicker, A., M. De Block, D. Inzé, M. Van Montagu, and J. Schell (1980b) Is-like element IS8 in RP4 plasmid and its involvement in cointegration. Gene 10:329-338.
26. Depicker, A., S. Stachel, P. Dhaese, P. Zambryski, and H.M. Goodman (1982) Nopaline synthase: Transcript mapping and DNA sequence. J. Mol. Appl. Genet. 1:561-574.
27. Deuring, R., U. Winterhoff, F. Tamanoi, S. Stabel, and W. Doerfler (1981) Site of linkage between adenovirus type 12 and cell DNAs in hamster tumour line CLAC3. Nature, London 293:81-84.
28. De Vos, G., M. De Beuckeleer, M. Van Montagu, and J. Schell (1981) Restriction endonuclease mapping of the octopine tumor inducing pTiAch5 of Agrobacterium tumefaciens. Plasmid 6:249-253.
29. Drummond, M.H., M.P. Gordon, E.W. Nester, and M.-D. Chilton (1977) Foreign DNA of bacterial plasmid origin is transcribed

in crown gall tumours. Nature, London 269:535-536.
30. Drummond, M.H., and M.-D. Chilton (1978) Tumor-inducing (Ti) plasmids of Agrobacterium share extensive regions of DNA homology. J. Bacteriol. 136:1178-1183.
31. Ellis, J.G., and P.J. Murphy (1981) Four new opines from crown gall tumours - Their detection and properties. Mol. Gen. Genet. 181:36-43.
32. Ellis, J.G., A. Kerr, J. Tempé, and A. Petit (1979) Arginine catabolism: A New function of both octopine and nopaline Ti-plasmids of Agrobacterium. Mol. Gen. Genet. 173:263-269.
33. Engler, G., M. Holsters, M. Van Montagu, J. Schell, J.P. Hernalsteens, and R.A. Schilperoort (1975) Agrocin 84 sensitivity: A plasmid determined property in Agrobacterium tumefaciens. Mol. Gen. Genet. 138:345-349.
34. Engler, G., A. Depicker, R. Maenhaut, R. Villarroel-Mandiola, M. Van Montagu, and J. Schell (1981) Physical mapping of DNA base sequence homologies between an octopine and a nopaline Ti-plasmid of Agribacterium tumefaciens. J. Mol. Biol. 152:183-208.
35. Engler, G. R. Villarroel-Mandiola, M. De Vos, M.Van Montagu, and J. Schell. Electron microscope heteroduplex analysis of base sequence relations among plasmids of Agrobacterium tumefaciens. (In preparation).
36. Firmin, J.L. and G.R. Fenwick (1978) Agropine - A major new plasmid-determined metabolite in crown gall tumours. Nature, London 276:842-844.
37. Garfinkel, D.J., and E.W. Nester (1980) Agrobacterium tumefaciens mutants affected in crown gall tumorigenesis and octopine catabolism J. Bacteriol. 144:732-743.
38. Garfinkel, D.J., R.B. Simpson, L.W. Ream, F.F. White, M.P. Gordon, and E.W. Nester (1981) Genetic analysis of crown gall: Fine structure map of the T-DNA by site-directed mutagenesis. Cell 27:143-153.
39. Gelvin, S.B., M.P. Gordon, E.W. Nester, and A.I. Aronson (1981) Transcription of Agrobacterium Ti plasmid in the bacterium and in crown gall tumors. Plasmid 6:17-29.
40. Gelvin, S.B., M.F. Thomashow, J.C. McPherson, M.P. Gordon, and E.W. Nester (1982) Sizes and map positions of several plasmid-DNA-encoded transcripts in octopine-type crown gall tumors. Proc. Natl. Acad. Sci. USA 79:76-80.
41. Genetello, Ch., N. Van Larebeke, M. Holsters, A. Depicker, M. Van Montagu, and J. Schell (1977) Ti-plasmids of Agrobacterium as conjugative plasmids. Nature, London 265:561-563.
42. Gordon, M.P. (1982) Reversal of crown gall tumors. In Molecular Biology of Plant Tumors, G. Kahl and J. Schell, eds. Academic Press, New York, pp. 415-426.
43. Gresshoff, P.M., M.L. Skotnicki, and B.G. Rolfe (1979) Crown gall tumour formation is plasmid and plant controlled. J. Bacteriol. 137:1020-1021.
44. Gurley, W.B., J.D. Kemp, M.J. Albert, D.W. Sutton, and J.

Callis (1979) Transcription of Ti plasmid-derived sequences in three octopine-type crown gall tumor lines. Proc. Natl. Acad. Sci. USA 76:2828-2832.

45. Guyon, P., M.-D. Chilton, A. Petit, and J. Tempé (1980) Agropine in "null type" crown gall tumors: Evidence for the generality of the opine concept. Proc. Natl. Acad. Sci. USA 77:2693-2697.

46. Hack, E., and J.D. Kemp (1977) Comparison of octopine, histopine, lysopine, and octopine acid synthesizing activities in Sunflower crown gall tissues. Biochem. Biophys. Res. Commun. 78:785-791.

47. Hack, E., and J.D. Kemp (1980) Purification and characterization of the crown gall-specific enzyme, octopine synthase. Plant Physiol. 65:949-955.

48. Hernalsteens, J.P., H. De Greve, M. Van Montagu, and J. Schell (1978) Mutagenesis by insertion of the drug resistance transposon Tn7 applied to the Ti plasmid of Agrobacterium tumefaciens. Plasmid 1:218-225.

49. Hernalsteens, J.P., F. Van Vliet, M.De Beuckeleer, A. Depicker, G. Engler, M. Lemmers, M. Holsters, M. Van Montagu, and J. Schell (1980) The Agrobacterium tumefaciens Ti plasmid as a host vector system for introducing foreign DNA in plant Cells. Nature, London 287:654-656.

50. Holsters, M., A. Silva, C. Genetello, G. Engler, F. Van Vliet, M. De Block, R. Villarroel, M. Van Montagu, and J. Schell (1978) Spontaneous formation of cointegrates of the oncogenic Ti-plasmid and the wide-host-range P-plasmid RP4. Plasmid 1:456-467.

51. Holsters, M. B. Silva, F. Van Vliet, C. Genetello, M. De Block, P. Dhaese, A. Depicker, D. Inzé, G. Engler, R. Villarroel, M. Van Montagu, and J. Schell (1980) The functional organization of the nopaline A. tumefaciens plasmid pTiC58. Plasmid 3:212-230.

52. Holsters, M. R. Villarroel, M. Van Montagu, and J. Schell (1982) The use of selectable markers for the isolation of plant-DNA/T-DNA junction fragments in a cosmid vector. Mol. Gen. Genet. 185:283-289.

53. Hooykaas, P.J.J., P.M. Klapwijk, M.P. Nuti, R.A. Schilperoort, and A. Rörsch (1977) Transfer of the Agrobacterium tumefaciens Ti-plasmid to avirulent Agrobacteria and to Rhizobium ex planta. J. Gen. Microbiol. 98:477-484.

54. Hooykaas, P.J.J., G. Ooms, and R.A. Schilperoort (1982) Tumors induced by different strains of Agrobacterium tumefaciens. In Molecular Biology of Plant Tumors, G. Kahl and J. Schell, eds. Academic Press, New York, pp. 374-388.

55. Joos, H., D. Inzé, A. Caplan, M. Sormann, M. Van Montagu, and J. Schell. Genetic analysis of T-DNA transcripts in nopaline crown galls. Cell (in press).

56. Keane, P.J., A. Kerr, and P.B. New (1970) Crown gall of stone fruit. II. Identification and nomenclature of Agrobacterium isolates. Austr. J. Biol. Sci. 23:585-595.

57. Kemp, J.D., D.W. Sutton, and E. Hack (1979) Purification and characterization of the crown gall specific enzyme nopaline synthase. Biochem. 18:3755-3760.
58. Kerr, A. (1969) Transfer of virulence between isolates of Agrobacterium. Nature, London 223:1175-1176.
59. Kerr, A. (1971) Acquisition of virulence by non-pathogenic isolation of Agrobacterium radiobacter. Physiol. Plant Pathol. 1:241-246.
60. Kerr, A., P. Manigault, and J. Tempé (1977) Transfer of virulence in vivo and in vitro in Agrobacterium. Nature, London 265:560-561.
61. Kerr, A., and C.G. Panagopoulos (1977) Biotypes of Agrobacterium radiobacter var. tumefaciens and their biological control. Phytopath. Z. 90:172-179.
62. Kersters, K. J. De Ley, P.H.A. Sneath, and M. Sackin (1973) Numerical taxonomic analysis of Agrobacterium. J. Gen. Microbiol. 78:227-239.
63. Kersters, K., and J. De Ley (1982) Family Rhizobiaceae - Genus Agrobacterium. In Bergey's Manual of Determinative Bacteriology, 9th ed. Wilkins and Wilkins, Baltimore (in press).
64. Klapwijk, P.M., T. Scheulderman, and R.A. Schilperoort (1978) Coordinated regulation of octopine degradation and conjugative transfer of Ti-plasmids in Agribacterium tumefaciens: Evidence for a common regulatory gene and separate operons. J. Bacteriol. 136:775-785.
65. Klee, H.J., M.P. Gordon, and E. W. Nester (1982) Complementation analysis of Agrobacterium tumefaciens Ti plasmid mutations affecting oncogenicity. J. Bacteriol. 150:327-331.
66. Koekman, B.P., G. Ooms, P.M. Klapwijk, and R.A. Schilperoort (1979) Genetic map of an octopine Ti-plasmid. Plasmid 2:347-357.
67. Koekman, B.P., P.J.J. Hooykaas, and R.A. Schilperoort (1980) Localization of the replication control region of the physical map of the octopine Ti plasmid. Plasmid 4:184-195.
68. Leemans, J., C. Shaw, R. Deblaere, H. De Greve, J.P. Hernalsteens, M. Maes, M. Van Montagu, and J. Schell (1981) Site-specific mutagenesis of Agrobacterium Ti plasmids and transfer of genes to plant cells. J. Mol. Appl. Genet. 1:149-164.
69. Leemans, J., R. Deblaere, L. Willmitzer, H. De Greve, J.P. Hernalsteens, M. Van Montagu, and J. Schell (1982) Genetic identification of functions of TL-DNA transcripts in octopine crown galls. EMBO Journal 1:147-152.
70. Lemmers, M., M. De Beuckeleer, M. Holsters, P. Zambryski, A. Depicker, J.P. Hernalsteens, M. Van Montagu, and J. Schell (1980) Internal organization, boundaries and integration of Ti-plasmid DNA in nopaline crown gall tumours. J. Mol. Biol. 144:353-376.
71. Lipetz, J. (1966) Crown gall tumorigenesis. II. Relation between wound healing and the tumorigenic response. Cancer Res. 26:1597-1604.

72. Liu, S.-T., K.L. Perry, C.L. Schardl, and C.I. Kado (1982) Agrobacterium Ti plasmid indoleacetic acid gene is required for crown gall oncogenesis. Proc. Natl. Acad. Sci. USA 79:2812-2816.
73. Loper, J.E., and C.I. Kado (1979) Host-range conferred by the virulence-specifying plasmid of Agrobacterium tumefaciens. J. Bacteriol. 139:591-596.
74. Marton, L., G.J. Wullems, L. Molendijk, and R.A. Schilperoort (1979) In vitro transformation of cultured cells from Nicotiana tabacum by Agrobacterium tumefaciens. Nature, London 277:129-130.
75. Matzke, A.J.M., and M.-D. Chilton (1981) Site-specific insertion of genes into T-DNA of the Agrobacterium tumor-inducing plasmid: an approach to genetic engineering of higher plant cells. J. Mol. Appl. Genet. 1:39-49.
76. McPherson, J.C., E.W. Nester, and M.P. Gordon (1980) Proteins encoded by Agrobacterium tumefaciens Ti plasmid DNA (T-DNA) in crown gall tumors. Proc. Natl. Acad. Sci. USA 77:2666-2670.
77. Meins, F. Jr., and A. Binns (1977) Epigenetic variation of cultured somatic cells: evidence for gradual changes in the requirement for factors promoting cell division. Proc. Natl. Acad. Sci. USA 74:2928-2932.
78. Meins, F., Jr. (1982) Habituation of cultured plant cells. In: Molecular Biology of Plant Tumors, G. Kahl and J. Schell, eds. Academic Press, New York, pp. 3-31.
79. Ménagé, A., et G. Morel (1964) Sur la presence d'octopine dans les tissus de crown gall. C. R. Acad. Sci., Paris 259:4795-4796.
80. Ménagé, A., et Morel, G. (1965) Sur la présence d'un acide aminé nouveau dans le tissu de crown-gall. C. R. Soc. Biol., Paris 159:561-562.
81. Merlo, D.J., and E.W. Nester (1977) Plasmids in avirulent strains of Agrobacterium. J. Bacteriol. 129:76-80.
82. Merlo, D.J., R.C. Nutter, A.L. Montoya, D.J. Garfinkel, M.H. Drummond, M.-D. Chilton, M.P. Gordon, and E.W. Nester (1980) The boundaries and copy numbers of Ti plasmid T-DNA vary in crown gall tumors. Mol. Gen. Genet. 177:637-643.
83. Moore, L., G. Warren, and G. Strobel (1979) Involvement of a plasmid in the hairy root disease of plants caused by Agrobacterium rhizogenes. Plasmid 2:617-626.
84. Murai, N., and J. Kemp (1982a) T-DNA of pTi-15955 from Agrobacterium tumefaciens is transcribed into a minimum of seven polyadenylated RNAs in a sunflower crown gall tumor. Nucl. Acids Res. 10:1679-1689.
85. Murai, N., and J.D. Kemp (1982b) Octopine synthase mRNA isolated from sunflower crown gall callus is homologous to the Ti plasmid of Agrobacterium tumefaciens. Proc. Natl. Acad. Sci. USA 79:86-90.
86. Murphy, P.J., and W.P. Roberts (1979) A basis for agrocin 84 sensitivity in Agrobacterium. J. Gen. Microbiol. 114:207-213.

87. Ooms, G., P.M. Klapwijk, J.A. Poulis, and R.A. Schilperoort (1980) Characterization of TN904 insertions in octopine Ti plasmid mutants of Agrobacterium tumefaciens. J. Bacteriol. 144:82-91.
88. Ooms, G., P.J. Hooykaas, G. Moleman, and R.A. Schilperoort (1981) Crown gall plant tumors of abnormal morphology, induced by Agrobacterium tumefaciens carrying mutated octopine Ti plasmids; analysis of T-DNA functions. Gene 14:33-50.
89. Ooms, G. P.J.J. Hooykaas, R.J.M. Van Veen, P. Van Beelen, T.J.G. Regensburg-Tuink, and R.A. Schilperoort (1982) Octopine Ti plasmid deletion mutants of Agrobacterium tumefaciens with emphasis on the right side of the T-region. Plasmid 7:15-29.
90. Otten, L., H. De Greve, J.P. Hernalsteens, M. Van Montagu, O. Schieder, J. Straub, and J. Schell (1981) Mendelian transmission of genes introduced in plant by the Ti plasmids of Agrobacterium tumefaciens. Mol. Gen. Genet. 183:209-213.
91. Panagopoulos, G.G., and P.G. Psallidas (1973) Characterisation of Greek isolates of Agrobacterium tumefaciens. Commun. Appl. Bacteriol. 36:233-240.
92. Petit, A., S. Delhaye, J. Tempé, et G. Morel (1970) Recherches sur les guanidines des tissus de crown gall. Mise en évidence d'une relation biochimique spécifique entre les souches d'Agrobacterium et les tumeurs qu'elles induisent. Physiol. vég. 8:205-213.
93. Petit, A., J. Tempé, A. Kerr, M. Holsters, M. Van Montagu, and J. Schell (1978) Substrate induction of conjugative activity of Agrobacterium tumefaciens. Nature, London 271:570-572.
94. Rogler, C.E. (1980) Plasmid-dependent temperature-sensitive phase in crown gall tumorigenesis. Proc. Natl. Acad. Sci. USA 77:2688-2692.
95. Ruvkun, G.B., and F.M. Ausubel (1981) A general method for site-directed mutagenesis in prokaryotes. Nature, London 289:85-88.
96. Schell, J., and M. Van Montagu (1977a) The Ti plasmid of Agrobacterium tumefaciens, a natural vector for the introduction of NIF genes in plants? In Genetic Engineering for nitrogen fixation, A. Hollaender, ed. Plenum Press, New York, pp. 159-179.
97. Schell, J. and M. Van Montagu (1977b) On the transfer, maintenance and expression of bacterial Ti-plasmid DNA in plant cells transformed with A. tumefaciens. Brookhaven Symp. Biol., 29:36-49.
98. Schell, J., M. Van Montagu, M. De Beuckeleer, M. De Block, A. Depicker, M. De Wilde, G. Engler, C. Genetello, J.P. Hernalsteens, M. Holsters, J. Seurinck, B. Silva, F. Van Vliet, and R. Villarroel (1979) Interactions and DNA transfer between Agrobacterium tumefaciens, the Ti-plasmid and the plant host. Proc. R. Soc. Lond. B 204:251-266.
99. Schröder, J., G. Schröder, H. Huisman, R.A. Schilperoort, and J.Schell (1981) The mRNA for lysopine dehydrogenase in plant

tumor cells is complementary to a Ti-plasmid fragment. FEBS Lett. 129:166-168.
100. Schröder, G., and J. Schröder (1982) Hybridization selection and translation of T-DNA encoded mRNAs from octopine tumors. Mol. Gen. Genet. 185:51-55.
101. Sciaky,D., A.L. Montoya, and M.-D. Chilton (1978) Fingerprints of Agrobacterium Ti plasmids. Plasmid 1:238-253.
102. Scott, I.M., J.L. Firmin, D.N. Butcher, L.M. Searle, A.K. Sogeke, J. Eagles, J.F. March, R. Self, and G.R. Fenwick (1979) Analysis of a range of crown gall and normal plant tissues for Ti-plasmid-determined compounds. Mol. Gen. Genet. 176, 57-65.
103. Simpson, R.B., P.J. O'Hara, W. Kwok, A.M. Montoya, C. Lichtenstein, M.P. Gordon, and E.W. Nester (1982) DNA from the AGS/2 crown gall tumor contains scrambled Ti plasmid sequences near its junctions with plant DNA. Cell 29:1005-1014.
104. Skoog, F., and C.O. Miller (1957) Chemical regulation of growth and organ formation in plant tissue cultures in vitro. Symp. Soc. Exp. Biol. 11:118-131.
105. Stonier, T. (1962) Normal, abnormal and pathological regeneration in Nicotiana. In Regeneration, D. Rudnick, ed. Ronald Press, New York, pp. 85-115.
106. Tate, M.E., P.J. Murphy, W.P. Roberts, and A. Kerr (1979) Adenine N^6-substituent of agrocin 84 determines its bacteriocin-like specificity. Nature, London 280:697-699.
107. Tate, M.E., J.G. Ellis, A. Kerr, J. Tempé, K. Murray, and K. Shaw (1982) Agropine: a revised structure. Carbohyd. Res. 104:105-120
108. Tempé, J., P. Guyon, D. Tepfer, and A. Petit (1979) The role of opines in the ecology of the Ti-plasmids of Agrobacterium. In Plasmids of medical, environmental and commercial importance, K. Timmis and A. Püler, eds. Elsevier, Amsterdam 353-363.
109. TempéJ., A. Petit, M. Holsters, M. Van Montagu, and J. Schell (1977) Thermosensitive step associated with transfer of the Ti-plasmid during conjugation: possible relation to transformation in crown gall. Proc. Natl. Acad. Sci. USA 74:2848-2849.
110. Tepfer, D.A., and J. Tempé (1981) Production d'agropine par des racines formées sous l'action d'Agrobacterium rhizogenes, souche A4. C. R. Acad. Sc. Paris 292, Série III, 153-156.
111. Thomashow, M.F., R. Nutter, K. Postle, M.-D. Chilton, F.R. Blattner, A. Powell, M.P. Gordon, and E.W. Nester (1980a) Recombination between higher plant DNA and the Ti plasmid of Agrobacterium tumefaciens. Proc. Natl. Acad. Sci. USA 77:6448-6452.
112. Thomashow, M.F., R. Nutter, A.L. Montoya, M.P. Gordon, and E.W. Nester (1980b) Integration and organisation of Ti-plasmid sequences in crown gall tumors. Cell 19:729-739.
113. Thomashow, M.F., C.G. Panagopoulos, M.P. Gordon, and E.W. Nester (1980c) Host range of Agrobacterium tumefaciens is determined by the Ti plasmid. Nature, London 283:794-796.
114. Thomashow, M.F., V.C. Knauf, and E.W. Nester (1981) Relation-

ship between the limited and wide host range octopine-type Ti plasmids of Agrobacterium tumefaciens. J. Bacteriol. 146:484-493.

115. Van Haute, E., H. Joos, M. Van Montagu, and J. Schell. Broad-host range exchange recombination of pBR322-cloned DNA fragments - application to the reversed genetics of Ti plasmids of Agrobacterium tumefaciens. (In preparation).

116. Van Larebeke, N., G. Engler, M. Holsters, S. Van den Elsacker, I. Zaenen, R.A. Schilperoort, and J. Schell (1974) Large plasmid in Agrobacterium tumefaciens essential for crown gall-inducing ability. Nature, London 252:169-170.

117. Van Larebeke, N., C. Genetello, J.P. Hernalsteens, A. Depicker, I. Zaenen, E. Messens, M. Van Montagu, and J. Schell (1977) Transfer of Ti-plasmids between Agrobacterium strains by mobilization with the conjugative plasmid RP4. Mol. Gen. Genet. 152:119-124.

118. Warren, G.J., A.J. Twigg, and D.J. Sherratt (1978) ColE1 plasmid mobility and relaxation complex. Nature, London 274:259-261.

119. Watson, B., T.C. Currier, M.P. Gordon, M.-D. Chilton, and E.W. Nester (1975) Plasmid required for virulence of Agrobacterium tumefaciens. J. Bacteriol. 123:255-264.

120. White, L.O. (1972) The taxonomy of the crown-gall organism Agrobacterium tumefaciens and its relationship to rhizobia and other agrobacteria. J. Gen. Microbiol. 72:565-574.

121. White, F.F., and E.W. Nester (1980a) Hairy root: plasmid encodes virulence traits in Agrobacterium rhizogenes. J. Bacteriol. 141:1134-1141

122. White, F.F., and E.W. Nester (1980b) Relationship of plasmids responsible for hairy root and crown gall tumorigenicity. J. Bacteriol. 144:710-720.

123. White, F., G. Ghidossi, M. Gordon, and E. Nester (1982) Tumor induction by Agrobacterium rhizogenes involves the transfer of plasmid DNA to the plant genome. Proc. Natl. Acad. Sci. USA 79:3193-3197.

124. Willmitzer, L., M. De Beuckeleer, M. Lemmers, M. Van Montagu, and J. Schell (1980) DNA from Ti-plasmid is present in the nucleus and absent from plastids of plant crown-gall cells. Nature, London 287:359-361.

125. Willmitzer, L., W. Schmalenbach, and J. Schell (1981a) Transcription of T-DNA in octopine and nopaline crown gall tumours is inhibited by low concentrations of α-aminitin. Nucl. Acids Res. 9:4801-4812.

126. Willmitzer, L., L. Otten, G. Simons, W. Schmalenbach, J. Schröder, G. Schröder, M. Van Montagu, G. De Vos, and J. Schell (1981b) Nuclear and polysomal transcripts of T-DNA in octopine crown gall suspension and callus cultures. Mol. Gen. Genet. 182:255-262.

127. Willmitzer, L., G. Simons, and J. Schell (1982a) The TL-DNA in octopine crown gall tumours codes for seven well-defined poly-

adenylated transcripts. EMBO Journal 1:139-146.
128. Willmitzer, L., J. Sanchez-Serrano, E. Buschfeld, and J. Schell (1982b) DNA from Agrobacterium rhizogenes is transferred to and expressed in axenic hairy root plant tissues. Mol. Gen. Genet. 186:16-22.
129. Willmitzer, L., P. Dhaese, P.H. Schreier, M. Schmalenbach, M. Van Montagu, and J. Schell. Size, location and polarity of T-DNA encoded transcripts in nopaline crown gall tumors: Evidence for common transcripts present in both octopine and nopaline tumors. Cell, in press.
130. Wood, H.N., A.N. Binns, and A.C. Braun (1978) Differential expression of oncogenicity and nopaline synthesis in intact leaves derived from crown gall teratomas of tobacco. Differentiation 11:175-180.
130a. Wöstemeyer, A., L. Otten, H. DeGreve, J.P. Hernalsteens, J. Leemans, M. Van Montagu, and J. Schell (1982) Regeneration of plants from crown gall cells. In Genetic Engineering in Eukaryotes, P. Lurquin, ed., Plenum Press, New York (in press).
131. Wullems, G.J., L. Molendijk, G. Ooms, and R.A. Schilperoort (1981) Retention of tumor markers in F1 progeny plants from in vitro induced octopine and nopaline tumor tissues. Cell 24:719-727.
132. Yadav, N.S., K. Postle, R.K. Saiki, M.F. Thomashow, and M.-D. Chilton (1980) T-DNA of a crown gall teratoma is covalently joined to host plant DNA. Nature, London 287:458-461.
133. Yadav, N.S., J. Vanderleyden, D.R. Bennett, W.M. Barnes, and M.D. Chilton (1982) Short direct repeats flank the T-DNA on a nopaline Ti Plasmid. Proc. Natl. Acad. Sci. USA 79:6322-6326.
134. Yang, F., J.C. McPherson, M.P. Gordon, and E.W. Nester (1980a) Extensive transcription of foreign DNA in a crown gall teratoma. Biochem. Biophys. Res. Commun. 92:1273-1277.
135. Yang, F., A.L. Montoya, D.J. Merlo, M.H. Drummond, M.-D. Chilton, E.W. Nester, and M.P. Gordon (1980b) Foreign DNA sequences in crown gall teratomas and their fate during the loss of the tumorous traits. Mol. Gen. Genet. 177:707-714.
136. Yang, F.M., A.L. Montoya, E.W. Nester, and M.P. Gordon (1980c) Plant tumor reversal associated with the loss of foreign DNA. In vitro 16:87-92.
137. Yang, F., and R.B. Simpson (1981) Revertant seedlings from crown gall tumors retain a portion of the bacterial Ti plasmid DNA sequences. Proc. Natl. Acad. Sci. USA 78:4151-4155.
138. Zaenen, I., N. Van Larebeke, H. Teuchy, M. Van Montagu, and J. Schell (1974) Supercoiled circular DNA in crown gall inducing Agrobacterium strains. J. Mol. Biol. 86:109-127.
139. Zambryski, P., M. Holsters, K. Kruger, A. Depicker, J. Schell, M. Van Montagu, and H.M. Goodman (1980) Tumor DNA structure in plant cells transformed by A. tumefaciens. Science 209:1385-1391.
140. Zambryski, P., A. Depicker, K. Kruger, and H. Goodman (1982) Tumor induction by Agrobacterium tumefaciens: analysis of the boundaries of T-DNA. J. Mol. Appl. Genet. 1:361-370.

IN VITRO PLANT TRANSFORMATION SYSTEMS USING LIPOSOMES AND BACTERIAL CO-CULTIVATION

Robb T. Fraley and Rob B. Horsch

Molecular Biology Department
Monsanto Co.
St. Louis, Missouri 63167

ABSTRACT

The development of efficient methods for introducing nucleic acids into plant cells is critical to the successful application of molecular genetic approaches to plant systems. Two methods have been developed for this purpose: 1) liposome-mediated delivery and 2) *in vitro* transformation of protoplasts by co-cultivation with *Agrobacterium tumefaciens* cells.

Liposome-mediated delivery of tobacco mosaic virus (TMV) RNA into protoplasts and resulting virus production has been used as an assay for determining incubation conditions which favor increased uptake of liposomal contents by plant cells. Under optimal conditions, the liposome method was found to be 10-1,000 times more efficient than other methods commonly used for introducing nucleic acids into plant cells.

The co-cultivation method, initially developed by Marton et al., (12) has been improved and extended to use with petunia protoplasts. *In vitro* transformants have been obtained with a variety of *A tumefaciens* strains at efficiencies near 10^{-1}. This method has been used to obtain transformants with avirulent bacterial strains which contain disarmed Ti plasmids. The implications of both these methodologies for plant genetic engineering is discussed.

INTRODUCTION

The development of efficient transformation systems for plant cells has been hampered by several problems, including: 1) a lack of suitable mutants, 2) inadequate vectors and selectable markers, 3) inefficient methods for introducing nucleic acids intracellularly and 4) difficulties in handling large numbers of plant cells and in efficiently selecting rare phenotypes. However, recent breakthroughs have occurred which should facilitate major advances in this area. For example, a number of biochemically characterized plant mutants have been isolated (1,2) which should be useful as recipients in complementation experiments. In addition, techniques for culturing protoplasts and for plating them at low cell densities have been developed by several laboratories (3,4) and these are starting to be incorporated into transformation and selection protocols. Studies on the Ti plasmid of A. tumenfaciens have resulted in increased knowledge of the molecular basis for crown gall disease and have resulted in the development of suitable methods for genetically engineering this large plasmid (5,6), thus enhancing the possibility for its use as a plant vector. Finally, a number of in vitro systems based on free DNA delivery (7,8), spheroplast fusion (9), liposome-mediated delivery (10,11) and co-cultivation of protoplasts with intact A. tumefaciens cells (12,13), have recently been reported for introducing the Ti plasmid and other nucleic acids into plant cells.

In this paper, we describe the use and application of two of these methods: 1) liposome-mediated delivery and 2) transformation of protoplasts by incubation with A. tumefaciens cells with respect to the development of efficient in vitro plant transformation systems.

Liposome Mediated Delivery

In the last few years, the use of phospholipid vesicles (liposomes) to introduce nucleic acids into plant cells has received considerable attention (14-24). Possible advantages perceived in using liposomes as a delivery system include: 1) low toxicity and applicability with many plant species, 2) protection of entrapped nucleic acids from degradation by nucleases present in the culture medium and 3) more efficient delivery of nucleic acids to plant protoplasts. However, despite the considerable amount of research on the interaction of liposomes with protoplasts (14-24), there has been little agreement among these reports regarding the conditions (i.e., vesicle lipid composition, incubation parameters, etc.) which actually favor the intracellular delivery of liposomal contents into plant cells. Much of this confusion has arisen because of the lack of suitable methods for measuring the introduction of liposome contents into cells.

We have previously demonstrated that viral nucleic acids (polio RNA, SV40 DNA) could be encapsulated in liposomes and under the appropriate incubation conditions could be introduced into cultured mammalian cells (25,26). Furthermore, the expression of the nucleic acids could be used as a sensitive biological assay for monitoring liposome-cell interactions and for determining those vesicle lipid compositions and incubation conditions which result in enhanced delivery.

For our preliminary studies on liposome-plant protoplast interactions, we elected to use a similar approach. TMV RNA was chosen for encapsulation in liposomes because of its availability in large quantities and the existence of several assays for monitoring its production by infected cells. A radioimmunoassay was used to monitor TMV levels in infected protoplasts (Fig. 1) because its sensitivity and rapidity facilitated the analysis of the numerous small samples generated during the optimization of the liposome delivery method.

In the initial experiments with tobacco protoplasts (10), TMV RNA (500 μg) was encapsulated in a variety of different liposome preparations using the reverse evaporation method developed by Szoka and Papahadjopoulos (27). This method is particularly suitable for the encapsulation of large macromolecules at high efficiency (∿40% of sample is entrapped) and offers the advantages of being applicable with most lipid compositions, adaptable to use with small sample volumes (∿50 μl), and can be performed under sterile conditions. The details of liposome preparation and incubation with protoplasts are given in references 10 and 25. When these various liposome preparations (∿5 μg RNA) were simply incubated with protoplasts (10^6) in buffer for 30 minutes and the protoplasts washed and resuspended in media, there was no detectable virus production observed after 48 hours (Tab. 1).

Control experiments demonstrated that: 1) the encapsulated TMV RNA was not degraded, 2) high levels of liposomes became associated with the protoplasts and 3) the viability of the protoplasts was not substantially reduced following exposure to liposomes. It seemed most likely that the lack of detectable virus production was a result of the failure of the liposomes to introduce a significant amount of the TMV RNA into protoplasts.

Since it had previously been shown that the inclusion of compounds such as glycerol, ethylene glycol and polyethylene glycol (PEG) during the incubation of mammalian cells with liposomes resulted in dramatic increases (∿1,000-fold) in the efficiency of liposome delivery, these and other polyalcohols were tested with plant protoplasts (10). Incubation of tobacco protoplasts with negatively-charged, phosphatidylserine liposomes containing TMV RNA

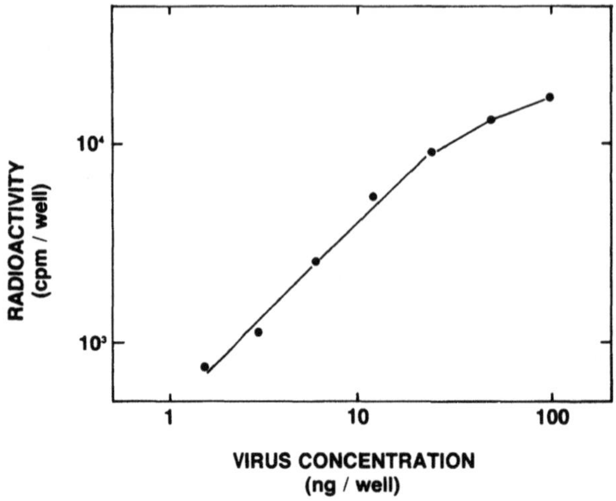

Fig. 1. Radioimmunoassay for monitoring TMV production in infected protoplasts. Microtiter plates (Dynatech Laboratories, Alexandria, VA) were precoated with unlabeled anti-TMV antibody (10 μg/ml in 50 mM phosphate buffer, pH 9.0) and washed three times with phosphate-buffered saline (containing 1% bovine serum albumin) before use. Dilutions of TMV were added to wells and incubated for 12-28 hr at the wells were washed three times with phosphate-buffer saline (containing 1% bovine serum albumin), and 125 μl of I^{125}-labeled anti-TMV antibody (approx. equal to 200,000 cpm) was added and allowed to incubate for 3.0 hr. The wells were washed five times with phosphate-buffered saline, separated and transferred to scintillation vials for the determination of radioactivity. The assay detects 1 ng of virus (at twice background levels); at saturating amounts of virus (> 1μg), approx. equal to 3-5% of the labeled antibody was bound.

virus production could be detected following incubations with neutral or positivelycharged liposome preparations. These later vesicle compositions also proved non-infectious when tested with petunia protoplasts (Tab. 1). An enhancement in infectivity ($\sim$5-fold) was observed when cholesterol was included in the vesicle preparation (Tab. 1) and this has been shown to correlate with its effect on reducing liposomal membrane permeability and decreasing leakage of vesicle contents which occur during cell incubations (26).

More recent studies with petunia protoplasts have revealed that the concentration of the polyalcohol solution used to stimulate liposome delivery is important. Both the PEG and PVA gave comparable results when tested at low concentrations, but PEG was found to be much more effective at higher concentrations (Fig. 2).

Table 1. Infectivity of TMV RNA encapsulated in various liposome preparations.

RNA PREPARATION	VIRUS YIELD (ng virus/10^5 protoplasts) Tobacco Protoplasts (-PVA)	(+PVA)	Petunia Protoplasts (+PVA)
PS liposomes	<1	85	13.1
PC liposomes	<1	<3	<1
PC-SA liposomes	<1	<1	<1
PS-Chol liposomes	<1	503	48.3
TMV, virions		6000	4150

LEGEND: TMV RNA was encapsulated in liposomes prepared by reverse-phase evaporation. A small amount of radioactive [^{3}H]poly(A) is included to permit precise calculations of encapsulation efficiency and TMV RNA concentrations. Unencapsulated TMV RNA was separated from liposomes by centrifugation (flotation) on discontinuous ficoll gradients (26). Ficoll solutions were prepared in 5 mM Tris, 0.5 M mannitol, 0.1 mM EDTA, pH 7.0. Rapidly growing suspension cultures were centrifuged (200 x g), resuspended at twice the original volume in 2% Cellulysin (Calbiochem), 1% Driselase (Kyowa Hakko Kogyo, Tokyo, Japan), 0.5% Macerase (Calbiochem), 0.5 M mannitol (pH 5.7), and incubated for 5-6 hr. The reulsting protoplasts were separated from debris and undigested cells by successive passage through 100, 150, and 200 mesh filters (Small Parts, Miami, FL) and were washed twice by centrifugation (100 x g for 5 min) with buffer (5 mM Tris, 0.5 mM $CaCl_2$, 0.5 M mannitol, pH 7.0). Protoplasts (10^6 in 0.5 ml of above buffer) were incubated with 5 μg of liposome encapsulated RNA for 5 min, prior to the addition of 4.5 ml of a polyvinyl alcohol solution (10% wt/v in the above buffer). After 20-30 min, buffer was added to dilute the viscous polymer solution and the protoplasts were centrifuged. The cells were washed again with buffer, once with medium and then resuspended in medium (1-2 x 10^5/ml) and incubated for 48 hr prior to analysis of virus production. PS = phosphatidylserine; PC = egg phosphatidylcholine; PC-SA = egg phosphatidylcholine-stearylamine (9:1); PS-chol = phosphatidylserine-cholesterol (1:1).

A second important parameter in liposome protoplast interactions was found to be the divalent metal ion concentration of the incubation buffer. High levels of $CaCl_2$ (5 mM) were found to stimulate delivery and virus production (Fig. 3). This enhancement is mediated both by increased liposome binding to protoplasts (Fig. 3)

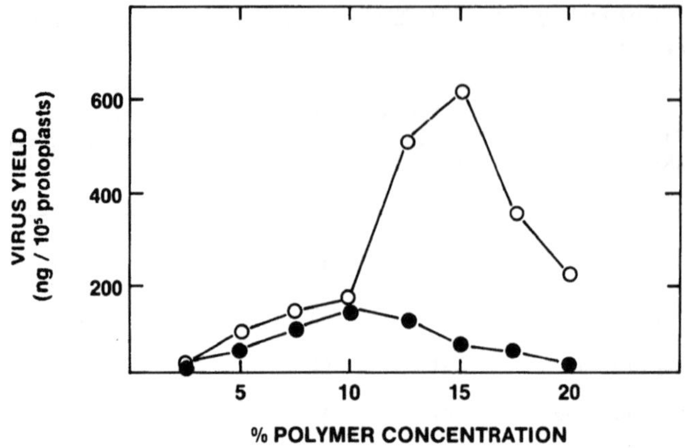

Fig. 2. The effect of polyalcohol concentration in the efficiency of liposome mediated delivery of TMV RNA to petunia protoplasts. The incubation of protoplasts (10^6) with liposomes (∿120 nmol vesicle lipid, 5 μg TMV RNA) were carried out in Tris-buffered mannitol (TMB) buffer containing 0.5 mM $CaCl_2$ for 5.0 min; 4.5 ml of either a PVA or PEG solution made in this buffer at the concentration shown in the figure was added, and the incubation allowed to continue for 20 min. The protoplasts were then washed, resuspended in medium and virus production was monitored after 48 hr. PVA treatment (-●-); PEG treatment (-○-).

and by stabilization of protoplast integrity by $CaCl_2$. The binding of Ca^{2+} to the negatively-charged liposome and protoplast membranes results in charge neutralization, which permits their close apposition and may also induce fusion at the sites of contact (28).

In examining other parameters for their effect on the efficiency of liposome-mediated delivery to petunia protoplasts, it was determined that maximal interactions occurred at neutral pH. The optimal incubation times (Fig. 4) were shown to be 5 minutes for the pre-incubation of liposomes with protoplasts (Fig. 4a) and 20-30 minutes for the length of exposure of protoplasts to PEG solutions (Figure 4b).

In order to establish the absolute efficiency of the liposome delivery method and to determine the effect of RNA content per lipsosome on the observed level of virus production, phosphatidylserine (PS-Chol) liposomes were prepared with various amounts of encapsulated RNA corresponding to 27, 3.0, 0.32 and 0.03 molecules of TMV RNA/vesicle in the final preparation. These were incubated with protoplasts at increasing lipid concentrations (0.5-300 nmol) under optimal conditions and the resulting levels of virus production determined (Fig. 5). For all preparations, the amount of virus

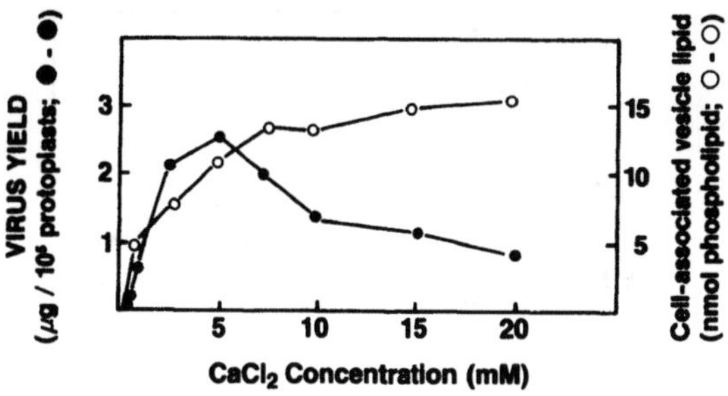

Fig. 3. The influence of buffer $CaCl_2$ concentration on the efficiency of liposome-protoplast interactions. Protoplasts (10^6) were washed twice with 10 ml of TBM buffer containing the level of $CaCl_2$ shown in the figure and then resuspended in 0.5 ml of the same buffer. Liposomes (∿125 nmol vesicle lipid) containing 5 µg of encapsulated TMV RNA were added and allowed to incubate with cells for 5.0 min prior to the addition of 4.5 ml of PEG solution (15% wt/v, made up in TBM at the appropriate $CaCl_2$ concentration). Following incubation with PEG, the cells were washed, resuspended in media and virus production was monitored after 48 hr. The association of vesicle lipid with protoplasts was determined using liposome containing [^{3}H]-dipalmitoylphosphatidyl choline (5 µCi/µmol of lipid) to label vesicle lipid as previously described (10). The incubation with radioactively labeled liposomes with protoplasts was carried out as described above, except that 2% (wt/v) ficoll was included in the buffers used in the post-incubation washes to facilitate the separation of unbound liposomes from protoplasts.

production increased roughly linearly with increasing concentrations of added liposomes. In addition, the specific infectivity (ng virus/nmol phospholipid) of the preparations increased with higher ratios of RNA/vesicle, indicating the delivery process has not been saturated even at RNA/vesicle >1. This observation emphasizes the importance of obtaining high efficiences of RNA encapsulation during liposome preparation, since the amount of delivery is proportional to the RNA content of the liposomes. It is noteworthy that measurable virus production (2.5 ng virus/10^5 protoplasts) could be observed when <0.5 ng of encapsulated TMV RNA was incubated with the cells (0.5 nmol of 0.32 RNA molecules/liposome preparation). When 0.25 µg of encapsulated RNA (10 nmol of 27 RNA molecules/liposome preparation) was incubated with protoplasts, the level of virus production (890 ng virus/10^5 protoplasts) was found to be only ∿40-fold lower than that observed in incubation with an equivalent amount of the intact virus (Tab. 1). At higher liposome concentrations (100-300 nmol of 27 RNA molecules/liposome preparation), the

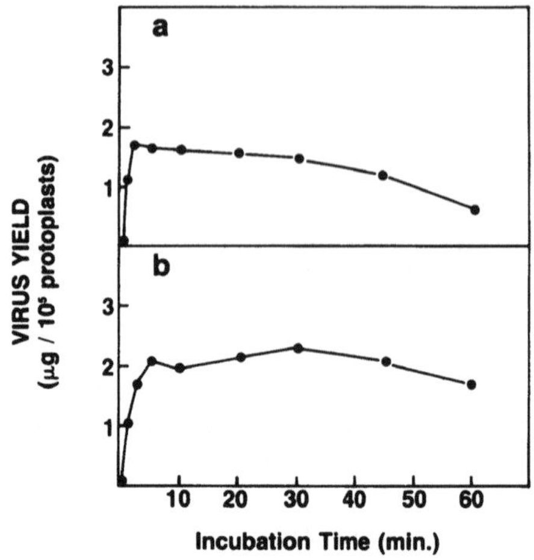

Fig. 4. Optimal incubation periods for the liposome-mediated delivery of TMV RNA into petunia protoplasts. The pre-incubation of protoplasts (10^6) with liposomes (∿120 nmol vesicle lipid, 5 μg RNA) was carried out in TBM containing 5 mM $CaCl_2$ for the times shown in the figure, prior to the addition of 4.5 ml of a PEG solution (15% wt/v) for the periods shown in the figure. The protoplasts were then washed, resuspended in medium and virus production was monitored after 48 hr. (a) pre-incubation with liposomes for various times (0-60 min prior to the addition of PEG for 20 min; (b) pre-incubation with liposomes for 5 min prior to the addition of PEG for various times (0-60 min).

level of virus production was >2,000 ng virus/10^5 protoplasts -- this corresponds to infection of >80% of the protoplasts as judged by staining of the protoplasts with FITC-labelled anti-TMV antibody (Fig. 6).

A direct comparison of the relative efficiency of the liposome delivery method with other techniques (7,8,29-32) used to introduce RNA and/or DNA molecules into plant protoplasts is impossible because of the many differences in plant sources, nucleic acids, and infectivity assays used in these various studies. We have attempted a rough comparison by repeating these procedures as closely as possible (Tab. 2), but substituting petunia protoplasts and TMV RNA for the protoplasts and nucleic acid used in the original studies.

It is emphasized these results should not be considered as a quantitative comparison because uncontrollable differences related to protoplast isolation methods or properties unique to protoplasts from certain plant species could dramatically influence the

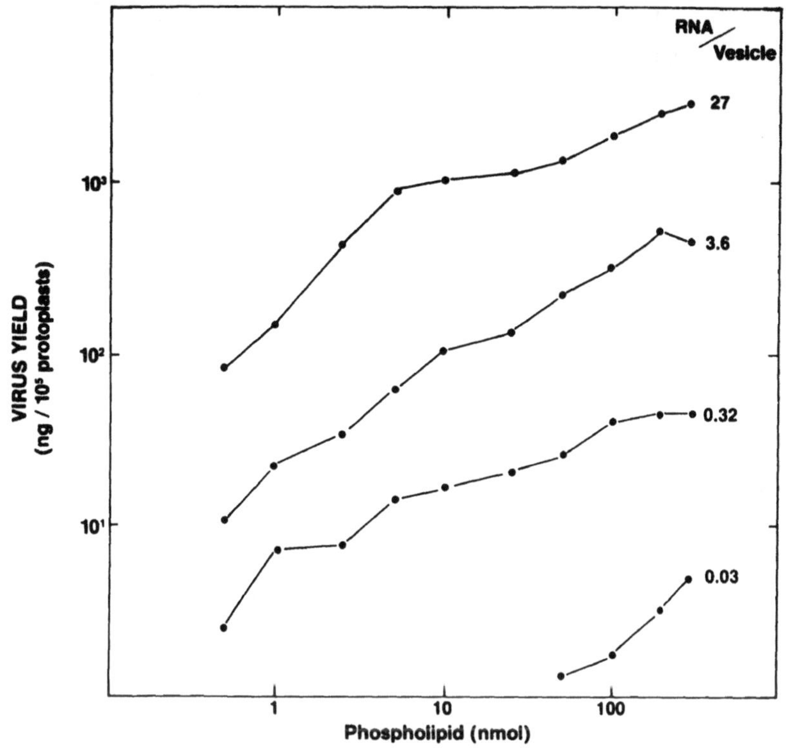

Fig. 5. Infectivity of TMV RNA encapsulated in liposomes at different RNA/vesicle ratios. Liposomes were prepared from 5 µmol each of phosphatidylserine and cholesterol and starting with 0.5, 5.0, 50 and 500 µg of TMV RNA. The encapsulation efficiency determined for each of these preparations was 46.6, 52.6, 60.5, and 45.8%, respectively. The number of RNA molecules per vesicle (0.03, 0.32, 3.6 and 27, respectively) was calculated assuming 4.5×10^{11} vesicles/µmol phospholipid (10) and a molecule weight of 2×10^6 daltons for TMV RNA. The incubation of protoplasts with liposomes (0.5-300 nmol) and the determination of virus production was as described in the text.

efficiency of a particular method. For example, in our hands a significant (>50%) loss of cell viability was associate with methods 3 and 5 (Tab. 2) and virus production was not detectable. All of the other methods tested (Tab. 2, line 1,2,4) resulted in varying levels of virus production -- all of which were significantly lower (10-1000 fold) than that obtained with liposomes using our procedure (line 7). It should be noted that the level of virus production obtained with method 4 (50 ng virus/10^5 protoplasts with 10 ng TMV

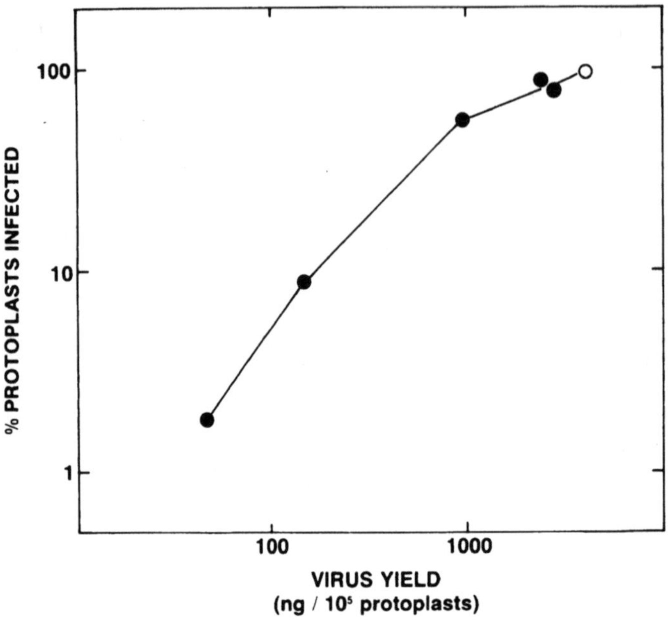

Fig. 6. Correlation between virus yield and the percentage of protoplasts infected following incubation with liposome-encapsulated TMV RNA. Duplicate samples from the experiment described in Fig. 5 (incubations with 0.5-300 nmol of the 27 RNA molecules/vesicle preparation) were prepared and stained with fluorescein-labeled anti-TMV antibody according to the method of Otsuki and Takebe (34). The stained preparations were examined using a Zeiss microscope equipped with fluorescence optics and epillumination. A minimum of 500 protoplasts were analyzed for each determination and the level of virus infection was expressed as the percentage of protoplasts which displayed discrete greenish fluorescence. The values for virus yield (ng virus/10^5 protoplasts) were obtained as described in the legend to Fig. 5. One sample obtained by infection of protoplasts with whole TMV using the poly-L-ornithine method is included for comparison. Values obtained from liposome delivery experiments (●), and infection with intact virus (○).

RNA) which would correlate with ~2% of the cells being infected, based on the data from Fig. 6, is much lower than reported by the original authors (31). Method 2, used previously for the introduction of Ti plasmid DNA into tobacco protoplasts (8), was found to be the most efficient method for introducing naked TMV RNA into petunia protoplasts. Slightly higher levels of virus production were obtained using liposomes (method 6) under conditions

Table 2. Comparison of various methods used to introduce nucleic acids into plant protoplasts.

Method	Brief Description And Application	Virus Yield[1] (1 μg RNA)	(10 μg RNA)	Reference
1	poly-L-ornithine co-precipitation; TMV RNA	1.8	14.6	Aoki and Takebe Virology 39:439-445, 1969
2	calcium phosphate co-precipitation PEG; Ti plasmid DNA	246.5	1205.2	Krens *et al.* Nature 296:72-74, 1982
3	protamine sulfate co-precipitation Brome Mosaic Virus RNA	<1	<1	Loesch-Fries and Hall J. Gen. Virol. 47: 323-332, 1980
4	alkaline pH, high KCl and $MgCl_2$; TMV RNA	3.8	52.6	Sarkar *et al.* Molec. Gen. Genet. 135:1-9, 1974
5	poly-L-ornithine co-precipitation, EDTA; Ti plasmid DNA	<1	<1	Davey *et al.* Plant Sci. Lett. 18:307-313, 1980
6	PS-Chol liposomes, PEG, alkaline pH, high $CaCl_2$; TMV RNA	80.3	894.8	Fukunaga *et al.* Virology 113:752-760 (1981)
7	this report	2087.3	4996.1	

[1]expressed as ng virus/10^5 protoplasts

LEGEND: Petunia protoplasts (10^6) and TMV RNA (1 and 10 μg were substituted for the cell type and particular nucleic acid used in the initial studies (7,8,29-32); otherwise the buffers and conditions for protoplast incubations were exactly those described in the original report. Calf thymus DNA was used as the carrier DNA in method 2. Following incubation for the appropriate periods, the protoplasts were washed twice, resuspended in medium ($\sim 10^5$ cells/ml) and virus production was monitored after 48 hrs. The values shown in the table are the averages of three separate experiments.

described by Fukanga et al. (29); however, these values were considerably less than obtained with the same liposome preparation using the conditions established in this paper (method 7).

From the success reported here using liposomes to introduce TMV RNA into protoplasts, it seems reasonable that liposomes will also be efficient carriers for DNA delivery in plant transformation experiments. In mammalian systems, it was found that the exact conditions (lipid composition, buffer, polymers, etc.) that resulted in the maximal introduction of SV40 DNA or the HSV-TK gene into cells (25,26), were identical to those conditions which produced maximal infection of cells by encapsulated polio RNA (R. Straubinger, unpublished results). We have had some success in preliminary experiments using liposomes to introduce encapsulated Ti plasmid DNA (pTiACH5) into tobacco protoplasts (S. Dellaporta and R. Fraley, unpublished results). Several calli have been obtained which produce octopine and contain T-DNA sequences; however, the efficiency of obtaining stable transformants was quite low in these experiments ($\sim 10^{-6}$) and the procedure is not yet practical for routine use. It is likely that liposomes will have their greatest application in transformation experiments utilizing non-Ti plasmid vectors and in experiments with protoplasts which are outside the host range of *A. tumefaciens*.

Co-cultivation of Protoplasts with A.tumefaciens Cells

Part of the reason for the very low frequency of transformation of protoplasts with liposome-encapsulated Ti plasmid may have been due to inefficient delivery or failure of the T-DNA to integrate. Alternatively, it seemed possible that our selection conditions (which were similar to those reported in references 7,12,13) were not suitable for identifying rare, hormone independent transformants against a large background of normal cells. Problems associated with cell plating and selection (e.g., minimal plating densities, cross feeding, cooperative cell death, etc.) could have interfered with the isolation of transformants, but unfortunately the low transformation frequency obtained using liposomes did not easily allow for the optimization of the selection protocol. Instead, we elected to determine conditions for selecting hormone autotrophic cells using a more efficient *in vitro* transformation procedure (10^{-4}-10^{-5}) than had been developed by Marton et al. (12). In their method, 3 day old tobacco protoplasts were cultivated in the presence of intact *A. tumefaciens* cells (for 30 hours) and during this period the bacteria bind to the protoplasts and transformation presumably results by a mechanism analogous to that occurring during *in planta* inoculations. We have adapted this technique to Mitchell petunia cells and combined it with a rapid protoplast culture system that permits early selection for transformed cells. The result is a rapid and efficient *in vitro* system for transforming plant cells with Ti plasmid. Its key components are: 1) the rapid culture method, 2) optimal conditions for co-cultivation of A. tumefaciens cells with protoplasts and 3) the selection for hormone autotrophy.

METHODS

We begin with standard enzymatic protoplast production, flotation isolation, and high density liquid culture as described by Ausubel et al. (33). We dilute the mannitol osmoticum more rapidly than reported, however, and within the first week, we transfer the growing cells from liquid to filter paper supports over tobacco nurse cultures on agar plates (feeder plates). The double filter paper plating system (34) permits rapid, density-independent growth of small cell clusters of single cells. In this way, we obtain individual colonies about 1mm in diameter within 15 days of protoplast production. The colonies can be grown close together or well isolated from each other. The filter paper substrate permits simple transfer to other media from the feeder plate.

We have adapted the procedure of Marton et al. (12) to our rapid culture system. Protoplasts are plated at 10^5 initial intact protoplasts per ml[1]. On the second day, A. tumefaciens cells are added to a titer of between 10^7 and 10^8 per ml. The cultures are incubated for 24 to 48 hours, then washed by centrifuging to remove free bacteria. The cells are replated in medium with 500 μg/ml carbenicillin and then plated at the normal time on feeder plates with reduced hormones to begin selection for tumor cells. The details of the culture regime for petunia will be published elsewhere.

Extensive aggregation is often observed during the co-cultivation period, but rapidly growing protoplast preparations survive the treatment quite well. In a typical control experiment, 5.6% of the initial intact protoplasts produced colony forming units[2]. In an identical sample of protoplasts treated with A. tumefaciens but not selected under hormone-independent conditions, 4.9% of the initial intact protoplasts produced colony forming units. Therefore, the absolute plating efficiency[3] was only reduced by 12.5% -- these were either killed by the treatment or not counted because of aggregation. Of the unselected colonies produced after treatment, 1.5% showed long-term growth on hormone-free medium (transformation frequency[4]). When selection was applied early (transfer to hormone-free medium) there was no reduction in the recovery of transformants, indicating very efficient selection.

Without defining the various parameters described above, it is extremely difficult to assess reports of low frequency transformation -- or failure to obtain transformants. In the absence of careful analysis one can not be sure of the ability to recover all

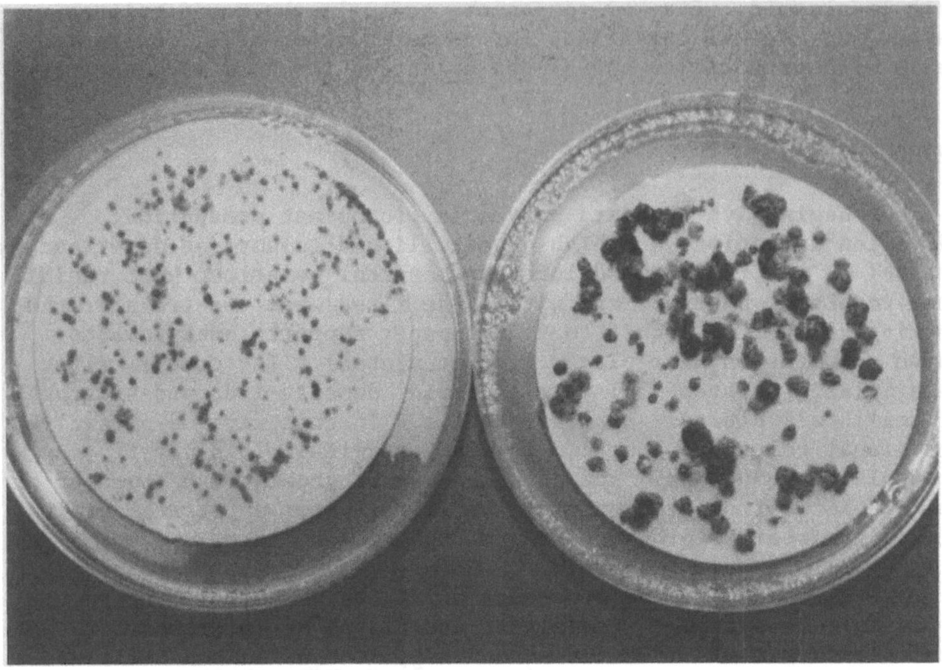

Fig. 7. Results of a co-cultivation experiment in which petunia protoplasts were incubated with *A. tumefaciens* stain B6S3. The conditions for protoplast isolation and incubation with bacteria were as described in the text. Following rapid culturing of control and bacterial treated cells on feeder plates for 10 days, the microcolonies on filters were transferred to selection medium without hormones. After two weeks, hormone-independent calli are observable on the plates treated with *A. tumefaciens* (right), whereas the colonies from control cells (left) have browned.

or most, or even any of the transformants present. Thus, to separate the efficiency of transformation from the ability to select for transformation from the ability to select for transformants, one must either validate the selection system or isolate colonies under non-selective conditions and then screen them for the transformed phenotype.

RESULTS

Using the co-cultivation technique adapted to our rapid plating system and early selection for hormone autotrophy, we have routinely observed transformation frequencies of 10^{-1} with *A. tumefaciens* strains carrying octopine, nopaline or agropine-type Ti plasmids

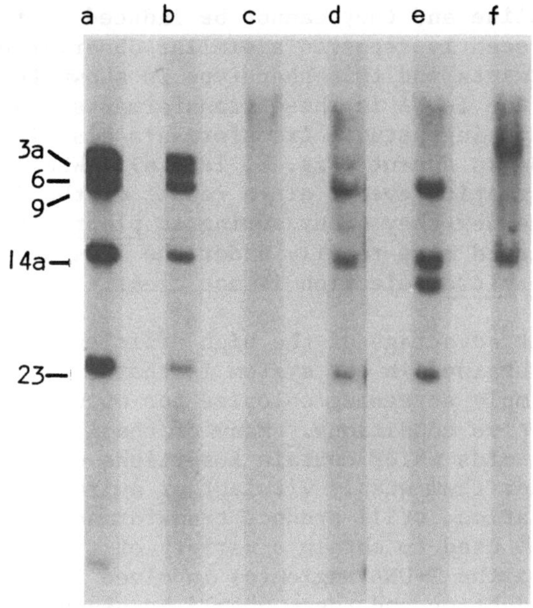

Fig. 8. Detection of T-DNA sequences in in vitro transformants obtained following incubation of protoplasts with A. tumefaciens strain A208. DNA from transformants was extracted and purified by CsCl centrifugation and 10 μg was digested with the restriction enzymes shown in the figure. Digestion products were fractionated on agarose gels, transferred to nitrocellulose filters and hybridized with a ^{32}P-labeled probe containing the entire T-DNA region in pBR322. Lane a and b contain 10 and 1 copy reconstructions, respectively. Lane c DNA from normal petunia callus. Lanes d,e, and f DNA isolated from in vitro transformants.

(Fig. 7). Frequencies as high as 80% have been observed. The majority of the hormone-independent calli (>90%) produce opines and Southern hybridization analysis has confirmed the presence of T-DNA in several of the transformants (Fig. 8). Control experiments demonstrate that habituation does not occur at a detectable frequency in the Mitchell petunia. Because of the early selection employed in this procedure, the majority of the in vitro transformants are clones comprised only of transformed cells.

A high percentage (∿1%) of the in vitro transformants have been observed to shoot spontaneously while being passaged on hormone-free medium. These shoots are transformed, producing high levels of octopine or nopaline and they cannot be induced to root. Wullems et al. (13) have recently reported a similar observation with in vitro tobacco transformants and this phenotype is shown to result from a loss of part of the T-DNA in these transformants. Analysis of spontaneously shooting petunia transformants also indicates that part of the T-DNA is absent (Fig. 8, lane e). Whether these aberrant transformation events are a result of the in vitro procedures or whether they occur during in planta infections but can simply be identified more readily under the less stringent conditions of in vitro selection is not clear.

An important advantage of the high efficiency of in vitro transformation obtained in our system is that transformants can be identified by simply screening colonies for opine production in the absence of selective conditions. Many of the A. tumefaciens strains harboring Ti plasmids which contain insertions or deletions in their T-DNA which render them weakly virulent or avirulent when tested by in planta inoculation, still produce transformants in vitro. This approach has been used to obtain a variety of transformants which have deletions in the T-DNA region(s) involved in regulating phytohormone production and these should be useful for correlating phytohormone levels with tumor phenotype and genetic analysis. This approach should also facilitate the routine regeneration of genetically-modified, phenotypically normal plants following the transformation of protoplasts with Ti plasmids which lack those genes responsible for tumorogenesis.

ACKNOWLEDGEMENTS

We wish to thank Ms. Darla Lam and Patsy Guenther for their assistance in the preparation of this manuscript, Annick De Framond for her gift of the T-DNA probe and Patricia Saunders for her help with the in vitro transformation experiments.

REFERENCES

1. Müller, A., and R. Grafe. 1978. Mol. Gen. Genet. 161:67-72.
1a. Somerville, C., and W. Ogren. 1981. Trends in Biochem. 7:171-176.
2. King, J., R. Horsch, and A. Savage. 1980. Planta 149:480-487.
3. Horsch, R., and G. Jones. 1980. In Vitro 16:103-108.
4. Caboche, M. 1980. Planta 149:7-18.
5. Matzke, A., and M.-D. Chilton. 1981. J. Mol. Appl. Genet. 1:39-49.
6. Leemans, J., C. Shaw, R. Deblaere, H. DeGreve, J. Hernalsteens, M. Maes, M. Van Montagu, and J. Schell. 1981. J. Mol. Appl. Genet. 1:149-164.
7. Davey, M., E. Cocking, J. Freeman, W. Pearce, and I. Tudor. 1980. Plant Sci. Lett. 18:307-313.

8. Krens, F., L. Molendijk, G. Wullems, and R. Schilperoort. 1982. Nature 296:72-74.
9. Hasezawa, S., T. Nagata, and K. Syono. 1981. Mol. Gen. Genet. 182:206-210.
10. Fraley, R., S. Dellaporta, and D. Papahadjopoulos. 1982. Proc. Natl. Acad. Sci. USA 79:1859-1863.
11. Nagata, T., K. Okada, I. Takebe, and C. Matsui. 1981. Mol. Gen. Genet. 184:161-165.
12. Marton, L., G. Wullems, L. Molendijk, and R. Schilperoort. 1979. Nature 277:129-131.
13. Wullems, G., L. Molendijk, G. Ooms, and R. Schilperoort. 1981. Proc. Natl. Acad. Sci. USA 78:4344-4348.
14. A. Cassells. 1978. Nature 275:760.
15. Lurquin, P. 1976. Planta 128:213-216.
16. Lurquin, P. 1981. Plant Sci. Lett. 21:31-40.
17. Ostro, M., D. Lavelle, W. Paxton, B. Matthews, and D. Giacomoni. 1980. Arch. Biochem. Biophys. 201:392-402.
18. Matthews, B., S. Dray, Widholm, J., and M. Ostro. 1979. Planta 145:37-44.
19. Rollo, F., M. Galli, and B. Parisi. 1981. Plant Sci. Lett. 20:347-354.
20. Matthews, B., and D. Cress. 1981. Planta 153:90-94.
21. H. Uchimiya. 1981. Plant Physiol. 67:629-632.
22. Lurquin, P., R. Sheehy, and N. Rao. 1981. FEBS Lett. 125:183-187.
23. Uchimiya, H., and H. Harada. 1981. Plant Physiol 68:1027-1030.
24. Lurquin, P., and R. Sheehy. 1982. Plant Sci. Lett. 25:133-146.
25. Fraley, R., S. Subramani, P. Berg, and D. Papahadjopoulos. 1980. J. Biol. Chem. 255:10431-10435.
26. Fraley, R., R. Straubinger, G. Rule, L. Springer, and D. Papahadjopoulos. 1981. Biochemistry 20:6978-6987.
27. Szoka, F., and D. Papahadjopoulos. 1978. Proc. Natl. Acad. Sci. USA 75:4194-4198.
28. Wilschut, J., N. Düzgünes, R. Fraley, and D. Paphadjopoulos. 1980. Biochemistry 19:6011-6021.
29. Fukunaga, Y., T. Nagata, and I. Takebe. 1981. Virology 113:752-760.
30. Aoki, S., and I. Takebe. 1969. Virology 39:439-445.
31. Sarkar, S., M. Upadhya, and G. Melchers. 1974. Mol. Gen. Genet. 135:1-9.
32. Loesch-Fries, L., and T. Hall. 1980. J. Gen. Virol. 47:323-332.
33. Ausubel, F., K. Bahnsen, M. Hanson, A. Mitchell, and H. Smith. 1980. PBM Newsletter 1:26-32.
34. Otsuki, Y., and I. Takebe. 1969. Virology 38:497-501.

APPENDIX

1. Initial intact protoplasts: the number of protoplasts at the beginning of the culture period, but after sufficient time to equilibrate to the culture medium, that are spherical, with evenly distributed cytoplasm, and containing a nucleus.
2. Colony-forming units: the number of propagules at the time of plating that will produce macroscopic colonies under the best conditions for growth.
3. Absolute plating efficiency: the ratio of colony forming units to initial intact protoplasts.
4. Transformation frequency: the ratio of transformed colonies to colony-forming units.

CHROMOSOMES FROM PROTOPLASTS -- ISOLATION, FRACTIONATION, AND UPTAKE

Russell L. Malmberg and Robert J. Griesbach*

Cold Spring Harbor Laboratory
Cold Spring Harbor, New York 11724
*U.S. Department of Agriculture
Beltsville, Maryland 20705

ABSTRACT

We have developed methods for the isolation of condensed mitotic and meiotic chromosomes from plant protoplasts. The chromosomes can be isolated from a variety of tissues: meiocytes, root tips, cell cultures, and from a variety of species: tobacco, tomato, lily, daylily, onion, and peas. The limiting steps are the ability to obtain synchronously dividing cells and conversion of those cells into protoplasts. We have also been able to demonstrate uptake of isolated chromosomes by recipient protoplasts with fluorescence microscopy at a frequency of 1%. We could partially fractionate chromosomes by size, but we also, unfortunately, were able to show that the *in vitro* size did not correspond well to the relative *in vivo* size, so the fractionation method was not noticeably useful. We have not obtained an actual genetic transformation as yet.

INTRODUCTION

Genetic transformation of plant cells either as a tool in understanding basic plant genetics or as a means of genetic engineering, has long been a desirable technique. As reported in other papers in this volume, tremendous progress has been made in transformation via *Agrobacterium* plasmids, and in liposome encapsulated DNA transformation. Here, we report our progress in using isolated mitotic chromosomes as vectors for plant transformations. We have developed methods for the isolation of mitotic and meiotic chromosomes from plant protoplasts (2,4), for their uptake by recipient protoplasts (3), and for their partial fractionation by size (1). We have not yet achieved a stable transformation with expression of the donor genetic material. Other researchers (6) have also been able to obtain mitotic chromosome preparations that are taken up by protoplasts. Chromosome mediated transformation of mammalian cell cultures has, of course, long been an established and well characterized fact (5). Thus, for certain applications, chromosome mediated gene transfer may succeed and be useful in plant genetic manipulations.

MANIPULATIONS

Chromosome Isolation (2,4)

Protoplasts were obtained from partially synchronous material, either cell cultures treated with a DNA synthesis inhibitor, or root tips, or synchronous meiocytes. The protoplasts were collected by centrifugation at 100 g for 15 min, and washed twice with 5 mM MES 0.6 M mannitol at pH 6.0. The washed protoplasts were resuspended in a small volume of 15 mM Hepes, 1 mM EDTA, 15 mM dithiothreitol (DTT), 0.5 mM spermine, 80 mM KCl, 20 mM NaCl, 300 mM sucrose, 500 mM hexylene glycol, all at pH 7.0 (chromosome buffer). This was then gently passed through a 27 gauge hypodermic needle about 3 times until the protoplasts were ruptured. Cellular debris was removed by centrifugation at 100 g for 15 min.; the chromosomes were collected after centrifugation at 2500 g for 10 min and resuspended in chromosome buffer.

Chromosome Uptake (3)

Isolated chromosomes were stained with 0.1% 4'6-diamidino-2-phenylindole (DAPI) for 1 hr in the dark. The chromosomes were then washed free of the non-bound stain via centrifugation. The uptake mixture contained chromosomes at 10^7 per ml, tobacco leaf protoplasts at 10^6 per ml, 35% PEG 4000, 2% mannitol, 12 mM $CaCl_2$ at pH 6.0. The mixture was incubated for 20 min at room temperature, and then diluted with 4 volumes of 50 mM $CaCl_2$, 10% mannitol at pH 8.5. The protoplasts were then collected by centrifugation at 100 g, washed in MES/Mannitol and examined by fluorescence microscopy (Zeiss) under dark field illumination, a UG-1 exciter filter, and various barrier filters.

Chromosome Fractionation (1)

Chromosomes were resuspended in 0.5 ml of chromosome buffer and then sedimented through a 5 ml 5-40% sucrose gradient at 2000 g for 30 to 45 min. Eleven fractions were collected for visual and hybridization analyses.

Hybridization to rRNA (1)

An equal volume of 0.5 M NaOH, 0.02 M EDTA, 0.1% Triton X-100 was added to chromosome fractions. After 1 hr at room temperature the solution was neutralized with 0.5 M HCl and made 2XSSC (0.3 M NaCl - 0.03 M Na_3 citrate, pH 7). The DNA was collected on nitrocellulose filters at a concentration of 0.1 μg per mm^2 of paper. The filter was washed in 2XSSC, dried, and baked in a vacuum oven. Ribosomal RNA (rRNA) was isolated from root tips, iodinated with I^{125} by standard techniques, and hybridized to the filters. The filters were blotted and submerged in 100 μg/ml of RNase (DNase free) for 1 hr at 37°. The filters are then washed in 2XSSC and counted.

RESULTS

Chromosome Isolation

Initially (4) we isolated mitotic chromosomes from suspension cultures by a very simple procedure. The cell cultures were treated with fluorodeoxyuridine to achieve some small degree of synchronization, 15% to 20%, the block in DNA synthesis was relieved with thymidine, and the cells were converted into protoplasts at the appropriate time. The protoplasts were collected and gently lysed, the large debris was spun out at low speed centrifugation, and the chromosomes collected by moderate speed centrifugation. Our yields were good, and the morphology of the chromosomes was acceptable, although not excellent. The key yield limiting steps were the degree of synchronization that could be obtained, and the yield of protoplasts. This led us to try the isolation of meiotic chromosomes from the naturally 100% synchronous meiocytes of *Lilium* and *Hemerocallus*. These tissues proved to be good starting material for the isolation of protoplasts, and hence also of chromosomes. Since the morphologies were not perfect, we performed two tests to show that the structures isolated were in fact chromosomes. First we labelled cells with tritiated thymidine, then isolated chromosomes, and performed autoradiography on the preparations. We were thus able to show that we had isolated DNA-containing structures, and some of the autoradiograms showed correct "X" shaped morphologies. Second, we took our isolated chromosome preparations and isolated histone proteins from them, as shown on an SDS polyacrylamide gel. Thus, we proved that we could isolate chromosomes.

We were able to improve the isolation process substantially by trying more than 200 different buffers and methods of protoplast lysis. The final method (ref. 2, and described in the Materials and Methods section herein) involves lysis by passage through a hypodermic needle. With this improved method we were able to isolate chromosomes with very good morphology and yield from several species and from root tips as well as meiocytes and cell cultures. The yield-limiting steps were still the degree of synchronization and the ability to make protoplasts. We calculated that our yields were approximately 50% from protoplasts containing condensed chromosomes. Figures 1 and 2 show fields of lily and daylily chromosomes isolated by this method.

Chromosome Uptake

In order to assay chromosome uptake by recipient protoplasts we developed a fluorescence assay method (3). Chromosomes were prestained with DAPI, a DNA specific stain that causes yellow or green fluorescence, and then mixed with tobacco leaf protoplasts under a variety of conditions. Since chlorophyll in the leaf protoplasts fluoresced a dull red, we were able to score the association of chromosomes with protoplasts by checking for the presence of the yellow-green chromosomes within red protoplasts under fluorescence microscopy. Several methods were used to indicate that a chromosome was inside a protoplast, and not above or below it. The cover slip could be tapped gently and the protoplast would then float and roll around, dislodging chromosomes only loosely associated with the surface. It is a great deal of fun to watch an object inside a recipient protoplast move as the protoplast rolls around. Also, to a limited extent, optical sectioning could be used to show that the plane of focus of the chromosome was within the protoplast.

Using this assay system, we optimized a standard PEG protoplast fusion regime for the uptake of *Hemerocallus* root chromosomes by tobacco leaf protoplasts. We found distinct optima for length of incubation in PEG, 20 min, and percent PEG, 35%. The ratio of chromosomes to protoplasts was an important variable which caused increasing uptake at higher ratios of chromosomes to protoplasts. The practical limit in terms of being able to isolate a large number of chromosomes was at 10:1 ratio of chromosomes to cells. The actual optimum, if it is not a monotonically increasing variable, is probably at a much higher ratio. Using these best conditions we were able to score a chromosome uptake event in 1% of the protoplasts. *Hemerocallus* chromosomes are much larger than typical tobacco chromosomes, so this uptake is perhaps an unrealistic model of transformation of tobacco protoplasts with tobacco chromosomes. However, we assayed the uptake of the much smaller *Pisum* chromosomes by the same method, and scored the same frequencies. Thus, the uptake method may be a useful starting point for experiments which attempt to achieve a complete chromosome transformation event including expression of a selectable forcing marker.

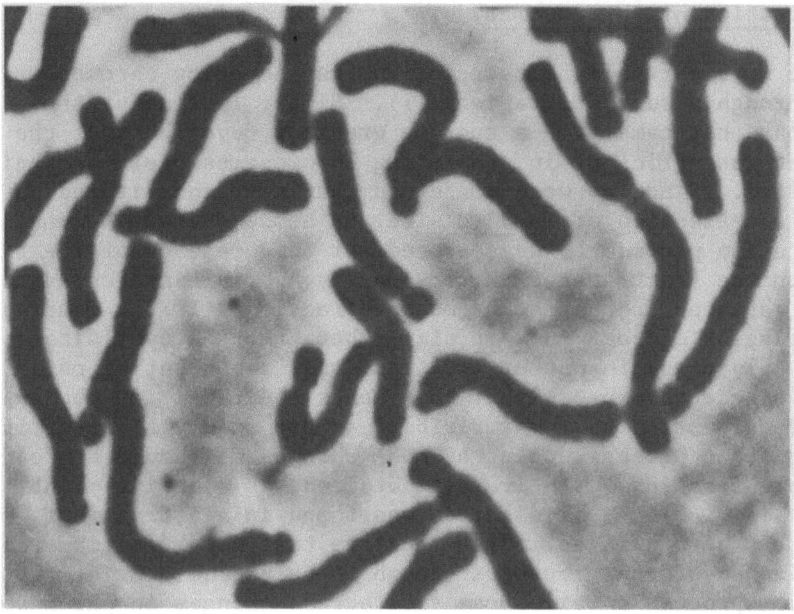

Fig. 1. Field of isolated *Lilium* (lily) chromosomes.

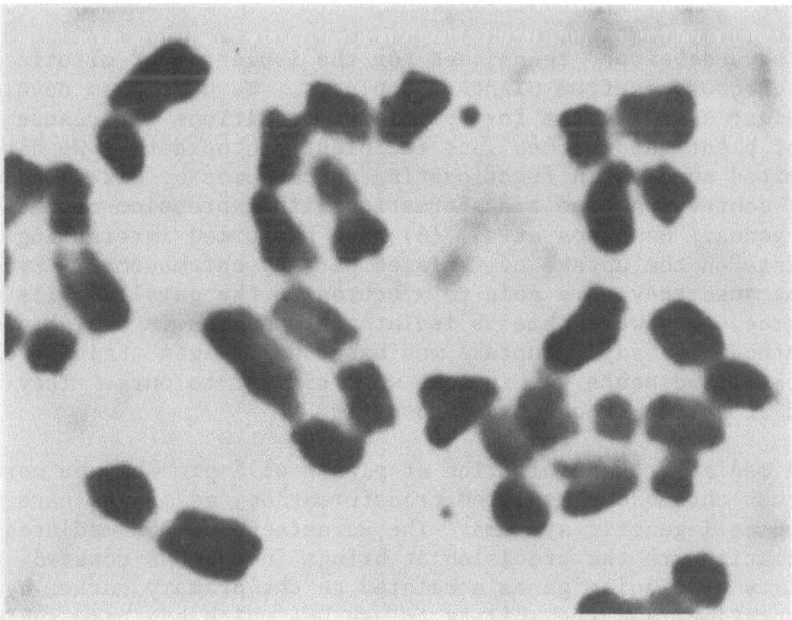

Fig. 2. Field of isolated *Hemerocallus* (daylily) chromosomes.

Fractionation of Chromosomes

We were partially successful in fractionating chromosomes by size through a sucrose gradient (1). Simply looking at the fractions that came from a sucrose gradient revealed that the gradient was efficient in separating chromosomes based on their *in vitro* size. Since we were concerned that this constituted an accurate reflection of *in vivo* size and hence would be a useful purification procedure, we searched for the presence of the rRNA genes on the fractionated chromosomes by hybridizing iodinated rRNA to DNA isolated from each chromosome fraction. If the gradient was separating by correct *in vivo* size, then the rRNA hybridization should only occur in one or two fractions, representing purification of the single chromosome pair that bears the rDNA. If the apparent size fractionation that was visible was due solely to an artifact of the isolation process, then the rDNA chromosome should be found among all fractions. The result obtained was annoyingly intermediate (Fig. 3). Hybridization was found in all fractions, but a broad peak also existed. This suggests that some fractionation by true size does occur, but not enough to be useful. Evidently the isolation process entails some artifactual changes in chromosome size and density.

DISCUSSION

We have developed techniques for the isolation of mitotic and meiotic chromosomes from plant protoplasts. We have also developed a convenient assay system for optimizing conditions that cause the uptake of plant chromosomes into recipient protoplasts. We have had only limited success in fractionating chromosomes by size, and we have not achieved a true transformation with expression of the donated genes. Szabados et al. (6) have performed interesting experiments on the uptake of isolated parsley chromosomes. Evidently because they were able to synchronize the parsley cells to a high degree, they were able to isolate chromosomes in a high yield. Their method of assaying uptake was based on Feulgen staining of the chromosomes, and hence was in some ways similar to ours. They reported an uptake frequency of about 10^{-5}.

DNA mediated transformation of plants will probably be more useful than chromosome mediated transformation, as is the case for mammalian cell genetic systems. The advantages of DNA mediated transformation are the precision it brings in the DNA donated, the possibility of cloning genes unrelated to the primary marker by complementation, and the ability to use DNA which has been subjected to *in vivo* mutagenesis techniques. The advantages of chromosome mediated transfer rest primarily on the possibility that genes linked to the primary selectable marker may be co-transferred even if there is no selection for these traits. Hypothetically this

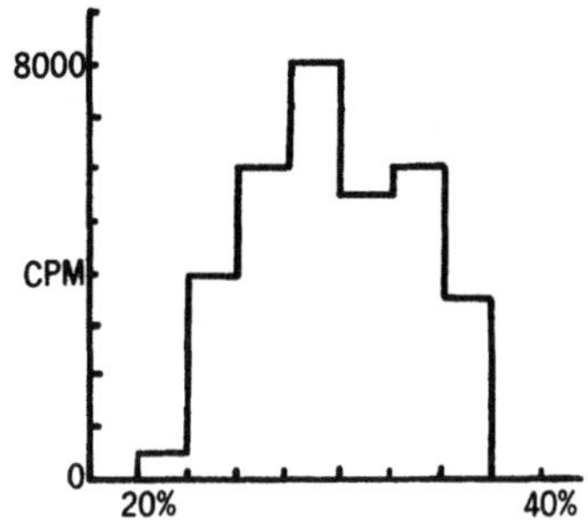

Fig. 3. Hybridization of daylily rRNA to DNA extracted from isolated chromosomes separated on a sucrose gradient.

means that if an agriculturally useful trait were linked to a simple marker, then it might be possible to transfer the trait to a new cell line or species. A variation on this theme would be if the linked traits were polygenic, of small effect, but could be transferred as a unit because of linkage. Since chromosome mediated gene transfer has not been achieved, this discussion is speculative.

REFERENCES

1. Griesbach, R.J. 1980. Ph.D. Thesis, Michigan State University, East Lansing, Michigan, U.S.A.
2. Griesbach, R.J., R.L. Malmberg, and P.S. Carlson. 1982. An improved technique for the isolation of higher plant chromosomes, Plant Sci. Letts. 24:55-60.
3. Griesbach, R.J., R.L. Malmberg, and P.S. Carlson. 1982. Uptake of isolated lily chromosomes by tobacco protoplasts, J. Hered. 73:151-152.
4. Malmberg, R.L., and R.J. Griesbach. 1980. The isolation of mitotic and meiotic chromosomes from plant protoplasts. Plant Sci. Letts. 17:141-147.
5. McBride, O.W., and J.L. Peterson. 1980. Chromosome mediated gene transfer in mammalian cells. Ann. Rev. Genet. 14:321-345.
6. Szabados, L., G. Hadlaczky, and D. Dudits. 1981. Uptake of isolated plant chromosomes by plant protoplasts. Planta 151:141-145.

COMPARISON OF GENOMIC CLONES DERIVED FROM THE *Sh* GENE IN *ZEA MAYS* L. AND OF TWO MUTANTS OF THIS GENE WHICH ARE CAUSED BY THE INSERTION OF THE CONTROLLING ELEMENT *Ds*

H.-P. Döring, M. Geiser, E. Weck, W. Werr,
U. Courage-Tebbe, E. Tillmann, and P. Starlinger

Institut für Genetik
Universität zu Köln
Weyertal 121
D-5000 Köln 41
Federal Republic of Germany

INTRODUCTION

Transposable DNA sequences have been studied extensively biochemically in the last 10 years, particularly in bacteria (17), and later in yeast (6), Drosophila (8), and vertebrates. However, transposable elements were first discovered and thoroughly investigated in maize, and were designated "controlling elements" by B. McClintock (10,11,12,15). The tremendous amount of genetic and physiological information accumulated by McClintock, Peterson, Brink, Rhoades, and other authors led to the understanding of the capabilities of the maize transposable elements.

Maize transposable elements were discovered by studies on unstable alleles. A controlling element inserted in a gene or its vicinity can give rise to a null mutation or to a different expression of the gene in question. However, mutations caused by controlling elements are known to be highly unstable; they can revert frequently to the normal phenotype. Reversion to normal phenotype can be achieved by transposition of the element to another position in the genome.

The transposition itself can be accompanied by different kinds of chromosomal rearrangements including deletions, duplications, and inversions. Controlling elements can cause not only a characteristic instability pattern of gene expression, but the elements are unstable in the sense that the type of the induced instability in gene expression can undergo considerable alterations (e.g., time of reversion or frequency of reversion).

Controlling elements usually fall into two main groups: some elements display their capabilities (e.g., transposition, "changes in state") by themselves; other elements can do so only in the presence of a second trans-acting element.

Since the genetically well-characterized controlling elements are not understood up to now on the molecular level, it might be helpful to isolate these elements. The experiments described in this article are concerned with the controlling element Ds, which can be transposed only in the presence of the trans-acting element Ac.

Ac and Ds belong to the most extensively investigated controlling elements (16). Several mutations at the Shrunken locus which were generated by the insertion of transposable element Ds at the Sh locus were isolated by McClintock (14,13). The Sh locus encodes the enzyme endosperm sucrose synthase in maize (3,4). In homozygotic sh kernels the enzyme is missing or highly reduced, giving rise to collapsed endosperm tissue. As these mutants are caused by the insertion of Ds at the Sh locus, they might enable us to isolate Ds-DNA. Isolation and comparison of genomic sequences of the Sh region of the wild type and of one of the mutants might give us the ability to detect Ds sequences.

Several cDNA clones from sucrose synthase, which are a prerequisite for the identification of the corresponding genomic clones, were isolated (1,2,7,18). Restriction analysis of the Sh region in wild type DNA and DNA derived from different Ds-caused mutants was performed with a cDNA clone as a probe (5). The different restriction maps constructed revealed homology on one side of the map, but differences on the other side. This already suggested that the mutations were caused by DNA rearrangements rather than by point mutations.

RESULTS

A 16 kb BamHI fragment which hybridized to the sucrose synthase cDNA clone was found by cloning wild type Sh DNA into λ1059 (9). This clone was designated λ::Zm Sh. The restriction map constructed previously (5) by hybridization of the cDNA clone to genomic Sh DNA was confirmed with the cloned insert of λ::Zm Sh. The cDNA clone derived from the 3' region of the Sh gene hybridizes to a 6.0 kb BglII fragment. The neighboring BglII fragments towards the 5' region of 3.3 kb and 1.1 kb, respectively, could be located in the genomic clone (Fig. 1). Direction of transcription was determined by hybridizations of labeled subfragments with mRNA and subsequent S1 digestion. At least five introns were detected by S1 mapping. One intron was completely sequenced and showed intron-exon junctions identical to other eukaryotic genes.

Genomic DNA of the mutant sh-m5933 when hybridized to sucrose synthase cDNA revealed a 21 kb BclI band. A recombinant λ clone was isolated which carried a BclI fragment of this size (λ::Zm sh-m5933). Restriction analysis of the cloned fragment and DNA-DNA hybridization showed that a segment of 6 kb located at the right end of the insert of λ::Zm sh-m5933 is identical to a 6 kb segment in λ::Zm Sh. This 6 kb fragment extends from the rightmost BclI site to the point where the DNAs of λ::Zm Sh and λ::Zm sh-m5933 start to deviate from one another. The remaining 15 kb of the 21 kb BclI insert in λ::Zm sh-m5933 do not hybridize to the insert of λ::Zm Sh. This already indicates that some kind of DNA rearrangement must have occurred in the mutant-derived DNA. Subclones of the insert of λ::Zm Sh were prepared containing a 2.25 kb HindIII/BamHI fragment, a 2.7 kb BamHI fragment and a 4.4 kb BamHI/EcoRI fragment, respectively. The point of divergence (junction point) of the two inserts in λ::Zm Sh and λ::Zm sh-m5933 have been roughly determined by restriction analysis, and more accurately by S1 digestion of the appropriate heteroduplex molecules and by DNA sequencing. The junction point is located on the 2.25 kb HindIII/BamHI fragment. The distance from this point to the BamHI site was determined to be 184 nucleotide pairs. From the measurement of the Sh mRNA, the placement of the 3' terminus by ExoVII mapping, and the observation of several introns by means of S1 mapping, we placed the 5' terminus of the gene to the left of the junction point in λ::Zm sh-m5933. Therefore the junction to presumptive Ds-DNA must be located within the Sh gene.

A DNA segment of ∿ 600-800 bp adjacent to the point of divergence in λ::Zm sh-m5933 is repeated in inverted orientation approx. 1.3 kb away. A second pair of inverted repeats of several hundred bp could be detected whose right member is located in the 1.3 kb segment mentioned above and whose second member is found to the left of the first pair of inverted repeats (Fig. 1).

Hybridization of genomic sh-m5933 with a restriction fragment of λ::Zm Sh spanning the junction point revealed that no deletion has occurred in this mutant. Two bands representing a right and a left junction fragment can be detected upon hybridization of the junction-spanning fragment with BglII/EcoRI digested genomic DNA. Another Ds-caused mutation is sh-m6233. A genomic clone carrying an insertion of ∿ 17.5 kb was isolated. When compared with the cloned DNA of λ::Zm Sh and λ::Zm sh-m5933, it turned out that the junction point in λ::Zm sh-m6233 is located 2.5 kb apart from the junction point in λ::Zm sh-m5933. From the tentative placement of the 5' terminus of the Sh gene mentioned above, the junction point in

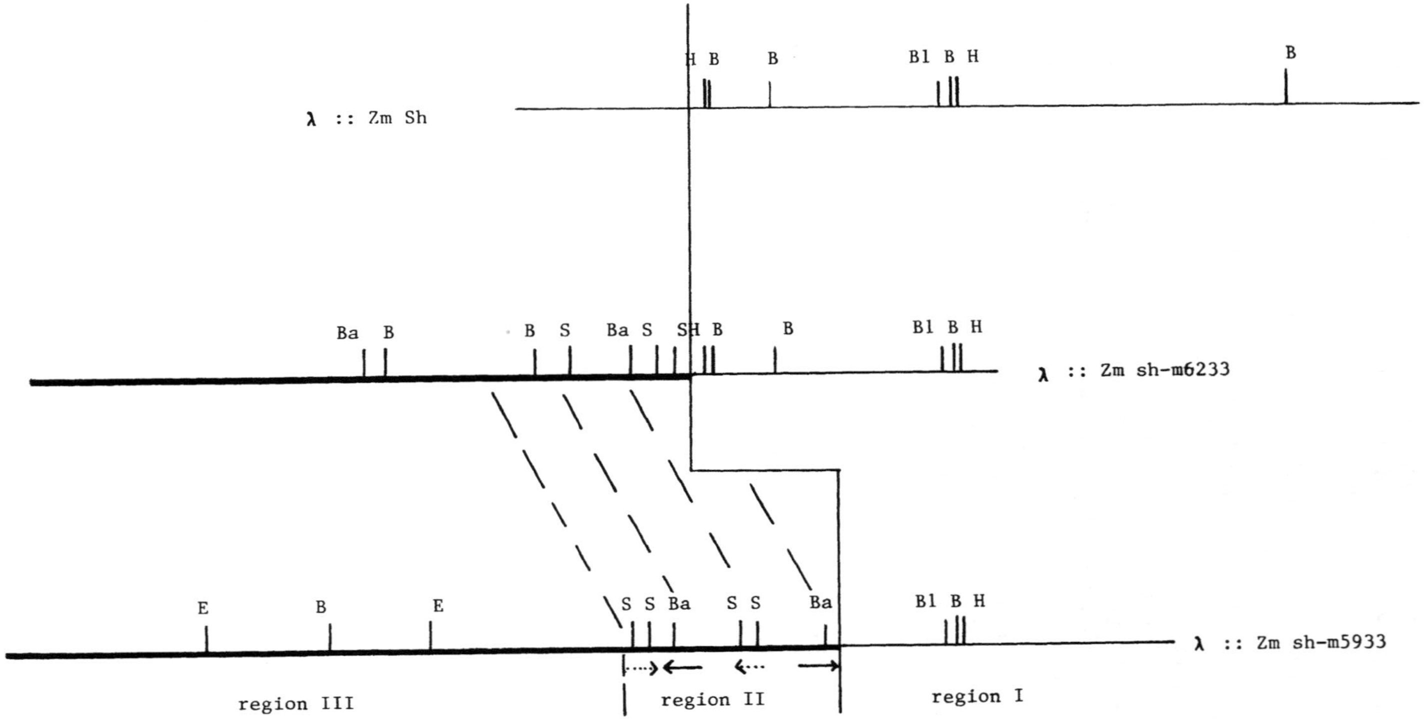

Fig. 1. Restriction maps of the inserts of the three recombinant phages λ::Zm Sh, λ::Zm sh-m6233, and λ::Zm sh-m5933. In region I all of the inserts show homology, region I represents the DNA segment where the Sh gene is located. In region II only partial homology could be detected between the two foreign DNAs present in the inserts of λ::Zm sh-m5933. The thick line represents the presumptive Ds DNA in both mutants. In region III no homology could be detected; this could be due to either different sequences in the internal part of the two Ds elements or due to different orientation of the two Ds elements. B = BglII, H = HindIII, S = SphI, Ba = BamHI, E = EcoRI, Bl = BclI.

λ::Zm sh-m6233 could be located outside the gene. If the junction point were outside the gene, this would mean that Ds insertion can inactivate the Sh gene from a distance.

The region of homology present at the junction point in λ::Zm sh-m5933 and λ::Zm sh-m6233 has a length of between 1.5 kb and 3 kb. Sequences near the junction point in λ::Zm sh-m6233, however, are not identical to λ::Zm sh-m5933. Whereas in sh-m5933 DNA a member of one pair of repeats terminated at the junction point, in sh-m6233 DNA sequences of the second pair of inverted repeats were located at the junction.

The remaining DNA sequences of the two mutant DNAs do not hybridize to each other. If the cloned DNA of sh-m5933 and sh-m6233 which is different to wild type DNA represents Ds-DNA, this could indicate that various Ds elements are of remarkable heterogeneity. However, as only two different Ds-caused mutants of Sh gene were cloned, it cannot be said with certainty whether the mutant-derived DNA sequences represent Ds-DNA or whether they correspond to DNA sequences which have been dislocated under the influence of the Ds element. The assumption that Ds elements can be very heterogeneous is supported by hybridization experiments with different subfragments (i.e., the 2.7 kb BamHI fragment and the 4.4 kb BamHI/EcoRI fragment mentioned above) of mutant-derived DNA of λ::Zm sh-m5933.

These subfragments containing one of the inverted DNA repeats showed similar band patterns when hybridized to restricted genomic maize DNA. The observation of a pattern of bands instead of a single band suggests that different Ds elements differ from one another to some extent. However, the similarity of the patterns indicate that all of these sequences share some DNA sequence, e.g., the inverted repeats. The number of bands obtained with DNA of different maize strains was between 20 and 40. It cannot be estimated whether the individual bands revealed by these hybridizations represent one or several copies of the hybridizing sequence. The number of Ds-like sequences is limited. Thus, the mutant-derived DNA subfragments of λ::Zm sh-m5933 do not represent highly repetitive DNA which would give rise to an unspecific smear rather than to a distinct pattern upon hybridization.

In summary, in two Ds-caused mutations at the sucrose synthase gene in maize, DNA sequences were detected which were not present in the wild type gene. It is possible that these new sequences represent Ds-DNA. If this is true, the two Ds elements share a DNA segment of about 1.5 kb to 3 kb near the junction point with wild type DNA. The remainder of the two Ds sequences were found to be different. The location of the insertion point in the two Ds sequences are different. The results are compatible with the occurrence of an insertion of more than 25 kb or with an insertion that has one breakpoint at the junction point.

The insertion point in sh-m5933 is located in the gene, while in sh-m6233 the insertion is possibly located upstream of the 5' terminus of the gene. The presence of DNA sequences in both mutants investigated, which belong to an inverted repeat, suggests that these sequences may play a role in transposition of the element.

ACKNOWLEDGEMENTS

This research was supported by the "Landesamt für Forschung des Landes Nordrhein-Westfalen" and by the "Kommission der Europäischen Gemeinschaften".

REFERENCES

1. Burr, B., and F. Burr. 1981. Controlling element events at the Shrunken locus in maize. Genetics 98:143-156.
2. Chaleff, D., J. Mauvais, S. McCormick, M. Shure, W. Wessler, and N. Fedoroff. 1981. Controlling elements in maize. Carnegie Inst. Wash. Yearbook 80:158-174.
3. Chourey, P., and O. Nelson. 1976. The enzymatic deficiency conditioned by the shrunken mutation in maize. Biochem. Genetics 14:1041-1055.
4. Chourey, P., and D. Schwartz. 1971. Ethyl methane sulfonate-induced mutation of the Sh protein in maize. Mutat. Res. 12:151-157.
5. Döring, H.P., M. Geiser, and P. Starlinger. 1981. Transposable element Ds at the shrunken locus in Zea mays, Mol. Gen. Genet. 184:377-380.
6. Fink, G., P. Farrabough, G. Roeder, and D. Chaleff. 1980. Transposable elements (Ty) in yeast. Cold Spring Harbor Symp. Quant. Biol. 45:575-580.
7. Geiser, M., H.P. Döring, J. Wöstemeyer, U. Behrens, E. Tillmann, and P. Starlinger. 1980. A cDNA clone from Zea mays endosperm surcose synthase mRNA. Nucleic Acids Res. 8:6175-6188
8. Green, M.M. 1980. Transposable elements in Drosophila and other Diptera. Ann. Rev. Genet. 14:109-120.
9. Karn, J., S. Brenner, L. Barnett, and G. Cesarini. 1980. Novel bacteriophage λ cloning vector. Proc. Natl. Acad. Sci. U.S.A. 77:5172-5176.
10. McClintock, B. 1946. Maize genetics. Carnegie Inst. Wash. Yearbook 45:176-186.
11. McClintock, B. 1947. Cytogenetic studies of maize and Neurospora. Carnegie Inst. Wash. Yearbook 46:146-152.
12. McClintock, B. 1948. Mutable loci in maize. Carnegie Inst. Wash. Yearbook 47:155-169.
13. McClintock, B. 1952. Mutable loci in maize. Carnegie Inst. Wash. Yearbook 51:212-219.

14. McClintock, B. 1953. Mutation in maize. Carnegie Inst. Wash. Yearbook 51:227-237.
15. McClintock, B. 1956. Controlling elements and the gene. Cold Spring Harbor Symp. Quant. Biol. 21:197-216.
16. McClintock, B. 1965. The control of gene action in maize. Brookhaven Symp. Biol. 18:162-184.
17. Starlinger, P. 1980. Review: IS elements and transposons. Plasmid 3:241-259.
18. Wöstemeyer, J., U. Behrens, A. Merckelbach, M. Müller, and P. Starlinger. 1981. Translation of *Zea mays* endosperm sucrose synthase mRNA *in vitro*. Eur. J. Biochem. 114:39-44.

PLANT GENE STRUCTURE

Joachim Messing, Daniel Geraghty, Gisela Heidecker*, Nien-Tai Hu, Jean Kridl, and Irwin Rubenstein

Department of Biochemistry and
Department of Genetics and Cell Biology
University of Minnesota
St. Paul, Minnesota 55108

SUMMARY

Techniques in molecular cloning and DNA sequencing have provided the tools to study the structure of genes at the nucleotide level. Most of these studies have been conducted on mammalian genes. From the comparison of individual genes much knowledge has been gained about organization and potential signal sequences. Although relatively little sequence data is available for plant genes, their number has grown to a degree where similar comparisons may be initiated. Using our studies of the zein storage protein and the data of other laboratories we may draw the following conclusions: 1) Like other plant genes, zein genes are organized in multigene families. Hybridization techniques and sequence data further subdivide the zein multigene family into subfamilies. 2) The sequence data obtained also allows us to determine the protein sequence and the sequence variation among the zein proteins which has occurred during evolution. 3) Comparing the zein genes and other known plant genes, we have identified potential signal sequences which can be distinguished from those of animal genes.

INTRODUCTION

Many laboratories are investigating gene primary structure in an attempt to answer the question of how genes are organized and regulated. Observed changes in primary structure can sometimes

*Permanent address: Department of Human Pathology, University of California, Davis, California 95616.

explain functional changes as well as evolutionary relationships. Although we strive to understand the basic mechanisms of biological systems, the data gathered and catalogued in the short-term is fundamental to the genetic engineer who intends to employ biological systems for industrial purposes.

The presence of multigene families and an abundance of non-coding sequences in the genomes of eukaryotic cells makes it necessary to sequence very long stretches of DNA. This requirement has made the purification and amplification of DNA fragments a very important part of DNA sequencing methodology. There are three major E. coli vector systems for purification of DNA fragments: one using plasmids (7), one using bacteriophage lambda (44,48,57), and one using the single-stranded bacteriophage M13 (41). The M13 system has been used to combine DNA sequencing and cloning into an integrated strategy for studying the primary structure of DNA (25,42,43). The cloned DNA segments generated in the M13 system do not serve only as templates for DNA sequencing, but also as a source of single-stranded DNA for single-strand specific hybridization (28) and site-directed mutagenesis (30,52).

Our current understanding of eukaryotic gene structure is based mainly on comparisons of the sequences of animal genes. One might ask if the same structural features which are thought to be important for the control of gene expression in animal systems are used in plant systems and, furthermore, if plant genes are organized in multigene families. Recently, data on some plant gene sequences have become available allowing tentative identification of structural features important for gene expression in plants. By studying the primary structure of plant genes of a common trait or function one can begin to look for sequences that are involved in expression and examine protein structure and its relationship to function.

THE ZEIN MULTIGENE FAMILY

The Maize Storage Protein as a Model System

One plant gene system that shares common traits and functions is the zein storage proteins of maize. These gene products are found in the endosperm tissue of the seeds and are made at a specific time during seed development. It is believed that these proteins function as a source of nitrogen and energy for the germinating seeds. Since zeins represent the most abundant proteins in endosperm during kernel development they have been of interest for two reasons: 1) as a model gene system for studying regulation of gene expression during development, and 2) for the improvement of the nutritional value of corn. Since the quality of corn as a food source for non-ruminant animals depends on the amino acid composition of the zeins, the analysis of their structure is directly related to a breeding program for the production of more nutritious corn.

Zein Protein Analysis

The analysis of zein and of future breeding steps is somewhat hampered by the complexity of the product. Zein is not a single protein, but is rather a family of possibly some one hundred related protein products. Zein isolated from the endosperm of maize kernels falls into two major size classes as determined by SDS gel electrophoresis (19,53). However, the complexity of zein goes beyond these sized classes. Protein sequencing indicates that there is microheterogeneity in amino acid sequences in addition to differences in molecular weight (3). This agrees with the analysis of zein proteins based on charge differences by isoelectric focusing (49). These data indicate that zein may be organized as a multigene family. Further support for this hypothesis was obtained by using variants in isoelectric focusing patterns to map the zein genes. It appears that the genes are located on at least three chromosomes (54). Based on two-dimensional gel electrophoresis data a minimum number of 25 zein genes has been estimated (23).

Microheterogeneity Within Zein mRNAs

A better understanding of the complexity of the zein family can be achieved by investigating the primary structure of individual genes. Since it is difficult to purify individual members of the zein protein family, it is useful to employ molecular cloning techniques to examine the genes and deduce the protein sequence from the nucleotide sequence. Endosperm tissue is a rich source of zein proteins and their mRNAs. The cDNAs prepared from this mRNA (6,26,46,62) were used to search for zein genes within a genomic library (38,46). The cDNA clones were characterized by hybrid-select translation (45). These studies confirmed that the charge heterogeneity observed at the protein level was not due to post-translational modification, but was due to the presence of many different mRNAs. The data support the hypothesis that the zein proteins are organized in a large multigene family. In addition, the hybridization properties of an individual member of the zein cDNA clones with respect to the entire mRNA population can be used to define the organization of zein genes into subfamilies. Although it is still unclear how many different mRNAs occur within the zein multigene family, the hybrization studies indicate that at least 3 to 5 subfamilies of zein mRNA exist (6,45), and that there are 50 to 100 genes per maize haploid genome (24).

If hybridization properties can be used to relate one cDNA clone to a group of other cDNAs or mRNAs then the same experiments can be used to relate genomic sequences to the subfamilies. Each subfamily has a characteristic restriction pattern that varies among inbred lines (38,59,63). Ultimately, the direct comparison of the nucleotide sequences of individual members of zein subfamilies will allow definition of the homology among subfamilies and the relationships of individual members of a subfamily to one another.

The Zein Family Tree

Based on current sequence and hybridization data the zein multigene family is divided into several subfamilies as shown in Fig. 1. Each subfamily is defined by sequence homology to a cDNA clone: A20, A30, B49, B59, or B36. Both B49 and B36 select mRNAs in hybrid-select translation studies that code for predominantly heavy class (23 kd) proteins, and A20, A30, and B59 select for predominantly the light class (19 kd) proteins (6). Where possible, chromosome location is also indicated. Other clones, cDNA or genomic, are related to the major cDNA representing the subfamily either by hybridization data or sequence data. Hybridization data are based on hybrid-select translation (45,62), dot hybridization (6) and genomic Southern blots (38,59). Nucleotide sequence data are from Pedersen et al. (46), Marks and Larkins (39), and our own laboratories. We have indicated that there may be joint members between any two subfamilies as suggested by the results of Viotti et al. (59).

ZEIN PROTEIN STRUCTURE

The Different Domains of the Zein Protein

In addition to providing information on the organization of a multigene family, a comparison of zein sequences provides information about the structure and variability of the zein proteins. We have drawn a consensus protein sequence (from 11 cDNA clones) for each of the A20, A30, and B49 subfamilies and have listed these in Figure 2. The protein sequences are divided into four distinct regions. Region I comprises the signal peptide present in most if not all zeins (5). Regions II and IV contain the amino and carboxyl termini respectively, and region III covers the remaining two thirds of the coding sequence including the tandem repeats of 20 amino acids first seen in A30 (16).

The Signal Peptide and the Terminal Regions

The signal peptide of the zeins is relatively invariant in length (21 amino acids), and is typical of that found in other eukaryotes (32). The salient features include a lysine near the

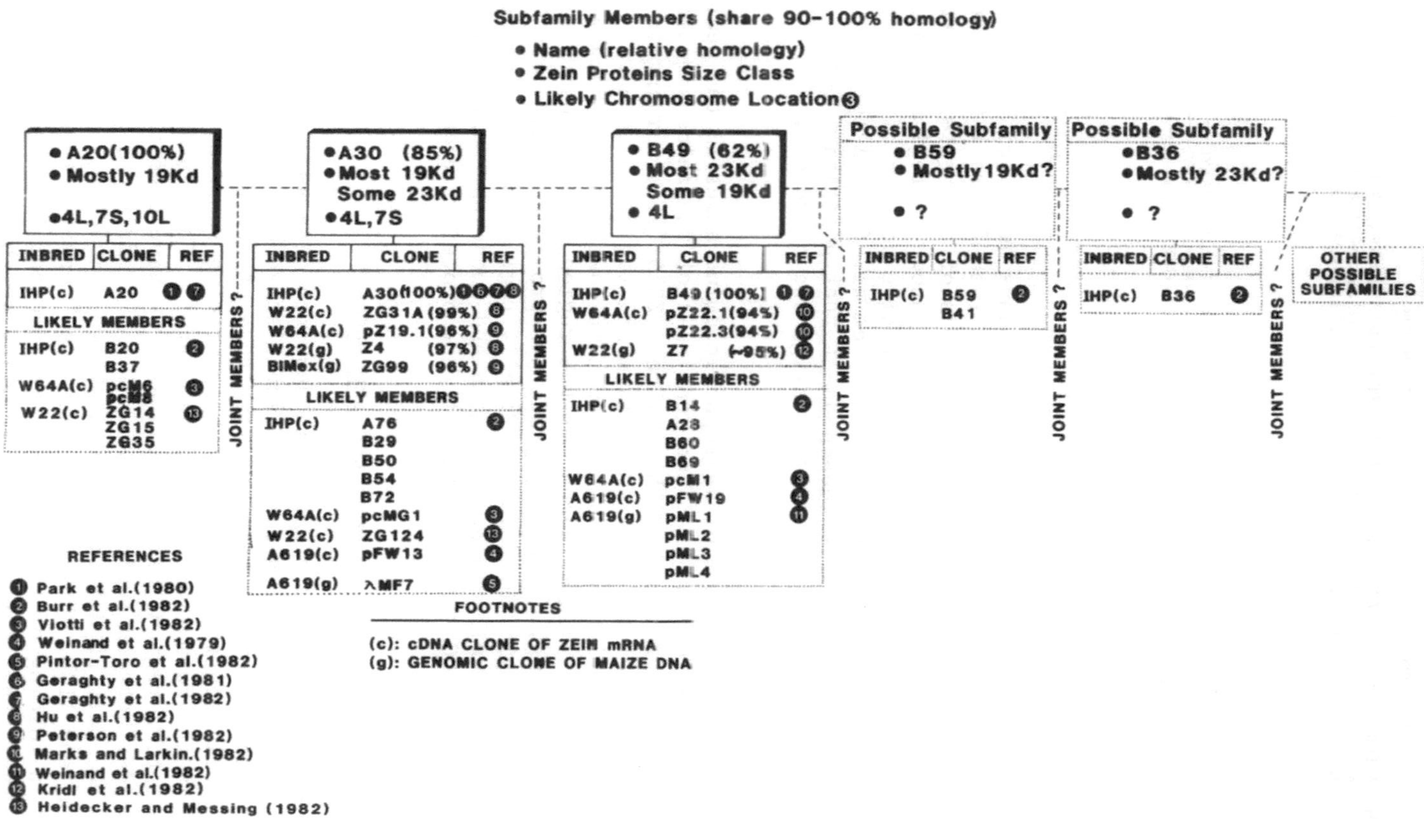

Fig. 1. The zein multigene family.

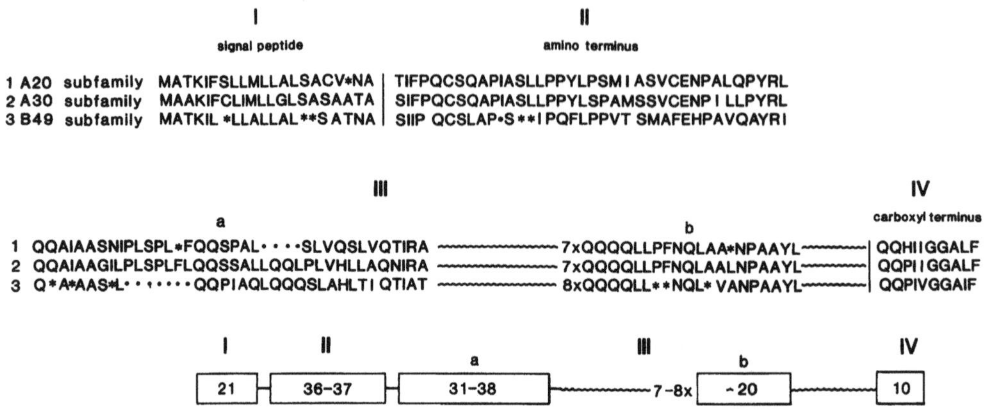

Fig. 2. Zein protein structure. A consensus amino acid sequence for each of the zein subfamilies (see figure 1 for references) is shown. The subdivisions of the protein structure are described in the text. Shown in region IIIb is a consensus sequence of the repetative portion of the zein proteins described in Geraghty et al. (17). Astericks indicate lack of a consensus at that position. Dots represent gaps inserted to align the sequences.

amino terminus, a sequence of hydrophobic amino acids in the center and an alanine residue at the junction. All of the zeins sequenced to date have these features with only one, ZG15 (26), varying in length (20 amino acids). The amino terminal regions also have little size variability and contain certain conserved sequences. Of possible structural significance for the region is the conservation, in similar positions, of the dipeptide pro-pro, a glutamic acid residue, and an arginine residue. Finally region IV, the carboxyl terminus, is very well conserved within and among the subfamilies. The sequence QQ*I*GGA*F terminates all the protein sequences.

The Central Region and its Repeated Sequences

Agros et al. (1) have used zein sequence data as well as physical data to develop a somewhat speculative but nonetheless intriguing model for the tertiary structure of the zein proteins. In this model each repeat, bounded by glutamine residues, is involved in the formation of a helical wheel. The glutamines constitute turn regions between helices. We have used this study to draw a border between regions II and III. In addition, we have drawn a secondary border to distinguish two domains within region III, IIIa, and IIIb. Their model has nine "repeats" that are involved in folding of the protein. It is clear from the sequence data, however, that the sequences in IIIa, which comprise the first two of the nine "repeats", have homology with the IIIb sequences limited to the presence of glutamines at the ends. While the IIIa sequences may indeed be involved in the formation of helical wheels, the high degree of homology among the IIIb sequences (17) and therefore the implied unique evolution of this region of the protein, necessitates their separate categorization.

It is in region III that one finds the greatest variability in size within and among the subfamilies. All three subfamilies show insertions and deletions relative to one another in region IIIa. The B49 subfamily can be considered to have an additional tandem repeat in IIIb. It must be pointed out, however, that the difference in protein size of the B49 subfamily and the A20 and A30 subfamilies is not simply due to an additional tandem repeat. There are what appear to be short insertions and deletions throughout the sequence (17,32). Within the A30 subfamily a rather interesting variation is found. Hu et al. (29) have sequenced a zein genomic clone, Z4, containing a duplication of 32 amino acids relative to A30. This duplication involves sequences overlapping regions IIIa and IIIb and has glutamines at its "ends". Since the structural model discussed above includes this same region as two helical structures, it may be reasonable to assume that the structure of the mature zein protein contains these helices as building blocks. If this is so, then A30 would contain 9 helices and Z4 would have 11.

Potential Region for Site-directed Alterations

The last region of interest occurs at the junction between IIIb and IV. There is a short sequence of marked variability which includes a tryptophan residue in B49 (17). Tryptophan and lysine are underrepresented in zein proteins and contribute to the inferior protein quality of corn as feed for nonruminant livestock. This region may be of interest to the genetic engineer for introducing alterations in the primary structure without violating any structural features of the zein storage proteins.

PLANT GENE STRUCTURE

DNA Sequence Comparisons

Comparisons of a multitude of eukaryotic genes have revealed a number of eukaryotic signal sequences potentially involved in the initiation of transcription, translation, and processing of the primary transcript. Relatively little of this sequence data has been derived from plant genes. Besides the zein multigene family, other genes such as the French bean phaseolin, the soybean leghemoglobin, maize alcohol dehydrogenase, and the soybean actin genes have been studied at the nucleotide sequence level (16,18,31,51,56). Also, the sequences of two cauliflower mosaic virus strains have been determined (14,15). Because cauliflower mosaic virus is transcribed in the plant nucleus (22), its sequences constitute an important contribution to the study of transcriptional control of plant genes.

We have used the putative regulatory sequences that have been described in animal genes as a guide in a search for similar plant homologies. The studies mentioned above have been extended recently with more information from the 5'-noncoding regions (17,26,39, 46,61). Figure 3 lists the pertinent homologies from a representative sample of these plant genes with a consensus sequence drawn for each.

Intervening Sequences

Like other eukaryotic genes, plant genes have intervening sequences. Nucleotide sequence data of the French bean phaseolin gene (56), soybean leghemoglobin genes (4,31,61), and a soybean actin gene (51), and R loop and S1 nuclease protection studies of a soybean glycinin gene and a rare class leaf gene (12) have shown that all these plant genes have 2 to 3 introns. In contrast, the maize storage protein zein gene family lacks introns (29,46,63) as do two other soybean protein genes and one rare class leaf gene as shown by R loop analysis (12).

Where intervening sequences are present, all plant genes are identical in the first and the last two bases of the introns, GT and AG, respectively. The conservation of the two dinucleotides at the intron/exon junctions in all eukaryotic genes (37) suggests that similar RNA splicing mechanisms are involved. Comparisons of plant storage proteins with animal storage proteins show that the former have simpler intergenic structures than the latter. Plant storage proteins have from zero (zein in maize) to three introns (phaseolin in French bean), whereas ovalbumin genes of chicken have seven introns (27) and vitellogenin genes have 33 introns (60). The significance of this difference is not clear.

The Polyadenylation Signal

Almost all of the animal genes studied thus far, including the animal viruses, have AATAAA located about 10-33 bases upstream from the poly A tail (2). This consensus sequence is not found in the unpolyadenylated histone genes (33), nor is it present in any yeast genes, which are nevertheless polyadenylated (64). Recent studies of a deletion mutant of the yeast iso-1-cytochrome c gene have revealed among yeast genes a consensus sequence different from AATAAA (64). The deletion of 38 bases, which included the yeast

Clone	Ref.	Agga Box	(36–59)	Tata Box	(29–33)	Cap Site	(11–68)	Translation Start	Introns	Stop	Poly A Addition Signals	3'
(g) pSAc3	50	TTTTAAGTGAATCCT	(51)	TCCATACAA TA	(?)	x	(?)	AAAAGATGG	3	TAA		
Lba	30			TCTATATAAACA	(32)	x	(49)	GAAATATGG	3	TAA	(29)AATAAA	GGATAAA(23/144)
Lbc1	30	AGCCAAGAGAAACTT	(40)	TCTATATAA CA	(32)	x	(57)	GAAATATGG	3	TAA	(29)AATAAA	GGATAAA(23/144)
Lbc2	59	AGCCAAGAGAAACTT	(40)	TCTATATAAACA	(33)	x	(54)	GAAATATGG	3	TAG	(29)AATAAA	TGATAAA(29/144)
Lbc3	59			TCTATATAAATA	(33)	x	(49)	GAAATATGG	3	TAG	(29)AATAAA	GGATAAA(* /131)
GmLb11	4	AGCCAAGAGAGACAT	(48)	TCTATATAAATA	(33)	x	(49)	GAAATATGG	3	TAG	(29)AATAAA	GGATAAA(* /126)
Cabb-S	13	TCCTAATTGAAATCC	(59)	CCTATTTAAACA	(33)	x	(11)	GAAATATGG	0	TGA		CAATAAA(34/295)
CMV1841	14	TCCTAATTGAAATCC	(58)	CCTATATAAACA	(32)	x	(11)	CAAGCATGG	0	TGA		CAATAAA(34/305)
ZG99	45	GCAAAATCGAAAATT	(39)	TGTATAAA TA	(30)	x	(57)	CAAGCATGG	0	TAG	(25)AATAAA	AAATAAG(23/87)
Z4	28	GCAAAATCGAAAATT	(39)	TGTATAAA TA	(29)	x	(57)	CAACAATGG	0	TAG	(25)AATAAA	AAATAAG(23/87)
Z7	34	CAAAAAGACAAAATC	(36)	TGTATGAA TA	(29)	x	(66)	CAATAATGG	0	TAG		AAATAAT(37/97)
(c)A20	16					x	(68)	CAACAATGG		TAG	(25)AATAAA	AAATAAG(23/87)
ZG14	25					x	(68)	GAACAATGG		TAG	(25)AATAAA	AAATAAC(23/87)
A30	15							CAATGG		TAG	(25)AATAAA	AAATAAG(23/87)
ZG124	25					x	(57)	CAACAATGG		TAG	(25)AATAAA	AAATAAG(16/80)
ZG19	25							CAACAATGG		TAG	(25)AATAAA(12-43)	
B49	16									TAG		AAATAAT(37/97)
pZ22.3	38							CAACAATGG		TAG		AAATAAT(17/77)
pZML84	17									TAG	(44)AATAAA	AATGAG(35/372)
pRC2.11.7	9									TAG	(30)AATAAG	AAATAAA(23/134)

	Agga Box	Tata Box	Translation Start	Poly A Addition Signals
Plant Consensus	$^{C}_{T}A_{2-5}{}^{G}_{T}NGA_{2-4}{}^{CC}_{TT}$	$T^{C}_{G}TATA^{T}_{A}A_{1-3}{}^{C}_{T}A$	$^{C}_{G}AANNATGG$	$^{G}_{A}ATAA_{1-3}$
Animal Consensus	$GG^{C}_{T}CAATCT$	$TATA^{T}_{A}A^{T}_{A}$	$^{A}_{G}NNATGG$	AATAAA

Fig. 3. Plant gene structure. The potential control sequences of plant genes are aligned. Bold face type indicates agreement with the consensus sequences listed at the bottom of the figure. The numbers in parentheses are the distances in base pairs between these potential control regions. Numbers on the far right of the figure indicate the distance of the second poly A addition signal to the 3'-end of the message followed by the total length of the 3' noncoding region. Sequence data information has been evaluated using the computer software for the Apple II microcomputer (Larson and Messing, 36).

consensus sequence, made the mutant synthesize longer messages than wild type, all of which were polyadenylated. This implies that the yeast consensus sequence is important for the correct position of transcriptional termination and not polyadenylation. No equivalent studies have been done with other animal genes. However, in vivo studies of SV40 late genes did suggest that the AATAAA at the 3'-end is required for polyadenylation of the viral messages and also determines the position of the poly A tail (13).

Is the structure of the 3'-end of plant genes similar to animal or yeast genes? By comparing gene sequences from the zein multigene family with other published plant gene sequences, we have obtained the following picture. The animal consensus sequence AATAAA is found in all the plant genes examined, with the exception of the B49 subfamily of zein. This subfamily, which includes Z7, B49, and pZ22.3, has the sequence AATAAT instead of the normal AATAAA. Even though most plant genes have AATAAA, its location with respect to the poly A tail differs from that in animal genes. All of the leghemoglobin genes, most of the zein genes and the alcohol dehydrogenase genes of maize have their AATAAA closer to the stop codon of the protein coding sequences than to their poly A tail. Variants of AATAAA, with a consensus sequence of (A/G)ATAA1-3, are found in these genes located approximately where AATAAA is found in animal genes. An exception is ZG19, a zein cDNA clone (26). This clone has high sequence homology to the zein cDNA clones A30 and ZG124 and the zein genomic clone Z4. However, the 3'-noncoding region of ZG19 is shorter than that of the other A30-like zeins. Poly A of the ZG19 message is added 15 bases downstream from the normal polyadenylation signal, AATAAA. The transcription of the other A30 subfamily members seems to have bypassed this AATAAA and terminated after the second signal. While it is not likely that the difference at the 3'ends of ZG19 and the other A30 members resulted from the cDNA cloning procedures, we cannot eliminate the possibility that sequence rearrangement occurred during propagation of the plasmid. The difference between A30 and ZG124 in the distances of their poly A tails to the same putative signal, AATAAG, could reflect the true lengths of different messages, which in turn would suggest that the position of poly A may vary within a few bases. The possibility of the difference being a result of artifacts has not been ruled out completely, but interestingly, such a difference is also observed between another pair of closely related B49-like zein cDNA clones, B49 and pZ22.3 (17,39).

In addition to the A30- and A20-like zeins, most of the other plant genes discussed here have two putative polyadenylation signals. The leghemoglobin and alcohol dehydrogenase genes are similar to the zeins in having AATAAA close to the stop codon and another variant of this sequence close to the poly A. The leghemoglobin genes have GATAAA and the alcohol dehydrogenase genes have AATGAG. In contrast, the legumin gene has AATAAG near the stop codon and AATAAA near the poly A tail (9). The inclusion body protein gene of cauliflower mosaic virus seems to be more similar to animal genes than to plant genes at the 3'-terminus. Only one AATAAA is located 34 bases upstream from poly A. There is no other variant. In summary, it would appear from the available data that multiple polyadenylation signals occur more often in plant genes than in animal genes, where only a few multiple polyadenylation sites have been observed (11,50,58), and that poly A signals in plant genes show considerably more variation in sequence.

The ATG Initiation Codon

All of the plant genomic and cDNA clones discussed here (20 sequences) have sequences surrounding the ATG initiation codon which are consistent with the consensus sequence derived by Kozak (34) from a large number of mRNAs. The homology can be extended in plant genes to include two additional nucleotides as seen in Fig. 3. Although the number of examples we have available is admittedly quite small, the possibility that there may be additional sequences important for proper recognition of the translation initiation site remains open.

The TATA Box

Of the sequences thought to be important in transcription initiation, the TATA-box or Goldberg Hogness-box (47) is the best characterized. This sequence is required for correct expression of eukaryotic genes *in vitro* (20) and accurate, efficient initiation *in vivo* (40). Since plant genes are also transcribed by RNA polymerase II, it is not surprising that all the plant genes described have a sequence analogous to the TATA box. The consensus sequence we have drawn includes some sequences surrounding a sequence similar to the animal consensus, giving a somewhat extended homology. The position of the transcription initiation site or cap site has been determined for cauliflower mosaic virus inclusion body mRNA (8), and for the leghemoglobin gene GmLb11 (4). We have estimated the length of the leader sequences of the other leghemoglobins using this information since all the members of this gene family share significant homology. The zein leader sequences have been estimated using cDNA clones containing what we believe are complete copiers of the corresponding mRNA (26). The combined information gives distances from the TATA box (measured from the second T) to the cap site of 29-33 nucleotides, typical of that found in other eukaryotes.

The sequence of the TATA homology in the soybean actin gene pSAc3 (51) is somewhat ambiguous. There are no data available on the transcription initiation site and there are two potential homologies upstream of the translation start codon. The first sequence, TGTAAATG, and subsequent sequences share homology with a similar region of the yeast actin gene (51). The second potential homology occurs about 30 nucleotides downstream of the first. We have included this sequence in Fig. 3 as it more closely resembles the analogous sequence of the other soybean genes and the other plant genes. Clearly, experimental data will be required to determine which sequence (if either) is necessary for proper initiation.

The AGGA Box

Another sequence that may be involved in regulation of transcription of some eukaryotic genes is the consensus sequence GG(C/T)CAATCT or "CAAT-box" (2). This sequence appears 80-100 nucleotides upstream of the cap site. Deletion of the CAAT box significantly reduces the transcription of the rabbit B-globin gene *in vivo* (21) although its sequence alteration had little effect on transcription of the thymidine kinase gene *in vivo* (40). Two zein genomic clones, Z4 and Z7 (29,35), have sequences with limited homology to the CAAT box. The leghemoglobin genes (4) have three sequences upstream of the coding region, all having some homology to the animal sequence. Only one of these, however, has a feature common to the other plant genes; the presence of a short tract of As two nucleotides after the CAAT homology. Indeed, all of these plant sequences have an interesting symmetry of adenines surrounding the trinucleotide (G/T)NG. We have named this possible regulatory sequence the AGGA box to differentiate it from its animal counterpart and have drawn a consensus sequence shown in Fig. 3. Comparison of future plant gene sequences will indicate if this homology is significant. It may be that despite the similarities of animal and plant systems at the molecular level the latter has its own characteristic features.

After the preparation of this manuscript we received a preprint of the DNA sequences of the octopine and nopaline synthase genes (10,55). These genes are encoded by *Agrobacterium tumefaciens* Ti plasmids and are expressed in plants. Both of these sequences have the dinucleotide TC preceding the TATA box, consistent with the other plant sequences discussed here. Both also contain polyadenylation signals which vary in position and sequence. The nopaline 3'-noncoding region is 175 nucleotides long with AATAAA 42 nucleotides after the stop codon and AATAAT 53 nucleotides preceding the 3'-end. The 3'-noncoding sequence of the octopine message can apparently vary in length as two sizes of 165 and 189 nucleotides are seen for this region. The ends of both of the messages are preceded by AATATA at 21 and 33 nucleotides from the poly A tail and the longer predominant message includes AATAAT 12 nucleotides before the poly A. The nopaline sequence has a CAAT homology 80 nucleotides upstream of the cap site with 7 of 9 nucleotides in agreement with the animal consensus sequence. No sequence resembling the plant AGGA box can be found. Conversely, the octopine gene contains two sequences resembling the AGGA box, GAAAGTTAAAGG and CAAGTCAATA, 43 and 25 nucleotides respectively, upstream of the TATA homology.

ACKNOWLEDGEMENTS

We thank Kris Kohn for aid in preparing the manuscript. This research was supported by grants from the National Institutes of Health, GM24756; from the USDA/SEA Competitive Grant Program-Genetic Mechanisms for Crop Improvement, 59-2271-0-1401-0; from the Department of Energy, DE-AC02-81ER 10901; from the Minnesota Agricultural Experiment Station, MN-15-030; and from the National Institutes of Health, ST32 GM07467-05.

REFERENCES

1. Argos, P., K. Pedersen, M.D. Marks, and B. Larkins. 1982. A structural model for maize zein proteins. J. Biol. Chem. (in press).
2. Benoist, C., K. O'Hare, R. Breathnach, and P. Chambon. 1980. The ovalbumin gene-sequence of putative control regions. Nucl. Acids Res. 8:127-142.
3. Bietz, J.A., J.W. Paulis, and J.S. Wall. 1979. Zein subunit homology revealed through amino-terminal sequence analysis. Cereal Chem. 56:327-332.
4. Brisson, N., and D.P. Verma. 1982. Soybean leghemoglobin gene family: Normal pseudo, and truncated genes. Proc. Natl. Acad. Sci. USA 79:4055-4059.
5. Burr, B., F.A. Burr, I. Rubenstein, and M.N. Simon. 1978. Purification and translation of zein messenger RNA from maize endosperm protein bodies. Proc. Natl. Acad. Sci. USA 75:696-700.
6. Burr, B., F.A. Burr, T.P. St. John, M. Thomas, and R.W. Davis. 1982. Zein storage protein gene family of maize. J. Mol. Biol. 154:33-49.
7. Cohen, S.N., A.C.Y. Chang, H.W. Boyer, and R.B. Helling. 1973. Construction of biological functional bacterial plasmids *in vitro*. Proc. Natl. Acad. Sci. USA 70:3240-3244.
8. Covey, S.N., G.P. Lomonossoff, and R. Hull. 1981. Characterization of cauliflower mosaic virus DNA sequences which encode major polyadenylated transcripts. Nucl. Acids Res. 24:6735-6747.
9. Croy, R.R., G. Lycett, J.A. Gatehouse, J.N. Yarwood, and D. Boulter. Cloning and analysis of cDNAs encoding plant storage protein precursors. Nature 295:76-78.
10. De Greve, H., P. Dhaese, H. Seurinck, M. van Montagu, and J. Schell. 1982. Nucleotide sequence and transcript map of

Agrobacterium tumefaciens Ti plasmid-encoded octopine synthase gene. J. Mol. Appl. Genet. (in press).
11. Early, P., J. Rogers, M. Davis, K. Calame, M. Bond, R. Wall, and L. Hood. 1980. Two mRNAs can be produced from a single immunoglobulin u-gene by alternative RNA processing pathways. Cell 20:313-319.
12. Fischer, R.L., and R.B. Goldberg. 1982. Structure and flanking regions of soybean seed protein genes. Cell 29:651-660.
13. Fitzgerald, M., and T. Shenk. 1981. The sequence 5'-AAUAAA-3' forms part of the recognition site for polyadenylation of late SV40 mRNAs. Cell 24:251-260.
14. Frank, A., H. Guilley, G. Jonard, K. Richards, and L. Hirth. 1980. Nucleotide sequence of cauliflower mosaic virus DNA. Cell 21:285-294.
15. Gardner, R.C., A.J. Howarth, P. Hahn, M. Brown-Luedi, R.J. Shepherd, and J. Messing. 1981. The complete nucleotide sequence of an infectious clone of cauliflower mosaic virus by M13mp7 shotgun sequencing. Nucl. Acids Res. 9:2871-2888.
16. Geraghty, D., M.A. Peifer, I. Rubenstein, and J. Messing. 1981. The primary structure of a plant storage protein: Zein. Nucl. Acids Res. 9:5163-5174.
17. Geraghty, D., J. Messing, and I. Rubenstein. 1982. Sequence analysis and comparison of cDNAs of the zein multigene family. (submitted for publication).
18. Gerlach, W.L., Pryor, A.J., E.S. Dennis, R.J. Ferl, M.M. Sachs, and W.J. Peacock. 1982. cDNA cloning and induction of the alcohol dehydrogenase gene (ADH1) of maize. Proc. Natl. Acad. Sci. USA 79:2981-2985.
19. Gianazza, E., P.G. Rhigetti, F. Pioli, E. Galante, and C. Soave. 1976. Size and charge heterogeneity of zein in normal and opaque-2 maize endosperms. Maydica 21:1-17.
20. Grosveld, G.C., C.K. Shewmaker, P. Jat, and R.A. Flavell. 1981. Localization of DNA sequences necessary for transcription of the rabbit beta-globin gene in vitro. Cell 25:215-226.
21. Grosveld, G.C., E. de Boer, C.K. Shewmaker, and R.A. Flavell. 1982. DNA sequences necessary for transcription of the rabbit beta-globin gene in vivo. Nature 295:120-126.
22. Guifoyle, T.J. 1980. Transcription of cauliflower mosaic virus genome in isolated nuclei from turnip leaves. Virology 107:71-80.
23. Hagen, G., and I. Rubenstein. 1980. Two dimensional gel analysis of the zein proteins in maize. Plant Sci. Lett. 19:217-223.
24. Hagen, G., and I. Rubenstein. 1981. Complex organization of zein genes in maize. Gene 13:239-249.
25. Heidecker, G., J. Messing, and B. Gronenborn. 1980. A versatile primer for DNA sequencing in the M13mp2 cloning system. Gene 10:69-73.

26. Heidecker, G., and J. Messing. 1982. Construction of a maize endosperm cDNA library by a new mRNA cloning technique (manuscript in preparation).
27. Heilig, R., F. Perrin, F. Gannon, J.L. Mandel, and P. Chambon. 1980. The ovalbumin gene family: Structure of the x gene and evolution of duplicated split genes. Cell 20:625-637.
28. Hu, N.-T., and J. Messing. 1982. The making of strand specific M13 probes. Gene 17:271-277.
29. Hu, N.-T., M.A. Peifer, G. Heidecker, J. Messing, and I. Rubenstein. 1982. Primary structure of a zein genomic clone (submitted for publication).
30. Hutchinson, III, C.A., S. Phillips, M.H. Edgell, S. Gillam, P. Jahnke, and M. Smith. 1978. Mutagenesis at a specific position in a DNA sequence. J. Biol. Chem. 253:6551-6560.
31. Hyldig-Nielsen, J., E. Jensen, K. Paludan, O. Wiborg, R. Garret, P. Jorgensen, and K. Marcker. 1982. The primary structure of two leghemoglobin genes from soybean. Nucl. Acids Res. 10:689-701.
32. Inouye, M., and S. Halegoua. 1980. Secretion and membrane localization of protein in Escherichia coli. Crit. Rev. Biochem. 7:339-371.
33. Kedes, L.H. 1979. Histone genes and histone messengers. Ann. Rev. Biochem. 48:837-870.
34. Kozak, M. 1981. Possible role of flanking nucleotides in recognition of the AUG initiator codon by eukaryotic ribosomes. Nucl. Acids Res. 9:5233-5252.
35. Kridl, J., J. Vieira, I. Rubenstein, and J. Messing. 1982. Further analysis of zein genes in maize (manuscript in preparation).
36. Larson, R., and J. Messing. 1982. Apple II software for M13 shotgun DNA sequencing. Nucl. Acids Res. 10:39-49.
37. Lerner, M.R., J.A. Boyle, S.M. Mount, S.L. Wolin, and J.A. Steitz. 1980. Are snRNPs involved in splicing. Nature 283:220-224.
38. Lewis, E.D., G. Hagen, J.I. Mullins, P. Mascia, W.D. Park, W.D. Benton, and I. Rubenstein. 1981. Cloned genomic segments of Zea mays homologous to zein mRNAs. Gene 14:205-215.
39. Marks, M.D., and B. Larkins. 1982. Analysis of sequence microheterogeneity among zein messenger RNAs. J. Biol. Chem. (in press).
40. McKnight, S.L., and R. Kingsbury. 1982. Transcriptional control signals of a eukaryotic protein coding gene. Science 217:316-324.
41. Messing, J., B. Gronenborn, B. Muller-Hill, and P.H. Hofschneider. 1977. Filamentous coliphage M13 as a cloning vehicle: Insertion of a Hind II fragment of the lac regulatory region in the M13 replicative form in vitro. Proc. Natl. Acad. Sci. USA 75:3642-3646.
42. Messing, J., R. Crea, and P.H. Seeburg. 1981. A system for shotgun DNA sequencing. Nucl. Acids Res. 9:309-321.

43. Messing, J. 1982. New M13 vectors for cloning, In Recombinant DNA, Part B, Methods In Enzymo Vol 71. R. Wu, ed. Academic Press, New York (in press).
44. Murray, N.E., and K. Murray. 1974. Manipulations of restriction targets in phage lambda to form receptor chromosomes for DNA fragments. Nature 251:476-481.
45. Park, W.D., E.D. Lewis, and I. Rubenstein. 1980. Heterogeneity of zein mRNA and protein in maize. Plant Physiol. 65:98-106.
46. Pedersen, K., J. Devereux, D.R. Wilson, E. Sheldon, and B.A. Larkins. 1982. Cloning and sequence analysis reveal structural variation among related zein genes in maize. Cell 29:1015-1026.
47. Proudfoot, N.J. 1979. Eukaryotic promoters? Nature 279:376.
48. Rambach, A., and P. Tiollais. 1974. Bacteriophage lambda having EcoRI sites only in nonessential region of the genome. Proc. Natl. Acad. Sci. USA 71:3927-3931.
49. Rhigetti, P.G., E. Gianazza, A. Viotti, and C. Soave. 1977. Heterogeneity of storage proteins in maize. Planta 136:115-123.
50. Setzer, D.R., M. McGrogan, J.H. Nunberg, and R.T. Schimke. 1980. Size heterogeneity in the 3' end of dihydrofolate reductase messenger RNA in mouse cells. Cell 22:361-370.
51. Shah, D.M., R.C. Hightower, and R.B. Meagher. 1982. Complete nucleotide sequence of a soybean actin gene. Proc. Natl. Acad. Sci. USA 79:1022-1026.
52. Shortle, D., and D. Nathans. 1978. Local mutagenesis: A method for generating viral mutants with base substitutions in preselected regions of the viral genome. Proc. Natl. Acad. Sci. USA 75:2170-2174.
53. Soave, C., P.G. Rhigetti, C. Lorenzoni, E. Gentinetta, and F. Salamini. 1976. Expressivity of the opaque-2 gene at the level of zein molecular components. Maydica 21:61-75.
54. Soave, C., N. Suman, A. Viotti, and F. Salamini. 1978. Linkage relationships between regulatory and structural gene loci involved in zein synthesis in maize. Theor. Appl. Genet. 52:263-268.
55. Stachel, S., A. Depicker, P. Dhaese, and H. Goodman. 1982. The nopaline synthase gene mapping and DNA sequence. J. Mol. Appl. Genet. (submitted for publication).
56. Sun, S.M., J.L. Slightom, and T.C. Hall. 1981. Intervening sequences in a plant gene - comparison of the partial sequence of cDNA and genomic DNA of French bean phaseolin. Nature 289:37-41.
57. Thomas, M., J.R. Cameron, and R.W. Davis. 1974. Viable molecular hybrids of bacteriophage lambda and eukaryotic DNA. Proc. Natl. Acad. Sci. USA 71:4579-4583.
58. Tosi, M., R.A. Young, O. Hagenbuchle, and V. Schibler. 1981. Multiple polyadenylation sites in a mouse alpha-amylase gene. Nucl. Acids Res. 9:2313-2323.

59. Viotti, A., D. Abildsten, N. Pogna, E. Sala, and V. Pirrotta. 1982. Multiplicity and diversity of cloned zein cDNA sequences and their chromosomal localization. EMBO J. 1:53-58.

60. Wahli, W., J.B. Dawid, G.U. Ryffel, and R. Weber. 1981. Vitellogenesis and vitellogenin gene family. Science 212:298-304.

61. Wiborg, O., J. Hyldig-Nielsen, E. Jensen, K. Paludan, and K. Marcker. 1982. The nucleotide sequences of two leghemoglobin genes from soybean. Nucl. Acids Rec. 10:3487-3494.

62. Wienand, U., C. Bruschke, and G. Felix. 1979. Cloning of double-stranded DNAs derived from polysomal mRNA of maize endosperm: Isolation and characterization of zein clones. Nucl. Acids Res. 6:2707-2715.

63. Wienand, U., P. Langridge, and G. Feix. 1981. Isolation and characterization of a genomic sequence of maize coding for a zein gene. Mol. Gen. Genet. 182:440-444.

64. Zaret, K.S., and F. Sherman. 1982. DNA sequence required for efficient transcription-termination in yeast. Cell 28:563-573.

CHROMOSOME STRUCTURE IN CEREALS: THE ANALYSIS OF REGIONS CONTAINING REPEATED SEQUENCE DNA AND ITS APPLICATION TO THE DETECTION OF ALIEN CHROMOSOMES INTRODUCED INTO WHEAT

R. Appels

Division of Plant Industry
CSIRO, P.O. Box 1600
Canberra City 2601, Australia

INTRODUCTION

The concept of chromosome engineering of plants is well established in plant breeding. As early as 1937 Blakeslee and Avery commented in their article on chromosome doubling by colchicine that "with increasing knowledge of the constitution of chromosomes and of methods whereby their structure and behavior may be altered, there arises an opportunity for the genetic engineer who will apply knowledge of chromosomes to building up to specification forms of plants adapted to the surroundings in which they are to grow, and suited to specific economic needs." The colchicine doubling technique which inspired this comment was taken up by plant breeders who had been interested in producing wheat-rye hybrids. Routine production of colchicine doubled wheat-rye hybrids with some fertility allowed intensive plant breeding for improved fertility, yield, seed quality, etc. (14,29,46). The breeding effort has resulted in recent years in many farmers around the world growing triticale as a crop. More directed chromosome engineering has been carried out for introducing alien chromosome segments carrying disease resistance (37,55,56).

In this contribution to the proceedings, I wish to summarize studies carried out on those regions of the chromosomes which contain high concentrations of repeated sequence DNA. Cytologically these regions often appear as heterochromatin. Studies of *Drosophila* chromosomes have provided evidence that manipulation of these regions with respect to amount and position in the genome can have marked effects in certain genotypes (for a recent discussion see Hilliker and Appels, ref. 32). The heterochromatic regions within a genome have pleiotropic effects on different genes and, considering the complex characters often dealt with in plant breeding, the modification of these regions may prove of interest. This possibility has been specifically considered in relation to the heterochromatic regions of rye chromosomes in triticale (11) and will be discussed further in this paper. Two other areas will also be discussed, namely, the mechanisms which may alter the tandem arrays of repeated DNA sequences and the recent application of some of the repeated DNA sequence probes for detecting alien chromosomes introduced into wheat.

CYTOLOGICAL ASPECTS OF CHROMOSOME STRUCTURE

The cytological studies of pachytene chromosomes in the 1940s and 1950s provided detailed analyses of the complexity of chromosome structure in plants. Figure 1 summarizes the main findings of these studies with reference to chromosome 1R of rye (39), namely, the presence of heterochromatin, the chromomere structure in the euchromatic region, the symmetrical structure of the centromere region and the nucleolus with its associated heterochromatin. In mitotic chromosomes the site from which the nucleolus originates is seen as a secondary constriction, the nucleolus organizer. Various staining techniques using fluorescent compounds such as Hoechst 33258 (50) and chromomycin (52), or Giemsa C-banding (reviewed for cereals in ref. 29), reveal substructure within mitotic chromosomes. The C-banded regions usually correlate with large chromomeres and heterochromatic regions observed in pachytene chromosomes. The fluorescent compounds often have well defined specificities for certain DNA sequences (44,53) and can be more selective in the regions they reveal; for example, in rye, chromomycin counter-stained with distamycin A and DAPI reveals only the rDNA region (53).

MOLECULAR ANALYSIS OF CYTOLOGICAL "LAND-MARKS" IN CEREAL CHROMOSOMES

The rDNA Region

The ribosomal RNA gene (rDNA) region of chromosome 1R (Fig. 1) is characterized by large chromomeres (in pachytene chromosomes) from which a nucleolus originates, while in mitotic chromosomes a secondary constriction with a proximal C-band (29) defines the same region. The C-band and large chromomeres most likely define the same chromosome segment. In wheat, rye, and barley a significant fraction of the rDNA is located in the C-band proximal to the secondary constriction of mitotic chromosomes (ref. 3; Fig. 2) and thus it is likely that the large chromomeres observed at the base of

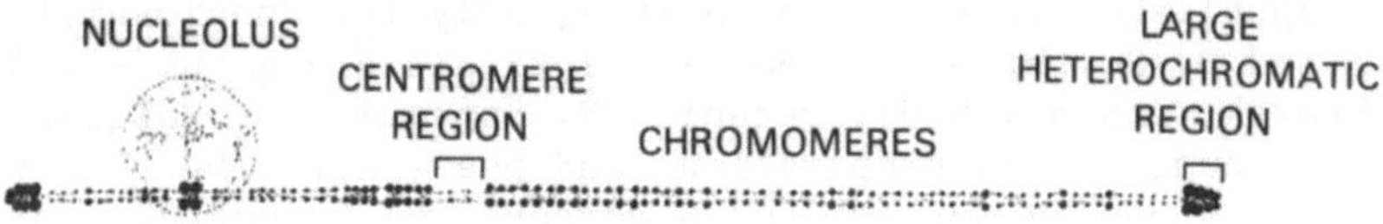

Fig. 1. Drawing of a chromosome 1R of rye at pachytene of meiotic prophase, modified from Lima-de-Faria (39), to emphasize the features which are readily observed cytologically.

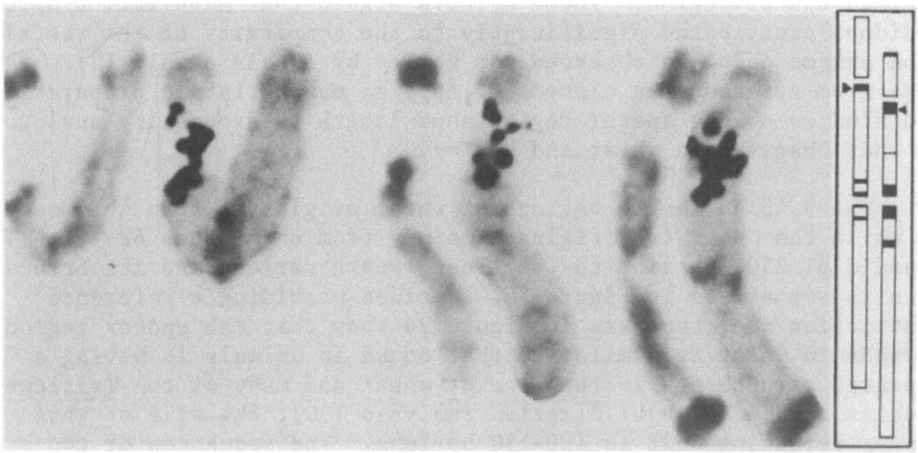

Fig. 2. I^{125}-rRNA hybridized to mitotic chromosomes of barley to show the distribution of rDNA relative to the secondary constriction (modified from Appels, et al., 3). Barley is shown because the secondary constrictions are more easily visualized; identical distributions of rDNA relative to secondary constrictions were observed in rye and wheat mitotic chromosomes. The inset shows the secondary constriction chromosomes of barley after C-banding (Linde-Laursen, 40).

the nucleolus (Fig. 1) contain rDNA. This correlation is also seen in maize where a proportion of rRNA genes have been argued to be in the nucleolus associated heterochromatin observed in pachytene chromosomes (19,27).

The DNA structure of the region has been well defined. In rye, wheat, and barley the repeating unit of rDNA is 9.0-10.0 kb in length as determined from the spacing of EcoRI sites occurring only once in the 26S rDNA genes (3); this length of repeating unit seems to be typical for plants since rDNA units in the 8.0-11.0 kb range are observed in soybean (23,57), flax (28), Panicum species (I.

Oliver, unpublished results), corn (60), lily (L. speciosum)(A.L. von Kalm and D. Smyth, pers. comm.), Elytrigia elongatum (R. Appels and J. Dvorak, unpublished), Petunia (P. Dunsmuir, J. Waldron, J. Bedbrook, manuscript in preparation) and pea (34). The rDNA region of wheat was shown by Gerlach and Bedbrook (24) to contain three major length variants--9.0, 9.15, and 9.45 kb--with the smallest two being cloned in the plasmid pACY184 by these authors. Gerlach and Bedbrook also cloned the rDNA from barley and showed that the main source of length variation in the rDNA units is in the non-transcribed spacer region. These authors argued that methylation of C residues contributed significantly to the complexity of the restriction enzyme patterns observed for barley by Appels et al. (3). The rDNA from rye has been cloned (R. Appels, manuscript in preparation) and studies on the spacer region show length heterogeneity analogous to that observed in wheat and barley.

The 9.45 kb length variant of wheat originates from chromosome 1B while the other two originate mainly from chromosome 6B (7). The plasmid pTA250 contains the smallest length variant and its structure is summarized in Figure 3a. Besides providing a reference restriction map, the data in Figure 3a show that the spacer region of rDNA in wheat is similar to that found in animals in having a repeated structure. In the case of wheat and many of the Triticum species studied by hybridization analyses (20), the size of this spacer repeating unit is 130-150 bp long. The sequences of two of the repeated "130 bp" units from pTA250 are shown in Figure 3b; from the cross-hybridization characteristics of a population of fourteen "130 bp" unit clones, the two shown in Figure 3b were maximally different in sequence. The sequences are seen to be identical except for differences at six positions. Although the spacer region is internally repeated, the "130 bp" units are unique to the rDNA region and do not occur in large blocks elsewhere in the wheat genome as shown by in situ hybridization and genomic DNA analyses (7).

At the chromatin level, the rDNA region in the cereals is similar to other eukaryotes in displaying a micrococcal nuclease sensitivity which results in the respective DNA being digested at locations separated, on average, by 170 bp (Fig. 3c). This is the same periodicity as that observed for the entire complement of DNA in the nucleus (Fig. 3c) and, as discussed later in relation to the analysis of telomeric heterochromatin, the periodicity is specifically related to the way DNA in the nucleus interacts with proteins (or other components of the nucleus). Thus it appears that the large chromomeres or heterochromatin characterizing the rDNA region (Fig. 1) are not correlated in a simple way with specific aspects of chromatin structure assayable by micrococcal nuclease. Higher levels of organization of the chromatin are under investigation.

The control of rDNA expression in cereals has been examined using indirect assays such as nucleolus formation and the presence of secondary constrictions (22). These studies suggest multi-genic control of rDNA expression as well as an effect on expression of the position of rDNA within the genome. The definition of the rDNA region in more detail, at a molecular level, should allow the development of probes for examining these questions more directly.

The 5S rDNA Region

The 5S rDNA of chromosomes 1R (rye) and 1B (wheat) is located in tandem arrays distal to the rDNA region (3) although it is not clearly associated with a heterochromatic band. In wheat the 5S rDNA is also located at site(s) other than on chromosome 1B, but these have not yet been localized. The different classes of 5S rDNA from wheat have been cloned and sequenced (25) and the sequence of the size variant from chromosome 1B, compared with the other size variant found in wheat, is shown in Figure 4. The data indicates that the spacer regions in the different size variants of 5S rDNA differ much more than the spacer regions of size variants in the 18S-26S rDNA region. This suggests that the variant of 5S rDNA may be characterized by a longer spacer located in a wheat genome different from the one in which the 1B site is found; both of the major 18S-26S rDNA sites, in contrast, are within the B genome.

Regions of Heterochromatin

The heterochromatic regions of the cereals are becoming well understood at a molecular level of 100-1000 bp (2,4,9,18,36) although relatively little is known about the organization over longer lengths of DNA (30 kb and longer). The chromatin of this region in rye chromosomes, like that of rDNA, is folded into a nucleosome-type structure as defined by micrococcal nuclease digestion (Fig. 5a). The 170 bp spacing is not a consequence of the sequence specificity inherent in the digestion of DNA by micrococcal nuclease (16,45); deproteinised DNA does not show the 170 bp "ladder" of bands although other bands are visible (Fig. 5a).

Thermal denaturation of the nucleosome structure can be assayed by short (1 min) heat pretreatments of nuclei prior to micrococcal nuclease digestion. The loss of nucleosome structure is highly co-operative (Fig. 5b) and for the heterochromatic sequences this consistently occurs at a temperature 2-4°C higher than that of the rDNA sequences. In Figure 5b, replicate samples of DNA from nuclei pretreated at the various temperatures and micrococcal nuclease digested were hybridized with either the heterochromatic sequence probe (pSc 7235) or an rDNA probe (pTA250.17). It is evident, particularly from the higher molecular weight bands, that the heterochromatic chromatin is more stable than the rDNA chromatin.

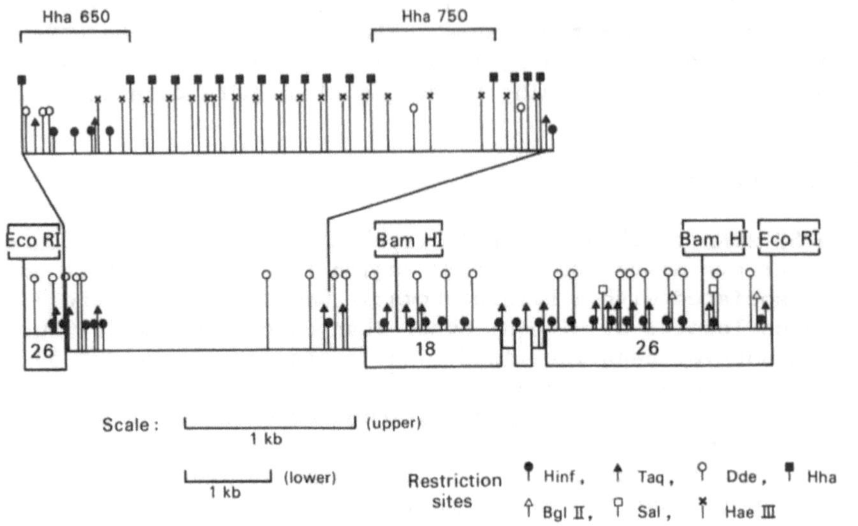

(a)

```
                                                    40
130.6   5'GGATCCGCGTCAACTTCCGTCCACGACGGCCGCCGGCCGTT
        3'CCTAGGCGCAGTTGAAGGCAGGTGCTGCCGGCGGCCGGCAA

                                                  A
130.8   5'GGATCCGCGTCAACTTCCGTCCACGACGGCCGCCGGGCCTT
        3'CCTAGGCGCAGTTGAAGGCAGGTGCTGCCGGCGGCCCGGAA
                                                  T ●

                                                    80
130.6   5'TTTTCCCGTGCCCACGTACAGCCCGTTCACGGCTCCGTGTA
        3'AAAAGGGCACGGGTGCATGTCGGGCAAGTGCCGAGGCACAT

                                        G
130.8   5'TTTTCCCGTGCCCACGTACAGCCCGTTCACGGGTCCGTGTA
        3'AAAAGGGCACGGGTGCATGTCGGGCAAGTGCCCAGGCACAT
                                        C  ●

                                                    120
130.6   5'AGGGTCCGTCTACGCGGCGTGTGCCACGTACGTGGTTTTGC
        3'TCCCAGGCAGATGCGCCGCACACGGTGCATGCACCAAAACG

130.8   5'AGGGTCCGTGTACGCGGCGTCTGCCACGTACGTGGTTTTGC
        3'TCCCAGGCACATGCGCCGCACACGGTGCATGCACCAAAACG
                   ●

130.6   5'CCGGTTTTCCATGGCGCGGATCC
        3'GGCCAAAAGGTACCGCGCCTAGG

                             C
130.8   5'CCGGTTTTCCATGGCGCGGATCC
        3'GGCCAAAAGGTACCGCGCCTAGG
                             G
```

(b)

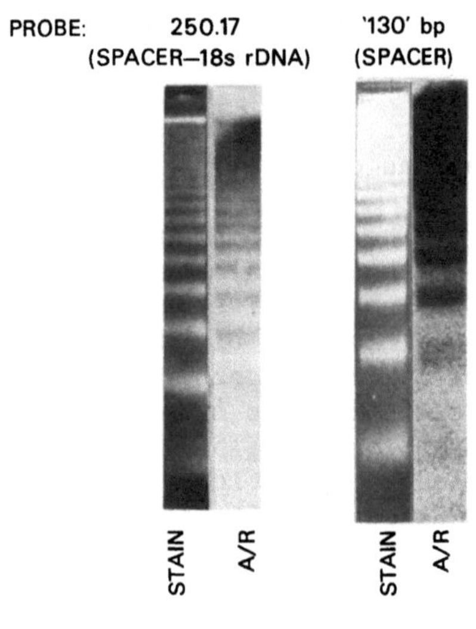

(c)

Fig. 3. Characteristics of the rDNA region of chromosomes.

(a) Restriction enzyme map of pTA250. In the lower map the positions of the restriction enzyme sites are ± 50-100 bp while in the upper map they are ± 10-50 bp. For the upper map the Hae III sites were not determined in the Hha 650 region.

(b) The nucleotide sequences of the 130-6 and 130-8 sequences. The region of the rDNA spacer, containing 130 bp repeating units was subcloned by isolating the Hha generated units, "filling-in" the ends, adding synthetic Bam HI linkers and inserting into the BAM HI site of pBR322. From a population of 130 bp subclones, the 130-6 and 130-8 were selected for sequencing since they were maximally different on the basis of cross-hybridization studies. Further details of the study are in Appels and Dvorak (7). The position of differences between 130-6 and 130-8 sequences is indicated (•). The difference at position 37 in the sequence results in the absence of a Hae III site in 130-8 relative to 130-6.

(c) Folding of the rDNA region in nuclei. Seeds from the triticale Satu were imbibed for 15-18 hrs (23°C) and nuclei isolated by chopping isolated embryos with a razor blade in 1M sucrose, 30 mM NaCl, 15 mM KCl, 15mM β-mercaptoethanol, 5 mM $MgCl_2$, 0.15 mM spermidine, 15 mM Tris-HCl (pH 7.5). After filtration through miracloth and adding TiX-100 to 0.1%, nuclei were isolated by centrifugation in a step gradient of metrizamide (Bedbrook _et al._, 9,10). Micrococcal nuclease (Boehringer) was added to nuclei resuspended in 0.34 M sucrose, 30 mM NaCl, 15 mM KCl, 15 mM β-mercaptoethanol, 1 mM $MgCl_2$, 1 mM $CaCl_2$, 0.15 mM spermidine, 15 mM Tris-HCl (pH 7.5), so that 1.4 units of enzyme was added per, approximately, 15 μg of nuclear DNA in a volume of 100 μl. Reaction was terminated after 2.5 min at 37°C by adding an equal volume of proteinase K (Boehringer, 0.1 mg/ml) in 0.2% SDS and incubating a further 60 min at 37°C. DNA was recovered for electrophoresis on 2% agarose by phenol/chloroform (1:1) extraction and ethanol precipitation. The DNA sample which was electrophoresed, transferred to Gene-Screen (New England Nuclear), prehybridized and then hybridized to ^{32}P-labelled insert from pTA250.17 (a subclone for pTA250 containing the 650 bp Hinf fragment which spans the transcribed spacer and start of the 18S rRNA gene - see map in [a]) in 3 x SSC/50% formamide/5 x Denhardt's/0.1% SDS at 37°C. The same pattern of hybridization was observed using a 130 bp probe from the repeated sequence region of the spacer.

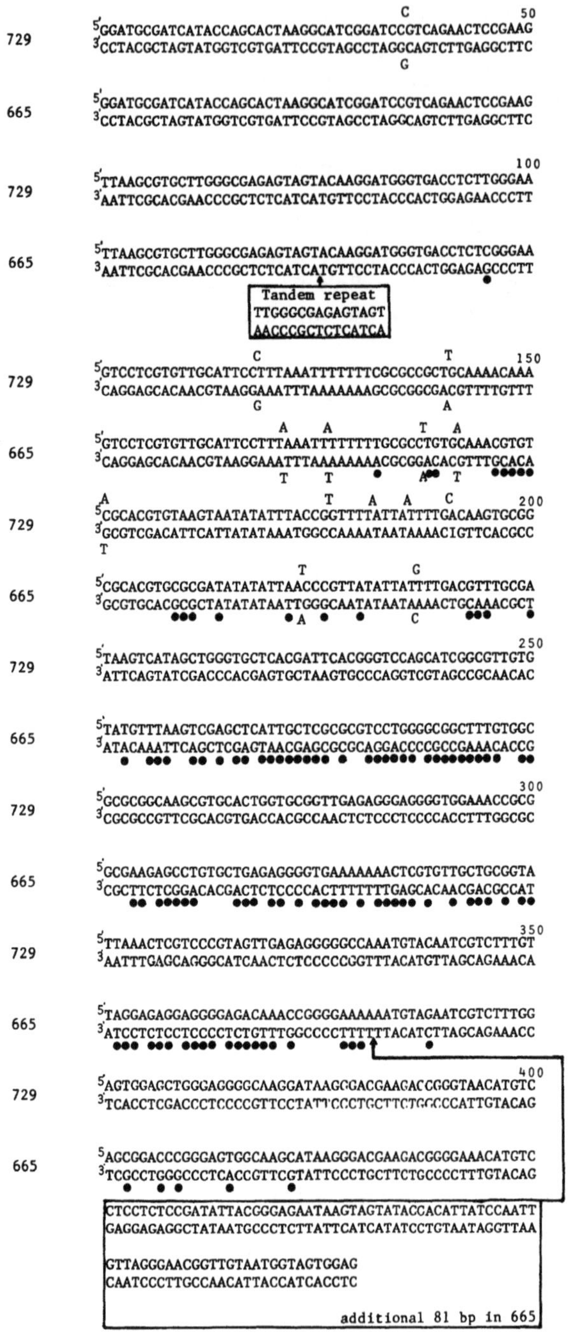
C 50
729 5'GGATGCGATCATACCAGCACTAAGGCATCGGATCCGTCAGAACTCCGAAG
3'CCTACGCTAGTATGGTCGTGATTCCGTAGCCTAGGCAGTCTTGAGGCTTC
G
665 5'GGATGCGATCATACCAGCACTAAGGCATCGGATCCGTCAGAACTCCGAAG
3'CCTACGCTAGTATGGTCGTGATTCCGTAGCCTAGGCAGTCTTGAGGCTTC
100
729 5'TTAAGCGTGCTTGGGCGAGAGTAGTACAAGGATGGGTGACCTCTTGGGAA
3'AATTCGCACGAACCCGCTCTCATCATGTTCCTACCCACTGGAGAACCCTT
665 5'TTAAGCGTGCTTGGGCGAGAGTAGTACAAGGATGGGTGACCTCTCGGGAA
3'AATTCGCACGAACCCGCTCTCATCATGTTCCTACCCACTGGAGAGCCCTT
Tandem repeat
TTGGGCGAGAGTAGT
AACCCGCTCTCATCA
C T 150
729 5'GTCCTCGTGTTGCATTCCTTTAAATTTTTTTCGCGCCGCTGCAAAACAAA
3'CAGGAGCACAACGTAAGGAAATTTAAAAAAAGCGCGGCGACGTTTTGTTT
G A
A A T A
665 5'GTCCTCGTGTTGCATTCCTTTAAATTTTTTTGCGCCTGTGCAAACGTGT
3'CAGGAGCACAACGTAAGGAAATTTAAAAAAACGCGGACACGTTTGCACA
T T A T
A T A A C 200
729 5'CGCACGTGTAAGTAATATATTTACCGGTTTTATTATTTTGACAAGTGCGG
3'GCGTCGACATTCATTATATAAATGGCCAAAATAATAAAACIGTTCACGCC
T
T G
665 5'CGCACGTGCGCGATATATATTAACCCGTTATATTATTTTGACGTTTGCGA
3'GCGTGCACGCGCTATATATAATTGGGCAATATAATAAAACTGCAAACGCT
A C
250
729 5'TAAGTCATAGCTGGGTGCTCACGATTCACGGGTCCAGCATCGGCGTTGTG
3'ATTCAGTATCGACCCACGAGTGCTAAGTGCCCAGGTCGTAGCCGCAACAC
665 5'TATGTTTAAGTCGAGCTCATTGCTCGCGCGTCCTGGGGCGGCTTTGTGGC
3'ATACAAATTCAGCTCGAGTAACGAGCGCGCAGGACCCCGCCGAAACACCG
300
729 5'GCGCGGCAAGCGTGCACTGGTGCGGTTGAGAGGGAGGGGTGGAAACCGCG
3'CGCGCCGTTCGCACGTGACCACGCCAACTCTCCCTCCCCACCTTTGGCGC
665 5'GCGAAGAGCCTGTGCTGAGAGGGGTGAAAAAAACTCGTGTTGCTGCGGTA
3'CGCTTCTCGGACACGACTCTCCCCACTTTTTTTGAGCACAACGACGCCAT
350
729 5'TTAAACTCGTCCCGTAGTTGAGAGGGGGCCAAATGTACAATCGTCTTTGT
3'AATTTGAGCAGGGCATCAACTCTCCCCCGGTTTACATGTTAGCAGAAACA
665 5'TAGGAGAGGAGGGGAGACAAACCGGGGAAAAAATGTAGAATCGTCTTTGG
3'ATCCTCTCCTCCCCTCTGTTTGGCCCCTTTTTTACATCTTAGCAGAAACC
400
729 5'AGTGGAGCTGGGAGGGGCAAGGATAAGGGACGAAGACCGGGTAACATGTC
3'TCACCTCGACCCTCCCCGTTCCTATTCCCTGCTTCTGGCCCATTGTACAG
665 5'AGCGGACCCGGGAGTGGCAAGCATAAGGGACGAAGACGGGGAAACATGTC
3'TCGCCTGGGCCCTCACCGTTCGTATTCCCTGCTTCTGCCCCTTTGTACAG
CTCCTCTCCGATATTACGGGAGAATAAGTAGTATACCACATTATCCAATT
GAGGAGAGGCTATAATGCCCTCTTATTCATCATATCCTGTAATAGGTTAA
GTTAGGGAACGGTTGTAATGGTAGTGGAG
CAATCCCTTGCCAACATTACCATCACCTC
additional 81 bp in 665

Although the biochemical basis for this greater stability is under investigation, the data suggests that it may be possible to correlate certain molecular features of a DNA-protein complex with cytological characteristics.

The major rye heterochromatic sequence assayed in Figure 5a is distributed throughout the large heterochromatic blocks observed at pachytene, as shown by *in situ* hybridization (Fig. 6a). This observation, taken together with the observations that several different sequences are located in heterochromatin and that each of these sequences are arranged in homogeneous arrays of at least 30 kb long (4,9), suggests that there exists some form of interspersion of the large blocks of sequences in heterochromatin. This suggestion contrasts with the sequence arrangement within telomeric heterochromatin suggested by Flavell (21).

The nucleotide sequence of the basic repeating unit cloned from rye is shown in Figure 6b. The nucleotide sequence includes the 240 bp unit of the major heterochromatic DNA sequence cloned by Bedbrook *et al.* (9); other heterochromatic sequences have been cloned (4,9) but have not yet been sequenced. It is evident from Figure 6b that a regular, internally repeated structure is not present in the major heterochromatic sequence although stretches of $\ldots\begin{smallmatrix}\mathrm{AAAA}\\ \mathrm{TTTT}\end{smallmatrix}\ldots$ are common. This contrasts with the sequence structure of the major heterochromatic sequence present in wheat and barley (18,47). This heterochromatic sequence, which is also present as a minor heterochromatic component in rye (2), is characterized by a simple sequence structure dominated by the occurrence of the triplets $\ldots\begin{smallmatrix}\mathrm{GAA}\\ \mathrm{CTT}\end{smallmatrix}\ldots$ and $\ldots\begin{smallmatrix}\mathrm{GGA}\\ \mathrm{CCT}\end{smallmatrix}\ldots$. These differences in complexity of heterochromatic sequences are also found in other species (reviewed in refs. 1, 15).

The basic repeating unit of sequences in the class shown in Figure 6b appears to be 350 bp, the periodicity of the "ladder" observed in genomic DNA digested with EcoRI, Pst, and Bam HI (ref. 4; Fig. 7). The analysis of Bam HI "trimer", cloned in the vector λ1059, confirms this and indicates the 350 pb unit may often consist of 120 bp and 230 bp units as defined by Hae III (Fig. 7). The "480 bp" family defined by Bedbrook *et al.* (9) belongs to the class of sequence shown in Figures 6 and 7, and it may prove useful to redefine this family as the "350 bp" family.

Fig. 4. The 5S rDNA sequences from two size variants in wheat cloned in pTA729 and pTA665, as determined by Gerlach and Dyer (25). Sequences are aligned to maximize similarities - occasional bases are "looped-out" to aid alignment. Differences are indicated (●).

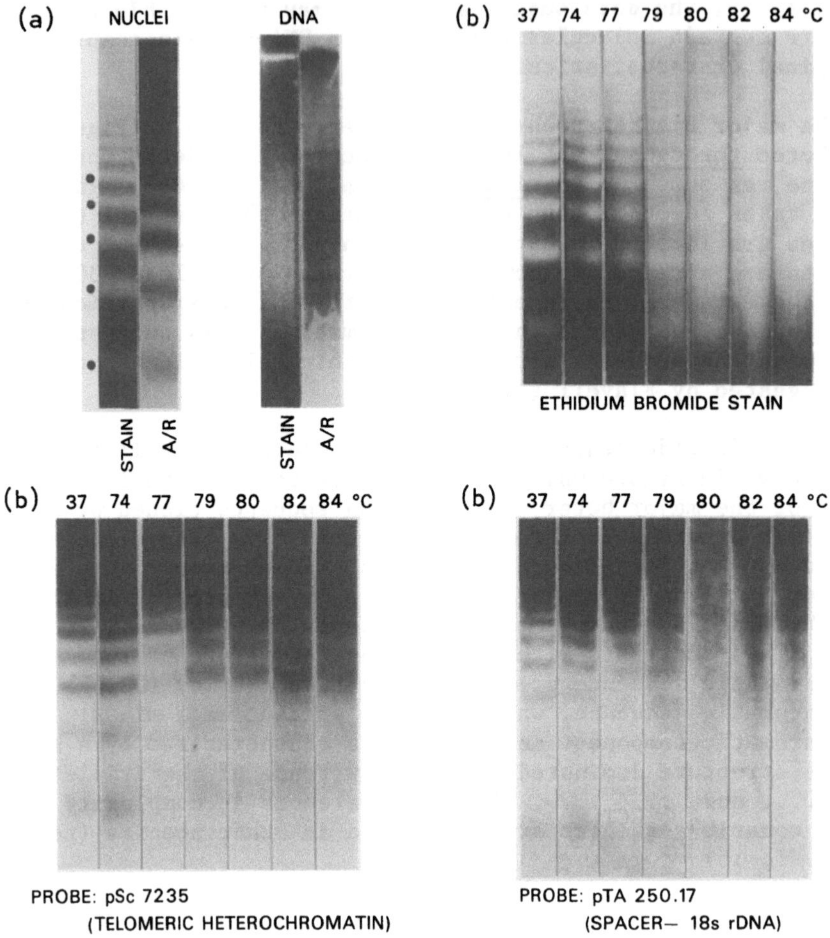

Fig. 5. Characteristics of the chromatin containing the major heterochromatic sequence of rye.

(a) Micrococcal nuclease digestion of embryo nuclei or deproteinized DNA from the triticale Satu$_2$ and assayed for the rye heterochromatic sequence using ^{32}P-labelled insert from the plasmid pSc7235. All relevant experimental details are given in the legend to Figure 3(c).

(b) Thermal denaturation of nucleosome structure. Micrococcal nuclease digestion was carried out as described in Figure 3(c) except that nuclei were pretreated for 1 min at the temperature indicated, plus 5 sec at 37°C (time needed to re-equilibrate to the temperature required for enzyme digestion) before adding the enzyme. Replicate DNA samples were assayed with either an rDNA probe (pTA250.17) or heterochromatic probe sequence (pSc7235).

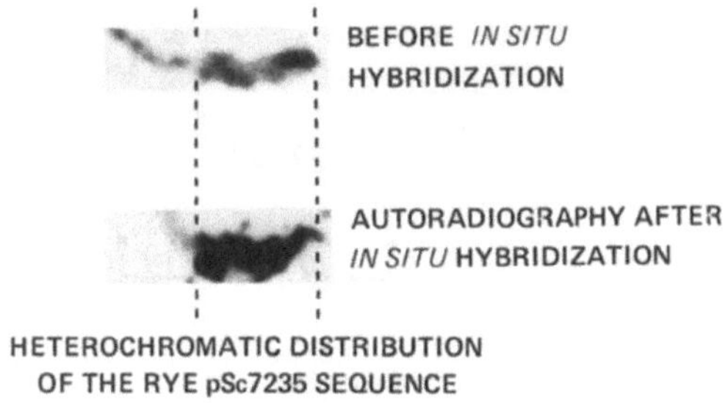

(a)

```
5'GAATTCAAACATTTCGAAACACACAAGCAATTTCTACAAATGGAAAAAA
3'CTTAAGTTTGTAAAGCTTTGTGTGTTCGTTAAAGATGTTTACCTTTTTT
50
TAAACTCCCCAAATAAAACTGAGAATACGAACATTTTTTGAATGGCAAAA
ATTTGAGGGGTTTATTTTGACTCTTATGCTTGTAAAAAACTTACCGTTTT
100
CAGATTTCGGATACGCGGACAGTTTTGAAAAATGGGAACTTTTTTTTAAT
GTCTAAAGCCTATGCGCCTGTCAAAACTTTTTACCCTTGAAAAAAAATTA
150
CTACCGAATAAAATTTGAAAACACCAACATTTTTTGAAACTCCAGAACAA
GATGGCTTATTTTAAACTTTTGTGGTTGTAAAAAACTTTGAGGTCTTGTT
200
AATTTGAAGCACGAACAATTTTTTAAATCCCAGATCAGAATTTGAAAACA
TTAAACTTCGTGCTTGTTAAAAAATTTAGGGTCTAGTCTTAAACTTTTGT
250
TGAACACTATATGAAAATTTTCCACGCACGTAATCCTGTGTGTGCATGTA
ACTTGTGATATACTTTTAAAAGGTGCGTGCATTAGGACACACACGTACAT
300
TGTAAAATCCTCCAGATCGTAAAAACTAGAAGTCCTATCACCTCAATTCC
ACATTTTAGGAGGTCTAGCATTTTTGATCTTCAGGATAGTGGAGTTAAGG
350
CCAATAGCTTTCCAACGCCTATGAAAACGACGAAATCGGTGCTTGTTCTC
GGTTATCGAAAGGTTGCGGATACTTTTGCTGCTTTAGCCACGAACAAGAG
400
ACTTGCTTTGAGAGTCTCGATCAATTCGAACTCTAGGTTGATTTTTGTAT
TGAACGAAACTCTCAGAGCTAGTTAAGCTTGAGATCCAACTAAAAACATA
450
TTTCTTTGATCACCGTTTCTTCGCGAGAGTTGGCCTAACACCCTATTGTA
AAAGAAACTAGTGGCAAAGAAGCGCTCTCAACCGGATTGTGGGATAACAT
500
TCCATACATGGGTGGGGCGCCCAGGACCTGAACACCAAAGTGATATGCCG
AGGTATGTACCCACCCCGCGGGTCCTGGACTTGTGGTTTCACTATACGGC
550
GCTCATCAAAATTTGAGCTGTAATTGCGATAACTTTCATTTTTTCCAATC
CGAGTAGTTTTAAACTCGACATTAACGCTATTGAAAGTAAAAAAGGTTAG
600
CTGCAGGCCATTTTTGAGGGCCAAGACGACCCTTTTTGGGCTCAGAATTC3'
GACGTCCGGTAAAAACTCCCGGTTCTGCTGGGAAAAACCCGAGTCTTAAG5'
```

(b)

Fig. 6. Characteristics of DNA in rye telomeric heterochromatin.
(a) In situ hybridization of a ^{3}H-cRNA probe synthesized from pSc7235 to chromosomes from pachytene of meiosis (Appels et al., 4). The sequence assayed is seen to be distributed along the entire length of cytologically defined telomeric heterochromatin.
(b) Nucleotide sequence of the rye heterochromatic DNA in pSc7235. The region underlined is the 120 bp Hae III segment indicated in Figure 7.

Overview of Repeated DNA Sequences in Cereal Chromosomes

Studies on defining the cytological "landmarks" of cereal chromosomes are summarized in Figure 8. This summary identifies

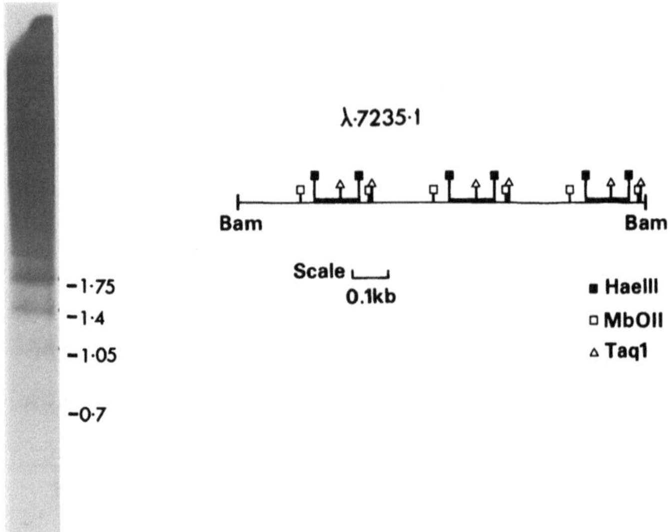

Fig. 7. The left panel shows a long (X-ray) exposure of rye DNA digested with BAM HI and hybridized with the pSc7235 sequence to show bands with a 350 bp periodicity. The right panel is the restriction map of a Bam HI "trimer", 1050 bp long, recovered in the bacteriophage 1059 (the bacteriophage λ.7235.1 was recovered by screening all plaques obtained in a cloning experiment with the pSc7235 sequence and contained the 1050 bp insert added to the normal complement of DNA in the bacteriophage).

some specific regions of repeated sequence DNA which would be of great interest in further developing our understanding of cereal chromosome structure.

(a) The repeated DNA sequences in and around the centromere region merit further investigation. In rye the major families of repeated DNA sequences are not found near the centromere region although this region is C-banded and thus an important (although minor) class of DNA sequences may be located here. In wheat and barley the major heterochromatic sequences are located near the centromere; the long range organization of these sequences is, however, complex (discussed in ref. 5) even though their short range structure is dominated by simple sequences such as ...$\genfrac{}{}{0pt}{}{GAA}{CTT}$.... . It would be of interest to know if certain complexities of the organization of these sequences in wheat and barley are related to the centromere region.

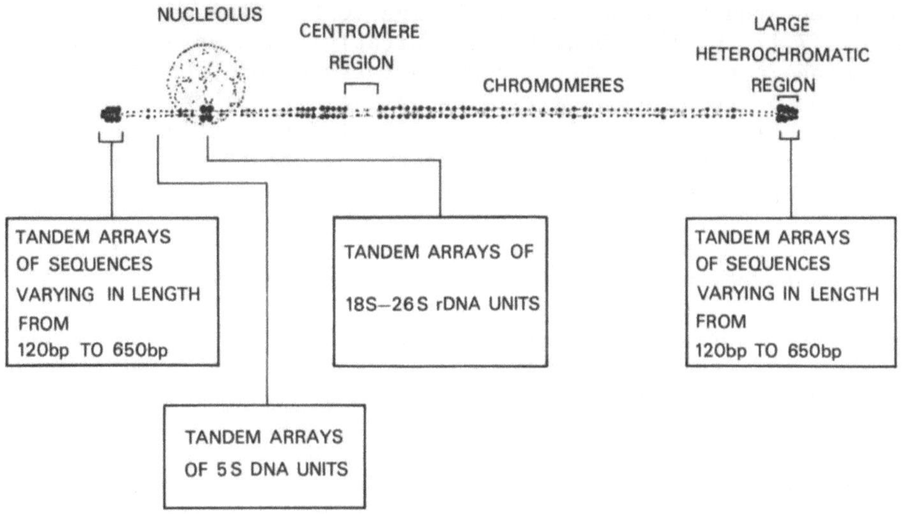

— % OF GENOME AS REPEATED SEQUENCE DNA IS 70–80% ($\approx 6\times10^6$ kb)
— APPROXIMATELY 0.3×10^6 kb OF DNA IS DEFINED AT A DNA SEQUENCE OR RESTRICTION ENZYME MAP LEVEL

Fig. 8. Summary of the cytological "land-marks" defined at a molecular level.

(b) Some repeated DNA sequences which are distributed at sites along the chromosome arms of rye have been characterized (2,9) but the relationship between these regions and those containing genes remains to be analyzed. The "120 bp family" of repeated DNA sequences in rye occurs as tandem arrays of 120 bp units as well as in linkage to a different (repeated) sequence (10). This latter finding is consistent with earlier findings that repeated DNA sequences of different families in cereal chromosomes are interspersed (49). If the chromomeres of pachytene chromosomes in Figure 1 (or Fig. 8) generally correspond to regions of repeated DNA sequences, linkage between a given repeated sequence and other unique (or low copy number) sequences is expected and the analysis of these boundary regions is of considerable interest.

RELATIVE TIME-SCALE OF CHANGES WHICH CAN OCCUR IN CHROMOSOME REGIONS CONTAINING REPEATED SEQUENCE DNA

Although it is difficult to establish an absolute time-scale of DNA sequence change within chromosomes, recent studies have shown that in several species, major changes must be occurring rapidly within repeated DNA sequences (relative to the time of duration of a species)(7,17,20,51). Dramatic quantitative changes have been demonstrated in heterochromatic regions (assayed by C-banding) of the genomes of polyploids classified as belonging, for example, to the D genome by chromosome pairing analyses (20). A detailed analysis of evolutionary changes in the rDNA regions of species in the Triticeae has been carried out by J. Dvorak and R. Appels (above references).

The "evolutionary divergence" profile of a specific region such as the rDNA region has been determined (Fig. 9) using eleven different probes from within the rDNA region. The significance of the asymmetric nature of divergence of the spacer sequences has been considered by Appels and Dvorak (8). These considerations are summarized here since they may have a bearing on mechanisms which can operate in individuals of a species to maintain homogeneity within an array of repeated sequences. It was suggested that although selection for promoter/processing signals, combined with genetic exchange processes, very likely contributes to the pattern of sequence conservation observed in Figure 9, the participation of RNA as a co-factor in gene conversion may also contribute. The mechanism envisaged is based on the in vitro observation that RNA can invade a double helix to form an R-loop (58,59). The displaced DNA strand could then invade the homologous DNA double helix and form a heteroduplex which is then subjected to excision repair processes leading to gene conversion (48). The stochastic nature of this process would result in the conservation of those sequences that are most often involved in heteroduplex formation. The participation of RNA as a co-factor would provide a bias toward those DNA sequences which produce RNA sequences in highest concentration. In the rDNA system this corresponds to the gene sequences. Transcribed spacer sequences, producing short-lived RNA molecules, are represented by RNA sequences at effectively lower concentrations. The model thus predicts that gene sequences would show the highest degree of conservation, transcribed spacer regions (e.g., those preceding the 18S rRNA gene and between the 18S and 26S genes) at a lower degree, and the non-transcribed spacer, the lowest degree of conservation. Although speculative, the model of RNA participating as a co-factor in democratic gene conversion, perhaps in combination with other processes such as unequal exchange and selection, may well have wide application.

The molecular mechanisms involved in the acquisition of new variation in a repeated array of genes such as the rDNA region is not clear since such variation appears to be "fixed" by stepwise processes within the hierarchy of arrays of repeated units, as discussed below.

"Fixation" of 130 bp Units within the Repeating Units of the Spacer Region

This phenomenon is demonstrated by the comparison of Triticeae

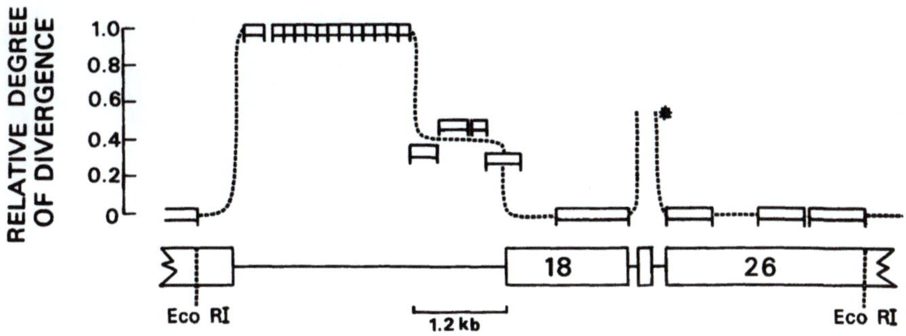

Fig. 9. Evolutionary divergence "profile" of the rDNA region. Utilizing the thermal denaturation of probes for eleven different regions of the rDNA unit hybridized to the DNA from six Triticum species, an estimate of relative evolutionary divergence along the length of the unit was obtained by Appels and Dvorak (8). The Triticum species utilized were T. aestivum, T. speltoides, T. tauschii, T. searsii, T. dichasians and T. urartu. Since the probes originated from T. aestivum, this species was used as the standard against which Δ Tm's of respective hybrids were determined. As expected the greatest Δ Tm was observed using spacer sequences hybridized to T. urartu (closely related to the A genome of T. aestivum) since its genome is well differentiated from the B genome of T. aestivum (containing the rDNA of T. aestivum). The Δ Tm's of the sequences were plotted against an arbitrary "evolutionary distance" scale with the relative distribution of the Triticum species being arranged in such a manner so that the data could be described by linear regressions. Although the slopes of the lines had no absolute meaning, the relative levels of divergence of the various sequences could be estimated by normalizing the data to assign the 130 bp sequence (showing maximum evolutionary change) a relative degree of divergence of 1.0 and gene sequences (showing minimum evolutionary change) a value of 0. These relative degrees of divergence are plotted in the Figure shown. The * denotes that the exact position, and degree of divergence, of the sequence in this region was not determined.

species. No simple relationship exists between the degree of sequence difference relative to Chinese Spring wheat (T. aestivum) as assayed by the Tm of hybrids formed between a ^{32}P-130 bp probe (see map in Fig. 3a) and various genomic DNAs, and the rDNA spacer structure as measured by the enzyme Hae III. Figure 10 shows, for example, that T. speltoides has a ΔTm of 3.5°C, T. dichasians 5.8°C, Elytrigia scirpea 10°C, and E. elongatum 10.5°C which indicates nucleotide sequence divergence levels of 4.2%, 7.0%, 12.0%, and

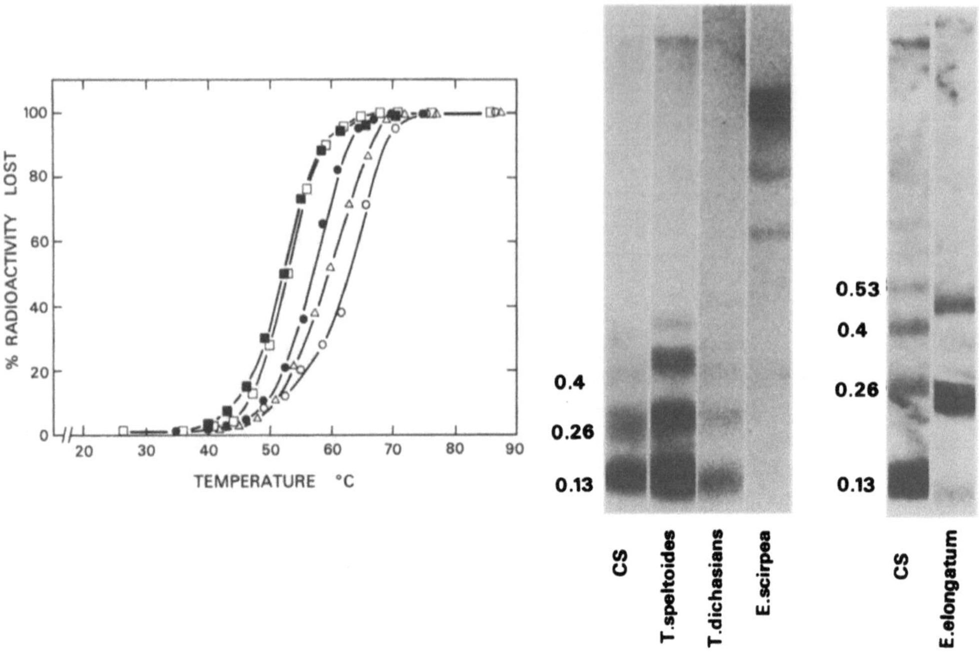

Fig. 10. The nature of sequence change in the rDNA spacer region of Triticeae species. The probe utilized was from the 130 bp repeating region of pTA250 [see Fig. 3(a)]. The left panel shows thermal denaturation profiles of hybrids formed between the 130 bp probe and DNA from T. aestivum (○-○, internal standard at 63°C), T. speltoides (△-△), T. uniaristatum (●-●), E. elongatum (■-■), E. scirpea (□-□). The right-hand panel shows Hae III digests of the DNAs hybridized with the 130 pb probe. A detailed analysis of rDNA sequence change in Triticeae species is presented in Dvorak and Appels (20).

12.6%, respectively. It is the spacer structure of T. dichasians (with respect to the distribution of Hae III sites), however, that is virtually indistinguishable from T. aestivum. The species T. speltoides is more closely related in overall sequence to T. aestivum but is distinguishable at the level of spacing of the Hae III sites (Fig. 10). The two Elytrigia species differ from T. aestivum to a similar degree, in overall sequence of the spacer region, but in one (E. scirpea) almost all Hae III sites are absent. The all or none differences in Hae III sites measured against a 3.5-12.6% difference in overall nucleotide sequence has been argued by Dvorak and Appels (20) to suggest a "fixation" of certain sequence variants within the entire array of 130 bp repeating units in a spacer region.

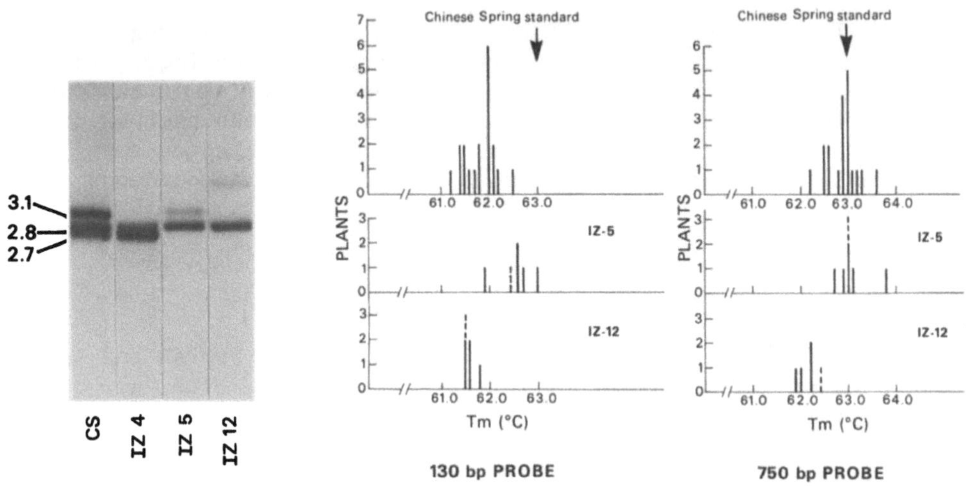

Fig. 11. Sequence heterogeneity in the rDNA region individuals in a single population of T. dicoccoides. Left panel shows Taq 1 digestions of individuals in Syrian population of T. dicoccoides called IZ (further details in Dvorak and Appels, 20). Taq I digestion, together with hybridization with a spacer region probe [in this case Hha 750, see Fig. 3(a)] allows length variation in the major spacer Taq I fragment [see Fig. 3(a)] to be measured. Right panel shows sequence heterogeneity in T. dicoccoides individuals as measured using two different rDNA spacer probes, the 130 bp sequence or the Ha 750 sequence. Thermal denaturation of hybrids between the probes and the respective DNAs immobilized on nitro-cellulose filters (for details see Appels and Dvorak, 7) show reproducible differences between individuals. The data has been summarized so that the length of a bar indicates the number of individuals with a Tm in a particular category. Sibling samples from IZ-5 and IZ-12 are shown to prove that differences between individuals was significant (the broken line indicates the position of the original measurement).

"Fixation" of the 9 kb rDNA Repeating Unit

The preceding level of "fixation" is accompanied by the spread of a new variant throughout the 1,000 or more 9 kb rDNA units found in the rDNA region of cereal chromosomes. The evidence for this is most striking from the Taq I digests of genomic DNA from individuals within populations of a species where the major Taq I spacer fragment is assayed using spacer sequence probes (Fig. 11). In any one individual, the arrays of 9 kb units at an allelic rDNA region were usually monomorphic with respect to the length of the Taq I fragment. This very likely applies to the sequence per se of the spacer since differences in Tm of the hybrids formed between the DNA of individuals and spacer sequence probes were also detected among individuals and thus the arrays must be relatively homogeneous (Fig. 11).

"Communication" between Non-allelic rDNA Regions

Most *Triticeae* species which were analyzed by Dvorak and Appels (20) have two chromosomes carrying rDNA (26,35,41). The "all or none" differences in the spacer regions (as assayed by Hae III) discussed in (a) above suggests that repeated sequences at these non-allelic sites are somehow kept similar to each other in a given species. In *T. aestivum*, the rDNA on chromosomes 1B and 6B are detectably different, but not as much as expected. Tm analysis shows the spacer regions on 1B differ from those on 6B by 1°C (1.2% nucleotide sequence difference) using the 130 bp sequences as probes (7). In contrast, a comparison of *T. araraticum* (which also has its major rDNA sites on chromosomes 1B and 6B) to *T. aestivum* show that by Tm analysis the rDNA of these species differ by 3.2% in nucleotide sequence. Since the two wheat lineages represented by *T. araraticum* and *T. aestivum* very likely had a common ancestor containing chromosomes 1B and 6B, the data suggests that some form of "communication" between these chromosomes has been occurring which prevents them from diverging from each other as much as expected.

The observation of rapid (at least at an evolutionary level) and differential "fixation" phenomena at the DNA level clearly has implications for a class of change which may occur in the DNA of plant cells passaged through tissue culture. This is discussed in detail elsewhere in this volume (see also review by Larkin and Scowcroft, ref. 38), but is mentioned here to reiterate the argument put forward in the Introduction that repeated DNA sequences can have measurable, sometimes subtle, effects on the phenotype of an organism. If large changes in the levels of particular repeated sequences of DNA occur in a given chromosomal region, there is good experimental evidence from the studies of *Drosophila* chromosomes to suggest that changes in gene expression will result (33).

DETECTING ALIEN CHROMOSOMES AND CHROMOSOME SEGMENTS INTRODUCED INTO WHEAT: THE ANALYSIS OF RYE CHROMOSOMES IN TRITICALE

In this section the use of repeated sequence DNA probes as tools for following alien chromosomes in plant breeding programs and some aspects of chromosome stability in foreign environments is discussed; these areas have been the subject of a recent review (5).

The introduction of chromosomes from rye, barley, *Elytrigia*, and *Haynaldia* into wheat results in alien chromosome/wheat translocations which appear to occur via centromere misdivision (references in 5). This type of "instability" of alien chromosomes in wheat occurs primarily as a result of the alien chromosome being present as a univalent during the course of manipulation of the genetic stocks. Once such translocations are homozygous, they appear to be stable and have potential value as breeding material (42). In situations where the source of the alien chromosome is clearly defined (e.g., rye chromosomes from wheat-rye addition lines introduced into commercial wheat varieties), the repeated DNA regions are stable components of the chromosome (or chromosome arm) at a DNA sequence organization level (4). In contrast, where the source of the alien chromosome is not clearly defined (e.g., in triticale production), different forms (with respect to the distribution of heterochromatic regions) of the alien chromosome have been "fixed". The fixation of the different forms of rye chromosomes (identified by differing distributions of heterochromatin) have been interpreted as indicating fine structure change after incorporation of the rye chromosomes into the triticale (for a recent discussion see ref. 54). Extensive quantitative change in the heterochromatic regions of chromosomes in the evolution of polyploids has been observed in the C-banding analyses of some *Triticum* polyploid species (20). As discussed before, extensive qualitative change can occur in the rDNA region. The interpretation of the different rye chromosomes in triticale in these terms (over the short period of time in which triticales have developed) is, however, complicated by the fact that the rye chromosomes used in triticale production are highly heterogeneous (data summarized in ref. 5). There is no convincing evidence to indicate that the different forms of rye chromosomes in triticale were generated subsequent to triticale formation. At present the simplest interpretation of the observed variants of rye chromosomes fixed in triticales is that they preexisted in *Secale cereale*. Whatever the source of the variant rye chromosomes, however, they clearly have different effects on, for example, seed phenotype (12,30). The identification of different rye chromosomes is thus of interest for the breeding of triticales.

Recognition of the different rye variants is particularly striking when the major rye heterochromatic sequence (shown in Fig. 6b) is used to probe the chromosome complement of triticales (6), as illustrated in Figure 12 (a more extensive list of the triticales analyzed is in the Appendix). This recognition of different variants allows genetically homologous chromosomes to be partitioned and thus has the potential for allowing the assignment of desirable agronomic characters to a certain variant.

The rye heterochromatic sequence probe has proven useful in analyzing the rye chromosomes in the triticales assessed in the first Australia-wide triticale trial conducted by Professor C.

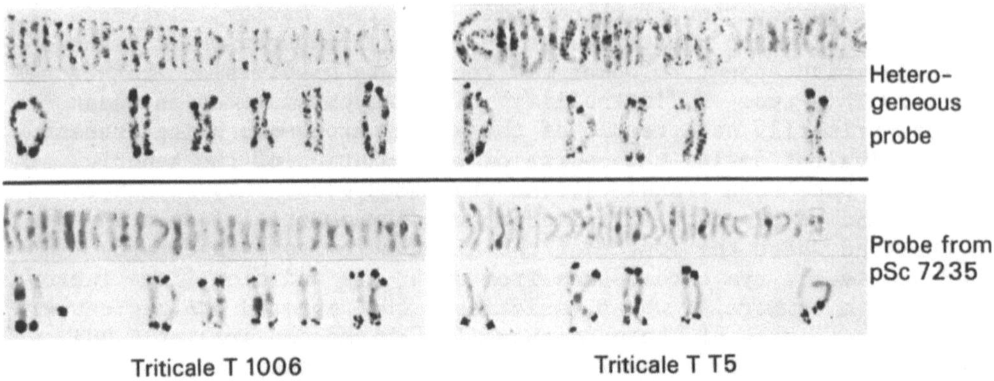

Fig. 12. Two triticales are shown, karyotyped after hybridization with either a heterogeneous probe for rye heterochromatin (i.e., assaying a range of repeated sequences which renature with a Cot of 0.02, Appels, Driscoll and Peacock, 1978) or a cloned probe (pSc7235). The comparison emphasized that the overall similarity of the rye chromosomes after hybridization with the heterogeneous probe is sharply resolved when hybridization with the cloned probe is carried out. The probes used were ^{3}H-cRNA copies of respective DNAs and the mitotic chromosomes were derived from root meristem tissue treated with colchicine (Appels, Driscoll and Peacock, 2).

Driscoll (Waite Agricultural Institute, South Australia). Furthermore, the probe is currently being used to study the rye chromosome segments present in back-cross progeny from a program to introduce speckled leaf blotch resistance from triticale into wheat varieties by Dr. C. E. May (N.S.W. Dept. of Agriculture, Wagga Wagga).

The usefulness of the rye chromosome probes in the analysis of triticale, and derived material (see also ref. 36), suggests that useful probes can be developed for other grasses as these are used to provide a source of new genetic material for wheat. The recent analysis of the rDNA region of species in the *Triticeae* has shown that, for example, the short arms of chromosomes 5E and 6E (from *Elytrigia elongatum*) are readily identifiable by virtue of the structure of their rDNA spacers (J. Dvorak, unpublished observation). Cytogenetic studies on a chromosome from *Aegilops sharonensis* which is preferrentially transmitted when it is in wheat has utilized a 1.7 kb segment of wheat DNA cloned in pACY184. *In situ* hybridization with ^{3}H-cRNA from this wheat DNA allows identification of *A. sharonensis* chromosomes (43).

PROSPECTS

The simple observation that 70-80% ($\cong 6 \times 10^6$ kb) of cereal chromosomes consists of repeated sequence DNA dictates that this class of sequence is important in the overall structure and function of chromosomes. At present there exists a reasonable understanding of only $\cong 0.3 \times 10^6$ kb of DNA in this class. As these regions of the chromosomes are delineated it seems likely that some regions may prove useful for flanking newly introduced genetic material, to enhance the stability or expression of such material. Furthermore, quantitative or qualitative modification of the regions per se (by unequal exchange or gene conversion events induced in systems possibly involving cycles of tissue culture) may provide the means of adjusting a particular pattern of gene expression to a given environment. Manipulation of the repeated DNA sequences can potentially effect large segments of the chromosomes to produce regions not necessarily found in nature but which are possibly more suited to the domesticated crop situation.

APPENDIX

The karyotypes of a selection of triticales analyzed using the cloned heterochromatic DNA sequence (pSc7235) from rye. Only the rye chromosomes are shown. Analyses were also carried out on the triticales shown using a heterogeneous probe for rye heterochromatin (Appels, Driscoll and Peacock, 2); although in every case the two probes were consistent in identifying the number of rye chromosomes present, the cloned sequence provided a more sensitive probe for the different rye chromosome variants (as discussed in the main paper). For each probe replicate slides were prepared and studied. Although the labelling of some rye chromosomes is low, this labelling is reproducible and is assayed against a background of no detectable labelling of the wheat chromosomes. The rye chromosomes are arranged in a manner to maximize similarities between the triticales, but as can be seen the variation between triticales is too great to allow assignments to homologous groups; the Imperial rye standard karyotype (Appels et al., 4), did not provide a useful guide for assigning chromosomes to homologous groups.

The author is grateful for Professor C. Driscoll for encouraging the analysis and providing the pedigrees, as far as they are known, of the triticales. The triticale originated from CIMMYT breeding material. Many triticales are based on crosses involving "Armidillo" and thus chromosome 2R is often absent (Gustafson and Zillinsky, 31).

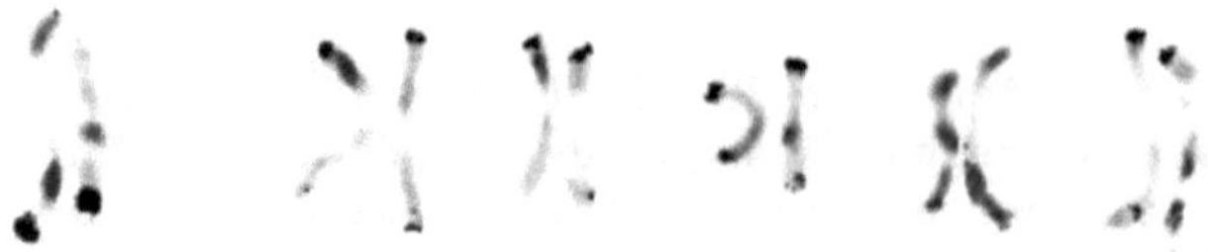

Mapache: Maya II-Arm "S" X2802-F-12M-1N-1M-OY

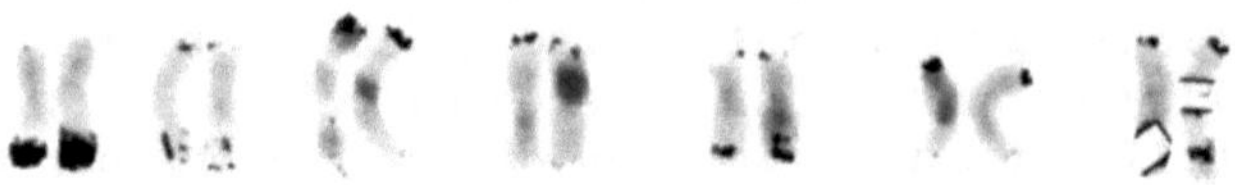

Drira: D-R44/3/Inia Rye 2// Arm "S" x 7110

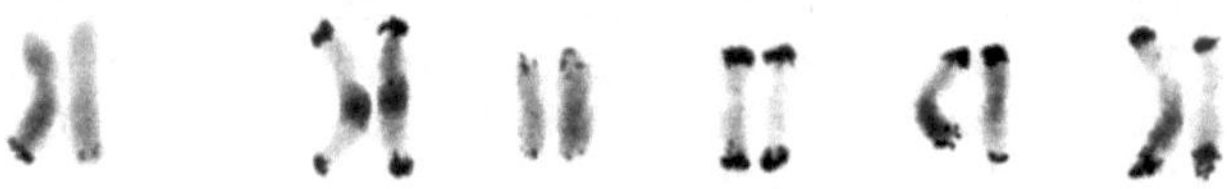

$F_2$181: M2A-Bg1 x Yoco
Bg1=UM "S"-Tc1 Bulk x 1530-A-12M-5N-1M-OY
Yoco=Inia rye-Arm "S"

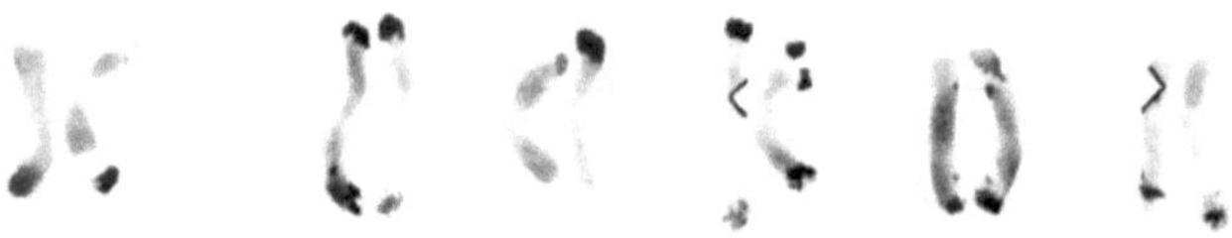

SW139: M2A x M2A

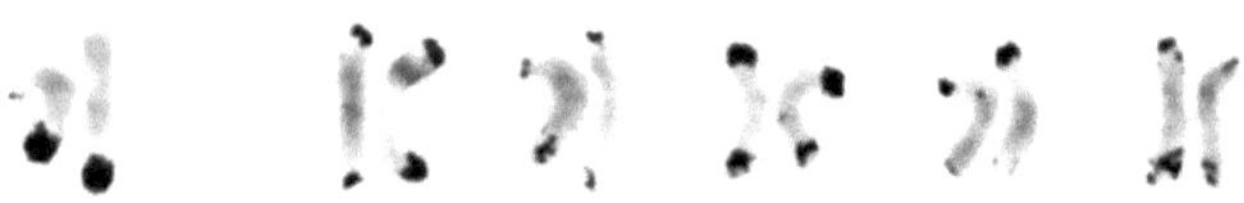

SW182: M2A-IGA x 1A

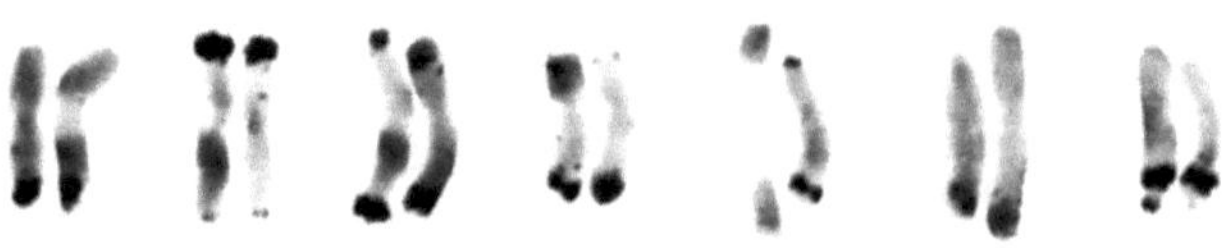

Growquick: Unknown pedigree

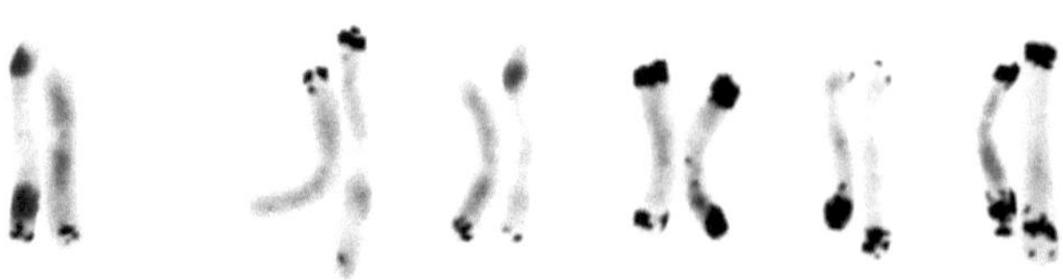

Satu: Tc1 M2A

Dua: Tcl M2A

TR1: DG-66 x 1648

TR3: DG-154

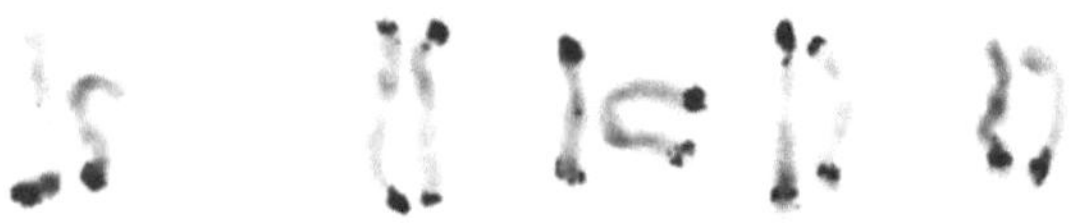

TT4: Koala x Tcl M2A-Arm "O"
Koala=[(Son64-P4 160 x T. Pol Mika/Son 64) Rye Merced]
(Tcl dur "Cilt" - Tcl dur Ghiza) Arm "S" x 2091-102Y-ON-1M-ON

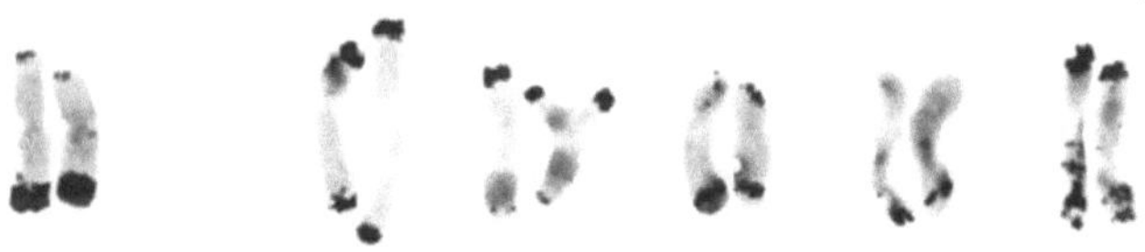

Tyalla: Maya II-Arm "S" x 2802-38N-5M-6N-6M-OY

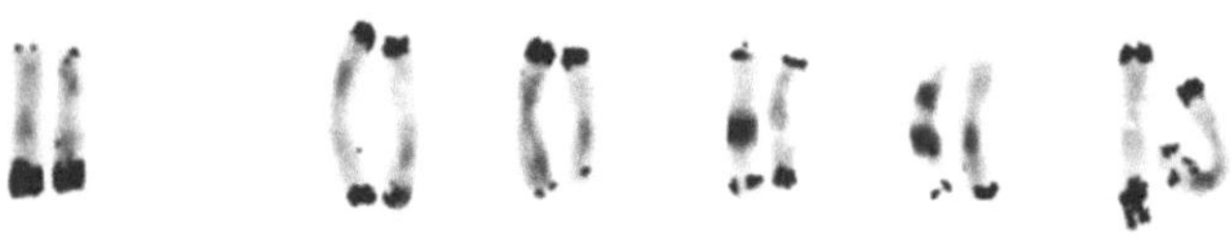

T1245: Maya II-Arm "S" x 2802-38N-5M-6N-6M-OY-1M-OY-OM

T1006: Mava II-Arm "S" x 2802-38N-5M-6N-6M-1Y-1M-OY

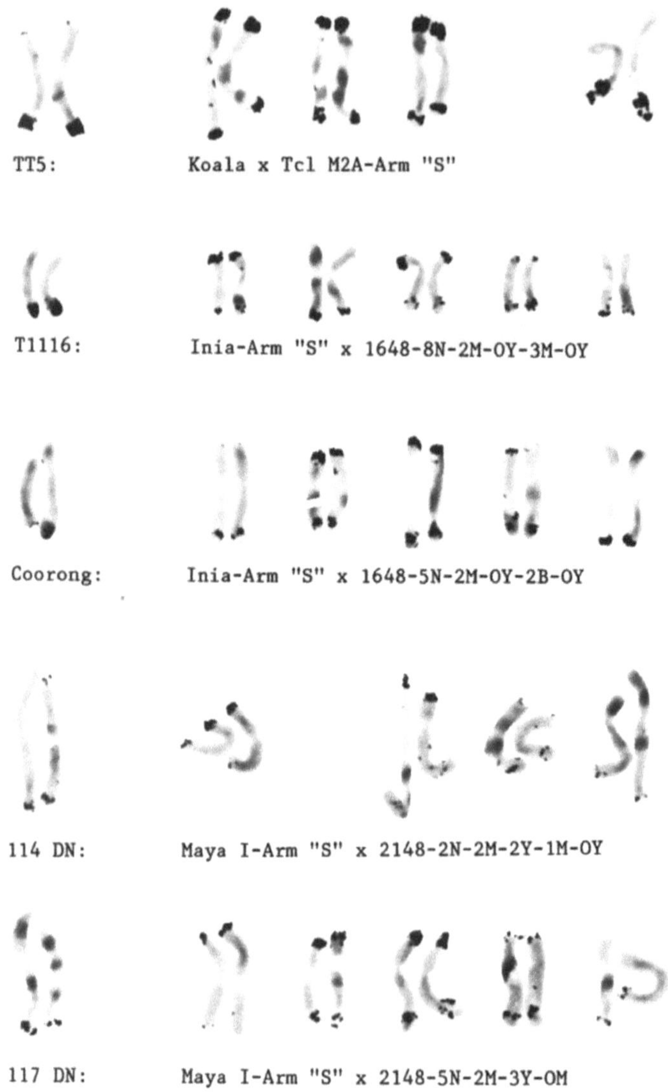
TT5:
Koala x Tcl M2A-Arm "S"
T1116:
Inia-Arm "S" x 1648-8N-2M-OY-3M-OY
Coorong:
Inia-Arm "S" x 1648-5N-2M-OY-2B-OY
114 DN:
Maya I-Arm "S" x 2148-2N-2M-2Y-1M-OY
117 DN:
Maya I-Arm "S" x 2148-5N-2M-3Y-OM

ACKNOWLEDGEMENTS

The author is grateful for the opportunity to have worked with several colleagues to generate the data and interpretations discussed. These colleagues include E.S. Dennis, C. Driscoll, J. Dvorak, W.L. Gerlach, J.P. Gustafson, C.E. May, W.J. Peacock, and D.R. Smyth. The experiments reported were made possible by the constant participation of Lyndall Moran, and more recently Joe Filshie in the research program. Research was in part supported by a grant from the Reserve Bank of Australia.

REFERENCES

1. Appels, R., and W.J. Peacock. 1978. The arrangement and evolution of highly repeated (satellite) DNA sequences with special reference to Drosophila. Int. Rev. Cytology Suppl. 8: 69-126.
2. Appels, R., C. Driscoll, and W.J. Peacock. 1978. Heterochromatin and highly repeated DNA sequences in rye (Secale cereale). Chromosoma 70:67-89.
3. Appels, R., W.L. Gerlach, E.S. Dennis, H. Swift, and W.J. Peacock. 1980. Molecular and chromosomal organization of DNA sequences coding for the ribosomal RNAs in cereals. Chromosoma 78:293-311.
4. Appels, R., E.S. Dennis, D.R. Smyth, and W.J. Peacock, 1981. Two repeated DNA sequences from the heterochromatin regions of rye (Secale cereale) chromosomes. Chromosoma 84:265-277.
5. Appels, R. 1982) The molecular cytology of wheat-rye hybrids. Int. Rev. Cytology (in press).
6. Appels, R. J.P. Gustafson, and C.E. May. 1982. Structural variation in the heterochromatin of rye chromosomes in triticales. Theor. Appl. Genet. (in press).
7. Appels, R., and J. Dvorak. 1982a. The wheat rDNA spacer region: Its structure and variation in populations and among species. Theor. Appl. Genet. (in press).
8. Appels, R., and J. Dvorak. 1982b. Relative rates of divergence of spacer and gene sequences within the rDNA region of species in the Triticeae: Implications for the maintenance of homogeneity of a repeated gene family. Theor. Appl. Genet. (in press).
9. Bedbrook, J.R., J. Jones, M. O'Dell, R.D. Thompson, and R.B. Flavell. 1980a. A molecular description of telomeric heterochromatin in Secale species. Cell 19:545-560.
10. Bedbrook, J.R., M. O'Dell, and R.B. Flavell. 1980b. Amplification of rearranged repeated DNA sequences in cereal plants. Nature 288:133-137.
11. Bennett, M.D. 1977. Heterochromatin, aberrant endosperm nuclei and grain shrivelling in wheat-rye genotypes. Heredity 39:411-419.

12. Bennett, M.D., and J.P. Gustafson. 1982. The effect of telomeric heterochromatin from Secale cereale L. on triticale (x Tritico secale. Wittmack). Can. J. Genet. Cytol. 24: 93-100.
13. Blakeslee, A.F., and A.G. Avery. 1937. Methods of inducing doubling of chromosomes in plants by treatment with colchicine. J. Heredity 28:393-411.
14. Briggle, L.W. 1969. Triticale--A review. Crop Sci. 9: 197-202.
15. Brutlag, D.L. 1980. Molecular arrangement and evolution of heterochromatic DNA. Ann. Rev. Genetics 14 (1980) 121-144.
16. Bryan, P.N., H. Hofstetter, and M.L. Birnstiel. 1981. Nucleosome arrangement of tRNA genes of Xenopus laevis. Cell 27:459-466.
17. Coen, E.S., J.M. Thoday, and G. Dover. 1982. Rate of turnover of structural variants in the rDNA gene family of Drosophila melanogaster. Nature 295:564-568.
18. Dennis, E.S., W.L. Gerlach, and W.L. Peacock. 1980. Identical polypyrimidine-polypurine satellite DNAs in wheat and barley. Heredity 44:349-366.
19. Doerschug, E.B. 1976. Placement of genes for ribosomal RNA within the nucleolar organizing body of Zea mays. Chromosoma 55:43-56.
20. Dvorak, J., and R. Appels. 1982. Chromosome and nucleotide sequence differentiation in genomes of polyploid Triticum species. Theor. Appl. Genet. (in press).
21. Flavell, R.B. 1981. The analysis of plant genes and chromosomes by using DNA cloned in bacteria. Phil. Trans. R. Soc. Lond. B (in press).
22. Flavell, R.B., and G. Martini. 1981. The genetic control of nucleolus formation with special reference to common wheat. Society Exp. Biol. Seminar Series. "Nucleolus." C.A. Cullis and E.G. Jordan, eds. Cambridge University Press (in press).
23. Fredrick, H., V. Hemleben, R.B. Meagher, and J.L. Key. 1979. Purification and restriction endonuclease mapping of soybean 18S and 25S ribosomal RNA genes. Planta 146:467-473.
24. Gerlach, W.L., and J.R. Bedbrook. 1979. Cloning and characterization of ribosomal RNA genes from wheat and barley. Nucleic Acids Res. 7:1869-1885.
25. Gerlach, W.L., and T.A. Dyer. 1980. Sequence organization of the repeating units in the nucleus of wheat which contain 5S rRNA genes. Nucleic Acid Res. 8:4851-4865.
26. Gerlach, W.L., T.E. Miller, and R.B. Flavell. 1980. The nucleolus organizers of diploid wheats revealed by in situ hybridization. Theor. Appl. Genetics 58:97-100.
27. Givens, J.F., and R.L. Phillips. 1976. The nucleolus organizer region of maize (Zea mays L.). Chromosoma 57: 103-117.
28. Goldsbrough, P.B., and C.A. Cullis. 1981. Characterisation of the genes for ribosomal RNA in flax. Nucleic Acids Res. 9: 1301-1309

29. Gustafson, J.P. 1976. The evolutionary development of Triticale: the wheat-rye hybrid. Evolutionary biology 9: 107-135.
30. Gustafson, J.P., and M.D. Bennett. The effect of telomeric heterochromatin from Secale cereale on triticale (x Triticosecale, Wittmack). Can. J. Genet. Cytol. 24:83-92.
31. Gustafson, J.P., and F.J. Zillinsky. 1973. Identification of D-genome chromosomes from hexaploid wheat in a 42-chromosome triticale. Proc. 4th Int. Wheat Genet. Symp.: 225-232.
32. Hilliker, A.J., and R. Appels. 1983. Pleiotropic effects associated with the deletion of heterochromatin surrounding rDNA on the X chromosome of Drosophila. Chromosoma (in press).
33. Hilliker, A.J., R. Appels, and A. Schalet. 1980. The genetic analysis of D. melanogaster heterochromatin. Cell: 607-619.
34. Jorgensen, R.A., R.E. Cuellar, and W.F. Thompson. 1982. Modes and tempos in the evolution of nuclear-encoded ribosomal RNA genes in legumes. Carnegie Institute Year Book (in press).
35. Hutchinson, J., and T. Miller. 1982. The nucleolar organizers of tetraploid and hexaploid wheats revealed by in situ hybridization. Theor. Appl. Genet. 61:285-288.
36. Jones, J.D.G., and R.B. Flavell. 1982. The mapping of highly-repeated DNA families and their relationship to C-bands in chromosomes of Secale cereale. Chromosoma (submitted for publication).
37. Knott, D.R. 1971. The transfer of genes for disease resistance from alien species to wheat by induced translocations. Mutation Breeding for Disease Resistance. Int. Atom. Energy Agency, Vienna: 67-77.
38. Larkin, P.J., and W.R. Scowcroft. 1981. Somaclonal variation- a novel source of variability from cell cultures for plant improvement. Theor. Appl. Genet. 60:197-214.
39. Lima-de-Faria, A. 1952. Chromomere analysis of the chromosome complement of rye. Chromosoma 5:1-68.
40. Linde-Laursen, I. 1978. Giemsa C-banding of barley chromosomes I: Banding pattern polymorphism. Hereditas 88:55-64.
41. Martini, G., M. O'Dell, and R.B. Flavell. 1982. Partial inactivation of wheat nucleolus organizers by the nucleolus organizer chromosomes from Aegilops umbellulata. Chromosoma 84:687-700.
42. May, C.E., and R. Appels. 1982. The inheritance of rye chromosomes in early generations of triticale x wheat hybrids. Can. J. Genet. Cytol. 24:285-291.
43. Miller, T.E., J. Hutchinson, and V. Chapman. 1982. Investigation of a preferentially transmitted Aegilops sharonensis chromosome in wheat. Theor. Appl. Genet. 61: 27-33.
44. Müller, W., and F. Gautier. 1975. Interaction of heterochromatin compounds with nucleic acids. Eur. J. Biochem. 54:385-394.

45. Musich, P.R., F.L. Brown, and J.J. Maio. 1982. Nucleosome phasing and micrococcal nuclease cleavage of African green monkey component αDNA. Proc. Natl. Acad. Sci. U.S.A. 79: 118-122.
46. O'Mara, J.G. 1953. The cytogenetics of Triticale. The Botanical Review 19:587-605.
47. Peacock, W.J., W.L. Gerlach, and E.S. Dennis. 1981. Molecular aspects of wheat evolution: repeated DNA sequences. Wheat Science--Today and Tomorrow. L.T. Evans and W.J. Peacock, eds. Cambridge University Press. 41-60.
48. Radding, C.M. 1978. Genetic recombination: Strand transfer and mismatch repair. Ann. Rev. Biochem. 47:847-880.
49. Rimpau, J.; D. Smith; and R.B. Flavell. 1978. Sequence organization analysis of the wheat and rye genomes by interspecies DNA/DNA hybridization. J. Mol. Biol. 123: 327-359.
50. Sarma, N.P., and A.T. Natarjan. 1973. Identification of heterochromatic regions in the chromosomes of rye. Hereditas 74:233-238.
51. Schopf, T.J.M. 1981. Evidence from findings of molecular biology with regard to the rapidity of genome change: Implication for species durations. Paleobotany, Paleocology and Evolution 1. K.J. Niklas, ed. 135-192.
52. Schweizer, D. 1979. Fluorescent chromosome banding in plants: Mechanisms, and implications for chromosome structure. Proc. 4th John Innes Symp.: 61-72.
53. Schweizer, D. 1981. Counterstain--Enhanced Chromosome Banding. Human Genetics 57:1-4.
54. Seal, A.G., and M.D. Bennett. 1981. The rye genome in winter hexaploid triticales. Can. J. Genet. Cytol. 23:647-653.
55. Sears, E.R. 1956. The transfer of leaf rust resistance from Aegilops umbellulatum to wheat. Brookhaven Symp. Biol. 9: 1-22.
56. Sears, E.R. 1972. Chromosome engineering in wheat. Stadler Symposium: 23-38.
57. Varsany-Breiner, A.; J.F. Gusella; C. Keys; D.E. Housman; D. Sullivan; N. Brisson; and D.P.S. Verma. 1979. The organisation of a nuclear DNA sequence from a higher plant: Molecular cloning and characterisation of soybean ribosomal DNA. Gene 7:317-334.
58. Wellauer, P.K., and I.B. Dawid. 1977. The structural organization of ribosomal DNA in Drosophila melanogaster. Cell 10:193-212.
59. White, R.L., and D.S. Hogness. 1977. R-loop mapping of the 18S and 28S sequences in the long and short repeating units of Drosophila melanogaster rDNA. Cell 10:181-192.
60. Zimmer, E., J. Swanson, C. Rivin, J. Bennetzen, S. Hake, V. Walbot. 1981. Rapid evolution of ribosomal and other repeated DNAs in corn. Fed Proc. 40:1752 (abstract).

CONSIDERATIONS OF DEVELOPMENTAL BIOLOGY FOR THE PLANT CELL GENETICIST

R.S. Chaleff

Department of Central Research & Development
Experimental Station
E.I. du Pont de Nemours and Company
Wilmington, Delaware 19898

ABSTRACT

Cultured plant cells offer many advantages for genetic manipulation of higher plants. But our efforts to select and characterize mutants *in vitro* remind us that cultured plant cells are not unicellular microorganisms, but components of highly complex developmental systems. First, many whole plant traits, especially agronomically important characteristics such as yield, are not expressed by cultured cells. Second, many novel phenotypes selected in cell culture are not expressed by regenerated plants. Epigenetic changes, the occurrence of developmentally regulated isozymes, and variation in the impact of selective conditions on plant cell growth and viability during development are considered as explanations for the failure of whole plants to manifest many phenotypic alterations that are expressed by cultured cells.

INTRODUCTION

One of the major advantages afforded by cell culture for genetic experimentation with higher plants is that it makes possible selection for novel phenotypes from large physiologically and developmentally uniform populations of cells grown under defined conditions. With the ability to manipulate plants like microbes came the expectation that all of the extraordinary feats of genetic experimentation performed with microbes would soon be realized with plants. However, to date, success has been limited, in part because of a unique feature of plant cell culture as a genetic system that is all too often overlooked in our haste to draw analogies to microbial systems: selection for a novel phenotype is conducted at a level of differentiation distinct from that at which expression of the phenotype is ultimately desired. This additional dimension of transposition between various levels of differentiation has several consequences of special significance for employing plant cell culture in the genetic modification of agronomic traits, which, for the most part, are products of differentiated cells, tissues, and organs present only in the whole plant.

Not All Traits Expressed by the Whole Plant are Expressed by the Cultured Cell

If a breeder identified a drought tolerant variety, he might look for deeper root penetration, altered control of stomatal closure, or a thicker cuticle as a basis for this phenotype. But these characteristics are functions not only of highly differentiated cells, but of the organization of such cells into complex organs and of interactions between these organs. At present, it is difficult to imagine how the expression of such traits could be elicited from single cells in culture. Accordingly, the somatic cell geneticist must accept that certain traits are exclusively whole plant functions and as such lie beyond his/her reach. Or, stated another way, certain difficulties and limitations are inherent in an approach that relies on manipulation of the component parts to modify the properties of a whole that is greater than the sum of its parts. This is not to say that cell culture cannot be used to modify whole plant traits, such as drought tolerance, but only that this technique restricts one to approaches that involve selection for alterations of basic cellular functions. Thus, in applying cell culture to the problem of drought tolerance, one could not expect to select mutants with an altered root architecture. However, it might be possible to select cells capable of regulating their osmotic potential by production of osmotically active solutes. Protoplasts or cells possessing this capability could preferentially multiply in a hypertonic medium.

That a given trait can be effected by several different mechanisms--some acting at the cellular level and others only at the whole plant level--is illustrated by comparative studies on the susceptibility of whole plants and callus cultures to salt. Thus, callus cultures of the halophyte glasswort (*Salicornia*) and of cabbage, sweet clover, and sorghum are equally sensitive to NaCl (32). In contrast, the relative degrees of salt tolerance of callus cultures of two barley species (*Hordeum vulgare* and *H. jubatum*) seem to correspond to those of the whole plants (26). Most encouraging with respect to the feasibility of selecting salt tolerance *in vitro* was the report (24) that plants regenerated from salt tolerant tobacco cell lines survived irrigation with a solution containing a NaCl concentration lethal to normal plants. Although progeny obtained by self-fertilization of these regenerated plants also proved tolerant to NaCl, the possibility remains that tolerance results from adaptive changes (such as an altered membrane composition) that are transmitted maternally in crosses rather than from a true mutation.

Other whole plant traits of agronomic importance, such as yield, grain quality, and many types of pest resistance, may prove less amenable to an in vitro approach. Not only are these traits not expressed by cultured cells, but our poor understanding of their molecular and cellular bases prevents us from identifying correlative cellular functions for which in vitro selection schemes might be devised. Ironically, cell culture grants us the general ability to select for mutant types, but precludes selection for many specific agronomically desirable features that are not expressed by cultured cells.

Not All Traits Expressed by Cultured Cells are Expressed by the Whole Plant

It is now generally recognized that successful selection for a novel phenotype in cell culture does not guarantee expression of that phenotype by the regenerated plant. But the casual indifference displayed by the regenerated plant toward the most ardent desires of the investigator is, after all, to be expected of a developmental system. Epigenetic changes and developmental control of expression of a mutant locus are frequently advanced as explanations of differences in phenotypic expression between callus and plant (5). In addition to briefly reviewing these two possibilities, I would like to consider a third: that the degree of physiological stress imposed by a given selective condition varies as a function of the demand placed on the affected metabolic processes at different developmental stages.

Epigenetic Events

In evaluating the basis of an altered phenotype, especially in a case in which expression of that phenotype appears to be developmentally dependent, it is useful to ascertain whether the alteration has a genetic or nongenetic origin. In contrast to genetic events, which involve changes in nucleotide sequence or in chromosome number, structure, or arrangement, epigenetic events reflect changes in gene expression (resulting from the operation of nonmutant control mechanisms, rather than from mutations in regulatory sequences, such as promoters, or in structural genes of regulatory proteins). In functional terms, epigenetic changes are characterized by their relative stability through mitotic divisions and their reversibility by the processes of differentiation and meiosis. Thus, novel phenotypes with an epigenetic basis may persist indefinitely throughout cellular divisions following removal of the inducing conditions, but are not expressed in regenerated plants or their sexual progeny (4,25). For the present, transmission through sexual crosses provides an acceptable criterion by which to distinguish genetic from epigenetic changes. However, this distinction provides us only with an operational definition that should not be applied too rigidly. Certain types of genetic change, such as gene amplification, are unstable even through mitotic divisions in the absence of selection. Perhaps future discoveries will reveal altered patterns of gene expression that are transmitted through meiosis.

The phenomenon of habituation is the most thoroughly studied example of epigenetic variation in plant cell culture. Habituation refers to the loss of a requirement for exogenous supplies of hormones--either auxin or cytokinin or both--that enables cultured cells to proliferate in the absence of these substances. The habituated state is extremely stable in that daughter cells produced by succeeding mitotic divisions exhibit the same phenotype, and hormone dependence is rarely regained in culture (14). Since habituated cells contain elevated endogenous concentrations of the hormone for which they are habituated (12,13), it is presumed that such changes result from an altered rate of hormone synthesis or degradation.

That habituation has an epigenetic basis was first suggested by observations of the reversibility of the hormone autotrophic phenotype by differentiation. That is, callus cultures initiated from plants that had been regenerated from habituated cell lines were found to require hormones for growth (29,30). The repetition of these experiments with cloned habituated cell lines demonstrated that restoration of the hormone requirement did not result from preferential regeneration of plants from normal cells in a heterogeneous population, but from the true reversibility of the hormone autotrophic state (4).

An epigenetic origin has been suggested for several other novel phenotypes expressed by cultured plant cells. Thus, although selected cell lines of Nicotiana sylvestris displayed enhanced tolerance to chilling injury, even following propagation at a nonselective temperature, such tolerance was not observed in cell cultures established from regenerated plants (10,11).

In the cases of selection for resistance to picloram and to hydroxyurea in N. tabacum cell cultures, plants regenerated from a majority of the resistant cell lines isolated gave rise to resistant callus cultures. However, several picloram-resistant and hydroxyurea-resistant cell lines gave rise only to plants that produced sensitive secondary callus cultures, even though the original isolates proved stably resistant following prolonged periods of maintenance on non-selective medium (7).

But production of normal secondary callus cultures from plants regenerated from a variant cell line is in itself insufficient evidence to establish an epigenetic basis for a novel phenotype. This question can be addressed only by studying plants regenerated from cloned derivatives of the original variant isolate. Thus, in the examples of resistances to chilling injury, picloram, and hydroxyurea, in which such cloning was not performed, the possibility remains that regenerated plants arose from normal or revertant cells contained within a mixed population.

Perhaps at this point it is appropriate to review studies of cycloheximide-resistance in N. tabacum conducted by Maliga and his associates (19), which are often cited as a possible example of epigenetic variation. In these experiments, expression of drug-resistance, even by cloned cell lines, was very unstable in the absence of selection. Interestingly, two observations suggested an association between cycloheximide-resistance and organogenesis. First, resistant callus cultures formed shoots on a medium that completely inhibited differentiation of sensitive callus. Second, the peroxidase isozyme banding pattern of resistant cells cultured on callus maintenance medium supplemented with cycloheximide was distinct from that of sensitive cells grown on callus maintenance medium, but resembled the banding pattern of sensitive callus cultured on a medium that promotes organogenesis. This evidence led Maliga and his colleagues (19) to conclude that resistance to cycloheximide was accomplished by the activation of genes that are expressed normally in differentiated tissues and not in callus cultures. However, since nonselected differentiating tissue is sensitive to cycloheximide (18), this interpretation seems unlikely. Instead, resistance may have resulted from mutation of such a developmentally regulated gene. In contrast, normal embryos and plantlets of carrot are considerably less sensitive to cycloheximide than are cultured cells (33). A cycloheximide-resistant carrot cell line was isolated. Although somatic hybrids formed by fusing protoplasts of the resistant cell line with normal protoplasts were sensitive to the drug, the segregation of resistant cell lines from these hybrids suggested that resistance had a genetic basis (17). Because resistance of both normal plantlets and mutant cell cultures was accomplished by inactivation of the drug, it was proposed that the resistant cell line carried a mutation altering regulation of synthesis of the cycloheximide-inactivating enzyme (33).

Genetic Events

An altered phenotype that appears in cultured cells, yet is not detectable in the whole plant, can also result from a genetic (mutational) event. A genetic basis is inferred from phenotypic expression of the alteration both by secondary callus cultures established from regenerated plants and by callus cultures initiated from seedlings produced by sexual crosses with regenerated plants. (Of course, expression of a mutant phenotype by progeny-derived cell cultures will depend upon the seedling genotype and the degree of dominance of the mutant allele.) Thus, transmission through sexual crosses demonstrates that the alteration results from a mutation and confinement of expression to a specific tissue or developmental stage indicates only that the mutated gene (or other genes encoding activities required for expression of the phenotype) is (are) under developmental control or, as we shall see later, that other developmentally controlled processes modify phenotypic expression.

Before proceeding, let us consider what is meant by "expression in the whole plant". Most often this phrase refers to germination of seeds plated on a medium containing the selective agent and subsequent seedling growth on this medium. In practical terms, it is difficult to imagine evaluating certain phenotypes at the whole plant level by other means. For example, in the case of a mutation that enabled cultured tobacco cells to utilize glycerol as sole carbon source, failure of extensive efforts to identify the altered gene product (Chaleff, unpublished) precluded monitoring gene expression in the mature plant by direct biochemical assay. It is also far too cumbersome to accomplish heterotrophic growth of a mature plant on glycerol. Therefore, to determine if the mutant trait were expressed in the whole plant, seeds were germinated in the dark on media containing various carbon sources and seedling growth was measured. No differences were observed between growth of normal and mutant seedlings under these conditions (8). In contrast, the PmR mutations conferred increased tolerance for the herbicide picloram upon both callus and seedling growth (6,7).

As mentioned above, failure to observe an altered phenotype at the whole plant level may simply reflect normal developmental control of the mutated gene. Alternatively, synthesis of proteins encoded by other genes required for function of the product of the mutant gene (e.g., other enzymes in a metabolic pathway or appropriate electron acceptors or structural elements) may be developmentally regulated. A mutant trait might also disappear in a plant regenerated from a mutant cell line as a result of the activation of genes that were previously repressed, as would be the case in the masking of an auxotrophic requirement by the opening of alternate biosynthetic routes, or in the modification or elimination of certain nutritional requirements of the cultured cell by the ability of progeny seedlings to mobilize nutrient reserves stored in specialized plant organs such as the endosperm or cotyledons.

But an aspect of plant development that is perhaps invoked most frequently to explain developmentally specific expression of a heritable mutant trait is the occurrence of isozymes. Isozymes are multiple forms of enzymes that are encoded by distinct genes yet that possess similar catalytic activities. Plants possess large numbers of isozymes and in many cases it has been demonstrated that the various forms of an enzyme are synthesized at different rates during development. For example, distinct forms of aspartokinase are produced by roots and by cultured tissue of carrot. Aspartokinase activity extracted from fresh carrot root tissue is inhibited greatly by threonine and only slightly by lysine. In contrast, the major activity present in root tissue slices after three days of culture and in well established suspension cultures is relatively sensitive to lysine and insensitive to threonine (22,31). Assuming that these two aspartokinase activities are products of distinct genes rather than of post translational modification of a single gene product, how will the existence of two aspartokinase isozymes influence in vitro mutant selection experiments?

Because the gene encoding the lysine sensitive form of aspartokinase is expressed in cultured cells, supplementation of the medium with lysine will cause inhibition of the enzyme and a consequent deficiency of threonine, methionine, and isoleucine. However, the occurrence of a mutation in the structural gene that renders the enzyme insensitive to lysine inhibition will enable cells to grow in the presence of excess lysine. Yet, because the threonine-sensitive isozyme is produced in the plant, plants regenerated from such lysine resistant cell lines will appear phenotypically normal. That is, mutant and nonmutant seedlings will respond similarly to exogenously supplied lysine and threonine. But unlike normal seedlings, mutant seedlings will give rise to lysine-resistant callus cultures

In addition to the structural gene mutation, a regulatory mutation that permits synthesis of the threonine-sensitive isozyme by cultured cells would also confer lysine resistance. An epigenetic event could produce the same altered regulatory pattern. But, as described earlier, a genetic basis for resistance would be demonstrated by the production of lysine resistant callus cultures from both regenerated plants and progeny seedlings.

Thus, developmental dependence of phenotypic expression can be readily explained by presuming the existence of isozymes. Considering the major role played by polyploidization in plant evolution, this explanation is not unreasonable. But perhaps another factor, still developmental in nature, should be considered when trying to rationalize the failure of plants to express a mutant trait that is readily visible in cultured cells.

As the demands on particular metabolic pathways vary during the course of development and differentiation, the adequacy of protection afforded by a mutation to a given stress may also vary. For example, a compound may inhibit cell growth by reducing the activity of a key enzyme to 10% of its normal level. Let us say that in the presence of the inhibitor a mutant form of this enzyme that has a lower affinity for the inhibitor possesses 50% of normal activity. If at this reduced level of activity the mutant enzyme is able to provide 80-90% of the amount of end product required for normal callus growth, callus growth will still proceed in the presence of the inhibitor, albeit at a slightly lower rate. But if the enzyme in question is a component (especially a rate-limiting step) of a pathway that is maximally utilized at a particular developmental stage (e.g., respiration or DNA, RNA, or protein synthesis during seedling growth) only 50% of the end product levels required at this point will be produced. Thus, the mutant enzyme may be synthesized at two developmental stages, but the same concentration of inhibitor produces a greater deficiency at one stage than at another.

Such an explanation of developmentally dependent expression of a mutant phenotype may be appropriate in the case of tobacco mutants that were selected *in vitro* on the basis of resistance to isonicotinic acid hydrazide (INH), an inhibitor of glycine decarboxylation in the glycolate pathway of photorespiration. Germination of seeds produced by plants regenerated from INH-resistant cell lines was as sensitive to INH as was germination of normal seeds. However, callus cultures established from plants regenerated from these resistant cell lines and from their progeny were resistant. Thus, INH-resistance has a genetic basis and its expression is restricted to the cellular level (3). By direct biochemical assay, glycine decarboxylase activities in both resistant callus cultures and in leaves of mutant plants were shown to be less sensitive to inhibition by INH than were the activities in the corresponding normal tissues. Cosegregation of INH-resistance and an altered glycine decarboxylase activity in sexual crosses strongly suggests that the reduced sensitivity of this enzyme complex to INH is the basis for the resistance phenotype (34). But although this biochemical alteration appears in both callus and plant, only callus and not seedling growth displays resistance to INH.

Perhaps, then, it is not a coincidence that resistance is manifested only at the callus stage by all mutants isolated to date by selection in culture for resistance to pyrimidine antagonists (21, Chaleff and Keil, unpublished results). Levels of pyrimidine biosynthetic enzymes and the endogenous concentrations of intermediates of this pathway vary during the cell cycle (16,27) and are higher in embryogenic than in nonembryogenic cells (1). This regulatory pattern provides greater amounts of purine and pyrimidine nucleotides at periods of active DNA and RNA synthesis. Thus, because of the increased demand for nucleotides during embryogenesis and seedling growth, a mutation affording partial reversal of enzyme inhibition by a pyrimidine antagonist may not confer resistance at these developmental stages.

Cell lines of Nicotiana tabacum resistant to 5-bromodeoxyuridine (BUdR) (20), methotrexate (MTX), and hydroxyurea have been isolated. MTX is a folic acid analogue that inhibits dihydrofolate reductase (9), an enzyme involved in production of the cofactor utilized in the conversion of deoxyuridylate to thymidylate. Hydroxyurea inhibits ribonucleotide reductase (15), which catalyzes the formation of deoxyribonucleotides from ribonucleotides. In all cases, germination of seeds produced by plants regenerated from resistant cell lines was as sensitive to the selective agent as was germination of normal seeds and progeny phenotypes were evaluated by the growth responses of derivative callus cultures. Although crosses with plants regenerated from one BUdR-resistant cell line demonstrated a genetic basis for resistance, the actual mode of inheritance was not established conclusively (21). The results of crosses with plants regenerated from one MTX-resistant cell line indicate that in this isolate resistance results from a single semidominant nuclear mutation for which the regenerated plants are homozygous (Table 1). Plants regenerated from hydroxyurea-resistant cell lines were first crossed with normal plants to minimize interference in the subsequent genetic analysis by other genetic events that may have accumulated during cell culture. Resistant, and presumably heterozygous, F_1 plants were then self-fertilized and crossed reciprocally with normal plants. In all cases, resistance segregated as a single dominant nuclear mutation (Table 2). This result was confirmed by further selfings of hydroxyurea-resistant F_2 progeny obtained from self-fertilization of HuR1/+ and HuR9/+ F_1 plants. By this procedure it was shown that, among 15 resistant isolates obtained from an HuR1/+ plant, 4 were homozygous and 11 were heterozygous. Similarly, a population of 18 resistant F_2 individuals derived from an HuR9/+ plant consisted of 5 homozygotes and 13 heterozygotes (Keil and Chaleff, unpublished results).

Because the biochemical mechanisms of the resistances to BUdR, hydroxyurea, and MTX have not yet been identified, it is not known if the mutant genes conferring these resistances are expressed in the seedling, where the altered phenotypes are not manifest, as well as in the callus or, for that matter, if the mutant genes are indeed in the pyrimidine biosynthetic pathway. Only with such evidence can it be decided whether the observed developmental dependence of the resistance phenotypes results from changes in the demand for products of the affected pathway(s) or from the synthesis of distinct isozymes in the seedling and callus stages.

CONCLUSION

The suitability of in vitro mutant selection as a means for modifying whole plant characteristics must be evaluated independently in each case. If morphogenetically competent cell

Table 1. Crosses with a regenerated plant (R) obtained from a MTX-resistant cell line.

Cross	Number of Individuals Resistant	Intermediate	Sensitive
R selfed	20	0	0
R X Normal	0	26	0
Normal X R	0	28	0
(R X Normal) selfed	7	20	13
(R X Normal) X Normal	0	10	10

Table 2. Progeny obtained from self-ferbilization and crosses with hydroxyurea-resistant F_1 plants.

Cross	Number of Individuals			
	Resistant		Sensitive	
	obs	(exp)	obs	(exp)
HuR1/+ selfed	43	(47)	20	(16)
HuR1/+ x +/+	19	(16)	13	(16)
+/+ x HuR1/+	15	(16)	17	(16)
HuR9/+ selfed	53	(48)	11	(16)
HuR9/+ x +/+	15	(14.5)	14	(14.5)
+/+ x HuR9/+	14	(16)	18	(16)
HuR12/+ selfed	63	(58)	14	(19)
HuR12/+ x +/+	17	(16)	15	(16)
+/+ x HuR12/+	15	(16)	17	(16)
HuR13/+ selfed	33	(36)	15	(12)
HuR13/+ x +/+	16	(16)	16	(16)
+/+ x HuR13/+	13	(16)	19	(16)
HuR17/+ selfed	52	(53)	19	(18)
HuR17/+ x +/+	19	(16)	13	(16)
+/+ x HuR17/+	15	(15)	15	(15)
HuR30/+ selfed	46	(49)	19	(16)
HuR30/+ x +/+	18	(15.5)	13	(15.5)
+/+ x HuR30/+	13	(15)	17	(15)
HuR32/+ selfed	51	(52)	18	(17)
HuR32/+ x +/+	10	(13)	16	(13)
+/+ x HuR32/+	20	(16)	12	(16)

SOURCE: KEIL AND CHALEFF, UNPUBLISHED RESULTS.

cultures can be obtained for the species in question, one must then ask if the trait that one seeks to alter (or a correlated function) is expressed in such cell cultures and if an appropriate procedure for selecting mutants of the desired phenotype can be devised. Because this phenotype may be produced by any of several disparate mechanisms, many isolates should be characterized. In some cases, expression of the desired phenotype will have resulted from an epigenetic event. In those cases in which a genetic basis for the altered phenotype can be established, it must be determined if expression occurs at the whole plant level and in the appropriate organ and developmental stage. These considerations are illustrated by a brief survey of efforts to isolate mutants displaying increased tolerance for herbicides.

Although the herbicides bentazon and phenmedipham bleach leaves of intact tobacco plants, they do not affect growth of cultured cells. Therefore, Radin and Carlson (28) turned to leaves of haploid plants for populations of cells in a differentiated state that was sensitive to these herbicides. After spraying plants with the herbicides, occasional green islands of resistant cells were visible in the otherwise yellow leaf tissue. These sectors were excised, placed into culture, and induced to form plants. By this procedure, eight bentazon-resistant and two phenmedipham-resistant isolates were obtained. In all cases, resistance was inherited as a recessive trait. Thus, an ingenious alternative was applied successfully to select for resistance to herbicides to which differentiated cells of the whole plant, but not cultured cells, were sensitive.

Because paraquat is converted to phytotoxic agents by a series of reactions initiated by its reduction to the free radical form by electrons supplied by photosynthetic electron transport, one might also expect this herbicide to be without effect on heterotrophic cell cultures. But apparently the initial reduction of paraquat can be accomplished in the dark, albeit at a slower rate, via mitochondrial respiration (2). Presumably it is for this reason that cultured tobacco cells proved as sensitive to paraquat as whole plants and resistant cell lines could be selected _in vitro_ (23). Plants regenerated from all but one resistant cell line gave rise to resistant callus cultures. But of these plants, those derived from fewer than half of the variant cell lines displayed increased tolerance for the herbicide (as assayed by the rate of bleaching of leaves floated on paraquat solutions). Thus, it appears that paraquat resistance can be achieved by several mechanisms, some of which are expressed in the whole plant and others only in cultured cells. Studies are now being conducted to determine if any of these resistance mechanisms has a genetic basis.

Although the mode of action of picloram is unknown, the sensitivity of tobacco cell cultures to this herbicide permitted resistant cell lines to be isolated. Plants were regenerated from six of seven cell lines initially selected from diploid cell cultures. Callus cultures established from plants originating from five cell lines proved resistant to the herbicide and plants obtained from four of these five cell lines produced seeds that germinated in the presence of picloram concentrations inhibitory to normal seed germination. In genetic crosses with the four mutants producing resistant seeds and with a fifth such mutant isolated subsequently, three mutations (PmR1, PmR2, and PmR7) behaved as dominant alleles and two (PmR6 and PmR85) as semidominant alleles of single nuclear genes (6,7). PmR1 and PmR7 were genetically linked and PmR6 and PmR85 defined two additional linkage groups (6). Increased tolerance for picloram was expressed by two month old plantlets homozygous for the PmR1 mutation (the only mutant to be so tested) (5). Thus, experiments designed to isolate picloram-tolerant mutants yielded an assortment of different types. In one case resistance was not expressed by callus cultures derived from regenerated plants, and in another case resistance was not expressed by germinating seeds. But characterization of a number of different isolates resulted in the identification of at least one genetically stable mutation that had been selected at the cell level, yet conferred a higher degree of tolerance for picloram on the whole plant.

REFERENCES

1. Ashihara, H., T. Fujimura, and A. Komamine. 1981. Pyrimidine nucleotide biosynthesis during somatic embryogenesis in a carrot cell suspension culture. Z. Pflanzenphysiol. 104:129-137.
2. Ashton, F.M., and A.S. Crafts. 1981. Mode of Action of Herbicides. New York: John Wiley and Sons.
3. Berlyn, M.B. 1980. Isolation and characterization of isonicotinic acid hydrazide-resistant mutants of *Nicotiana tabacum*. Theor. Appl. Genet. 58:19-26.
4. Binns, A., and F. Meins. 1973. Habituation of tobacco pith cells for factors promoting cell division is heritable and potentially reversible. Proc. Natl. Acad. Sci. U.S.A. 70:2660-2662.
5. Chaleff, R.S. 1981. Genetics of Higher Plants: Applications of Cell Culture. New York: Cambridge University Press.
6. Chaleff, R.S. 1980. Further characterization of picloram-tolerant mutants of *Nicotiana tabacum*. Theor. Appl. Genet. 58:91-95.
7. Chaleff, R.S., and M.F. Parsons. 1978a. Direct selection *in vitro* for herbicide-resistant mutants of *Nicotiana tabacum*. Proc. Natl. Acad. Sci. U.S.A. 75:5104-5107.
8. Chaleff, R.S., and M.F. Parsons. 1978b. Isolation of a glycerol-utilizing mutant of *Nicotiana tabacum*. Genetics 89:723-728.
9. Crosti, P. 1981. Effect of folate analogues on the activity of dihydrofolate reductases and on the growth of plant organisms. J. Exp. Bot. 32:717-723.
10. Dix, P.J. 1977. Chilling resistance is not transmitted sexually in plants regenerated from *Nicotiana sylvestris* cell lines. Z. Pflanzenphysiol. 84:223-226.

11. Dix, P.J., and H.E. Street. 1976. Selection of plant cell lines with enhanced chilling resistance. Ann. Bot. (London) 40:903-910.
12. Dyson, W.H., and R.H. Hall. 1972. N^6-(Δ^2-Isopentenyl) adenosine: Its occurrence as a free nucleoside in an autonomous strain of tobacco tissue. Plant Physiol. 50:616-621.
13. Fox, J.E. 1963. Growth factor requirements and chromosome number in tobacco tissue cultures. Physiol. Plant. 16:793-803.
14. Gautheret, R.J. 1946. Comparaison entre l'action de l'acide indole-acetique et celle du Phytomonas tumefaciens sur la croissance des tissus vegetaux. C.R. Soc. Biol. 140:169-171.
15. Hovemann, B., and H. Follmann. 1979. Deoxyribonucleotide synthesis and DNA polymerase activity in plant cells (Vicia faba and Glycine max). Biochim, Biophys. Acta 561:42-52.
16. Kanamori-Fukuda, I., H. Ashihara, and A. Komamine. 1981. Pyrimidine nucleotide biosynthesis in Vinca rosea cells: Changes in the activity of the de novo and salvage pathways during growth in a suspension culture. J. Exp. Bot. 32:69-78.
17. Lazar, G.B., D. Dudits, and Z.R. Sung. 1981. Expression of cycloheximide resistance in carrot somatic hybrids and their segregants. Genetics 98:347-356.
18. Maliga, P. 1978. Resistance mutants and their use in genetic manipulation. In Frontiers of Plant Tissue Culture 1978, pp. 381-392. T.A. Thorpe, ed. Calgary: International Association for Plant Tissue Culture.
19. Maliga, P., G. Lazar, Z. Svab, and F. Nagy. 1976. Transient cycloheximide-resistance in a tobacco cell line. Mol. Gen. Genet. 149:267-271.
20. Maliga, P., L. Marton, and A. Sz. Breznovits. 1973. 5-Bromodeoxyuridine-resistant cell lines from haploid tobacco. Plant Sci. Lett. 1:119-121.
21. Marton, L., and P. Maliga. 1975. Control of resistance in tobacco cells to 5-bromodeoxyuridine by a simple Mendelian factor. Plant Sci. Lett. 5:77-81.
22. Matthews, B.F., and J.M. Widholm. 1978. Regulation of lysine and threonine synthesis in carrot cell suspension cultures and whole carrot roots. Planta 141:315-321.
23. Miller, O.K., and K.W. Hughes. 1980. Selection of paraquat-resistant variants of tobacco from cell cultures. In Vitro 16:1085-1091.
24. Nabors, M.W., S.E. Gibbs, C.S. Bernstein, and M.E. Meis. 1980. NaCl-tolerant tobacco plants from cultured cells. Z. Pflanzenphysiol. 97:13-17.
25. Nanney, D.L. 1958. Epigenetic control systems. Proc. Natl. Acad. Sci. U.S.A. 44:712-717.
26. Orton, T.J. 1980. Comparison of salt tolerance between Hordeum vulgare and H. jubatum in whole plants and callus cultures. Z. Pflanzenphysiol. 98:105-118.

27. Parker, N.F., and J.F. Jackson. 1981. Control of pyrimidine biosynthesis in synchronously dividing cells of Helianthus tuberosus. Plant Physiol. 67:363-366.
28. Radin, D.N., and P.S. Carlson. 1978. Herbicide-tolerant tobacco mutants selected in situ and recovered via regeneration from cell culture. Genet. Res. 32:85-89.
29. Sacristan, M.D., and G. Melchers. 1969. The caryological analysis of plants regenerated from tumorous and other callus cultures of tobacco. Mol. Gen. Genet. 105:317-333.
30. Sacristan, M.D., and M.F. Wendt-Gallitelli. 1971. Transformation to auxin-autotrophy and its reversibility in a mutant line of Crepis capillaris callus culture. Mol. Gen. Genet. 110:355-360.
31. Sakano, K., and A. Komamine. 1978. Change in the proportion of two aspartokinases in carrot root tissue in response to in vitro culture. Plant Physiol. 61:115-118.
32. Strogonov, B.P. 1970. Structure and Function of Plant Cells in Saline Habitats. 1973 Israel Program for Scientific Translations. New York: Halsted Press.
33. Sung, Z.R., G.B. Lazar, and D. Dudits. 1981. Cycloheximide resistance in carrot culture: A differentiated function. Plant Physiol. 68:261-264.
34. Zelitch, I., and M.B. Berlyn. 1982. Altered glycine decarboxylation inhibition in isonicotinic acid hydrazide-resistant mutant callus lines and in regenerated plants and seed progeny. Plant Physiol. 69:198-204.

PROTOPLAST FUSION: AGRICULTURAL APPLICATIONS OF SOMATIC HYBRID PLANTS

David A. Evans and C.E. Flick

DNA Plant Technology Corporation
2611 Branch Pike
Cinnaminson, NJ 08077

INTRODUCTION

Interspecific sexual hybridization has been important both in the evolution of cultivated crops (1) and for the development of new cultivated varieties. Several cultivated crops are allopolyploids that were originally derived through sexual hybridization, chromosome doubling, and subsequent diploidization (2). Hence, the historical value of interspecific hybridization is well documented. Interspecies hybrids have been extremely useful for transfer of genes into cultivated crops (3). Several plant varieties have been released that express traits derived from wild species. These include varieties of tomato, tobacco, barley, potato, wheat, etc. For example, resistance to several diseases has been transferred from *Solanum demissum* into cultivated potatoes (4). One variety of tobacco has resistance to three diseases derived from three different wild *Nicotiana* species (5). Disease resistance often may be controlled by a single gene and is relatively easy to transfer by hybridization. However, traits traditionally viewed as more complex have also been transferred into cultivated crops using sexual hybridization. For example, interspecies hybrids of oats (6) and tobacco (7) have been identified with improved yield. Novel variation, not expressed in either parent species, has also been observed in some interspecies sexual hybrids, such as cytoplasmically controlled male sterility in tobacco and improved fruit pigmentation in tomato.

Despite the value for crop improvement, limitations of interspecies hybridization encourage the development of new technologies to create genetic variation and to transfer genes.

Most cultivated crops have a narrow genetic base and have limited variation available for new variety development. Several crops are highly inbred (e.g., tomato, lettuce) and many known cultivated crops were derived from a very narrow genetic base (e.g., soybeans, corn). In recent years, efforts have been made to collect and preserve wild species with interesting and potentially useful traits. However, many sexual crosses between new accessions of wild species and related cultivated crops are not possible using conventional methods. It is hoped that embryo culture or protoplast fusion can be used to circumvent limits to successful hybridization. Protoplast fusion methods offer unique opportunities to introduce variation and to transfer genes between different varieties or species.

ISOLATION AND IDENTIFICATION OF SOMATIC HYBRIDS

The techniques for isolation, culture, and fusion of protoplasts have been well documented (8,9). Protoplasts can be isolated from cell cultures or intact plants using commercially available enzymes. To facilitate observations of fusion products, leaf-derived protoplasts of one species are often fused with cell culture-derived protoplasts of the second species.

Polyethylene glycol (PEG) is most frequently used to induce protoplast fusion, and its use was first reported by Kao and Michayluk (10). High molecular weight PEG, 1540-6000, when added at 25-30%, causes agglutination of protoplasts. When PEG is diluted at high pH in the presence of a high Ca^{++} concentration, a high frequency of protoplast fusion is obtained. In some cases, up to 100% fusion has been reported using PEG (11). The mechanism of action of PEG has been discussed, but appears to be complex. It has been suggested that PEG acts as a molecular bridge, dissociating the plasmalemma, resulting in intercellular connections. Several other methods, both physical and chemical, have been used to facilitate protoplast fusion, including electrical stimulation and treatment with polyvinyl alcohol.

Following the fusion treatment, protoplasts that are cultured in the appropriate liquid medium regenerate cell walls and undergo mitosis. A typical fusion experiment results in a mixed population of parental cells, homokaryotic fusion products, and heterokaryotic fusion products or hybrids. Hybrid cells and regenerated hybrid plants must be distinguished from the other cells present. Identification and recovery of protoplast fusion products has been based on the general observation that hybrid cells display genetic complementation for recessive mutations and physiological complementation for <u>in vitro</u> growth requirements. Carlson et al. (12) first successfully used complementation to isolate auxin autotrophic somatic hybrids. Following fusion of two <u>Nicotiana</u>

species, each with an auxin requirement for cell growth, somatic hybrids were isolated by growth on auxin-free culture medium. Auxin autotrophy was expressed as a result of the specific genetic combination of the two parental species used.

Melchers and Labib (13) first used albino gene complementation to isolate green somatic hybrids following fusion of two distinct homozygous recessive albino mutants of Nicotiana tabacum. This has developed as the most frequently used method to isolate somatic hybrids. A population of protoplasts isolated from a genetically recessive albino is fused with either a population of protoplasts isolated from a second non-allelic albino mutation, or with a population of normal green mesophyll protoplasts. For example, Schieder (14) fused protoplasts of two diploid homozygous albino mutations of Datura innoxia, A1/5a and A7/1s, that had been induced by X-ray treatment. Intraspecific somatic hybrids were selected by isolating young, green regenerating shoots. Similarly, Douglas et al. (15) isolated green interspecific hybrid shoots following fusion of chlorotic N. rustica protoplasts with albino N. tabacum protoplasts.

We have used a semidominant albino mutation (Su) to successfully identify somatic hybrids in the genus Nicotiana (16). The light green Su/su, when selfed, produces dark green, light green and aurea (albino) seedlings in a 1:2:1 ratio. A cell suspension culture derived from aurea leaves of the mutant (Su/Su) was established and has maintained stable chromosome number (2n=48) for five years. Protoplasts isolated from Su/Su cultures have been fused to mesophyll protoplasts isolated from leaves of several wild Nicotiana species. While regenerated dark green and albino shoots are derived from the parental species, light green shoots are derived from hybrid cells. Somatic hybrids have been produced between N. tabacum (Su/Su) and N. sylvestris, N. otophora, N. glauca, N. nesophila, and N. stocktonii.

The development of more powerful selection methods, utilizing mutants induced in vitro, may be necessary to isolate interspecific and intergeneric hybrids between more distantly related species. Utilization of two amino acid analogue resistance mutants (17) or two nitrate reductase deficient mutants (18) has been proposed, but has not yet resulted in the recovery of mature hybrid plants. On the other hand, some variants have been successfully used to recover somatic hybrid plants. Maliga et al. (19) used a kanamycin resistant variant of N. sylvestris, KR103, isolated from cultured cells, as a genetic marker to recover fusion products between N. sylvestris and N. knightiana. Similarly, the SR1 streptomycin resistant mutant of N. tabacum, also isolated from cultured cells, was used to recover intraspecific hybrids with N. tabacum (20), interspecific hybrids with N. sylvestris (21), and interspecific hybrids with N. knightiana (22). The SR1 mutation is encoded in cytoplasmic DNA, and those somatic hybrids that contained *N. tabacum* chloroplast DNA, as determined by restriction analysis, expressed streptomycin resistance. Variants isolated in vitro from carrot, *Daucus carota*, have also been used to identify somatic hybrid plants. Cycloheximide (CHX) resistant plants of carrot were isolated using cultured cells. Regenerated CHX-resistant plants have abnormal leaves. When these resistant lines were fused with albino lines of *D. carota*, somatic hybrids could be identified as being CHX-resistant, green, and having normal, dissected leaves (23). Similarly, Kameya et al. (24) used the C123 cell line of *D. carota* that simultaneously expressed 5-methyltryptophan (5MT) and azetidine-2-carboxylate (A2C) resistance to identify interspecific hybrids between *D. carota* and *D. capillifolius*. While selection for hybrid cells was based on resistance to 5MT, callus reinitiated from somatic hybrid plants expressed only intermediate resistance to 5MT and complete resistance to A2C.

Once hybrids are selected, hybridity can be verified using numerous criteria (16). In most cases, characteristics of somatic hybrids are intermediate between the two parents. These traits include both vegetative and floral characters for several hybrids.

In many instances, the somatic hybrids with intermediate characters have been favorably compared to sexual hybrids when available. The genetic basis for most of these morphological traits has not been elucidated, but the intermediate behavior in hybrids suggests the traits are controlled by multiple genes. Based on the sexual and somatic hybrids produced, intermediate morphology is the most frequently cited criterion to verify hybridity. Some traits, though, behave as dominant single gene traits as they are present in only one parent, but are also expressed in the somatic hybrids. Such traits include stem anthocyanin pigment (Fig. 1), flower pigment (25), heterochromatic knobs in interphase cells (26), and leaf size (25). Consequently, intermediate morphology is not observed for all characters in somatic hybrids. When possible, additional genetic data should be presented to support hybridity.

Several characters are not intermediate in somatic hybrids. Pollen viability is usually dependent on the taxonomic closeness of the two parental species used for hybridization; consequently, more distant somatic hybrids have lower pollen viability. Chromosome number, on the other hand, should equal the sum of somatic chromosome numbers of the two parents, but has been quite variable in most somatic hybrids reported to date.

Isoenzyme analysis has been used extensively to verify hybridity (27). Enzymes that have exhibited unique banding patterns for somatic hybrids versus either parental species include esterase, aspartate aminotransferase, amylase, and isoperoxidase (see 16). Isoenzymes, though, are extremely variable within plant tissues, and zymograms should be interpreted cautiously.

Fig. 1. Anthocyanin pigmentation, a N. glauca-derivedtrait, is prominent at the base of the petiole of this N. glauca + N. tabacum somatic hybrid plant.

VARIABILITY IN POPULATIONS OF SOMATIC HYBRID PLANTS

Interspecific or intraspecific sexual hybrids are usually genetically uniform. In most instances, particularly if hybrids are unilaterally incompatible, only certain nuclear-cytoplasmic combinations are possible. On the other hand, genetic variability is common among somatic hybrids as it is among plants regenerated from leaf explants (28) and from mesophyll protoplasts (29). As both agronomic and non-agronomic characteristics are variable, traits important for crop improvement may be expressed in somatic hybrids and not in comparable sexual hybrids. This phenomenon emphasizes the potential value of somatic hybrids even when produced between sexually compatible species. Evans et al. (30) have compared somatic and sexual hybrids between N. nesophila and N. tabacum and reported increased phenotypic variability in the somatic hybrids.

The phenotypic variability observed in somatic hybrid plants is a reflection of genetic phenomena that occur prior to plant regeneration. Four potential sources of variability have been identified: nuclear incompatibility, mitotic recombination, somaclonal variation, and organelle segregation. 1) Nuclear genetic instability in wide species combinations may result in chromosome elimination. Directional chromosome elimination has been documented in several intergeneric plant cell hybrids (31) and this instability has resulted in recovery of aneuploid plants resembling one or the other parental type for all intergeneric hybrid plants (32). Aneuploid plants have also been recovered for the more closely related interspecific somatic hybrids (16) although some of this chromosome variation may be due to the use of protoplasts isolated from long term aneuploid cell cultures. 2) Intergenomic mitotic recombination (translocation) results in unique mixtures of genetic information. Recombination could be observed in several somatic hybrids (33) when the fate of genetically controlled spot formation was monitored on leaves of somatic hybrids. If mitotic recombination occurs prior to shoot regeneration, this nuclear genetic variation could be detected when regenerated clones are compared. It is likely that chemical or physical agents could be used to increase the frequency of intergenomic genetic exchange. 3) Somaclonal variability has been observed in plants regenerated from mesophyll protoplasts (29). This phenomenon is the result of either preexisting genetic variation in leaf cells or is induced by undefined components of the protoplast, callus or plant regeneration media. It is as yet impossible to measure the effects of somaclonal variation on interplant variation of somatic hybrids; however, it should be noted that leaf mesophyll protoplasts, where somaclonal variation has been detected, were used to produce most somatic hybrid plants. 4) Evidence from a number of somatic hybrids suggests that although two types of cytoplasm are initially mixed during protoplast fusion, cytoplasmic segregation usually occurs (16). Segregation may not occur until after meiosis, but when it does, genetic information from the second cytoplasm is apparently irreversibly lost. This phenomenon produces different classes of hybrid plants based on organelle DNA content. Recent evidence suggests that recombinant plastomes, particularly mitochondrial genomes, may be recovered following fusion (34). Hence, with organelle segregation and recombination, a wider range of nuclear-cytoplasmic genetic combinations is possible in somatic hybrids than in comparable sexual hybrids. These four factors may all contribute to recovery of unique variability in somatic hybrid plants.

GENETIC ANALYSIS OF SOMATIC HYBRIDS

Value of Genetic Analysis

A role for somatic hybrids in crop improvement cannot be

realized unless the hybrids can be integrated into conventional breeding programs. For example, although disease resistance may be expressed in a somatic hybrid, undesirable characters are also expressed in the hybrid. These undesirable traits may effect quality or depress yield of the crop. Elimination of deleterious traits is best achieved through repeated backcrossing of a somatic hybrid to the cultivated crop, while continually selecting for desirable characteristics. Hence, the somatic hybrid provides a novel pool of genes as a basis for new variety development.

When somatic hybrids are produced between distantly related species, the use of these hybrids in conventional breeding is complicated. For example, if an interspecific hybrid is produced in order to incorporate disease resistance from a wild species into a cultivated crop, it would be most advantageous to eliminate all of the wild species genome except the few genes of interest. Two factors complicate incorporation of desirable genes from a wild species into a crop. First, if the trait of interest is polygenic, then an intact functional block of genes, rather than a single gene, must be transferred in sexual crosses. Transfer of such a block of genes will also entail incorporation of closely linked genes. Some of these closely linked genes may depress yield or have other undesirable characteristics. Second, incorporation of a single gene trait from a wild species into a cultivated crop will necessitate interspecific recombination between genomes. The frequency of such recombination is reduced in hybrids between distantly related species. Double spots, composed of adjacent green and albino sectors, on heterozygous light green Su/su plants, have been interpreted as reciprocal genetic events and verified in *N. tabacum* as the result of mitotic recombination (35,36). Similar single and double spots have been observed on light green leaves of somatic hybrids of albino *N. tabacum* Su/Su combined with *N. tabacum* (su/su), *N. sylvestris*, *N. otophora*, *N. glauca*, *N. nesophila*, and *N. stocktonii*. The presence of double spots on leaves of somatic hybrids of widely divergent species such as *N. nesophila* and *N. tabacum* (Fig. 2) is encouraging if these result from mitotic recombination between distantly related genomes. Indeed, one somatic hybrid between *N. tabacum* and *N. sylvestris* has a greatly increased frequency of spot formation relative to *N. tabacum* Su/Su. The high spotting frequency has been transmitted to sexual progeny of this somatic hybrid. It should be noted, though, that spots associated with the Su locus, particularly the non-reciprocal single spots, may be produced by genetic phenomena other than mitotic recombination (37).

Fate of Nuclear Genes

Several nuclear traits have been monitored in progeny of interspecific somatic hybrids. In some instances where detailed

Fig. 2. Spots on the leaf of light green *N. nesophila* + *N. tabacum* somatic hybrid plant obtained by fusing albino Su/Su *N. tabacum* protoplasts with normal green *N. nesophila* protoplasts.

data were collected, progeny of somatic hybrids could be compared to sexual hybrids. However, in distant interspecies somatic hybrids, comparable amphiploid hybrids were not available. Schieder and Vasil (38) reported segregation of albino seedlings in self-fertilized progeny of *Datura innoxia* and *D. stramonium* or *D. discolor* somatic hybrids, and Evans et al. (39) reported segregation of light green and albino seedlings in backcrossed and self-fertilized progeny of *N. nesophila* and *N. tabacum* somatic hybrids. Segregation ratios in the sexual progeny of these hybrids were as expected for hybrids between divergent species. Morphological traits, though often not ascribed to a single gene, have also been monitored in the progeny of several somatic hybrids. *Petunia hybrida* and *P. parodii* somatic hybrids were self-fertilized and found to segregate for flower color (40). Schieder (41) monitored flower color in the sexual progeny of several *Datura* somatic hybrids. In both of these cases, the segregation patterns obtained were identical in somatic hybrids and comparable sexual hybrids.

We have monitored morphological characters in three generations of our *N. tabacum* + *N. glauca* somatic hybrid plants. Data for morphological characters are summarized in Table 1. We have adopted the terminology suggested by Chaleff (42). Somatic hybrids that have been regenerated from fused protoplasts are called R or R_0 plants. The progeny of self-fertilized progeny have increasing R numbers as subscripts. As previously reported (33), the somatic hybrids are intermediate between *N. tabacum* and *N. glauca* for all five morphological characters listed in Table 1. Morphological characters of somatic hybrids and R_1 plants are comparable to F_1 and F_2 *N. tabacum* x *N. glauca* sexual hybrids (unpublished). For all traits except corolla length, R_2 plants were intermediate between the two backcross populations. As expected, the backcrosses to *N. glauca* more closely resembled *N. glauca* while backcrosses to *N. tabacum* more closely resembled *N. tabacum*. Both R and R_1 plants contained the expected chromosome number (2n=72), while chromosome elimination was observed in the next generation (Table 2). Only two of eight backcrosses to *N. glauca* had the expected chromosome number as all 24 other plants in this generation lost from 2 to 14 chromosomes. No mosaic plants were detected. No single morphological character in the backcrosses could be directly correlated with the number of chromosomes lost. On the other hand, for the most characters, the R_2 plants seemed to resemble the backcrosses to *N. tabacum* more than the backcrosses to *N. glauca*. In this generation, several morphological abnormalities were detected, including a plant with very narrow leaves. This R_1 x *N. tabacum* plant had 55 chromosomes. Another R_1 x *N. tabacum* plant containing 55 chromosomes had several abnormal flowers with deformed corolla (Fig. 3). In these flowers, anthers were fused to the petals, and no pollen was formed. Pollen viability was extremely variable in the plants of this generation. Several of the backcrosses to *N. glauca* had no viable pollen, while no backcrosses to *N. tabacum* and only one R_2 plant had 0% pollen viability. In general, pollen viability was not correlated with number of chromosomes lost. Isoenzymes for aspartate aminotransferase and alanyl aminopeptidase were monitored for each plant. All R and R_1 plants had all the *N. tabacum* and *N. glauca* bands for these enzymes. Nearly all plants in the next (R_2) generation also retained all *N. glauca* and *N. tabacum* bands for both enzymes. Only one R_1 x *N. tabacum* plant was identified that lost the *N. glauca* bands for aspartate aminotransferase. However, even this plant retained the *N. glauca* band for alanyl aminopeptidase.

Fate of Cytoplasmic Genes

Due to the unique nuclear-cytoplasmic combinations that are possible via somatic hybridization, a great deal of interest has

Table 1. Morphological characters of sexual progeny of *Nicotiana glauca* + *N. tabacum* somatic hybrids.

Plant	Corolla Length	Corolla Diameter	Petiole Pigment*	Petiole Length	Trichome Density**	Chromosome Number
Nicotiana glauca	38.0 ± 0.3 mm	10.0 ± 0.2 mm	5	55.9 ± 0.36 mm	0	24
N. tabacum (JWB)	59.0 ± 0.7 mm	25.0 ± 0.6 mm	0	0	23.8 ± 1.35	48
N. glauca + *N. tabacum* somatic hybrid (R)	42.0 ± 0.4 mm	18.0 ± 0.4 mm	4	12.9 ± 0.52 mm	1.9 ± 0.44	72
R_1 (8 plants)	41.2 ± 7.7 mm	21.2 ± 0.57 mm	3-5	30.4 ± 2.15 mm	4.2 ± 0.54	72
R_1 (single plant used as parent of R_2 and backcross generation)	45.6 ± 0.68 mm	18.0 ± 0.55 mm	4	26.6 ± 9.2 mm	3.5 ± 0.23	72
R_2 (R_1 selfed, 7 plants)	41.4 ± 0.77 mm	19.4 ± 0.39 mm	3	19.6 ± 6.5 mm	11.0 ± 2.01	58-68
Backcross (R_1 x *N. tabacum*, 11 plants)	43.1 ± 0.72 mm	22.0 ± 0.52 mm	2	13.0 ± 6.6 mm	12.9 ± 0.75	55-68
Backcross (R_1 x *N. glauca*, 8 plants)	42.0 ± 1.18 mm	16.8 ± 0.66 mm	4	47.0 ± 1.96 mm	4.7 ± 1.49	44-48

* Scale of 0-5 for no pigment to most intense pigment.
** Density = # of trichome per mm^2.

Fig. 3. Abnormal flower from a R_2 x N. tabacum plant (left) and normal flower (right) from another comparable R_2 x N. tabacum plant.

Table 2. Chromosome number of third sexual generation following Nicotiana glauca + N. tabacum somatic hybridization.

Plants	Number of Plants	Expected	Observed Chromosome Number
R_2	7	72	58(2 plants), 64, 66, 68 (3)
Backcross, R_1 x N. tabacum	11	60	4, 55(2), 56(7), 58
Backcross, R_1 x N. glauca	8	48	44(3), 46(3), 48(2)

been expressed in the fate of mixed cytoplasms and genetic transmission of cytoplasmic traits. Chen et al. (43) examined the fate of the large subunit of fraction-1 protein (F-1 P) in N. glauca and N. langsdorffii somatic hybrids. Random chloroplast segregation that resulted in fixed langsdorffii or glauca-type F-1 P, was eventually observed in each clone analyzed. Seventeen of 18 somatic hybrid clones had already segregated by the time regenerated plants were analyzed (10 langsdorffii-type and 7 glauca-type), while the remaining clone segregated following self-fertilization. Chloroplast segregation was also complete prior to meiosis in all four N.

rustica and N. tabacum somatic hybrids examined for large subunit F-1 P, resulting in two tabacum-type and two rustica-type hybrids (44).

Segregation has been observed for other cytoplasmically controlled traits. Male sterile (cms) P. hybrida and normal P. axillaris segregated to male sterile and male fertile cytoplasm in the first and second meiotic cycle (45). Similar segregation of male sterility has been reported in other cybrids (34). Following analysis of three chloroplast characters (large subunit F-1 P, tentoxin sensitivity, and restriction enzyme patterns of organelle DNA), cosegregation was observed in Nicotiana cybrids for all three traits (46). Cosegregation implies that following cytoplasmic segregation, the cytoplasm contains all characters from one parent or the other, suggesting that cytoplasmic recombination does not occur. However, evidence from N. tabacum + N. sylvestris somatic hybrids suggests that chloroplasts and mitochondria assort independently (47). Hence, novel combinations of nuclear, chloroplast, and mitochondrial DNA are possible using protoplast fusion. Biochemical evidence from restriction analysis of organelle DNA also suggests that recombination occurs in both mitochondrial DNA (34) and chloroplast DNA (48) prior to cytoplasmic segregation.

Although it has been suggested that organelle segregation is random following protoplast fusion, non-random organelle segregation has been reported (46,49). Sexual hybrids of N. nesophila and N. tabacum can only be made by ovule rescue when N. nesophila is the female parent. Somatic hybrids between these two species reflect this incompatibility as they only contain N. nesophila chloroplasts as indicated by tentoxin sensitivity (49). When unilateral sexual incompatibility is due to post fertilization events, as in the N. nesophila x N. tabacum hybrids, both somatic and sexual hybrids may exhibit the same nuclear-cytoplasmic incompatibilities. Random organelle segregation resulting in novel nuclear-cytoplasmic combinations more likely occurs when there is no apparent incompatibility or when the sexual incompatibility is due to a pre-fertilization event. For example, while N. glauca x N. langsdorfii sexual hybrids can be easily produced, the reciprocal hybrid is difficult to produce due to insufficient growth of the N. glauca pollen tube (50), a pre-fertilization incompatibility. However, the equivalent of reciprocal hybrids are randomly produced following protoplast fusion of these two species (43). While interspecies incompatibility may be most important in determination of non-random organelle segregation, other factors may influence the segregation of organelles. Defective chloroplasts, resulting from albino nuclear or plastid mutations, may be at a selective disadvantage during segregation. Also, although no supporting data has been reported, the larger number of chloroplast units present in leaf protoplasts compared to the number of proplastids in protoplasts isolated from cell culture proplastids may afford a selective advantage for chloroplasts of leaf protoplasts during segregation.

AGRICULTURAL APPLICATIONS

Limitations to Success

While protoplast fusion has a great deal of promise for crop improvement, no new varieties have yet been produced using these methods. The successful agricultural application of somatic hybridization is dependent on overcoming several limitations. First, application of protoplast methodology requires efficient plant regeneration from protoplasts. Protoplasts from any two species can be fused; however, production of somatic hybrid plants has been mostly limited to the Solanaceae. More recently, several other plant species have been regenerated from protoplasts, suggesting that novel fusion products may soon be recovered for other agronomically important crops.

Second, somatic hybrids must be capable of sexual reproduction. In all cases reported to date, somatic hybrids containing a mixture of genes from two species must be backcrossed to the cultivated crop to develop new varieties. All wide intergeneric somatic hybrids reported to date are sterile (9), and are therefore useless for new variety development. It may be necessary to use back-fusion or embryo culture to produce gene combinations that are sufficiently stable to permit incorporation into a breeding program.

Third, in order to transfer useful genes from a wild species into a cultivated crop, it is necessary to achieve intergenomic recombination or chromosome substitution between the two species. While meiotic chromosome pairing and recombination frequently occur between varieties of a species or closely related species, recombination would be minimal in divergent hybrids. Also, in hybrids between distantly related species, chromosome substitution, in which large blocks of genes are transferred from the wild species into the cultivated crop may be more likely than intergenomic recombination. Depression of yield would most likely result from transfer of whole chromosomes between divergent species. It may be necessary to induce mitotic recombination prior to plant regeneration to insure intergenomic gene transfer.

Fourth, as protoplast fusion results in plants with the summation chromosome number, it may be necessary to manipulate chromosome number before release of new varieties. While variation of chromosome number can be tolerated in several asexually propagated crops, such as potato or sugarcane, most sexually propagated crops, particularly fruits and vegetables, will not tolerate variation of chromosome number. Androgenesis has been proposed as the most practical way to manipulate ploidy level. Either the initial fusion can be done with haploid plants derived from cultured anthers (13), or anthers of somatic hybrids can be cultured to recover the original ploidy level (51). It may be difficult to extend anther culture methodology to distantly related interspecies hybrids.

Unique Applications Using Somatic Hybrids

Protoplast fusion results in unique mixtures of genetic information that may facilitate transfer of nuclear and cytoplasmic genetic information between plant species. Fusion would be most useful for transfer of single genes or small blocks of closely linked genes. Incorporation of disease resistance is particularly exciting (39). Transfer of unlinked polygenic traits using products of protoplast fusion is probably not likely to occur as early as transfer of single genes. Chromosome elimination occurs frequently following protoplast fusion (31) and most hybrid plants recovered following fusion have been aneuploid (16). This loss of chromosomes after fusion may advance a backcrossing program to reduce the time required for new variety release.

Novel somatic hybrids may also be valuable as bridges in development of distant gene combinations. For example, we have produced somatic hybrids of *N. nesophila* with cultivated tobacco (39), but did not recover hybrids using *N. repanda*, a well-characterized species which is a unique source of disease resistance. However, we have been able to sexually hybridize the *N. nesophila* + *N. tabacum* somatic hybrid with pollen of *N. repanda* to produce a bridge hybrid that contains genetic information of all three species (Fig. 4). This new hybrid may be used as a novel source of disease resistance after backcrossing to cultivated tobacco.

Cybridization or cytoplasmic transfer using protoplast fusion has been explored in several laboratories as a possible method to transfer genes between species in one step, thereby circumventing a lengthy backcross program. Use of X-rays to inactivate the nucleus prior to protoplast fusion (e.g., 46) and introduction of selectable cytoplasmic mutants (e.g., 21) may increase the efficiency of interspecies or intergeneric cytoplasmic transfer. Several agriculturally useful traits are cytoplasmically encoded, including some types of male sterility and some herbicide and disease resistance factors. As new traits are ascribed to organelle DNA, the value of this technique should rapidly increase.

In addition to the several unique applications of somatic hybrids described above, there are several unexplored potential applications of protoplast fusion for crop improvement. 1. While somaclonal variation in protoplast-derived plants has been observed, the impact of this phenomenon on crop improvement has not yet been assessed. It is likely that somaclonal variation will also be observed among clones of somatic hybrids (30). 2. Organelle transfer might be facilitated by using cytoplasts, incapable of mitosis, as organelle donors (52). This more direct method of transfer would eliminate nuclear contamination. As methodology is perfected, it may eventually be possible to directly isolate and

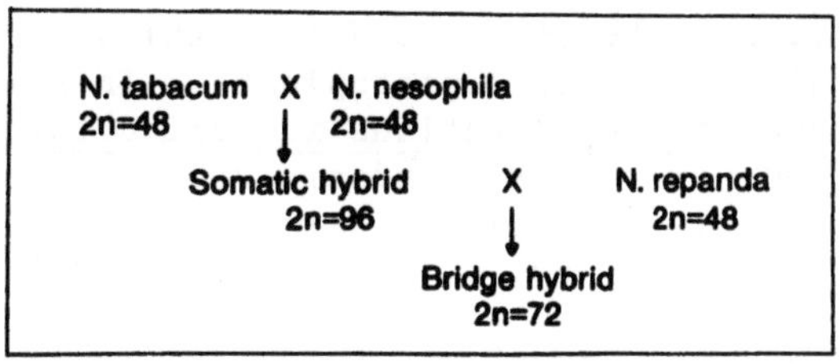

Fig. 4. Bridge crossing scheme successfully used to combine N. repanda with N. tabacum using the N. nesophila + N. tabacum somatic hybrid as an intermediate parent.

transfer chloroplasts or mitochondria. If developed, such organelle uptake and transfer could lead to the development of organelles as gene vectors. 3. As eukaryotic gene transfer systems are developed, it is expected that protoplasts will be very important for plant molecular biology. While current methods using the Agrobacterium Ti-plasmid as a vector system rely on reinfection of the plant with the bacterium to achieve gene transfer, recent reports of gene transfer by fusing isolated Ti-plasmids with plant protoplasts are encouraging (20).

It is expected that as the limitations outlined above are successfully addressed, protoplast fusion will continue to evolve into a tool that will be useful for crop improvement. The unique gene combinations possible using protoplast fusion insure that new plant varieties will soon be derived from somatic hybrid plants.

REFERENCES

1. Simmonds, N.W. 1976. Evolution of Crop Plants. London: Longman.
2. Brown, W.V. 1972. Textbook of Cytogenetics. St. Louis: Mosby.
3. Bates, L.S., and C.W. Deyoe. 1973. Wide hybridization and cereal improvement. Econ. Bot. 27:401-412.
4. Ross, H. 1979. Wild species and primitive cultivars as ancestors of potato varieties. In Broadening the Genetic Base of Crops, pp. 237-245. A.C. Zeven and A.M. van Harten, eds. Wageningen, Netherlands: PUDOC.
5. Collins, G.B., C.C. Litton, P.D. Legg, and J.H. Smiley. 1978. Registration of Kentucky 15 tobacco. Crop Sci. 18:694.
6. Langer, I., K.G. Frey, and T.B. Bailey. 1978. Production response and stability characteristics of oat cultivars developed in different eras. Crop Sci. 18:938-942.
7. Oupadissakoon, S., and E.A. Wernsman. 1977. Agronomic performance and nature of gene effects in progenitor species-derived genotypes of tobacco. Crop Sci. 17:843-847.

8. Gamborg, O.L., J.P. Shyluk, and E.A. Shahin. 1981. Isolation, fusion and culture of plant protoplasts. In Plant Tissue Culture: Methods and Applications in Agriculture, pp. 115-153. T.A. Thorpe, ed. New York: Academic Press.
9. Evans, D.A. 1982. Protoplast fusion. In Handbook of Plant Cell Culture. D.A. Evans, et al., eds. MacMillan Press (in press).
10. Kao, K.N., and M.R. Michayluk. 1974. A method for high-frequency intergeneric fusion of plant protoplasts. Planta 115:355-367.
11. Vasil, I.K., V. Vasil, W.D. Sutton, and K.L. Giles. 1975. Protoplasts as tools for the genetic modification of plants. In Proceedings IV International Symposium on Yeast and Other Protoplasts, p. 82. Nottingham, U.K.: Univ. of Nottingham.
12. Carlson, P.S., H.H. Smith, and R.D. Dearing. 1972. Parasexual interspecific plant hybridization. Proc. Natl. Acad. Sci. 69:2292-2294.
13. Melchers, G., and G. Labib. 1974. Somatic hybridization of plants by fusion of protoplasts I. Selection of light resistant hybrids of "haploid" light sensitive varieties of tobacco. Mol. Gen. Genet. 135:277-294.
14. Schieder, O. 1977. Hybridization experiments with protoplasts from chlorophyll-deficient mutants of some Solanaceous species. Planta 137:253-257.
15. Douglas, G.C., L.R. Wetter, C. Nakamura, W.A. Keller, and G. Setterfield. 1981. Somatic hybridization between Nicotiana rustica and N. tabacum. Can. J. Bot. 59:220-227.
16. Evans, D.A. 1982. Protoplast fusion and plant regeneration in tobacco. In Plant Regeneration and Genetic Variation, pp. 303-323. Praeger Press.
17. White, D.W.R., and I.K. Vasil. 1979. Use of amino acid analogue-resistant cell lines for selection of Nicotiana sylvestris somatic cell hybrids. Theor. Appl. Genet. 55:107-112.
18. Glimelius, K., T. Eriksson, R. Grafe, and A.J. Muller. 1978. Somatic hybridization of nitrate-deficient mutants of Nicotiana tabacum by protoplast fusion. Physiol. Plant. 44:273-277.
19. Maliga, P., G. Lazar, F. Joo, A.H. Nagy, and L. Menczel. 1977. Restoration of morphogenetic potential in Nicotiana by somatic hybridization. Mol. Gen. Genet. 157:291-196.
20. Wullems, G.J., K.J. Molendij, and R.A. Schilperoort. 1980. The expression of tumor markers in intraspecific somatic hybrids of normal and crown gall cells from Nicotiana tabacum. Theor. Appl. Genet. 56:203-208.
21. Medgyesy, P. L. Menczel, and P. Maliga. 1980. The use of cytoplasmic streptomycin resistance: Chloroplast transfer from Nicotiana tabacum into Nicotiana sylvestris and isolation of their somatic hybrids. Mol. Gen. Genet. 179:693-698.
22. Menczel, L., F. Nagy, Z.R. Kiss, and P. Maliga. 1981. Streptomycin resistant and sensitive hybrids of Nicotiana

tabacum + Nicotiana knightiana: Correlation of resistance with N. tabacum plastids. Theor. Appl. Genet. 59:191-195.
23. Lazar, G.B., D. Dudits, and Z.R. Sung. 1981. Expression of cycloheximide resistance in carrot somatic hybrids and their segregants. Genetics 98:347-356.
24. Kameya, T., M.E. Horn, and J.M. Widholm. 1981. Hybrid shoot formation from fused Daucus carota and D. capillifolius protoplasts. Z. Pflanzenphysiol. 104:459-466.
25. Schieder, O. 1978. Somatic hybrids of Datura innoxia Mill. + Datura discolor Bernh. and of Datura innoxia Mill. + Datura stramonium L. var. tatula L. I. Selection and characterization. Mol. Gen. Genet. 162:113-119.
26. Maliga, P., Z.R. Kiss, A.H. Nagy, and G. Lazar. 1978. Genetic instability in somatic hybrids of Nicotiana tabacum and Nicotiana knightiana. Mol. Gen. Genet. 163:145-151.
27. Wetter, L.R. 1977. Isoenzyme patterns in soybean--Nicotiana somatic hybrid cell lines. Mol. Gen. Genet. 150:231-235.
28. Sibi, M. 1978. Multiplication conforme, non conforme. Le Selectionneur Francais 26:9-18.
29. Shepard, J.F. 1982. The regeneration of potato plants from leaf cell protoplasts. Sci. Am. 246:154-166.
30. Evans, D.A., C.E. Flick, S.A. Kut, and S.M. Reed. 1982. Comparison of Nicotiana tabacum and Nicotiana nesophila hybrids produced by ovule culture and protoplast fusion. Theor. Appl. Genet. 62:193-198.
31. Kao, K.N. 1977. Chromosomal behavior in somatic hybrids of soybean--Nicotiana glauca. Mol. Gen. Genet. 150:225-230.
32. Hoffman, F., and T. Adachi. 1981. "Arabidobrassica": Chromosomal recombination and morphogenesis in asymmetric intergeneric hybrid cells. Planta 153:586-593.
33. Evans, D.A., L.R. Wetter, and O.L. Gamborg. 1980. Somatic hybrid plants of Nicotiana glauca and Nicotiana tabacum obtained by protoplast fusion. Physiol. Plant. 48:225-230.
34. Belliard, G., F. Vedel, and G. Pelletier. 1979. Mitochondrial recombination in cytoplasmic hybrids of Nicotiana tabacum by protoplast fusion. Nature 281:401-403.
35. Evans, D.A., and E.F. Paddock. 1976. Comparisons of mitotic crossing-over frequency in Nicotiana tabacum and three other crop species. Can. J. Genet. Cytol. 18:57-65.
36. Carlson, P.S. 1974. Mitotic crossing-over in plants. Genet. Res. 24:109-112.
37. Malmberg, R.L., P.J. Koivuniemi, and P.S. Carlson. 1980. Plant cell genetics--stuck between a phene and its genes. In Plant Cell Cultures: Results and Perspectives, pp. 15-30. F. Sala, et al., eds. N. Holland: Elsevier.
38. Schieder, O., and I.K. Vasil. 1980. Protoplast fusion and somatic hybridization. In International Review of Cytology Suppl. 11B, pp. 21-46. I.K. Vasil, ed. New York: Academic Press.

39. Evans, D.A., C.E. Flick, and R.A. Jensen. 1981. Somatic hybrid plants between sexually incompatible species of the genus Nicotiana. Science 213:907-909.
40. Power, J.B., K.C. Sink, S.F. Berry, S.F. Burns, and E.C. Cocking. 1981. Somatic and sexual hybrids of Petunia hybrida and Petunia parodii. J. Hered. 69:373-376.
41. Schieder, O. 1980. Somatic hybrids of Datura innoxia Mill. + Datura discolor Berth. and Datura innoxia Mill. + Datura stramonium L. var. tatula L. II. Analysis of progenies of three sexual generations. Mol. Gen. Genet. 139:1-4.
42. Chaleff, R.S. 1981. Genetics of Higher Plants. Cambridge: Cambridge Univ. Press. See pp. 94-95.
43. Chen, K., S.G. Wildman, and H.H. Smith. 1977. Chloroplast DNA distribution in parasexual hybrids as shown by polypeptide composition of fraction-1 protein. Proc. Natl. Acad. Sci. 74:5109-5112.
44. Iwai, S., T. Nagao, K. Nakata, N. Kawashima, and S. Matsuyama. 1980. Expression of nuclear and chloroplast genes coding for fraction-1 protein in somatic hybrids of Nicotiana tabacum + rustica. Planta 147:414-417.
45. Izhar, S., and Y. Tabib. 1980. Somatic hybridization in Petunia Part II. Heteroplasmic state in somatic hybrids followed by cytoplasmic segregation into male sterile and male fertile lines. Theor. Appl. Genet. 57:241-245.
46. Aviv, D., R. Fluhr, M. Edelman, and E. Galun. 1980. Progeny analysis of the interspecific somatic hybrids: Nicotiana tabacum (cms) + Nicotiana sylvestris with respect to nuclear and chloroplast markers. Theor. Appl. Genet. 56:145-150.
47. Galun, E., P. Arzee-Gonen, R. Fluhr, M. Edelman, and D. Aviv. 1982. Cytoplasmic hybridization in Nicotiana mitochondrial NDA analysis in progenies resulting from fusion between protoplasts having different organelle constitutions. Mol. Gen. Genet. 186:50-56.
48. Conde, M.R. 1981. Chloroplast DNA recombination in Nicotiana somatic parasexual hybrids. Genetics 97:s26.
49. Flick, C.E., and D.A. Evans. 1982. Evaluation of cytoplasmic segregation in somatic hybrids in the genus Nicotiana: Tentoxin sensitivity. J. Hered. 73:264-266.
50. Kostoff, D. 1943. Cytogenetics of the Genus "Nicotiana". Sofia, Bulgaria: State Printing House.
51. Schieder, O. 1978. Genetic evidence for the hybrid nature of somatic hybrids from Datura innoxia Mill. Planta 141:333-334.
52. Maliga, P., H. Lorz, G. Lazar, and F. Nagy. 1982. Cytoplast-protoplast fusion for interspecific chloroplast transfer in Nicotiana. Mol. Gen. Genet. 185:211-215.

SOMACLONAL VARIATION AND CROP IMPROVEMENT

Philip J. Larkin and William R. Scowcroft

CSIRO, Division of Plant Industry
Canberra, Australia

INTRODUCTION

There has been a remarkable escalation of interest in tissue-culture derived plant variation (somaclonal variation) in the last few years. Earlier authors were aware to some extent that abnormalities could result from a tissue culture cycle (42,65,67). However it is only more recently that the thought has been seriously entertained that some of this variation may be useful for varietal improvement (95,100). It is our contention that the lateness of this realization was a consequence of the fact that so few tissue culturists were engaged in careful analysis of the regenerated plants and also failed to see cell culture manipulation in a genetic context. We were committed to the idea of variation but envisaged it would only happen after specific manipulations (somatic hybridization or DNA-mediated transformation). Underlying the development of these important means of modifying the plant genome was the presupposition that tissue culture was cloning.

Significantly the earliest awareness of the potential for somaclonal variation came when tissue culturists were closely associated with breeders. This was the case in the Hawaiian Sugar Planters' Association Experimental Station (46,71), in the tobacco industry (30,74), and the floricultural industry (96).

It is not our intention here to provide an exhaustive compendium of somaclonal variation. There have been many new examples of the phenomenon since our recent review (57). Our intention is rather to review the most significant and substantial cases. We believe the examples chosen demonstrate the phenomenon to be accessible, manageable, and applicable to many crops and their problems. However, only when many breeders undertake to evaluate the approach with the thoroughness they apply already to varietal evaluation will we be in a position to judge the utility of this technique.

VARIANT PLANTS WITHOUT DEFINED GENETICS

The literature contains many substantial examples of variant plants regenerated from tissue cultures where the genetic nature of the variation has not been defined. The lack of definition can result for a number of reasons: the inability or impracticality of performing sexual crosses in vegetatively-propagated species; the unimportance of sexual crossing in vegetatively-propagated species; the sexual sterility of the regenerants; the publication of results before sexual cross data become available.

In potato there are now a number of independent reports of somaclonal variation. Some potato cultivars are sterile, others can be crossed only with difficulty and in all cases vegetative stability is the major practical issue. Shepard and his colleagues (89,90,91) have worked largely with the sterile cultivar Russet Burbank and have derived many thousands of "protoclones" from leaf protoplasts. From their earliest experiments 1,700 protoclones were evaluated for generally favorable horticultural characters and reduced to 396. General evaluation of these in a subsequent year reduced their number to 65 which were then carefully examined in randomized complete blocks (9.1 m rows) with six replications. Statistical analysis of 35 characters showed significant variation in 22 characters and no significant variation in 9 characters. The other 4 characters showed variation but were not amenable to statistical analysis. Two of the characters showing variation were the number of seed balls (7 protoclones produced at least 50X more seed balls than Russet Burbank), and the weight of USDA #2 tubers (14 protoclones with higher weights). None of the protoclones had a significantly higher total tuber weight and only 11 had significantly less.

Three clones varied in only 1 character, 1 clone varied in 17 characters and the mode class (15 clones) varied in 4 characters. Perhaps the most exciting results of their work relate to disease resistance. They screened 500 protoclones for resistance to *Alternaria solani* (early blight) toxin preparation and 5 were found to be resistant. Four of these also showed resistance to the pathogen. They screened 800 protoclonal plantlets for resistance to late blight (*Phytophthora infestans*) and found 20 resistant to race 0. Most of these also showed some resistance to race 1,2,3,4. These resistances appeared stable through subsequent tuber generations and partially stable through a second protocloning cycle (91).

Wenzel et al. (106) failed to find extensive variation among 192 protoclones derived from dihaploids. However Behnke (9,10) seems to have recovered tolerance to Phytophthora infestans toxic culture filtrates from similar potato dihaploid cultures. More recently Wenzel and Uhrig (107) observed 13 self compatible clones out of 600 doubled monohaploids (obtained by microspore culture of self-incompatible potato dihaploids). Some of these showed segregation in their progeny. One must assume culture-induced variation either to derive self-compatiblity (if the segregation is from unreduced microspores, as they suggest) or culture-induced variation to produce segregating characters in the progeny of doubled monoploids.

High frequencies of potato variation have also been observed by Van Harten et al. (103) in plants derived by adventitious shoots from cultured non-irradiated rachis, petiole and leaflet pieces, and also by Thomas et al. (99,101) in protoclones of cv. Maris Bard examined for 10 morphological characters.

Sugarcane is also vegetatively propagated. It is a highly polyploid complex hybrid and sexual crosses can only be interpreted with difficulty. The early examples of sugarcane somaclonal variants for disease resistance and yield components are well known (47,57) and will not be reiterated here. There is a recent report from Liu (59) describing the recovery of smut (Usitilago scitaminea) resistance by tissue culture.

Our own work with sugarcane eyespot (Helminthosporium sacchari) serves to illustrate the high frequency of variation, the additional possibilities of in vitro selection, and the possibility of character stacking in subsequent culture cycles. Complex callus cultures were initiated from sugarcane cultivar Q101 leaf base and maintained for 6-18 months before regeneration. Plants were then assayed for their reaction to the cultivar-specific pathotoxin at a defined concentration (58). The reactions are expressed as leaf electrolyte leakage rates in Fig. 1. When toxin is included in the final culture phases there is a further bias of the variation toward resistance. It is however striking that the distribution of variation is already biased toward resistance without toxin selection. It is possible that there is some inadvertent link between culturability and toxin tolerance or culturability and the variation-generating phenomenon. This is consistent with the observation that the bias toward resistance increases with the length of time in culture.

A total of 480 somaclones have been examined in this way and 85 have been reassayed in subsequent asexual generations. The majority (70%) have proved stable in the subsequent generations. The remaining somaclones either partially reverted or underwent somatic segregation. These reversions (10%) may represent initial misclassifications or tolerance due to physiological effects. The somatically segregating somaclones (20%) may be chimeric for resistant and

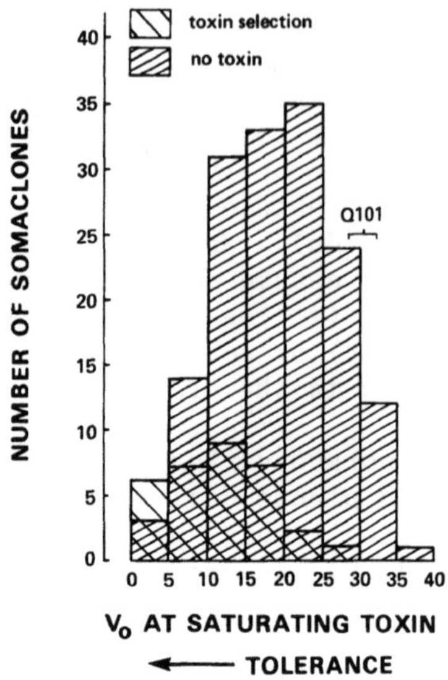

Fig. 1. Data from Larkin and Scowcroft (unpublished). Shows the distribution of reactions to eyespot toxin of Q101 sugarcane somaclones regenerated following tissue culture with and without in vitro toxin selection. The horizontal axis, "V_o at saturating toxin," shows initial rate of electrolyte leakage (μmho $cm^{-1} \cdot h^{-1}$) at saturating toxin concentration.

susceptible cells. A number of the stably resistant somaclones have been subjected to a second tissue culture cycle for 3-6 months. Out of 60 second cycle somaclones, 24 (40%) had a similar tolerance to the corresponding primary somaclone, 13 (22%) were more tolerant and 23 (38%) were more susceptible. We suggest that this indicates the possibility of "character stacking" as first proposed by Shepard et al. (91). One may screen for modification of a second characteristic following a second culture cycle and expect a useful proportion of these to have retained the desirable first characteristic selected in the first cycle.

Reisch and Bingham (84) have presented data on variant alfalfa plants recovered from tissue culture. They examined somaclones derived with and without EMS and with and without ethionine treatments. The plants were propagated vegetatively and grown in a randomized complete block design with 4 replications. Yield, plant height, longest shoot length and number of primary branches varied dramatically. The presence of an ethionine selection step in the culture increased the frequency of variants. However variants were obtained without overt mutagenesis and without ethionine. Indeed the highest yielding clone was from such a control culture and significantly outyielded all other clones and the parents (both diploid and tetraploid). The authors suggest that ethionine is mutagenic. However it would seem more consistent with their results that the ethionine selection step is simply enriching for cells having undergone the variation phenomenon.

PLANTS WITH CONSPICIOUS CHROMOSOME MODIFICATION

There are a great number of reports concerning the variation in chromosome numbers in plant cell cultures and regenerated plants (26). Most of these reports concern only mitotic chromosome counts. Indeed aneuploidy and polyploidy are well known to occur in somaclones of crops such as sugarcane (45,56,61,102), potato (17), rice (76,77), barley (36,37,64), sorghum (14), ryegrass (1,54), peanut (7), and wheat (33).

In mixaploid plants regenerated from barley anther culture, Mix et al. (64) found evidence in root tip mitoses not only for mixaploidy (haploid, diploid, triploid, tetraploid and aneuploid cells) but also for chromosome breakage. Abnormalities such as fragments, multicentric chromosomes and mitotic anaphase bridges were observed. Acentric and centric fragments, ring chromosomes, bridges and micronuclei have also been found in the root cells of mixaploid regenerants from garlic (*Allium sativum*) tissue cultures (72) and Arabidobrassica somatic hybrids (49).

In the large chromosomes of *Haworthia setata* Ogihara (73) was able to observe translocations and deletions in somaclone mitotic preparations. These were confirmed as reciprocal translocations by the observance of quadrivalents in metaphase I. The study of anaphase I also revealed bridges and fragments or bridges alone interpreted as paracentric inversions and subchromatid aberrations. Subchromatid exchange bridges always have symmetrically placed arms and side arms (see 13). This serves to illustrate the greater power of meiotic studies to reveal chromosome rearrangements especially where the chromosomes are not large or distinctive.

Nakamura et al. (70) have reported chromosome loss and meiotic abnormalities in *Triticum crassum* somaclones from immature inflorescence cultures. The donor was 2n = 42 and all 9 somaclones were 2n = 35. An examination of meiosis revealed multivalents, anaphase I bridges, and micronuclei. The frequency of univalents was much higher than expected from the chromosome number suggesting some suppression of homologue pairing either mechanistically or by extensive rearrangements.

McCoy et al. (63) have examined meiotic chromosomes of oat somaclones. They observed a high frequency of heteromorphic bivalents which they interpreted as the result of chromosome segment deletions. Monosomy, trisomy, fragments, translocation (quadrivalents), and other aberrations were also seen in this useful and extensive study. All but 1 of 61 plants examined showed the same chromosome constitution in the 2 or 3 panicles studied. The chromosome aberrations of 3 somaclones were found to segregate as expected in their progeny.

Heteromorphic bivalents have also been reported, albeit less frequently, in metaphase I of some somaclones of maize (44), wheat (3), and ryegrass hybrids (1,2). Reciprocal translocations and inversions have also been implicated in wheat (3) and ryegrass (1,2) and garlic (72).

When plants were regenerated from *Hordeum vulgare* x *H. jubatum* callus cultures, some showed a dramatic enhancement of bivalent and multivalent formation. Their hybrid donor was virtually asynaptic (78). This enhanced meiotic association is presumambly the consequence either of translocations or deletion of pairing suppressor genes.

McCoy et al. (63) have argued that the high frequency of large deletions in oat somaclones may well allow the selection and recovery of recessive mutations in diploid cultures. When the locus of the recessive mutation is matched by a deletion in the homologue the mutation will express. Siminovitch (94) referred to this as "culture-induced hemizygosity".

The ability to detect aberrations in chromosomes can also be increased using banding techniques. The Giemsa C-banding study of *Crepis capillaris* chromosomes by Ashmore and Gould (6) illustrates the resolving power possible and also how dramatic the internal rearrangements can be. The rearrangements in culture were so extensive that it was difficult to relate the culture chromosomes to root tip chromosomes with any certainty. Giemsa C-banding alterations have also been seen in *Vicia faba* cultures (51).

Greater use of banding techniques and *in situ* hybridization techniques (e.g., using radioactively labelled ribosomal DNA probes) will undoubtedly yield valuable information on somaclonal variation.

PLANTS WITH DEMONSTRABLY HERITABLE VARIATION

Possibly the most extensive statistical analysis and field experimentation applied to somaclonal variation has been with tobacco dihaploids. A number of studies have compared dihaploid progeny lines to donor inbred cultivars for a range of agronomic characters and found significant differences between lines and donors but not within lines (21,22,27,29,30,74). One suggestion has been that the colchicine used to double the haploids was mutagenic. However, De Paepe et al. (27) and Burk and Matzinger (22) and Deaton et al. (29) also used spontaneously-doubled haploids (no colchicine) to discount this explanation. Another suggestion is that some variant androgenetic plants may derive from microspores which have biased or inadequate sampling of cytoplasmic organelles. Counter to this is the failure to find maternal inheritance effects for most of the variant characters (21,29).

The most controversial suggestion has been that the inbred tobacco cultivars do in fact have residual heterozygosity and the variant dihaploids represent elimination of this heterozygosity. Some studies do show trends among the variants such as reduced vigor (5,22,74), or crumpled leaf phenotype (27). Such trends would be consistent with elimination of a small amount of residual heterozygosity. However Brown and Wernsman (21) argue that the inbreeding depression caused by enforced homozygosity (in dihaploids) should at least partially be dispelled by hybridizing among the yield depressed dihaploids (a heterosis effect). They did not observe this and conclude that residual heterozygosity was not involved in the phenomenon. The most compelling data against the residual heterozygosity explanation is that from De Paepe et al. (27) showing that a second cycle of androgenesis from a doubled haploid generates yet further variation. Indeed 5 subsequent cycles of androgenesis generated more variation following each cycle. The variations were sometimes homozygous (did not segregate upon selfing) and were sometimes heterozygous (segregating 3/1 upon selfing). They have proposed the unusual hypothesis that variants are regenerants from the vegetative microspore nucleus or fusions of the generative and vegetative nuclei. They argue that there are some systematic genetic differences between these nuclei. This is an interesting concept and deserving of further investigation. However this explanation would appear to be insufficient to explain the diversity of variation in some of the dihaploid studies (22).

Heritable variations have also been found in tobacco somaclones regenerated from protoplasts and callus cultures. Barbier and Dulieu (8) observed heritable reversions and deletions of mutant genes at two specific loci which effect chloroplast development. Chaleff and Keil (23) selected among cultured tobacco cells for picloram resistance and regenerated plants. The picloram resistance was conditioned by single nuclear genes showing dominance or semi-dominance to wild type. They discovered to their surprise that 3 out of 5 picloram resistant mutants also carried dominant alleles for hydroxyurea resistance (a single nuclear gene). The donor plants were sensitive to hydroxyurea. In 2 of these plants the hydroxyurea resistance gene assorted independently from the picloram resistance gene. The recovering of hydroxyurea resistance without overt selection and without linkage to the locus at which selection was focused is indeed remarkable. The appearance of simply inherited mutations at high frequencies in plant cell cultures is indicative of widespread genetic flux.

In Brassica napus there have been some interesting examples of heritable somaclonal variation. Hoffman et al. (50) produced 45 dihaploids by androgenesis from a highly inbred donor cultivar. Of these, 28 gave uniform vigorous progeny but each line differed for one of a number of morphological traits. Eight gave uniform progeny but with reduced vigor and partial sterility; 9 gave progeny segregating for a number of traits. The donor line was low in glucosinolates but there was a general trend for the derived dihaploids to have much higher glucosinolates. The stem embryogenesis phenomenon in rape also seems to generate variation. A single microspore was cultured to give one haploid plantlet which was maintained in culture for a number of subculturing cycles until the swollen stem gave rise to embryoids. These embryoids were cultured to give about 50 spontaneous diploids. The selfed progeny of these plants showed a high degree of segregation (50). Sacristan (86) was able to regenerate rape plants from suspension cultures of a haploid donor. The cultures were selected by exposure to blackleg (Phoma lingam) culture filtrates. Out of 29 somaclones, 20 were susceptible, 7 were tolerant, and 2 resistant to the pathogen. The 2 resistant somaclones were sterile. The 7 tolerant variants, when selfed, segregated for tolerance.

Heritable somaclonal variation has also been observed in the cereals. Deambrogio and Dale (28) found various abnormalities among the progeny of barley somaclones derived from immature embryo callus cultures. Oono (76,77) has observed dramatic variation in rice (Oryza sativa L.) somaclones. The donor plant was absolutely homozygous (being a dihaploid derived from the inbred cv. Norin 8). Seed calli were initiated from 75 seeds and 1121 somaclones regenerated. The selfed progeny of these were examined over 3 generations for many characters including plant height, chlorophyll mutations, heading date, panicle length, grain yield, and 1,000-grain weight. Many of the variants observed in the primary somaclones (D_1) segregated in the selfed progeny (D_2) and by D_3 many variant lines had been obtained in a pure breeding state. Among them were some with increased grain number, panicle number, and 1,000-grain weight. When the D_1 were analysed for just 5 characters only 28.1% of the somaclones were normal, and 28.0% of them were variant in 2 or more characters.

Oono (76,77) has also selected diploid rice cultures for tolerance to NaCl. Tolerance was retained in some regenerants and was heritable to D_3. Among the regenerants from these experiments were also plants which were true-breeding (up to D_4) for reduced height. Reciprocal crosses showed the mutation was not maternally inherited. If these last results are confirmed our hypotheses to explain somaclonal variation will have to take into account that "homozygous mutations" can occur in diploid tissue.

Maize somaclones also can display remarkable variation which is heritable. Edallo et al. (34) initiated cultures from immature embryos of maize inbreds and regenerated plants after various subcultures up to the 8th. Among 110 regenerates almost all were 2n = 20 (one was 2n = 40; one, 2n = 21). Pollen sterility in some did however suggest genetic or cytogenetic abnormalities. In the R_2 generation derived from a total of 77 fertile somaclones (R_0) they scored for the presence of simply inherited mutations (R_1 was not used because of low seed numbers). Both endosperm and seedling mutations were scored (17 phenotypes listed) provided they showed 3:1 segregation (i.e., from a heterozygous R_1 plant). The average frequency of observable, simply-inherited mutations was 1.2 for each R_0 plant from donor genotype W64A and 0.8 for each R_0 plant from donor genotype S65. Thus the overall average frequency was 1.0 observed, simply-inherited mutation for every somaclone. When more than one plant was regenerated from the one subculture they showed different mutations.

Wheat tissue cultures are also proving amenable to these types of studies. When Ahloowalia (3) regenerated 46 plants from cultures of 4 cultivars, only 28 of these were self-fertile. A number of morphological variants were observed among the R_0 plants but there has been no report on their heritability. In our own work we are discovering simply inherited mutations in the selfed progeny (R_1 generation) of wheat somaclones regenerated from immature embryo cultures. One of the frequently encountered mutations is one effecting awn development. Using the highly inbred line Yaqui 50E which is tip-awned and dwarfed we have recovered simply inherited variants which are fully awned or entirely awnless. A few of these appeared in the R_0 plants and have remained pure breeding ("homozygous mutation"?), most appeared as segregants in the R_1 plants. Segregation was also observed in R_1 for both increased height and reduced height relative to the very uniform parental material. Many other heritable morphological variants are also being recovered. We are in the process of analyzing wheat somaclones for mutations at specific rust resistance loci and loci affecting GA_3 responsiveness. We are also attempting to use *in vitro* selection to recover herbicide tolerance and *Septoria tritici* resistance in important Australian cultivars.

Somaclonal variation appears to be a phenomenon which affects cytoplasmic genomes as well as nuclear genomes. In a series of reports beginning with Gengenbach et al. (39) and Brettell et al. (15) it has been shown that mitochondrial DNA (mtDNA) changes occur in plants as a result of maize tissue culture. Changes have been detected in the expression of *Drechslera maydis* T toxin reaction and male sterility in somaclones from a T cytoplasm, cytoplasmic male sterile maize. Both with and without in vitro T toxin selection, T toxin resistant and male fertile mutants have been recovered (15,39-41). Analyses of mtDNA restriction patterns have shown a widespread occurrence of rearrangements. For example all 9 toxin resistant lines (from T cytoplasm donor) had mtDNA patterns distinguishable from each other and the T and N patterns (40,41). Kemble et al. (54) in a similar study found all somaclones from a T cytoplasm donor and an N cytoplasm donor to closely resemble their parents in mtDNA general characteristics. However all somaclone mtDNA patterns were distinguishable from each other and their donor by minor modifications.

Gengenbach and Umbeck (41) found that 14 out of 15 toxin resistant variants (ex T cytoplasm cultures) were missing a XhoI 6.6 kb fragment. None out of 34 toxin sensitive regenerants was missing this fragment. Remarkably Brettell et al. (16) have also found the 6.6 kb Xho I mtDNA fragment missing in all 4 toxin resistant variants analyzed in an independent series of somaclones. They also isolated the 6.6 kb fragment, nick translated it, and used it as a probe on Southern transfers. Three of the variants showed homology at the 6 kb position, the fourth showed homology at the 20 kb position.

There is a close correlation between a particular 13,000 M_r polypeptide synthesized by T cytoplasm mitochondria and male sterility and T toxin sensitivity. In maize T cytoplasm lines which have been restored to fertility by nuclear restorer genes, the synthesis of the 13,000 M_r polypeptide is depressed. Dixon et al. (31) have shown that toxin resistant somaclones recovered from T cytoplasm maize cultures both with and without toxin selection have decreased 13,000 M_r polypeptide synthesis. In addition four lines derived from a tissue culture regenerant showing varying degrees of fertility/sterility and toxin resistance/sensitivity were analyzed. It appeared that the critical level of mitochondrial synthesis of the 13,000 M_r polypeptide was about 30% relative to the T cytoplasm parental level. Below this, fertility was restored and toxin resistance evident.

In all of these experiments, variations in toxin reaction, male fertility, mtDNA restriction patterns, and 13,000 M_r polypeptide synthesis were all transmitted maternally to progeny.

It is worth noting that in two studies of mtDNA in Nicotiana somatic hybrid plants (11,69), dramatic restriction pattern variation has been observed. The somatic hybrids contain fragments from both parents but also non-parental fragments. In both cases the investigators favored the explanation that mtDNA recombination had occurred. However, non-parental fragments may equally well be the consequence of culture-induced variation. There are some indications of cytoplasmic inheritance of lettuce somaclonal variants (92).

THE NATURE OF THE VARIATION

Having reviewed a large body of data concerning somaclonal variation in a diversity of plant systems, it is appropriate to ask a number of general questions.

Is the Variation Preexisting?

This question is of practical significance mainly as it relates to the possible control of variation. If all the variation preexisted in somatic cells of the plants, and cell culture merely allowed a variant cell to form a new plant, then the only chance to eliminate somaclonal variation would be to initiate cultures from explant tissues in which such genetic variations had not occurred. There are two reports indicating a change in the frequency of somaclonal variation depending on the explant source for initiating the culture. These involve pineapple (105) and potato (103). However, these differences may be as easily explained by the degree of dedifferentiation achieved during the brief culture phase.

Certainly some studies imply that some of the variation may preexist. Barbier and Dulieu (8) analyzed the frequency of variation in tobacco plants regenerated from culture after varying durations of culture from 30-255 days. By applying fluctuation test analyses they concluded that many of the mutations either preexisted or were generated very early in the culture process. Kasperbauer et al. (53) have reported the recovery of a mutant tobacco plant with ruffled leaves controlled apparently by a single dominant allele. A plant was discovered with a small sector of ruffled leaf. Tissue culture allowed the isolation of plants "solid" for the mutation. This may be an easily visualized example of a widespread phenomenon, namely the accumulation of mutations in somatic cells.

Radin and Carlson (83) isolated herbicide resistant mutants by a technique known as "*in situ* selection". Young plants were γ-irradiated and sprayed with herbicide following a period for further growth. Unaffected "green islands" of cells on the leaves were excised and cultured and regenerated. Some of the regenerants (21% for bentazon, 13.5% for phenmedipham) retained the resistance. Out of 22 of these, 10 had simply inherited resistance (one or two genes). It would be valuable to attempt similar "green island selection" without overt mutagenesis.

The appearance of variants in a number of systems gives good evidence that some variation occurs in the culture phase and does not preexist. When T cytoplasm maize was cultured (with toxin selection) the early regenerants were still toxin susceptible and male sterile. Only after further culture cycles could resistant, fertile plants be regenerated (40). The frequency of variants among garlic somaclones increased with duration of culture (72). Probably the best documented increase of variant frequency with increased culture duration is from the oat culture studies of McCoy et al. (63). The frequency of regenerants with meiotic abnormalities increased dramatically with the duration of the culture phase for both cultivars studied (Fig. 2).

Thomas et al. (101) regenerated a number of plants from a single potato protoplast. The fact that there was variation between those plants confirms its occurrence during the culture phase.

Oono (76) reports that diploid regenerants from rice microspore culture show mutations which prove sometimes to be homozygous and sometimes heterozygous. If all the variation preexisted in the microspores all the mutant diploid regenerants would be homozygous. The heterozygous mutants clearly derive from mutational events following the spontaneous chromosome doubling during the culture phase.

Similarly Hoffmann et al. (50) found that 9 out of 36 microspore-derived rapeseed dihaploids were heterozygous for some characters. Since the donor was highly inbred and true breeding the mutational event giving rise to the segregation must have occurred after spontaneous diploidization otherwise they would be homozygous for the mutation. Therefore, the mutational events observed must be during the culture phase. Similarly many of the variant *N. sylvestris* dihaploids were also heterozygous (27).

Sectorial analysis of calli and the variant plants derived from them also indicates the culture phase origin of mutations. This is the case with rice diploid cultures (75) and also maize diploid cultures (34). In this maize work different somaclones regenerated from related cell culture lineages seemed no more likely to carry the same heritable mutation than somaclones of entirely unrelated cell lineages. This is suggestive of new mutational events occurring continuously during cell culture.

Do the Mutations have Simple or Complex Inheritance?

It will by now be apparent that at least some of the mutations are simply inherited (one or two genes) (8,23,34). Some of the heritable mutations have shown complex inheritance. For example,

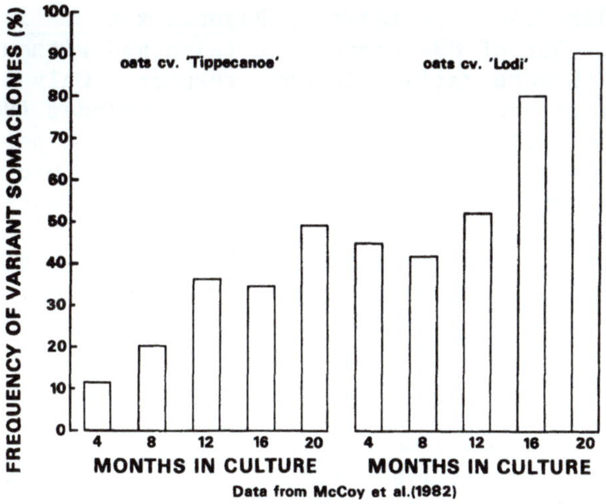

Fig. 2. Data from McCoy et al. (63). Shows the frequency of chromosomally variant oat plants regenerated from cultures of two cultivars, Lodi and Tippecanoe. In both cultures the frequency of variants increases with the duration of the culture phase. Lodi somaclones are more variable than Tippecanoe somaclones.

complex (multigenic) inheritance was apparent for heading date and plant height in rice (75). Inheritance data are not available for the many disease resistant variants in sugarcane somaclones (47) but it seems likely they represent multigenic changes. In the disease resistant potato somaclones (90) the indications are also suggestive of complex genetics. Diallel analysis of tobacco somaclonal variants (21,29) has shown significant general combining ability mean squares. This is indicative of quantitative inheritance of the variant parameters studied. Analysis of the types of inheritance has not yet been extensive enough to draw general conclusions. However, it seems we should expect both simple and complex characters to be affected.

Are the Mutants Solid?

This question concerns the issue of whether the regenerants are chimeric and is related to the issue of the single cell origin of regenerants. The pragmatist is more concerned with the solidity of the variant because if the somaclonal mutant is solid it can be used irrespective of how it came to be solid. The evidence available should be fairly encouraging to the pragmatist. Chimeric variants are only rarely reported from tissue culture. Roest et al. (85)

specifically addressed this issue in Begonia x hiemalis culture-derived plants. Out of 894 somaclones (with and without X-irridation) 266 were variant in some respect. Only 4 of these showed any indications of chimericity, the remainder (98.5%) were apparently solid variants. In Chrysanthemum, where periclinal chimeras are common, the variants arising from culture were nevertheless all apparently solid (100% of 168 variants) (18,19).

In potato somclones derived from adventitious buds of cultured explants a high frequency of stable mutation occurred. 360 lines (from the no irradiation control) were examined through 3 tuber generations, 104 were variant but only 1 proved to be chimeric (103).

The fact that so many somaclonal variants are proving heritable also suggests they are at least sufficiently solid to be fixed or made solid in subsequent generations. Of course even a chimeric variant may be recoverable in a solid form by a second tissue culture cycle (53).

The issue of whether this preponderance of solid variants is a consequence of the single cell origin of in vitro regenerated plants is not so easily settled. Histological studies do in some cases suggest the origin of shoot primordia or embryos from a small number rather than a large number of cells (45,60,68). In Sorghum immature embryo cultures the initiation of divisions was both within and on the surface of the scutellum. Beginning probably from single cells the dividing centers often formed embryo-like structures (32). Shoots appeared to arise from these or directly from scutellar cells. Development was similar in maize but without clearly identifiable embryo-like structures (43). King et al. (55) have emphasized that regenerable cereal cultures are composed of many repressed shoot primordia or embryos. They have argued that new structures arise only by cleavage of preexisting meristematic centers. The observations of these histological studies could be just as easily interpreted as new structures developing from single cells of the original scutellum or of the secondary organized structures in the culture. The histological study of regenerable wheat cultures (80) supports this idea. The compact callus, formed from the scutellum, itself gave rise to leaf-like structures which themselves were not shoot primordia. Upon lowering the 2,4-D level shoot primordia and shoots arose from cells of these structures.

Two further types of observation argue strongly for single cell origin of regenerants even from complex cereal cultures. Firstly, the variants arising from such cultures appear to be solid and are often simply inherited. Secondly, different regenerants from the one subculture of rice or maize callus were not often mutationally related (34,75). If cleavage of preexising structures was the only origin of shoot primordia then one might expect a solid variant (however it might arise) to produce many identical sister regenerants. For the purpose of somaclonal variation it may not be important whether or not a cereal or grass somaclone develops from a cell which is within an organized structure. Provided the cells of origin of regenerants are subject to the mutational phenomenon the pragmatist will be satisfied. The experimenter interested in DNA-mediated transformation will not be so easily satisfied. He is most attracted to protoplast systems in which each free-floating cell has the capacity to grow to a plant. In cereals this is not yet possible though there are some very promising approaches developed in Pennisetum (104) and Panicum (62) which have resulted in plants from protoplasts.

Are There Mutational "Hot Spots"?

A few of the somaclonal variation reports appear to indicate a non-random trend toward a particular type of variation, e.g., the crumpled leaf phenotype in tobacco (27), reduced vigor in tobacco (21), sterility and reduced chlorophyll in cereals (76,88), and increased glucosinolates in rape (50). Reduced height, chlorophyll, vigor, and fertility can be caused by very many independent mutations and may be expected from older cultures as deleterious mutations accumulate. The very high frequency of the crumpled leaf phenotype in dihaploids from Nicotiana sylvestris anther culture is not easily explained. This phenotype appears to result from mutation at one locus as shown by simple Mendelian segregation in selfings and test crosses. De Paepe et al. (27) argue that if this is mutation then it is some sort of "systematic mutation" resulting from mutational "hot-spots" in the genome.

In our own work with sugarcane the frequency of appearance of eyespot toxin tolerance (even without in vitro toxin selection) is sufficiently high to be suggestive of "systematic mutation". Chaleff and Keil (23) found that more than half of their picloram-resistant tobacco cell lines were also hydroxyurea resistant. The genes responsible for these resistance assorted independently. It can be argued that there is some unintentional selection for eyespot toxin tolerance, hydroxyurea resistance, and the crumpled leaf phenotype in their respective cultures. The toxin and antimetabolite resistances may result from some generalized stress tolerance since the tissue culture environment may be highly stressful. Alternatively the loci affecting these phenotypes may indeed be "hot spots" for controlling element mutagenesis triggered by culture.

MECHANISMS

In a previous paper we have discussed some hypotheses to explain somaclonal variation (57). It is not our intention here to reiterate these speculations but rather to reemphasize a couple of possibilities.

It is abundantly clear, particularly from the work of McCoy et al. (63) and Ogihara (73), that tissue culture can result in chromosome deletions, translocations, and other minor rearrangements. If these modifications can occur at a level gross enough to be cytologically observed they may also be occurring at a fine-structural level.

Variants may be the consequence of heritably altered expression of genes. Expression can be altered by amplification of gene copy, depletion of gene copy, translocation to positions under different control signals, movement of transposable elements or "controlling elements" to positions influencing expression of the gene, or heritable modification of methylation patterns (20). It is possible, for example, that the apparent "systematic mutation" or "hot spot" mutation is a consequence of transposable element movements which have a predilection for the "hot-spot" loci. Transpositional events in mitochondrial DNA have recently also been implicated in altered plant cell culture morphology. Chourey and Kemble (24) observed that the S_1 and S_2 plasmid-like DNA molecules in maize (S cytoplasm) mitochondria were still present in 10 plants regenerated after 6 months tissue culture. After 2 years culture there were two types of cultures. The callus with some organization still has S_1 and S_2 molecules but in reduced amounts. The friable cell line had no detectable S_1 and S_2, however Southern blotting data showed they had integrated into the high molecular weight mtDNA.

De Paepe et al. (27) have proposed that variant tobacco dihaploids derive either from the vegetative nucleus of the microspore or fusions of the generative and vegetative nucleus. They argue that there is a systematic genetic difference between the two nuclei explaining the apparent systematic mutation of the variants. This proposal does not seem to adequately account for the breadth of variation even among tobacco dihaploid and certainly can be no explanation for somaclonal variation when microspores are not involved.

There has been a series of reports over the last 20 years of variants induced in sorghum and barley following colchicine treatment (38,87,93). The coleoptiles of germinating seeds were treated with 0.5% colchicine in lanolin. A tumor-like growth occurs and plant expansion is arrested for a couple of weeks, after which a vigorous tiller grows from the arrested rosette. In sorghum many of these treated plants (10-67%) were mutant and most of these mutants (78%) proved to be true-breeding. Some (12%) were segregating and some (10%) had both segregating and true-breeding variant phenotypes (87). To explain the preponderance of true-breeding variant mutations the authors propose mechanisms involving cycles of polyploidization and reduction in the cells of the colchicine- induced swelling. For example Simantel and Ross (93) used as parental material an F_1 heterozygous for 2 reciprocal-translocation chromosome pairs (4 cytologically marked chromosomes). Two out of 90 surviving treated plants were true breeding and had all normal chromosomes. It was proposed that the chromosome complement was reduced to n = 10 (separating the normal from translocated chromosomes in some cells) and then doubled in some cells of the colchicine swelling. The diploid tiller then arose from such cell(s).

Sanders and Franzke (87) colchicine treated the inbred, highly uniform sorghum line "Experimental #3". The frequency of variants under certain environmental conditions was as high as 67% of surviving seedlings. The majority (78%) of these proved to be true-breeding for the complex mutations which included height, color markings, and awning. They proposed that sorghum is an allopolyploid and that the variants arise by first doubling the chromosomes and then reducing back to diploid level but with homoeologous substitutions resulting in compensated (and therefore fertile) nulli-tetrasomics.

Whether or not these proposed mechanisms for colchicine-induced variation are correct there are certain similarities with some cases of the somaclonal variation. It may well be that the phenomenon of the colchicine-tumor is similar to a cell culture. Some incidents of somaclonal variation appear to give rise to homozygous mutants. Homozygous mutants may arise if the cultured cells are passing through cycles (however brief) of haploidy. Alternatively many culture cell lineages may have periods of hemizygosity caused by monosomy for particular chromosomes. Mutation occurring before or during haploidy or monosomy can result in homozygosity for the mutation when euploidy is restored. Most somaclonal mutants are proving to be heterozygous as would be expected.

A heterozygous mutant can behave in a true-breeding fashion if the variant plant is also apomictic. We are aware of no evidence for this, however, the potential value of apomoxis in cereals warrants some attention to this possibility. By apomixis the high yield of a heterotic hybrid could be fixed.

It should soon be possible to investigate the causes of somaclonal variation at the molecular level. If a somaclonal mutant can be found at a locus for which a DNA probe is available, it will be possible to explore the causes of the altered gene expression. Questions could be asked and answered concerning amplification, insertions, deletions, flanking sequences, and nucleotide sequence.

DISADVANTAGES, ADVANTAGES, AND INTEGRATION

Some of the features of somaclonal variation which are attractive and unattractive will already have been evident in this manuscript. The most outstanding disadvantage is the lack of control by the investigator or breeder over the characters being varied. Classical mutagenesis has proved not to be entirely random. It remains to be seen to what extent somaclonal variation will be random and whether we can expect all agronomic characters to be modifiable by the phenomenon.

Another question not yet satisfactorily answered relates to the extent of variation within an individual somaclone. When somaclonal variation occurs, are too many characters altered at once for the plant to be of any interest for varietal improvement?

Some of the sugarcane variants chosen for amelioration of one trait have also been reevaluated in field trials for variation in the other agronomic parameters. In at least some cases general performance and quality was as good as the parental cultivars (47,59). For example an independent reevaluation of the yield of Fiji-disease resistant variants of sugarcane cultivar Pindar (56) showed no significant decline (4). Secor and Shepard (89) examined 65 potato somaclones in replicated field trials analyzing 35 parameters. Three clones varied in only 1 character from the parent, 1 clone varied in 17 of the parameters, and the mode class (15 clones) varied in only 4 parameters.

Somaclonal variation has the advantage of being accessible in most crop species. Even when the only regenerable culture systems available are the so-called "complex" cultures (e.g., cereal cultures), these are subject to the phenomenon. The evidence is as yet scant but it seems that complex characters (quantitatively inherited) can be affected by somaclonal variation as well as simply-inherited genes. If this is the case this would augur well for applicability since so many agronomic traits are complex.

Many possible applications can already be envisaged and many more may emerge in time as our understanding of the phenomenon improves. A crop cultivar may be cultured with the aim of recovering a variant with some specific improvement. The efficiency of this recovery may be improved dramatically if appropriate selection can be imposed during the culture phase. Some systems lend themselves readily to selection strategies such as pathotoxins, herbicides, heavy metal toxicity, and salinity. Other systems will require more ingenuity in order to devise appropriate selection such as those for recovering tobacco plants which support only slow tobacco mosaic virus development (66), or the use of lysine plus threonine inhibition for selection of variants with potentially increased nutritional value (48).

The combination of mutagenic treatments and tissue culture may prove advantageous. It is hoped, however, investigators will include non-mutagenized controls so that we may continue to learn more about this phenomenon in isolation from the mutagen effects.

Since deletion is one of the frequently reported cytological consequences of tissue culture we may envisage applications based on exploiting deletion. Bennett (12) has argued that the grain shrivelling associated with triticale is a consequence of the telomeric heterochromatin on the rye chromosomes. He has some preliminary evidence that the deletion of the telomeric heterochromatin on two rye chromosomes reduces the grain shrivelling and increases yield. Tissue culture of triticale may enable the rapid recovery of variants with telomeric heterochromatin deleted from other rye chromosomes.

Commercial sugarcane is derived from hybrids between Saccharum officinarum and S. spontaneum. The desirable chromosome number is 2n = 100-125, involving a preferential loss of many of the spontaneum chromosomes which is normally achieved by many years of backcrossing (so-called "nobilization"). Tłaskal (102) has argued that tissue culture of F_1 and BC_1 hybrids may enable the desirable chromosome complement to be derived rapidly without the traditional nobilization process.

We are convinced that one of the most exciting prospects for usefully integrating somaclonal variation into existing breeding programs will be by the tissue culture of wide hybrids. A significant proportion of modern plant breeding is aimed at the introgression of alien genes or gene complexes into crop cultivars. Since tissue culture has been strongly implicated in causing chromosomal rearrangements and exchanges, it may aid in the introgression process. Introgression is often hampered by a lack of association between the crop and alien genome which itself may be the consequence of spatial separation (35).

Orton (79) passaged Hordeum vulgare x H. jubatum hybrid tissue through a tissue culture cycle and recovered regenerants which were morphologically like H. vulgare and, although mixoploid (6-9 chromosomes), most cells had 7 chromosomes (the haploid H. vulgare number). However these regenerants showed some evidence of jubatum gene introgression in their perenniality and in esterase isozymes. Cooper et al. (25) initiated cultures from Hordeum distichum x Secale cereale hybrid embryos and subsequently regenerated plants. Without an intervening callus phase the normal hybrid plantlets died before they could be transferred to soil. However a number of somaclonal hybrids (from an intervening callus phase) grew and tillered vigorously and appeared to possess 8 rye and 6 barley chromosomes. Thus the intervening callus phase can also help to recover hybrid plants.

This approach to introgression breeding is also being applied to Lolium perenne x L. multiflorum (1,2), Lolium multiflorum x Festuca arundinacea (52), Triticum crassum x Hordeum vulgare (70), various Hordeum x Secale species hybrids (81), Trifolium pratense x T. sarosiense (82), Saccharum x Zea (97), and Lycopersicon esculentum x L. peruvianum (98). The results of this type of approach will be difficult to analyze but may prove to be the most fruitful in the short term.

REFERENCES

1. Ahloowalia, B.S. 1976. Chromosome changes in parasexually produced ryegrass. In Current Chromosome Research. K. Jones and P. Brandham, eds. Elsevier/North Holland, Amsterdam. pp. 115-122.
2. Ahloowalia, B.S. 1978. Fourth Intl. Cong. Plant Tissue Cell Culture, (abst.). Calgary, Canada. p. 162.
3. Ahloowalia, B.S. 1982. Plant regeneration from callus culture in wheat. Crop Sci. 22:405-410.
4. Anonymous. 1976. Yield trial on Fiji disease resistant sub-clones of Pindar. David North Plant Research Centre, April 1976 research report. p. 26.
5. Arcia, M.A., E.A. Wernsman, and L.G. Burk. 1978. Performance of anther-derived dihaploids and their conventionally inbred parent as lines, in F_1 hybrids, and in F_2 generations. Crop Sci. 18:413-418.
6. Ashmore, S.E., and A.R. Gould. 1981. Karyotype evolution in a tumour-derived plant tissue culture analysed by giemsa C-banding. Protoplasma. 106:297-308.
7. Bajaj, Y.P.S., A.K. Ram, K.S. Labana, and H. Singh. 1981. Regeneration of genetically variable plants from the anther derived callus of Arachis hypogaea and Arachis villosa. Pl. Sci. Letts. 23:35-39.
8. Barbier, M., and H.L. Dulieu. 1980. Effets génétiques observés sur des plantes de Tabac régénérées à partir de cotylédons par culture in vitro. Ann. Amelior. Plantes. 30: 321-344.
9. Behnke, M. 1979. Selection of potato callus for resistance to culture filtrates of Phytophthora infestans and regeneration of resistant plants. Theor. Appl. Genet. 55:69-71.
10. Behnke, M. 1980. Selection of dihaploid potato callus for resistance to the culture filtrate of Fusarium oxysporum. Z. Pflanzenzüchtg 85:254-258.
11. Belliard, G., F. Vedel, and G. Pelletier. 1979. Mitochondrial recombination in cytoplasmic hybrids of Nicotiana tabacum by protoplast fusion. Nature. 281:401-403.
12. Bennett, M.D. 1981. Nuclear instability and its manipulation in plant breeding. In The Manipulation of Genetic Systems in Plant Breeding. Phil. Trans. R. Soc. Lond. B292:475-485.
13. Brandham, P.E. 1970. Chromosome behaviour in the Aloineae. 3. Correlations between spontaneous chromatid and sub-chromated aberrations. Chromosoma 31:1-17.
14. Brar, D.S., S. Rambold, O. Gamborg, and F. Constabel. 1979. Tissue culture of corn and sorghum. Z. Pflanzenphysiol. 95:377-388.
15. Brettell, R.I.S., E. Thomas, and D.S. Ingram. 1980. Reversion of Texas male-sterile cytoplasm maize in culture to give fertile, T-toxin resistant plants. Theor. Appl. Genet. 58:55-58.

16. Brettell, R.I.S., M.F. Conde, and D.R. Pring. 1982. Analysis of mitochondrial DNA from four fertile maize lines obtained from a tissue culture carrying Texas cytoplasm. Maize Genet. Coop. Newsl. 56:13-14.
17. Bright, S., V. Jarrett, R. Nelson, G. Creissen, A. Karp, J. Franklin, P. Norbury, J. Kueh, S. Rognes, and B. Miflin. Modification of agronomic traits using in vitro technology. In Plant Biotechnology. S. Mantell and H. Smith, eds. Cambridge Uni. Press, Cambridge (in press).
18. Broertjes, C., S. Roest, and G.S. Bokelmann. 1976. Mutation breeding of Chrysanthemum moriflorum Ram. Using in vivo and in vitro adventitious bud techniques. Euphytica. 25:11-19.
19. Broertjes, C. and A. Keen. 1980. Adventitious shoots: Do they develop from one single cell? Euphytica. 29:73-87.
20. Brown, D.D. 1981. Gene expression in eukaryotes. Science. 211:667-674.
21. Brown, J.S., and E.A. Wernsman. 1982. Nature of reduced productivity of anther-derived dihaploid lines of flue-cured tobacco. Crop Sci. 22:1-5.
22. Burk, L.G., and D.F. Matzinger. 1976. Variation among anther-derived doubled haploids from an inbred line of tobacco. J. Hered. 67:381-384.
23. Chaleff, R.S., and R.L. Keil. 1981. Genetic and physiological variability among cultured cells and regenerated plants of Nicotiana tabacum. Molec. Gen. Genet. 181:254-258.
24. Chourey, P.S., and R.J. Kemble. 1982. Transposition event in tissue cultured cells of S-cms genotype of maize. Maize Genet. Coop. Newsl. 56:70.
25. Cooper, K.V., J.E. Dale, A.F. Dyer, R.L. Lyne, and J.T. Walker. 1978. Hybrid plants from the barley x rye cross. Plant Sci. Letts. 12:293-298.
26. D"Amato, F. 1978. Chromosome number variation in cultured cells and regenerated plants. In Frontiers of Plant Tissue Culture. T.A. Thorpe, ed. IAPTC/University of Calgary, Calgary. pp. 287-295.
27. De Paepe, R., E. Bleton, and F. Gnangbe. 1981. Basis and extent of genetic variability among doubled haploid plants obtained by pollen culture in Nicotiana sylvestris. Theor. Appl. Genet. 59:177-184.
28. Deambrogio, E., and P.J. Dale. 1980. Effect of 2,4-D on the frequency of regenerated plants in barley and on genetic variability between them. Cereal Res. Comm. 8:417-423.
29. Deaton, W.R., P.D. Legg, and G.B. Collins. 1982. A comparison of burley tobacco doubled-haploid lines with their source inbred cultivars. Theor. Appl. Genet. 62:69-74.
30. Devreux, M., and V. Laneri. 1974. Anther culture, haploid plants, isogenic lines and breeding research in Nicotiana tabacum L. In Polyploidy and Induced Mutations in Plant Breeding. IAEA, Vienna. pp. 101-107.
31. Dixon, L.K., R.I.S. Brettell, B.G. Gengenbach, and C.J. Leaver.

1982. Mitochondrial sensitivity in Drechslera maydis T-toxin and the synthesis of a variant mitochondrial polypeptide in plants derived from maize tissue cultures with Texas male-sterile cytoplasm. Theor. Appl. Genet. (63:75-80).
32. Dunstan, D.I., K.C. Short, and E. Thomas. 1978. The anatomy of secondary morphogenesis in cultured scutellar tissues of Sorghum bicolor. Protoplasma. 97:251-260.
33. Dutrecq, A. 1981. Studies of the effects of toxic preparations of Helminthosporium sativum P.K. and B. on barley and wheat and perspectives of in vitro selection of these cereals. Agronomie 1:167-176.
34. Edallo, S., C. Zucchinali, M. Perenzin, and F. Salamini. 1981. Chromosomal variation and frequency of spontaneous mutation associated with in vitro culture and plant regeneration in maize. Maydica. 26:39-56.
35. Finch, R.A., J.B. Smith, and M.D. Bennett. 1981. Hordeum and Secale mitotic genomes lie apart in a hybrid. J. Cell Sci. 52:391-403.
36. Forche, E., K.H. Neumann, B. Foroughi, and G. Mix. 1979. Untersuchung zur ploidieverteilung in gerstenpflanzen (Hordeum vulgare L.) aus antherenkulturen. Z. Pflanzenzüchtg. 83:222-235.
37. Foroughi-Wehr, B., G. Mix, and W. Friedt. 1979. Fertility of microspore derived plants over three successive generations. Barley Genet. News. 9:20-22.
38. Franzke, C.J., and J.G. Rose. 1952. Colchicine induced variants in sorghum. J. Hered. 43:107-115.
39. Gengenbach, B.G., C.E. Green, and C.M. Donovan. 1979. Inheritance of selected pathotoxin resistance in maize plants regenerated from cell cultures. Proc. Nat. Acad. Sci. (USA). 74:5113-5117.
40. Gengenbach, B.G., Connelly, D.R. Pring, and M.F. Conde. 1981. Mitochondrial DNA variation in maize plants regenerated during tissue culture selection. Theor. Appl. Genet. 59:161-167.
41. Gengenbach, B., and P. Umbeck. 1982. Characteristics of T-cytoplasm revertants from tissue culture. Maize Genet. Coop. Newsl. 56: 140-142.
42. Green, C.E. 1977. Prospects for crop improvement in the field of cell culture. Hort. Sci. 12:7-10.
43. Green, C.E. 1978. In vitro plant regeneration in cereals and grasses. In Frontiers of Plant Tissue Culture 1978. T.A. Thorpe, ed. IAPTC, Calgary. pp. 411-418.
44. Green, C.E., R.L. Phillips, and A.S. Wang. 1977. Cytological analysis of plants regenerated from maize tissue cultures. Maize Genet. Coop. Newsl. 51:53-54.
45. Haccius, B. 1978. Question of the unicellular origin of non-zygotic embryos in callus cultures. Phytomorphology. 28:74-81.
46. Heinz, D.J., and G.W.P. Mee. 1971. Morphologic, cytogenetic, and enzymatic variation in Saccharum species hybrid clones derived from callus tissue. Am. J. Bot. 58:257-262.
47. Heinz, D.J., M. Krishnamurthi, L.G. Nickell, and A. Maretzki.

1977. Cell, tissue and organ culture in sugarcane improvement. In Applied and Fundamental Aspects of Plant, Cell, Tissue and Organ Culture. J. Reinert and Y.P.S. Bajaj, eds. Springer, Berlin. pp. 3-17.

48. Hibberd, K.A., and C.E. Green. 1982. Inheritance and expression of lysine plus threonine resistance selected in maize tissue culture. Proc. Nat. Acad. Sci. (USA). 79:559-563.

49. Hoffmann, F., and J. Adachi. 1981. "Arabidobrassica": Chromosomal recombination and morphogenesis in asymmetric intergeneric hybrid cells. Planta 153:586-593.

50. Hoffmann, F., E. Thomas, and G. Wenzel. 1982. Anther culture as a breeding tool in rape. II. Progeny analysis of androgenetic lines and induced mutants from haploid cultures. Theor. Appl. Genet. 61:225-232.

51. Jelaska, S., D. Papes, B. Pevalek, and Z. Devide. 1978. Development and karyological studies of Vicia faba callus cutlures. In Fourth Intl. Cong. Plant Tissue Cell Cult. (abst.) Calgary, Canada. p. 101.

52. Kasperbauer, M.J., R.C. Buckner, and L.P. Bush. 1979. Tissue culture of annual ryegrass x tall fescue F_1 hybrids: Callus establishment and plant regeneration. Crop Sci. 19(4):457-460.

53. Kasperbauer, M.J., T.J. Sutton, R.A. Anderson, and C.L. Gupton. 1981. Tissue culture of plants from a chimeral mutation of tobacco. Crop Sci. 21:588-590.

54. Kemble, R.J., R.B. Flavell, and R.I.S. Brettell. 1982. Mitochondrial DNA analyses of fertile and sterile maize plants from tissue culture with the Texas male sterile cytoplasm. Theor. Appl. Genet. 62: 213-217.

55. King, P.J., I. Potrykus, and E. Thomas. 1978. In vitro genetics of cereals: Problems and perspectives. Physiol. Vég. 16:381-399.

56. Krishnamurthi, M., and J. Tlaskal. 1974. Fiji disease resistant Saccharum officinarum var. Pindar subclones from tissue cultures. Proc. Int. Soc. Sugar Cane. Technol. 15:130-137.

57. Larkin, P.J., and W.R. Scowcroft. 1981. Somaclonal variation - A novel source of variability from cell cultures for plant improvement. Theor. Appl. Genet. 60:197-214.

58. Larkin, P.J., and W.R. Scowcroft. 1981. Eyespot disease of sugar cane. Host-specific toxin induction and its interaction with leaf cells. Plant Physiol. 67:408-414.

59. Liu, M.-C. 1981. In vitro methods applied to sugar cane improvement. In Plant Tissue Culture. T.A. Thorpe, ed. Academic Press, New York. pp. 299-323.

60. Liu, M.-C., and W.-H. Chen. 1976. Histological studies on the origin and process of plantlet differentiation in sugarcane callus mass. Proc. Int. Soc. Sugarcane Technol. 1:118-128.

61. Liu, M.-C., and W.-H. Chen. 1976. Tissue and cell culture as aids to sugarcane breeding. I. Creation of genetic variation through callus culture. Euphytica. 25:394-403.

62. Lu, C.-Y., V. Vasil, and I.K. Vasil. 1981. Isolation and culture of protoplasts of Panicum maximum Jacq. (Guinea grass): Somatic embryogenesis and plantlet formation. Z. Pflanzenphysiol. 104:311-318.
63. McCoy, T.J., R.L. Phillips, and H.W. Rines. 1982. Cytogenetic analysis of plants regenerated from oat (Avena sativa) tissue cultures; high frequency of partial chromosome loss. Can. J. Genet. Cytol. 24:37-50.
64. Mix, G., H.M. Wilson, and B. Foroughi-Wehr. 1978. The cytological status of plants of Hordeum vulgare L. regenerated from microspore callus. Z. Pflanzenzüchtg. 80:89-99.
65. Morel, G. 1971. The impact of plant tissue culture on plant breeding. In The Way Ahead in Plant Breeding. F.G.H. Lupton et al., eds. Cambridge. 6th Congress Eucarpia. pp. 185-194.
66. Murakishi, H.H., and P.S. Carlson. 1982. In vitro selection of Nicotiana sylvestris variants with limited resistance to TMV. Plant Cell Reports 1:94-97.
67. Murashige, T. 1974. Plant propagation through tissue culture. Ann. Rev. Plant Physiol. 25:135-166.
68. Nadar, H.M., S. Soepraptopo, D.J. Heinz, and S.L. Ladd. 1978. Fine structure of sugarcane (Saccharum sp.) callus and the role of auxin in embryogenesis. Crop Sci. 18(2):210-216.
69. Nagy, F., I. Török, and P. Maliga. 1981. Extensive rearrangements in the mitochrondrial DNA in somatic hybrids of Nicotiana tabacum and N. knightiana. Molec. Gen. Genet. 183:437-439.
70. Nakamura, C., W.A. Keller, and G. Fedak. 1981. In vitro propagation and chromosome doubling of a Triticum crassum x Hordeum vulgare intergeneric hybrid. Theor. Appl. Genet. 60:89-96.
71. Nickell, L.G., and D.J. Heinz. 1973. Potential of cell and tissue culture techniques as aids in economic plant improvement. In Genes, Enzymes, and Populations. A.M. Srb, ed. Plenum, New York. pp. 109-128.
72. Novák, F.J. 1980. Phenotype and cytological status of plants regenerated from callus cultures of Allium sativum L. Z. Pflanzenzuchtg. 84:250-260.
73. Ogihara, Y. 1981. Tissue culture in Haworthia. Part 4: Genetic characterization of plants regenerated from callus. Theor. Appl Genet. 60:353-363.
74. Oinuma, T., and T. Yoshida. 1974. Genetic variation among doubled haploid lines of burley tobacco varieties. Jap. J. Breeding. 24:211-216.
75. Oono, K. 1978. Test tube breeding of rice by tissue culture. Trop. Agric. Res. Series 11:109-123.
76. Oono, K. 1981. In vitro methods applied to rice. In Plant Tissue Culture. T.A. Thorpe, ed. Academic Press, New York. pp. 273-298.
77. Oono, K. Genetic variability in rice plants regenerated from cell culture. In Potential of Cell and Tissue Culture Techniques in the Improvement of Cereal Plants. Science Press, Beijing (in press).

78. Orton, T.J. 1980. Chromosomal variability in tissue cultures and regenerated plants of Hordeum. Theor. Appl. Genet. 56:101-112.
79. Orton, T.J. 1980. Haploid barley regenerated from callus cultures of Hordeum vulgare x H. jubatum. J. Hered. 71:780-782.
80. Ozias-Akins, P., and I.K. Vasil. 1982. Plant regeneration from cultured immature embryos and inflorescences of Triticum aestivum L. (wheat): Evidence for somatic embryogenesis. Protoplasma 110:95-105.
81. Pershina, L.A., and V.K. Shumny. 1981. A characterization of clonal propagation of barley x rye and barley x wheat hybrids by means of tissue cultures. Cereal Res. Comm. 9:273-279.
82. Phillips, G.C., G.B. Collins, and N.L. Taylor. 1982. Interspecific hybridization of red clover (Trifolium pratense L.) with T. sarosiense Hazsl. using in vitro embryo rescue. Theor. Appl. Genet. 62:17-24.
83. Radin, D.N., and P.S. Carlson. 1978. Herbicide-tolerant tobacco mutants selected in situ and recovered via regeneration from cell culture. Genet. Res. 32:85-89.
84. Reisch, B., and E.T. Bingham. 1981. Plants from ethionine-resistant alfalfa tissue cultures: Variation in growth and morphological characteristics. Crop Sci. 21:783-788.
85. Roest, S., M. Van Berkel, G. Bokelmann, and C. Broertjes. 1981. The use of an in vitro adventitious bud technique for mutation breeding of Begonia x hiemalis. Euphytica 30:381-388.
86. Sacristan, M.D. 1982. Resistance responses to Phoma lingam of plants regenerated from selected cell and embryogenic cultures of haploid Brassica napus. Theor. Appl. Genet. 61:193-200.
87. Sanders, M.E., and C.J. Franzke. 1980. The effect of light on origin of colchicine-induced complex mutants in sorghum. J. Hered. 71:83-92.
88. Sears, R.G., and E.L. Deckard. 1982. Tissue culture variability in wheat; callus induction and plant regeneration. Crop. Sci. 22:546-550.
89. Secor, G., and J.F. Shepard. 1981. Variability of protoplast-derived potato clones. Crop Sci. 21:102-105.
90. Shepard, J.F. 1981. Protoplasts as sources of disease resistance in plants. Ann. Rev. Phytopath. 19:145-166.
91. Shepard, J.F., D. Bidney, and E. Shahin. 1980. Potato protoplasts in crop improvement. Science 208:17-24.
92. Sibi, M. 1976. La notion de programme génétique chez les végétaux supérieurs. II. Aspect expérimental obtention de variants par culture de tissues in vitro sur Lactuca sativa L. apparition de vigueur chez les croisements. Ann. Amelior. Plantes 26:523-547.
93. Simantel, G.M., and J.G. Ross. 1963. Colchicine-induced somatic chromosome reduction in sorghum. III. Induction of plants homozygous for structural chromosome markers in four pairs. J. Hered. 54:277-284.

94. Siminovitch, L. 1976. On the nature of heritable variation in cultured somatic cells. Cell 7:1-11.
95. Skirvin, R.M. 1978. Natural and induced variation in tissue culture. Euphytica. 27:241-266.
96. Skirvin, R.M., and J. Janick. 1976. Tissue culture-induced variation in scented Pelargonium spp. J. Amer. Soc. Hort. Sci. 101:281-290.
97. Sreenivasan, T.V., and N.C. Jalaja. 1982. Production of subclones from the callus culture of Saccharum-Zea hybrid. Pl. Sci. Lett. 24:255-259.
98. Thomas, B.R., and D. Pratt. 1981. Efficient hybridization between Lycopersicon esculentum and L. peruvianum via embryo callus. Theor. Appl. Genet. 59:215-219.
99. Thomas, E. 1981. Plant regeneration from shoot culture-derived protoplasts of tetraploid potato (Solanum tuberosum cv. Maris Bard). Pl. Sci. Letts. 23:81-88.
100. Thomas, E., P.J. King, and I. Potrykus. 1979. Improvement of crop plants via single cells in vivo - An assessment. Z. Pflanzenzüchtg. 82:1-30.
101. Thomas, E., S.W.J. Bright, J. Franklin, V.A. Lancaster, B.J. Miflin, and R. Gibson. 1982. Variation amongst protoplast derived potato plants (Solanum tuberosum cv. "Maris Bard"). Theor. Appl. Genet. 62:65-68.
102. Tlaskal, J.G. 1978. The cytology of sugarcane. M.Sc. thesis Uni. of New South Wales.
103. Van Harten, A.M., H. Bouter, and C. Broertjes. 1981. In vitro adventitious bud techniques for vegetative propagation and mutation breeding of potato (Solanum tuberosum L.). II. Significance for mutation breeding. Euphytica 30:1-8.
104. Vasil, V., and I.K. Vasil. 1980. Isolation and culture of cereal protoplasts. Part 2. Embryogenesis and plantlet formation from protoplasts of Pennisetum americanum. Theor. Appl. Genet. 56:97-99.
105. Wakasa, K. 1979. Variation in the plants differentiated from the tissue culture of pineapple. Jap. J. Breeding. 29:13-22.
106. Wenzel, G., O. Schieder, T. Przewozny, S.K. Sopory and G. Melchers. 1979. Comparison of single cell culture derived Solanum tuberosum L. Plants and a model of their application in breeding programs. Theor. Appl. Genet. 55:49-55.
107. Wenzel, G., and H. Uhrig. 1981. Breeding for nematode and virus resistance in potato via anther culture. Theor. Appl. Genet. 59:333-340.
108. Wernicke, W., I. Potrykus, and E. Thomas. 1982. Morphogenesis from cultured leaf tissue of Sorghum bicolor - The morphogenetic pathways. Protoplasma. 111:53-62.
109. Woo, S.C., and C.Y. Huang. 1982. Anther culture of Oryza sativa L. x Oryza spontaneous taiwania hybrids. Bot. Bull. Acad. Sinica. 23:39-44.

ON THE USE OF MICROSPORES FOR GENETIC MODIFICATION

N. Sunderland

John Innes Institute, Colney Lane

Norwich, NR4 7UH, United Kingdom

SUMMARY

The practicality of using microspores in mass culture for mutant selection at the haploid level is explored with reference to two culture systems, one giving high yields of embryos (Datura innoxia), the other, calluses (Hordeum vulgare). Both rely on a stress pretreatment to switch the spores into morphogenic competence before the anthers are dissected out. Removal of spores from the anthers is either mechanical (Datura) or by natural dehiscence (Hordeum). Up to 1.0% of spores give embryos in Datura, 4% give calluses in Hordeum. Ways of improving culture efficiency are discussed. It is estimated that with such improvements, mutant plants could be recovered in Datura at a rate of 10 per person per year. Use of the Hordeum system is restricted by a low frequency of green-plant regeneration from the calluses. Mutant selection would be feasible but costly in terms of both personnel and cultivation of plants. It is concluded that for many crop species, mutant selection at the microspore level would be impractical owing to the small size of anthers, difficulties of accumulating populations large enough and at a uniform developmental stage, and the need for plants grown under standardized conditions. A more general application is seen in the genetic manipulation of individual cells.

INTRODUCTION

In the field of higher plant somatic cell genetics, protoplasts are now established as one of the major tools for cell modification, somatic hybridization, cell cloning and mutant selection. At the haploid level they have been mainly exploited for mutant selection.

A classic example is that of Carlson (1) who used mutagenized protoplasts of haploid Nicotiana tabacum to select cells resistant to methionine sulfoximine, which mimics the action of the toxin associated with wildfire disease of tobacco. Plants regenerated from selected cells were resistant to the disease. The technique has been adapted for the isolation of a wide variety of biochemical and other types of mutants in different species, but in the case of important crops such as barley and wheat, application is frustrated by inability to regenerate plants from protoplasts. The entire procedure from protoplast to plant is lengthy and labour intensive. Gebhardt et al. (2), for instance, report having screened 29,000 mutagenized cell lines derived from mesophyll protoplasts of haploid Hyoscyamus muticus to obtain six amino-acid and two temperature-sensitive mutants. Shoots were regenerated from three of the mutants.

Microspores are an alternative, readily available source of free haploid cells, which can be obtained in fairly-well synchronized populations with respect to each phase of the cell cycle. Cultured inside the anther, microspores (or young pollen gametophytes derived from them) may be diverted from their normal program to develop as embryos or calluses from which plants can be regenerated. This facility applies to major cereals such as rice (3), barley (4), wheat (5), rye (6), and maize (7). Calluses are the main product of culture, but in rye and maize, embryos may also develop directly from the spores. Embryos are the main product of culture in other plant groups, such as Brassica (8) and the tuberous Solanums (9). Solanaceous plants, tobacco (10), Datura (11), and Hyoscyamus niger (12) are among the most productive in anther culture.

In view of the facility with which plants can be generated by anther culture, the question arises whether microspores could be used to advantage in programs of mutant selection. In the past, culture efficiencies, even with whole anthers, have been considered too low for the possibility to be entertained seriously (13). More recently, however, with the advent of float culture (14) (in which spores may be released from anthers by natural dehiscence), and with more stringent attention given to inoculation densities and pre-culture requirements, culture efficiencies have been markedly improved. Barley is a notable example (4). Following the lead given by Wernicke et al. (15) culture efficiency has also been improved with isolated tobacco pollen by density centrifugation on Percoll (16). In the light of these new developments, the use of microspores for mutant selection may be re-considered.

To illustrate the scale of operation required, two culture systems are here compared; one based on mechanically-isolated spores (Datura innoxia) modified from that originally described by Nitsch and Norreel (18), and the other on shed spores (Hordeum vulgare) released into liquid medium from floating anthers by natural dehiscence (17). Both these species, the one giving embryos and the other calluses, have the advantage of being true diploids whose microspores are highly responsive while still unicellular - the ideal state for mutagenesis and the formation of solid mutants.

DATURA INNOXIA

This species has the advantage of large anthers containing many thousands of spores. The average number is given as 53,000 (19). Not all spores are haploid. In the first flowers, about 1% of the spores are unreduced, but the frequency increases with plant age and by the end of the flowering period may be as high as 10% (19). Plants are readily raised in the glasshouse or growth rooms at temperatures around 20°C. At lower temperatures, meiotic irregularities leading to non-haploid spores become more prevalent. Seeds are difficult to germinate unless first soaked in a solution of gibberellic acid (11) followed later by removal of the testa.

Owing to cell shrinkage and loss of spores, sectioned anthers give little indication of the arrangement of spores in D. innoxia. However, with freeze-fractured anthers viewed in the scanning electron microscope (20) the arrangement is seen with striking clarity, just as it is in situ. The spores are arranged randomly in each loculus, completely free of each other and of the enveloping layer of nutritive tapetal cells - an important layer in normal pollen development. All available space is occupied by spores. In this situation, neither crushing (21) nor surgical opening (22,23) is necessary; one transverse cut followed by gentle pressure on each piece is sufficient to extrude the spores into medium without disruption of the anther tissues. Subsequent processing (18) of the isolated spores is straightforward (Table 1). By selective use of nylon mesh at this stage, non-haploid spores (which are larger than the haploid population) can be eliminated.

Table 1 indicates the sort of embryo yields that can be attained with isolated Datura spores cultured in a simple nutrient medium comprising only salts, vitamins and sucrose. Yields ranging from 200-300 embryos per anther are possible providing anthers are used from 5 to 6.5mm in length. Inoculation density does not appear to be critical in this species. Densities equivalent to the contents of one half to two anthers per ml are appropriate (Table 1). Higher densities tend to be inhibitory. With 15ml cultures in 90mm Petri dishes, recovery values are equivalent to about 1% of spores inoculated. This does not take into account spores lost during processing or those killed during isolation. Assuming a 1% recovery with this simple culture system and the chances of a mutational event to be 1 in 10^6, at least 100 dishes each containing 10^6 spores must be envisaged as a minimal

Table 1. Embryo yields compared in specimen float cultures (FC) of Datura innoxia anthers and microspore cultures (MC) prepared from anthers of the same flower buds.

Mean bud length mm	Mean anther length after pre-treatment mm	Number of anthers per culture	Volume of medium ml (vessel size mm)	Embryo yield per culture FC	MC	Approximate spore density ml^{-1}
34	4.9	8	4(50)	240	1280	10^5
35	5.3	8	4(50)	690	2570	10^5
36	5.8	8	4(50)	140	2060	10^5
31	5.9	8	4(50)	270	3000	10^5
39	5.8	6	6(50)	180	2150	5×10^4
39	6.3	6	15(90)	280	3140	2×10^4

Culture procedure: flower buds were pretreated for 7-10 d at 7°C in sealed plastic Petri dishes. Anthers (5) were removed from each bud and floated on N_6 medium (24) containing 3% w/v sucrose without hormones. One anther from each bud was used to check the developmental stage (see ref. 40). Length of remaining anthers was measured under a dissecting microscope. Paired anthers were combined from several buds of similar stage to initiate each float culture. Remaining pairs of anthers were used to initiate microspore cultures. Spores were extruded from anthers into medium after one transverse cut. The suspension was filtered through 30-40µ nylon and centrifuged at 100 x g for 5 min. The pellet was washed in medium and the suspension recentrifuged. Washed pellets were re-suspended in appropriate volumes of medium and transferred to culture vessels, either 50 or 90 mm plastic Petri dishes. Dishes were sealed with parafilm and incubated in darkness for 14 d at 28°C, thereafter in light at 25°C.

requirement for the possible isolation of one mutant. The whole operation would entail the processing of 2,000 anthers for which at least 400 buds are needed.

Float cultures of anthers (that is, cultures in which anthers are present throughout) may be less productive than microspore cultures (as in the selected cultures of Table 1). However, they are more reliable. In the experiment to which the data of Table 1 relate, roughly 50% of the float cultures gave higher yields than the corresponding microspore cultures. All float cultures were productive whereas 20% of the microspore cultures failed altogether.

Embryos derived by float culture are also easier to handle subsequently than embryos derived by microspore culture. They grow and develop more rapidly, forming well developed shoots and roots while still in contact with the anthers. Such plantlets readily transfer to soil particularly if they are first incubated in an agar medium containing reduced salts and sucrose to encourage further shoot and root growth. The Murashige and Skoog (MS) formulation (25) or the N_6 (24) at half strength, without hormones, is appropriate. In contrast, embryos derived by microspore culture grow less rapidly, the rate of growth being inversely related to the numbers present. In the cultures depicted by the data of Table 1, embryos developed no further than the torpedo stage. Neither roots nor shoots developed when embryos were transferred to half strength MS or N_6 agar. A few switched to callus formation, but in most instances, shoot formation occurred randomly over the entire surface of the embryos. There is clearly potential in this observation for cloning of individual embryos. However, it is evident that structures formed in isolation from anthers cannot be assumed to have the same regenerative potential as those formed in the presence of anthers. Hormonal control is different in the two systems. Vital growth regulation is provided by the anther tissues.

HORDEUM VULGARE

Barley anthers are much smaller than those of Datura and accordingly all the more tedious to dissect out in large numbers. They are about 2.0 x 0.5mm and contain on average 3,000 spores (26). The spores are less easily removed from the anthers than those of Datura. Freeze-fracturing shows the spores to be aligned against the tapetum in a single layer, each spore being oriented with the germ pore towards and abutting on to a tapetal cell. There are no spores free of tapetum within the lumen of the loculus. This highly ordered and regular positioning of spores against the tapetum has been described in several grasses from sectioned material (27,28,29) and is probably a characteristic of the family as a whole. Vigorous crushing or grinding of anthers is needed to break the physical association of the two tissues. Many spores are killed in the process. It is doubtful whether spores that survive are completely freed of tapetal debris.

The association between spores and tapetum can be broken by subjecting excised barley spikes to a temperature stress before the anthers are dissected out (Table 2). Freeze-fracturing shows the same positioning of spores at the end of the pretreatment, but a degenerate tapetum. With pretreated anthers, therefore, the spores come away more freely from the anthers. A simple method of isolation, which cuts down mechanical damage, is for the spores to be tapped out gently from individual anthers, each being held in turn in medium by means of fine forceps, while viewed under a dissecting microscope (17). The anther tissues are not disrupted. The procedure is not laborious if used on a micro-scale with appropriate reductions in volume of medium and vessel size. Both

Table 2. Callus yields in shed spore cultures of Hordeum vulgare cv Sabarlis (17). Values in brackets denote the percentage of multicellular pollen grains present in addition to calluses.

Medium	Days anthers in contact with medium	Number of calluses per culture
P-9H	0-1	7
	1-3	1
	3-7	0
	7-28 (+ anthers)	48 (2.7)
P-9H(C)	0-1	5097
	1-3	615
	3-7	78
	7-28 (+ anthers)	3282 (15.2)

Culture procedure: Tillers harvested as the flag leaf starts to emerge, 0-50 mm. Leaf laminae removed, leaf bases swabbed with alcohol and cut open to expose the spike. Spikes removed into a 90 mm Petri dish and awns cut off. Three anthers removed from one spikelet in the middle of the spike and the stage assessed by the acetocarmine test. Spikes having anthers at stage 2 (40) retained. Others rejected. Dishes sealed with parafilm and stored in darkness for 14 d at 7°C (stress pretreatment) (4). Upper- and lower-most spikelets discarded. Anthers from the rest dissected out and floated in groups of 80 on the surface of 1.0 ml aliquots of potato medium (P-9H) (43) containing 9% w/v sucrose, kinetin 0.5 mg l^{-1}, and 2,4-D 1.5 mg l^{-1} or conditioned potato medium (P-9H(C)) (4). Cultures incubated in darkness at 25°C. Anthers transferred to fresh medium after 1,3 and 7 d.

anthers and isolated spores of barley have been successfully cultured by a 'droplet procedure' (17,30) similar to that in use for culture of protoplasts.

However, isolation of barley microspores from pretreated anthers is not necessary. After a 14d pretreatment of spikes at 7°C, anthers already show signs of dehiscence as they are being dissected out. Within 24 h of being floated on medium they open fully and at least half of the spore population is shed into it (17). Removal of the anthers after 24h, followed by re-incubation of the cultures, leads to mass callus production. 'Droplet culture' is not then necessary. Cultures of shed spores are far more productive than cultures of isolated spores (17). This greater productivity is thought to be due to the release of key growth compounds from the anthers into the medium (conditioning factor) (4)

After removal of the anthers from the culture vessel, this factor is available for callus growth.

The data of Table 2 show one of the best results to date with shed spores. After 24 h, anthers were transferred to fresh medium and again after 3 and 7 d. Microscopic examination showed that fewer spores were shed into the medium between 1-3 d and between 3-7 d than are shed during the first 24 h. For high yields of calluses in the 24 h fraction, inoculation densities of 80-100 anthers per ml of medium are required (17). Both callus production and growth are improved by use of conditioned media (4,17) (Table 2) and by inclusion of myo-inositol (31). Assuming that half of the spore population is released, the recovery value for the 0-24 h culture period depicted in Table 2 is about 4%. With 1 ml cultures each inoculated with 70 anthers, 10 dishes would give approximately 10^6 spores. A total of 250 such dishes has to be envisaged to assure the possible isolation of one mutant. The operation would entail processing of 17,500 anthers. About 50 anthers can be used from one spike of Sabarlis barley (4). Hence, sufficient plants must be provided to give 350 spikes for one mutant.

The regenerative potential of calluses produced from shed spores has not been directly evaluated. Such calluses have the advantage of developing in the presence of conditioning substances whereas isolated spores of *Datura* are washed free of such substances during processing. Values have been derived from barley calluses produced in float cultures of anthers (32). These show that for every 100 anthers cultured a mean of 5 green plants is possible and also a mean of 9 albinos. On the assumption that the regenerative potential would not be altered by a mutation, the number of spikes deemed necessary for one green mutant needs to be increased 20 times. Thus the 350 spikes assumed previously must be increased to a value of 7,000 - clearly a colossal undertaking.

LIMITATIONS OF THE STRESS PRETREATMENT

The stress pretreatment (which is as much an integral part of the procedure with *Datura* as it is with barley) is a key determinant of culture yield (33,34). The pretreatment is an extension of the 'cold shock' used by Nitsch and Norreel (18). The 'shock', administered to excised flower buds kept in water for a few days at 3°-4°C, was then interpreted as a means of altering the polarity of the spores such that when given the stimulus of culture they divide by an equal rather than the unequal division that heralds the formation of the pollen gametophyte. The hypothesis has been considerably modified as a result of investigations on tobacco (35,36) and barley (26,37) in which equal divisions of haploid spores are a rarity even under prolonged periods of cold stress. In barley, with spikes kept in sealed containers at 7°C (Table 2), the spores develop slowly through the cycle, divide by the usual unequal division, but with subsequent constraints to division removed (34). The two daughter cells start to divide while the spores are still in the spike and with a sufficiently long pretreatment may undergo more than one division cycle. 'Desynchronization' of spores and tapetum is an important facet of the stress, and it has been suggested that this leads to failure in the triggering of genes essential to gametophytic development (34). A much broader concept of the stress pretreatment is that it induces the spores into a state of morphogenic competence and, in species such as barley, into division, before the anthers are dissected out. Counts made of dividing spores in pretreated barley spikes show that on average up to 60% may be switched and even more in individual anthers (34,38). Different methods of pretreatment have been examined in barley at different temperatures, the best results accruing from spikes kept in sealed Petri dishes at 4°C for 21-28 d (32). It is unlikely that the stress pretreatment can be further manipulated in barley significantly to alter culture efficiency.

With *D. innoxia*, however, only limited pretreatment is at present possible. Excised flower buds stay fresh in sealed Petri dishes for only a few days at 4°-7°C. By 7-9d, the calyx starts to turn brown and deteriorate, followed closely by corolla and finally the anthers. By 12 d, anthers and spores are deteriorating rapidly. In float cultures, embryo yields increase with increasing time of pretreatment at 7°C, reaching a peak at 9 d (Fig. 1), after which, with bud deterioration, embryo yields decline. Pretreatment is much more effective at 7° than at 4°C. Temperatures above 7°C (34) have been tested, also below 4°C with various methods of pretreatment, but to no avail. With the 9 d pretreatment, there is little movement of spores through the cell cycle, and if anthers are removed from the plant half way through the cycle, spores are still unicellular at the end of the pretreatment. Hence the number of spores switched into morphogenesis (if any) cannot be ascertained. On some occasions, buds have survived for periods longer than 12 d without severe browning. In these instances, culture yields were very high. There seems little doubt that culture efficiency in *Datura* would be improved if bud deterioration could be controlled. *Datura* anthers do not dehisce naturally after a 9 d pretreatment. This too might be influenced by longer pretreatments.

LIMITATIONS OF THE ISOLATION PROCEDURE

In the context of mutagenesis, the shed-spore approach for barley can be criticized on the grounds that the spores will be dividing or will start to divide during the initial flotation period. With the period of stress adjusted so that the spores are still unicellular when the anthers are floated, most will have divided within 24 h. This is also the case with *D. innoxia*.

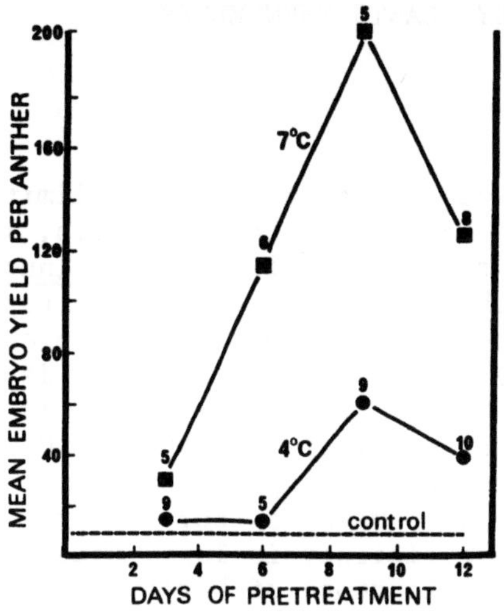

Fig. 1. Changes in embryo yield per anther in float cultures of Datura innoxia anthers with increasing days of bud pretreatment at 4° or 7°C. Medium and culture conditions as in Table 1. Figures denote the numbers of cultures for each pretreatment.

Nuclear fusion is common during the initial division cycles in both Datura (11) and barley (26). Mutagenesis of bicellular units in these species would be disadvantageous in that any recessive mutation might be masked. Survival of mutant lineages remaining distinct from others would depend upon their selective growth advantage. This latter comment is particularly appropriate to the tobacco pollen system (16) in which the unit isolated is bicellular.

Horner and Street (23) working with tobacco were the first to comment on the destructive element of mechanical isolation. It was largely because of such mechanical damage that the shed pollen procedure was adopted (14). By use of density centrifugation on Percoll (15) to remove spores damaged or killed during isolation, culture efficiencies could probably be improved in Datura. Spores degenerating during the stress pretreatment would likewise be eliminated. Such enrichment of shed-spore populations in barley is also feasible, though to maintain the benefit of the conditioning factor, the enriched population would need to be re-suspended in the original medium or in a medium specially conditioned for the purpose. Such enrichment in barley, however, is unlikely to alter culture efficiency so significantly that the number of spikes stipulated above for mutagenesis studies could be reduced to practical levels.

LIMITATIONS IMPOSED BY PLANTS THEMSELVES

Anther Size

Anther size is one of the main factors limiting the use of microspores on a mass scale. As has been seen, mechanical isolation is applicable mainly to species, like Datura innoxia, which have large anthers in which spores are free of the tapetum. Few species of economic importance have such easily manipulable anthers. Solanum tuberosum (9) is one, Brassica napus is another (39). Among the cereals (at which this paper is mainly aimed) rye and maize are two with reasonably-sized anthers. Culture efficiencies in both anther culture and microspore culture of rye are much lower than those in barley (9). Maize anther culture is still being developed and culture efficiency is low. Rice and wheat have anthers similar in size to those of barley. The shed spore procedure has been used to good effect with rice (3), but in neither species are culture efficiencies up to the level of that in barley. Other cereals such as the millets have even smaller anthers. Anthers of Setaria italica, for example, have ten times fewer spores than those of barley. Some of the Panicums have yet smaller anthers.

Rate of Flower Production

The rate of flower production is a limiting factor in the use of Datura microspores in mass. Flower buds are produced singly. An interval of several days elapses before another bud is ready for use. Though every bud is theoretically usable, some are lost because of the shortness of the microspore period. A bud slightly too young at the end of the working day may be too old the next morning. Accordingly, many more plants than are theoretically needed have to be provided to ensure a constant supply of buds at the microspore stage at any given time.

In Datura, the critical phase in the microspore cell cycle that would be needed for mutagenesis studies can be judged reasonably accurately by anther length. In the experiment to which the data of Table 1 relate, the most productive microspore cultures came from anthers having mean lengths of 5.6-5.8 mm after the pretreatment. Cytological examination showed most of these anthers to contain young pollen gametophytes; microspores were present only in a few instances. Examination of anthers of similar length before the pretreatment showed them to be either at the end of the microspore cell-cycle (stage 3) (40), at the mitotic stage (stage 4) or the immediate post-mitotic stage (stage 5). Spores in stage 3 anthers usually started to divide during the pretreatment. Simply as a means of producing plants, these stages are clearly the best, but to ensure that the microspores are still unicellular for mutagenesis, slightly shorter anthers (stage 2) are required. Under the plant-growth conditions used here, anthers about 5.0 mm would have been appropriate.

Complexity of Inflorescence Structure

The unit harvested in barley is the whole inflorescence or spike. A succession of tillers, each giving one spike, is produced. As with Datura buds, barley spikes are lost unless the plants are under constant harvesting supervision. Anthers from the earliest spikes are usually the most productive in culture; in practice, probably no more than 3-4 per plant can be used. Tillers have to be harvested by reference to the length of the emerging flag leaf. The actual length corresponding to the microspore stage is found by experience for the prevailing growth regime. Further spikes are lost because some of the harvested tillers are not at the desired anther stage, that is, stage 2. More than 7,000 plants may be required to give the 7,000 spikes stipulated for one mutant.

Restriction of anthers to stage 2 in barley is essential not so much from the point of view of the mutagenesis requirement but because high yielding cultures depend upon it. Anthers at stage 3 are less productive in barley than those at stage 2 (37); anthers at stage 1 (beginning of the cell cycle) are unproductive. Feulgen photometry shows that in a typical stage 2 anther of Sabarlis barley most of the spores have the 1C level of DNA (41). The data of Table 3, however, indicate that some have values slightly above the 1C level. This shows that the spore population is just starting to move into the S phase. Since stage 3 anthers are less productive than stage 2 anthers, it follows that the switch in program is in some way associated with the end of the G1 interphase and possibly with the transition into the S phase (34). With pretreatment of spikes at 7°C the spores in stage 2 anthers move into the S phase, then into G2, and by 21 d are just starting to divide (Table 3). Presumambly, this would also be the case in Datura if the pretreatment could be extended beyond 9 d.

In two-rowed barleys such as Sabarlis, spikelets (each with one floret) are borne on opposite sides of the spike axis, there being 20-30 spikelets depending on the genotype and the conditions of cultivation. Spikelets in the middle of the spike are slightly more advanced than those above and below. Anthers are too small for the length to be used as a criterion of stage. Each spike is accordingly tested by cytological examination of anthers removed from the middle spikelets before the pretreatment. Only those spikes with anthers at stage 2 are retained. Anthers from all spikelets along the spike axis can be used except those at the base and tip (4).

Table 3. Changes in the percentage of staining spores, and in Feulgen DNA values, with increasing pretreatment at 7°C of an excised spike of Hordeum vulgare cv Sabarlis having anthers at stage 2 (41). Spores stained with aceto-carmine. The 1C level of DNA determined on spore-tetrads = 9.8. In most cases, 300 spores or nuclei were scored.

Days of pretreatment	% Unicellular	% Bicellular	Empty or Degenerating	% G1	% S	% G2	Mean DNA value
0	99	0	1	84	16	0	9.7
7	96	0	4	84	16	0	9.9
14	76	0	24	24	59	17	13.8
21	61	2	37	59	37	4	10.7

With most other cereals and forage grasses, however, not all the florets from one spike can be used. In wheat and the ryegrasses, in which the spikelets each contain several florets at different developmental stages, the strategy is to restrict culture to the basal (or two basal) floret(s) in each spikelet as they pass through stage 2. In the loose panicles of rice, on the other hand, spikelets are borne on many branches of the spike axes. Infinite patience and careful dissection are needed to locate stage 2 anthers. With the close, contracted spike-like panicles of Setaria italica, location of stage 2 anthers is well nigh impossible. Maize presents yet another problem. In addition to there being only one complex male inflorescence per plant, there is no obvious morphological criterion, such as the length of the emerging flag leaf, by which the desired spikes can be harvested. Under growing conditions in Norwich, the stage 2 spike is well submerged in the leaf bases. Accordingly large numbers of plants have to be destroyed in order to find a spike at the correct stage.

THE NEED FOR PLANTS GROWN UNDER STANDARD CONDITIONS

To provide the large numbers of plants needed for microspore studies in the cereals, outdoor cultivation of plants is clearly indicated. In Norwich, however, cereals grown outdoors are usually infested with thrips. The insects crawl down into the leaf bases and infect the developing inflorescences. Anther cultures cannot be obtained free of infection. Glasshouse cultivation of cereals also has its problems. Plants are often over-heated during summer, which is deleterious to pollen fertility, and in winter they are particularly susceptible to fungal and other diseases. Fumigations and chemical spraying, now an accepted part of glasshouse routine, also have their toll in the reduction of pollen fertility and reduced culture yields. Notwithstanding these biotic and other problems, the main difficulty with glasshouse plants is the variability in response from one batch of plants to another. The experiments cited above on plant regeneration from barley calluses (32) were carried out on anthers from plants grown during spring through the progressively increasing day lengths of March to May. The same regenerative potential was not observed in experiments in which calluses were derived from plants grown in autumn during the decreasing daylengths of September to November. For consistent results, plants need to be grown under standard conditions - a costly process for growing large numbers of plants. An increasing daylength probably cannot be easily simulated. Good results are being obtained with barley plants grown in a 16 h day under mercury halide lamps at a temperature of 15°C day and night. Provided there is restricted access to growth rooms, and the rooms are fumigated and sprayed beforehand, diseases can generally be contained.

CONCLUSIONS

Some of the observations recorded here may seem trivial and obvious; others have dealt more with the creation of obstacles rather than their removal. This has been done purposely to rationalize the overstatement regarding the role of pollen as the future higher-plant equivalent of the microorganism (16,42). On present evidence, the number of species with which this might by feasible is distinctly limited. The *Datura* system is one possibility. It needs to be improved and conditions for plant regeneration defined. As has been seen, to initiate a culture of 10^6 spores, 4 buds are needed each having anthers at the same developmental stage. One individual worker could probably process 20 such buds per day. On the assumption that sufficient plants are available and that the total of 400 buds stipulated above as a basic requirement for the isolation of one mutant could be substantially reduced by: i) improved stress pretreatment, ii) removal of dead spores after isolation, and iii) use of better media, it would probably take about a month for sufficient cultures to be initiated for the one mutant. A turnover of 10 mutants per year by one worker seems realistic. The advantage of the *Datura* system is its simplicity in operation. The procedure is far easier and shorter than with protoplasts. Given a team of workers such as was presumably concerned in the isolation of the mutants of *Hyoscyamus muticus* cited at the beginning of the paper (2), a steady stream of mutants would probably be forthcoming. The major disadvantage of *Datura* is that, despite its alkaloid content, it cannot be described as an economic species. Little is known of the genetics of *D. innoxia*.

The tobacco system (16,42) is another possibility. It has not been considered here partly because of the amphidiploid origin of the species but also because it is essentially not a microspore system. Tobacco anthers at stage 2 are relatively unproductive (40). Tobacco has the advantage of having a well documented genetic background, and may be regarded as having some economic role. However, work on 'easy species' like tobacco and Datura now seems unrealistic. Guidelines for application to other more valuable species have been defined. With present financial constraints, effort should be directed at such species.

The barley system is not yet refined enough to make mutant selection a feasible proposition. It would take one individual more than a year, working consistently, to process the 7,000 spikes stipulated above as the minimal requirement for the possible isolation of one mutant. The cost of mounting a large-scale operation in terms of both plant cultivation and personnel would be heavy. The major block in the barley system is undoubtedly the albino problem and the poor potential of the microspore calluses for green-plant regeneration. Ability to regenerate green plants from every callus, or better still to obtain one green embryo from one spore, as is likely in Datura, would bring down the spike stipulation to much more acceptable levels. Detailed and systematic investigation of the albino problem and green-plant regeneration would be an amply rewarded investment, not only in respect of barley but also of other cereals and forage grasses. It is noted that better regeneration of green plants is afforded by certain genotypes of rice (44) than by Sabarlis barley. Development of a high-yielding shed spore system with such genotypes might be exploitable.

Culture efficiencies in the barley system could no doubt be slightly improved by resort to technological expediencies such as density gradient centrifugation. However, glib statements to the effect that mass spore populations could be attained simply by grinding up large quantities of inflorescence material regardless of anther stage, and by sieving off spores and pollen grains for subsequent separation by density centrifugation are unrealistic and evade the real issues. Whatever is done to improve efficiencies whether by stress pretreatments, medium improvement or the 'cleaning up' of spore preparations, the fact remains that the bulk of spores still degenerate during the first few days of culture. This is likely to remain so in the immediate future.

The findings of Z. H. Xu (4,30,31) on the synergistic interactions between hormones, conditioning factor, and myo-inositol point to endogenous constituents in both spores and anther tissues which hold the key, first, to the triggering of division in morphogenically competent spores, and second, to the maintenance of division and the formation of calluses. Identification of these compounds seems a more rational approach to the problem of low culture efficiencies than the purely empirical one of varying medium components in the hope that a few more calluses or embryos might be attained. Identification of conditioning compounds could also assist in application of the technique to other important plant groups such as the legumes, palms, woody species, and indeed to those cereals like oats, which have so far not responded in anther culture.

The use of cereal microspores generally for mass culture and mutant selection cannot be envisaged. The plants themselves present too many obstacles in respect of anther size, complexity of inflorescence structure, and growth-room cultivation. For most cereals, therefore, the major role of microspore studies in the future will still be in the rapid production of haploid and doubled haploid plants for exploitation in breeding programs. The float culture method with drops of medium offers a simple approach for application to species having minute anthers (30).

The main obstacles to the use of cereal microspores for mutant selection are at least clear and defined. With these out of the way, there may well be some justification for regarding certain species of microspores as the higher-plant equivalent of the micro-organism. However there are other areas in the context of genetic manipulation and modification in which microspores might be more generally used. With the present knowledge of triggering spores into morphogenic competence before culture, it is now possible to remove one spore from a population with some degree of certainty that it will respond in culture given the right medium and conditions. The microspore is thus clearly in the forefront for consideration as a cell for individual genetic manipulation. Perhaps a far more important role for the microspore in the future will be in the field of microinjection and the direct introduction of DNA into individual cells.

REFERENCES

1. Carlson, P.S. 1973. Methionine sulfoimine-resistant mutants of tobacco. Science 180:1366.
2. Gebhardt, C., V. Schnebli, and P.J. King. 1981. Isolation of biochemical mutants using haploid mesophyll protoplasts of Hyoscyamus muticus. II. Auxotrophic and temperature-sensitive clones. Planta 153:81.
3. Chen, Y.W., R.F. Wang, W.H. Tian, Q. Zuo, S. Zheng, D.Y. Lu, and G. Zhang. 1980. Studies on pollen culture in vitro and induction of plantlets in Oryza sativa subsp. Keng. Acta. Genet. Sin. 7:46.
4. Xu, Z.H., B. Huang, and N. Sunderland. 1981. Culture of barley anthers in conditioned media. J. Exp. Bot. 32:767.
5. Henry, Y., and J. De Buyser. 1981. Float culture of wheat anthers. Theor. Appl. Genet. 60:77.

6. Wenzel, G., F. Hoffmann, and E. Thomas. 1977. Increased induction and chromosome doubling of androgenetic haploid rye. Theor. Appl. Genet. 51:81.
7. Miao, S.H. 1980. Effects of different ammonium salts on the formation of maize pollen embryoids. Acta Bot. Sin. 22:356.
8. Keller, W.A., and G. Stringham. 1978. Production and utilization of microspore-derived haploid plants. In Frontiers of Plant Tissue Culture. T.A. Thorpe, ed. International Association for Plant Tissue Culture, Calgary.
9. Wenzel, G. 1980. Recent progress in microspore culture of crop plants. In The Plant Genome. D.R. Davies, and D.A. Hopwood, eds. The John Innes Charity, Norwich.
10. Nitsch, J.P. 1972. Haploid plants from pollen. Z. Pflanzenzuchtg. 67:3.
11. Sunderland, N., G.B. Collins, and J.M. Dunwell. 1974. The role of nuclear fusion in pollen embryogenesis of Datura innoxia Mill. Planta 117:227.
12. Raghavan, V. 1976. Role of the generative cell in androgenesis in Henbane. Science 191:388.
13. Sunderland, N., and J.M. Dunwell. 1977. Anther and pollen culture. In Plant Tissue and Cell Culture. H.E. Street, ed. Blackwell, Oxford.
14. Sunderland, N., and M. Roberts. 1977. New approach to pollen culture. Nature, Lond. 270:236.
15. Wernicke, W., C.T. Harms, H. Lorz, and E. Thomas. 1978. Selective enrichment of embryogenic microspore populations. Naturwissen. 65:540.
16. Rashid, A., and J. Reinert. 1980. Selection of embryogenic pollen from cold-treated buds of Nicotiana tabacum var. Badischer Burley and their development into embryos in cultures. Protoplasma 105:161.
17. Sunderland, N., and Z.H. Xu. 1982. Shed pollen culture in Hordeum vulgare. J. Exp. Bot. 33:1086.
18. Nitsch, C., and B. Norreel. 1973. Effet d'un choc thermique sur le pouvoir embryogene du pollen de Datura innoxia cultive dans l'anthere ou isole de l'anthere. C.R. Hebd. Seances Acad. Sci. Paris. 276:303.
19. Collins, G.B., J.M. Dunwell, and N. Sunderland. 1974. Irregular microspore formation in Datura innoxia and its relevance to anther culture. Protoplasma. 82:365.
20. Huang, B., and G.P. Hills. unpublished results.
21. Nitsch, C. 1974. Pollen culture - a new technique for mass production of haploid and homozygous plants. In Haploids in Higher Plants - Advances and Potential. K.J. Kasha, ed. University of Guelph Press, Guelph.
22. Wernicke, W., and H.W. Kohlenbach. 1977. Versuche zur Kultur isolierter Mikrosporen von Nicotiana und Hyoscyamus. Z. Pflanzenphysiol. 81:330.
23. Horner, M., and H.E. Street. 1978. Problems encountered in the culture of isolated pollen of a Burley cultivar of

Nicotiana tabacum. J. Exp. Bot. 29:217.
24. Chu, C.C. 1978. The N_6 medium and its application to anther culture of cereal crops. In Proceedings of Symposium on Plant Tissue Culture. Science Press. Peking.
25. Murashige, T., and F. Skoog. 1962. A revised medium for rapid growth and bioassays with tobacco tissue cultures. Physiol. Plant. 15:473.
26. Sunderland, N., M. Roberts, L.J. Evans, and D.C. Wildon. 1979. Multicellular pollen formation in cultured barley anthers. I. Independent division of the generative and vegetative cells. J. Exp. Bot. 30:1133.
27. Christensen, J.E., and T.H. Horner. 1974. Pollen pore development and its spatial orientation during microsporogenesis in the grass Sorghum bicolor. Am. J. Bot. 61:604.
28. Scoles, G.J., and L.D. Evans. 1979. Pollen development in male-sterile and cytoplasmic male-sterile rye. Can. J. Bot. 57:2782.
29. Steer, M.W. 1977. Differentiation of the tapetum in Avena. I. The cell surface. J. Cell Sci. 25:125.
30. Xu, Z.H., and N. Sunderland. 1982. Inoculation density in the culture of barley anthers. Scient. Sin. 25:961.
31. Xu, Z.H., and N. Sunderland. 1981. Glutamine, inositol and conditioning factor in the production of barley pollen callus in vitro. Plant. Sci. Lett. 23:161.
32. Huang, B., and N. Sunderland. 1982. Temperature-stress pretreatment in barley anther culture. Ann. Bot. 49:77.
33. Sunderland, N. 1978. Strategies in the improvement of yields in anther culture. In Proceedings of Symposium on Plant Tissue Culture. Sci. Press. Peking.
34. Sunderland, N. 1982. Induction of growth in the culture of pollen. In Differentiation In Vitro. M.M. Yeoman and D.E.S. Truman, eds. Cambridge University Press. Cambridge.
35. Sunderland, N., and M. Roberts. 1979. Cold pre-treatment of excised flower buds in float culture of tobacco anthers. Ann. Bot. 43:405.
36. Sunderland, N. 1979. Comparative studies of anther and pollen culture. In Plant Cell and Tissue Culture: Principles and Applications. W.R. Sharp, P.O. Larsen, E.F. Paddock, and V. Raghavan., eds. Ohio State University Press, Columbus.
37. Sunderland, N., and L.J. Evans. 1980. Multicellular pollen formation in cultured barley anthers. II. The A, B, and C pathways. J. Exp. Bot. 31:501.
38. Sunderland, N. 1981. The concept of morphogenic competence with reference to anther and pollen culture. In Proceedings of the Symposium on Plant Cell Culture in Crop Improvement. Calcutta. (in press).
39. Lichter, R. 1982. Induction of haploid plants from isolated pollen of Brassica napus. Z. Pflanzenphysiol. 105:427.
40. Sunderland, N. 1974. Anther culture as a means of haploid induction. In Haploids in Higher Plants: Advances and

Potential. Kasha, K.J., ed. University of Guelph Press, Guelph.
41. Xu, Z.H., and N. Sunderland. Unpublished observations.
42. Rashid, A., and J. Reinert. 1981. High-frequency embryogenesis in *ab initio* pollen cultures of *Nicotiana tabacum*, *Naturwissen*. 68:378.
43. Chuang, C.C., T.W. Ouyang, H. Chia, S.M. Chou, and C.K. Ching. 1978. A set of potato media for wheat anther culture. In *Proceedings of Symposium on Plant Tissue Culture*. Science Press. Peking.
44. Chaleff, R.S., and A. Stolarz. 1981. Factors influencing the frequency of callus formation among cultured rice (*Oryza sativa*) anthers. *Physiol. Plant*. 31:201.

CROP PRODUCTIVITY AND QUALITY: CHAIRMAN'S INTRODUCTION

Donald Rasmusson

Department of Agronomy and Plant Genetics
University of Minnesota
St. Paul, Minnesota 55108

If you ask a plant breeder about the major objectives of his/her program, the answer will likely include controlling pests, enhancing yield potential, reducing losses due to stress (such as salt and drought), and improving crop quality. This portion of the conference focuses on these four topics which are at the heart of plant breeding. They are pertinent topics from the standpoint of the plant breeder who is at times frustrated with slow progress, and uncertain about how to proceed in the breeding program. They are also pertinent to genetic engineers and others who desire to learn about plant breeding goals and new approaches in dealing with them. The four speakers consider selection strategies and genetic stocks that may have utility in improvement programs as well as important fundamental aspects of the traits. In considering selection strategies the authors do not limit themselves to molecular genetics, but rather emphasize a full range of possibilities which are emerging as an outgrowth of ongoing research in the four areas.

USE OF PATHOGEN-PRODUCED TOXINS IN GENETIC ENGINEERING OF PLANTS AND PATHOGENS

O.C. Yoder

Department of Plant Pathology
334 Plant Science Building
Cornell University
Ithaca, New York 14853

SUMMARY

Certain bacterial and fungal plant pathogens produce extracellular toxins that are known to be causally involved in disease. Some of these are required for pathogenicity (pathogenicity factors) whereas others contribute to virulence (virulence factors) of the producing organism. Both types of toxin are potentially useful in pest control. First, disease-resistant plants or cells can be efficiently selected both *in vitro* and *in vivo*. If a pathogenicity factor is used, a high level of resistance is expected; if a virulence factor is used an intermediate level of resistance is expected. Either level can be economically valuable. Toxins that have not been shown to be causally involved in disease are not expected to select any disease-resistance at all. Second, pathogen-produced toxins may provide an effective approach to pest control. For example, a gene or set of genes that controls production of a nonspecific toxin might be transferred to and expressed in a pathogen that is specific for a certain pest, such as a weed or an insect, making the pathogen capable of reducing the pest population to a low level. Before pathogens can be genetically engineered in this way, genes that control toxin production must be isolated and cloned. The isolation of most genes of this type will require complementation of toxinless recipient cells with a library of DNA fragments from a toxin-producing strain (these toxins are synthesized constitutively and no abundant mRNA is likely to be found). However, the genes in question will probably not function in easily transformable organisms such as *E. coli* or yeast. Therefore, genetic transformation systems must be developed for toxin-producing bacterial and fungal pathogens so that they can serve as hosts for the isolation of their own genes. Recent advances in several laboratories indicate that the technology for cloning pathologically important genes from bacteria is now available. Fungi are less tractable than bacteria, but a transformation system for the toxin-producing fungal pathogen Cochliobolus heterostrophus is under development. It is based on complementation of adenine-requiring fungal protoplasts with a cloned ADE gene from yeast. To date all transformants have aborted; efforts to construct a cloning vector that can be maintained indefinitely by Cochliobolus cells are in progress.

INTRODUCTION

Can gene cloning technology be used in combination with conventional methods to help manage agricultural pests such as insects, weeds and microbial pathogens? I shall respond to this question by describing ways that a model group of biostress molecules, pathogen-produced toxins, can be used to advantage. The focus will be on three areas in which it is reasonable to expect that advances can be readily made. These are: 1) the use of pathogen-produced toxins to select disease-resistant individuals in populations of genetically modified plants or plant parts, 2) the construction of toxin-producing pathogens for biological control of weeds or insects, and 3) the prospect of developing genetic transformation systems for fungal pathogens so that they can be used as hosts for the selection and cloning of their own pathologically-significant genes, including genes that control production of toxins.

PATHOGEN-PRODUCED TOXINS FOR SELECTION OF DISEASE-RESISTANT PLANTS

Genetic modification of plants for resistance to disease, whether by conventional or genetic engineering technology, requires that resistant individuals be identified among large populations of susceptible plants. This is usually done by inoculating plants with the pathogen itself, a procedure that is cumbersome and inefficient. Several authors have advocated the use of toxins produced by pathogens for more efficient identification of disease-resistant plants in vivo or plant parts such as tissues, cells, and protoplasts in vitro (4,14,15,18,22). There is merit in this approach, but only if adequate background information about the toxin in question is first obtained.

Before a toxin can be used reliably as a surrogate for a pathogen it must be convincingly shown to play a causal role in disease. For those toxins that are pathologically-important, the role in disease should be defined. Several classification schemes have been offered (10,23,45,47,48,60), none of which is completely satisfactory. For simplicity, it is convenient to think about toxins as either pathogenicity factors or virulence factors (62). A pathogenicity factor is required by the producing pathogen to cause any disease at all. A virulence factor is not necessary to initiate disease but, if present, changes the amount of disease that develops. A third possibility is that the material in question is not causally involved in disease but rather is fortuitously coincidental with disease or is the result of disease.

Approaches to distinguish among these three possibilities have been discussed previously (62). The most straightforward analysis involves induction of pathogen mutants that produce no toxin. If toxinless mutants cause no disease, the toxin is a pathogenicity factor. If toxinless mutants cause disease but an altered amount of it, the toxin is a virulence factor. If toxinless mutants cause the same level of disease as toxin-producing strains, the toxin has no apparent role in disease and is a spurious factor with respect to pathogenesis.

The foregoing distinction among types of toxin has value in predicting whether or not a toxin will be useful to screen for disease resistance, as well as in determining the quality of resistance that might be obtained. If the toxin has no causal role in disease, then plants selected for resistance to it will probably not be resistant to disease. If the toxin is a pathogenicity factor, plants highly resistant to the toxin will be completely resistant to disease. If the toxin is a virulence factor, plants resistant to the toxin will be partially resistant to disease. Partial resistance selected with a virulence factor can be fully satisfactory for disease control if the portion of the disease caused by the toxin is the economically important part.

Only a few toxins produced by fungal and bacterial pathogens have been investigated to the extent that their roles in disease are both convincingly established and clearly defined (62). All such toxins described so far have low molecular weights but are otherwise chemically diverse. Chemical classes include peptides (either cyclic or linear), terpenoids, sugars, and polyketols (12). Some of these toxins have been called "host-specific" because they affect only plants that are susceptible to the toxin-producing pathogen. Others are nonspecific and may be toxic not only to host plants but to nonhost plants, microorganisms, and animals as well (62). However, host-specificity is not an important criterion in the use of toxins to select disease resistant plants, as discussed below.

Use of Pathogenicity Factors to Select Disease-Resistant Plants

HV-toxin, produced by the fungus *Cochliobolus victoriae*, is a pathogenicity factor because isolates that lack ability to produce toxin cause no disease. The toxin is host-specific since it affects only plants that are susceptible to C. victoriae, ie. certain varieties of oats (62), and is a powerful tool for identifying rare resistant plants in populations that are mostly susceptible. Wheeler and Luke (59) were able to use HV-toxin to select 973 resistant oat plants from 100 bushels (4.5×10^7 seeds) of a susceptible seed lot. This study and others showed that all plants highly resistant to HV-toxin are fully resistant to the pathogen, whereas plants partially resistant to toxin are intermediate in response to the pathogen (36). Since HV-toxin is a pathogenicity factor, the level of resistance to toxin corresponds directly with the amount of disease caused by the pathogen. Other toxins known to be pathogenicity factors have also been used to select disease-resistant plants and results with them have been similar to those obtained with HV-toxin (63).

Use of Virulence Factors to Select Disease-Resistant Plants

Two examples, one a host-specific toxin, the other nonspecific, illustrate that virulence factors do indeed select plants with partial resistance to disease. One case involves T-toxin, a host-specific toxin produced by the fungus C. heterostrophus, which affects only corn that is highly susceptible to C. heterostrophus. The other is tabtoxin, a nonspecific toxin produced by several bacteria including Pseudomonas syringae pv. tabaci, a pathogen of tobacco. Tabtoxin itself is probably biologically inactive; its hydrolysis product, tabtoxinine-β-lactam, seems to be the actual toxin (13,57,63). Both of these toxins are virulence factors because neither is required by the producing pathogen to cause disease and each is responsible for only part of the symptoms of the respective diseases (62).

Callus derived from susceptible corn tissues was selected in vitro for resistance to T-toxin (19). Plants were regenerated from the toxin-resistant callus and tested for reaction to both T-toxin and a T-toxin-producing isolate of C. heterostrophus. The regenerated plants, which were highly resistant to T-toxin, were still susceptible to the pathogen but sustained much less damage than the original T-toxin-sensitive plants. Since T-toxin causes only part of the symptoms of disease, plant-resistance to toxin does not confer resistance to symptoms of disease caused by other, unknown, factors produced by the pathogen. In this case, the symptom caused by T-toxin is the economically-important portion of the disease and resistance to T-toxin provides effective disease control under field conditions.

Tabtoxinine-β-lactam also appears capable of selecting plants partially resistant to disease. Tobacco cells were selected for resistance to methionine sulfoximine, a compound which has biological activity similar to that of tabtoxinine-β-lactam, then regenerated into plants and tested for reaction to the bacterium.

The toxin-resistant plants were still susceptible to the bacterium, but did not develop the symptom the toxin causes and sustained less damage than toxin-sensitive plants (5). This observation, coupled with the results with T-toxin-resistance, supports the idea that partial resistance to disease can be predicted if selection involves use of a virulence factor. It also shows that even a nonspecific toxin can select disease resistant plants, provided it actually has a significant role in disease.

Danger of Using Pathologically Spurious Toxins in Selection Programs

Most toxins produced by plant pathogenic fungi and bacteria have not yet been rigorously evaluated for roles in disease. Such toxins cannot be used with confidence to select for disease resistance because they may be neither pathogenicity nor virulence factors and in fact may not be directly involved in disease at all. Although it is difficult to prove a negative, there is evidence that some pathogen-produced toxins are not important in pathogenesis. Victoxinine, a nonspecific toxin from *Helminthosporium* spp., is produced by both pathogenic and nonpathogenic strains (39). Fusaric acid, another nonspecific toxin produced by *Fusarium* spp., also seems unlikely to be pathologically significant; mutants that produce little or no toxin can be highly virulent, and high producers may be weakly virulent (31). Naphthazarine toxins produced by *Nectria haematococca*, a pea pathogen, have been rigorously examined for a role in disease, using genetic analysis of toxin production and pathogenicity (27). Segregation in progeny of crosses between a highly virulent, high toxin-producing strain and a weakly virulent, nontoxin-producing strain showed no correlation between toxin production and pathogenicity or high virulence. The toxins have no apparent role in disease caused by *N. haematococca*. Even though the foregoing toxins are produced by plant pathogens, they are pathologically irrelevant and plants selected for resistance to them would likely be fully susceptible to the respective pathogens themselves.

There are a few documented cases in which use of toxins with no defined pathological significance has failed to provide an adequate test for disease-resistant plants. For example, Straley et al. (55) proposed that a phytotoxic glycopeptide from *Corynebacterium insidiosum*, a bacterial pathogen of alfalfa, could be used to screen populations of alfalfa clones for disease resistance, although the correlation coefficient relating disease-reaction to toxin-sensitivity was only 0.62. When VanAlfen and McMillan (58) reexamined this claim using a more precise bioassay, they found no correlation at all and indicated that the toxin should not be used to screen for disease resistance. Similar circumstances surround the report that amylovorin, a phytotoxic glucan from *Erwinia amylovora*, can be used to screen apple, pear and quince trees for disease resistance (21). It was later found, using a larger sample size, that amylovorin can distinguish between plants which vary in transpiration rate but not between plants susceptible and resistant to disease (2; Beer et al., ms. in preparation).

It should not be assumed that uncharacterized crude culture filtrates from plant pathogens can be helpful in screening or selection programs. For example, Behnke (2a) regenerated potato plants from callus that had been selected for resistance to culture filtrate from the fungal pathogen Phytophthora infestans and tested the plants for resistance to the fungus itself. There was a slight, statistically significant, difference between the mean lesion sizes on selected and control plants, but this may have been fortuitous since the ranges overlapped and there were several clear exceptions to the correlation. The plants were not tested under field conditions. Crude filtrates would not necessarily be expected to select plants resistant to disease since growth media that have been colonized by microorganisms (whether pathogenic or not) contain many secondary metabolites which, in combination, can be phytotoxic but have nothing to do with disease development.

POTENTIAL OF TOXIN-PRODUCING PATHOGENS FOR WEED CONTROL

Precedents For Biological Control of Pests

Although biological control of weeds has had very limited success to date, the approach seems promising because of the striking accomplishments that have been achieved with other pests. Two well-known examples make the point and emphasize the importance of toxins in biological control.

Bacillus thuringiensis has been available for commercial use against certain insects since the 1950s (9). The mechanism by which B. thuringiensis kills insects is complex, but appears to involve the production of several toxins by the bacterium (11). Another well-documented case is the control of Crown Gall disease by Agrobacterium radiobacter (37). Again, control is mediated through a toxin, agrocin 84, which affects the target organism A. tumefaciens. There are many other pathogens which offer potential for pest control; at least some of these are known to produce toxins effective against the target pest (11,42).

Prospects For Genetic Engineering of Weed Pathogens

Use of microorganisms for weed control is an old idea, dating back to the 1800s (61). One approach is to introduce an exotic pathogen into a local population of weeds and hope that it becomes endemic. Another is to apply formulations of pathogen propagules directly to weeds in the same manner as chemical herbicides are used

(56). Neither of these tactics has been generally successful for weed control.

There are many reasons for failure of pathogens to adequately control populations of pests. In the case of weed control, these include poor dissemination of inoculum, spatial isolation of host plants, narrow environmental requirements for disease, and low virulence of the pathogen toward the weed host (56). The latter, low virulence, seems most likely to be amenable to manipulation by the genetic engineering techniques that are now available. As mentioned earlier, certain pathogen-produced toxins are known to increase the virulence of pathogens toward hosts. Specific genes controlling production of some toxins have been identified by conventional genetic analysis (62). If a gene or set of genes controlling production of a toxin could be cloned and transferred to a pathogen that is specific for a target weed, the enhanced virulence could achieve a satisfactory reduction in the weed population that would be environmentally safe and specific for the target organism. Since host-specific toxins for pests such as weeds may be difficult to discover, it may be more promising to focus on the cloning and transfer of genes that control production of nonspecific toxins.

It is anticipated that the transfer of genes in addition to those required for toxin production will be necessary. For example, with nonspecific toxins one or more genes for resistance to the toxin must be included so that the recipient cells do not self-destruct. In the case of phaseolotoxin, this would be straightforward since one mechanism of self-immunity is toxin-insensitive ornithine carbamoyltransferase, which is the site of toxin action (52). Other genes that may be desirable in the construction of a biocontrol agent include those that control traits such as the ability of inoculum to disseminate or the ability of the pathogen to adapt to a wide range of environments.

Advantages of Microorganisms Over Pathogen-Produced Chemicals Alone

The proposed biological control described above is mediated by toxic chemicals. It seems reasonable to predict that control could be achieved by dispensing with the pathogen and applying only the toxin. This would be analogous to the use of a herbicide and in fact may be useful in some cases. However, there are several advantages in applying the toxin-producing microorganism itself. 1) The microorganism may persist in the local population so that frequent reapplications may be unnecessary. 2) The microorganism provides an effective toxin-delivery system. This is important since some toxins do not readily enter intact plants (63); the microorganism ensures not only entry but entry at the proper location. 3) A weed-specific microbial pathogen permits the use of a nonspecific toxin in the construction of a biocontrol agent. As discussed below, there is reason to assume that nonhost plants will not be affected by a heterologous pathogen which produces a nonspecific toxin.

Assessment of Risk

It is important to consider whether or not the transfer of genes for control of toxin production across phylogenetic boundaries constitutes an acceptable risk with respect to the environment. Based on information now available concerning roles of toxins in disease, the risk seems to be very low. Host-specific toxins pose no apparent threat because of their extreme specificity for only those genotypes of plants that are sensitive. In fact, these remarkable substances would never have been discovered were it not for the unique susceptibility of certain plants that have been developed by humans for agricultural purposes (46). Nonspecific toxins require additional investigation. There are two such toxins whose roles in disease have been reasonably well defined (62). Tabtoxinine-β-lactam produces secondary chlorosis in diseases caused by several pathovars of *Pseudomonas syringae* (12). Phaseolotoxin causes chlorosis and ornithine accumulation in beans susceptible to *P. syringae* pv. *phaseolicola* (12). In both cases, the bacteria are highly host-specific even though the toxins are nonspecific. This means that before the toxins can be effective in producing symptoms, the bacteria must first initiate disease. Disease-initiating factors produced by the bacteria, unknown at this time, are necessary prerequisites for expression of symptoms due to toxin.

There is also experimental evidence which bears on the consideration of risk. As described earlier, Carlson (5) selected tobacco plants resistant to methionine sulfoximine, a biological surrogate for tabtoxinine-β-lactam, and compared them with plants that were either susceptible or resistant to *P. syringae* pv. *tabaci*. Plants susceptible to both the bacteria and the toxin developed necrotic lesions with chlorotic halos. Plants susceptible to the bacteria but resistant to the toxin produced necrotic lesions but no halos. Plants susceptible to toxin but resistant to the bacteria sustained no disease at all. Thus, if inoculated with a toxin-producing strain of the bacterium, plants must be susceptible to both the bacterium and the toxin to sustain damage caused by the toxin.

By extrapolation to other host/parasite interactions, it can be predicted that a pathogen specific for a particular pest and carrying a foreign gene for production of a nonspecific toxin will be highly virulent toward the pest but will cause no damage to nonhosts or to other organisms. However, since the available information on this issue is meager, experiments involving transfer of pathologically significant genes from one organism to another should be done under the containment conditions and with the usual precautions that are accorded pathogenic organisms.

For the cloning of Cochliobolus genes that control production of toxins (discussed below), we are using P2 level physical containment as prescribed by the NIH Recombinant DNA Advisory Committee (Federal Register, Vol. 46, No. 126, p. 34486, July 1, 1981). In addition, we are developing biological containment in anticipation of species-to-species transfer of genes for toxin production in the future. As a first precaution, only host-specific toxins are presently under consideration. Secondly, all recipients of recombinant DNA molecules are alb1, which is an allele for albinism (the wild type fungus produces a brown pigment). Isolates bearing alb1 cannot survive in the field, (W.E. Fry and O.C. Yoder, unpublished) presumably because they are known to be very sensitive to ultraviolet light. Another level of biological containment involves use of auxotrophic mutants. A series of field tests has shown that one auxotroph, ins2 (requires inositol), does not survive at all, whereas another, his1 (requires histidine), can maintain itself at a low level over the course of a single growing season (R. C. Garber, W.E. Fry, and O.C. Yoder, unpublished). We hope to find additional mutants which do not survive under field conditions. Thus, it seems possible to achieve adequate biological containment with built-in redundancy by combining alb1 with one or more auxotrophic alleles that debilitate the fungus in nature. On this point, it may be significant that all strains of the fungus isolated from the field and tested so far are prototrophic. This suggests that when auxotrophic mutations occur naturally, selection pressure is against them.

FEASIBILITY OF CLONING GENES FOR TOXIN PRODUCTION

To date there are no reports that genes controlling toxin production by plant pathogenic fungi or bacteria have been cloned, although a gene for production of the crystal protein toxin of the insect pathogen Bacillus thuringiensis has been isolated by virtue of its ability to function in E. coli (49). Other pathologically important genes that have been physically isolated are those on plasmid-borne T-DNA from Agrobacterium tumefaciens (38) and A. rhizogenes (8), and two genes controlling production of the plant hormone indole acetic acid, which are carried on a plasmid harbored by Pseudomonas syringae pv. savastanoi (38). Evidence now available indicates that genes controlling production of the bacterial toxins phaseolotoxin (28), and tabtoxin (40) are not located on plasmids, although the genes for the bacterial toxins syringomycin (20) and coronatine (J. V. Leary, personal communication) may be plasmid-borne . However, transfer of plasmid cloning vehicles into plant pathogenic bacteria has been demonstrated (24,32,38). Therefore, it should be possible to isolate chromosomal genes that control toxin production by complementing toxinless recipient cells with a library of chromosomal DNA fragments prepared from a toxin-producing strain, and then screening transformants for ability to produce toxin.

Although prospects for cloning toxin-production genes from bacteria appear promising, little attention has been directed toward the cloning of genes for toxin production from plant pathogenic fungi. One reason for this is that until recently it has not been possible to introduce cloned DNA into eukaryotic cells by transformation. However, with the development of an efficient transformation system for budding yeast (26) it has now become feasible to attempt development of similar systems for other eukaryotes. So far, fission yeast (1), *Neurospora crassa* (6), *Podospora anserina* (44), and *Chlamydomonas reinhardii* (51) have been transformed, in all cases using methods adapted from those developed for budding yeast. It seems likely that the technology can be extended to plant pathogenic fungi, thereby permitting the cloning and characterization of genes for toxin production and for other pathological traits.

DEVELOPMENT OF *COCHLIOBOLUS* FOR GENE CLONING EXPERIMENTS

Most organisms, as they are collected from nature, are unsuitable for efficient molecular and genetic analyses in the laboratory. Before meaningful work can be done, they must be developed so that it is possible to isolate useful mutants and to effect genetic recombination by sexual or asexual mechanisms (or both). For gene-cloning experiments, several additional tools must be available: a library of DNA fragments in a suitable cloning vehicle, large numbers of protoplasts that are competent for transformation and that will regenerate at high frequency, and one or more cloned genes which can be selected in appropriate recipient cells.

Cochliobolus heterostrophus was chosen as an object for genetic engineering because it has a sexual stage (62) and can also be manipulated asexually (33) and because it produces T-toxin, a linear polyketol (30) that enhances the virulence of the fungus to susceptible corn (62). Our immediate objective is to physically isolate a defined gene, *Tox1* (35), which is a single locus with two alleles that control production of T-toxin, and to then investigate the function of this gene both in *Cochliobolus* spp. and in other organisms. We assume that *Tox1* will not function in easily-transformable organisms such as *E. coli* or yeast and therefore must be isolated by complementation in *Cochliobolus* itself. For this, a transformation system effective with *Cochliobolus* is needed.

Efforts in my laboratory over the last two years have focused on the acquisition of materials required to transform C. heterostrophus. Auxotrophic mutants are available (34) and large numbers (10^8–10^9) of protoplasts can be conveniently prepared (33). The protoplasts can be regenerated at high frequency, they can be lysed for DNA isolation and for preparation of cell-free extracts for enzyme assays, or they can be treated with cloned DNA for transformation. Difficulties were encountered when we attempted to extract DNA from Cochliobolus using published procedures developed for other filamentous fungi. The initial preparations were highly degraded, but we now have protocols that yield good quality DNA from either mycelium or protoplasts.

To determine whether isolated Cochliobolus DNA was functional, it was tested for its ability to act as an origin of replication in yeast. Stinchcomb et al. (53) have described a straightforward procedure for isolating presumptive origins of replication from a variety of eukaryotes. These isolated DNA fragments are currently called ars's (autonomously replicating sequences) because they have been shown to function in yeast but not yet in the organisms from which they came. We used the Stinchcomb procedure to select ars's from Cochliobolus.

Ten μg of YIp5, an E. coli-yeast hybrid plasmid that contains the yeast URA3 gene but no yeast origin of replication, and 30 μg of total Cochliobolus DNA were digested with EcoRI. The two DNA preparations were mixed, ligated with T4 DNA ligase, and half of the ligation mixture was used to transform 8 x 10^8 spheroplasts of yeast strain TD1 ura3-52, which contains a gross rearrangement at the ura3 locus so that plasmids rarely integrate by homology. Transformation was done as described by Hinnen et al. (26), except that 75 μg E. coli chromosomal DNA was added to the reaction mixture; E. coli DNA will not function as ars's in yeast (53). About 750 transformants resulted from transformation with YIp5/Cochliobolus DNA.

Forty transformants were isolated and tested for stability by growing them in complete medium and then transferring to medium with or without uracil. Most of the transformants were unstable although a few were not. DNA was isolated from 10 unstable and 2 stable transformants, fractionated on an agarose (0.8%) gel, Southern blotted, and probed with ^{32}P-labeled YIp5. The probe hybridized to plasmid DNA from all unstable transformants (indicating autonomous replication) and to chromosomal DNA in the stable ones (indicating that the plasmid had integrated into a chromosome). Plasmid DNA from the unstable transformants was propagated in E. coli, digested with EcoRl, and run on an agarose gel along with size standards and EcoRI-digested YIp5. The Cochliobolus ars inserts ranged in size from 1.05 to 7.9 kb.

It is clear that Cochliobolus DNA is of high enough quality to function in yeast. Furthermore, we now have available a collection of defined Cochliobolus DNA sequences, each bearing an ars, which may be useful in the construction of a cloning vehicle.

The vehicle chosen for construction of a library of Cochliobolus DNA fragments was pTR262 (43), which permits easy selection of only those plasmids that receive inserts. The feature of pTR262 relevant to our work is the tetracycline-resistance (tet^R) gene from pBR322, which is under control of λ promoter but cannot function because the promotor is inactivated by the gene for λ repressor. The phenotype of E. coli cells carrying pTR262 is tet^S. However, by cloning into either the unique HindIII or BclI sites in the repressor gene, the repressor gene is inactivated, λ promoter functions, and the tet gene is expressed. Thus, by selecting for tet^R, only cells containing plasmids with inserts are recovered.

The library was prepared from Cochliobolus DNA that was partially digested with HindIII and size-fractionated on a sucrose gradient to give fragments of about 8 kb. Fragments were ligated into the HindIII site of pTR262, and the ligation mixture was used to transform E. coli strain 6507; selection was for tet^R. Plasmid DNA was prepared directly from the colonies that arose on the tetracycline plates and stored as a library of Cochliobolus genomic sequences.

The most likely approach to transformation of Cochliobolus is to complement a Cochliobolus auxotrophic mutant with the corresponding wild type gene isolated from the Cochliobolus library and cloned. We chose to isolate the Cochliobolus gene that codes for β-isopropylmalate dehydrogenase, an enzyme in the leucine biosynthetic pathway, because a Cochliobolus mutant deficient in that enzyme is available. In addition the corresponding yeast gene (LEU2) functions in E. coli (41) and the E. coli gene (leuB) functions in yeast (54). Therefore, it seemed reasonable to expect a Cochliobolus LEU gene to function in either E. coli or yeast or both.

E. coli leuB was transformed with the library of Cochliobolus genomic fragments and selected for leu^+. Two clones that resulted from the transformation were analyzed further. Plasmid isolated from either clone, designated pChLEU2-1 and pChLEU2-2, transformed E. coli leuB to leu^+ at high frequency. Furthermore, either of the plasmids transformed yeast leu2 (which corresponds to leuB in E. coli) to $LEU2^+$ at moderate frequency (100-1000 transformants/μg DNA).

To determine if the plasmids contained sequences of Cochliobolus genomic DNA and not contaminating E. coli chromosomal or yeast sequences, genomic DNA from Cochliobolus, E. coli, and yeast was digested with HindIII, separated on an agarose gel (0.8%), transferred to nitrocellulose paper and probed with ^{32}P-labeled pChLEU2-2. For comparison, pChLEU2-2 was digested with HindIII and run in parallel with the genomic samples. pChLEU2-2 hybridized with a HindIII fragment of Cochliobolus DNA that comigrated on the agarose gel with the HindIII insert in the plasmid. pChLEU2-2 also hybridized with a HindIII fragment of yeast DNA, but the yeast fragment was smaller than the Cochliobolus fragment. There was no hybridization to E. coli chromosomal DNA. These results support the interpretation that pChLEU2 contains a HindIII fragment which bears the Cochliobolus LEU2 gene; the fragment also has homology with the yeast fragment carrying the LEU2 gene. Other experiments showed that the cloned Cochliobolus and yeast LEU2 genes hybridize with each other nearly as well as each with itself.

The mechanism by which pChLEU2-1 and pChLEU2-2 transform yeast was of interest, since neither plasmid has any yeast sequences. There are two possibilities: 1) integration into yeast chromosomal DNA, either by homologous recombination at the yeast Leu2 locus or by spurious recombination elsewhere, and 2) autonomous replication. For the latter to occur, each Cochliobolus LEU2 fragment must also bear an ars, which functions as an origin of replication in yeast. This is not unlikely, since ars's apparently occur frequently in the Cochliobolus genome as they do in yeast (7). To test the two possibilities, we made use of the fact that in yeast autonomously replicating plasmids are unstable under nonselective conditions whereas plasmids integrated into chromosomes are stable (3). Eleven yeast transformants bearing pChLEU2-1 and twelve bearing pChLEU2-2 were grown for about 10 cell-doublings in complete medium, then plated on agar medium either containing or lacking leucine. For pChLEU2-1 there were 0.02% as many colonies on medium lacking leucine as on medium containing leucine; for pChLEU2-2 the value was 3%. Thus, both plasmids are unstable in yeast and appear to replicate autonomously, presumably because the Cochliobolus sequences contain ars's that function in yeast.

Several attempts have been made to transform Cochliobolus with either the cloned Cochliobolus LEU2 gene, yeast ADE genes (described below), genes for resistance to tricodermin (16) and cycloheximide (17) from yeast, or antibiotic G418 from E. coli (29). All attempts at transformation involved modifications of the yeast protocol (26) or the Neurospora crassa protocols (6,50). In general, the procedure is to prepare 10^7-10^8 protoplasts, wash them thoroughly in osmoticum followed by osmoticum containing $CaCl_2$, add DNA (plasmid plus salmon sperm carrier DNA) with or without temperature shock, incubate in the presence of polyethylene glycol 4000, wash in osmoticum, and plate in selective agar regeneration medium. The two basic controls are protoplasts treated with carrier DNA only (no plasmid) and plated in either selective (no colonies expected) or nonselective (many colonies expected) medium. No transformants have resulted from the trials with drug-resistance genes, and we have had difficulty interpreting results of experiments with the Cochliobolus LEU2 gene because the recipient leu2$^-$ cells are leaky, which causes an interferring background on the regeneration plates. Additional tests of LEU2 and drug-resistance genes are currently second in priority to promising experiments with the yeast ADE3 gene.

The rationale for attempting to transform Cochliobolus with yeast ADE genes was 3-fold: 1) the yeast ADE8 gene appears to be highly conserved across phylogenetic boundaries since the Drosophila gene corresponding to it functions in yeast (25), 2) five of the genes in the yeast adenine synthesis pathway have been cloned (25) and 3) we have in our collection eight Cochliobolus ade mutants, at seven different loci, which represent at least half of the steps in the adenine biosynthetic pathway. Therefore, it seems likely that one or more of the cloned yeast genes will complement a Cochliobolus mutant. Dr. Steve Henikoff has provided the five yeast genes that he has cloned (ADE1, ADE3, ADE4, ADE5,7, and ADE8), each in plasmid YEp13.

In a preliminary experiment, the five genes were pooled and used in an attempt to transform each of six of the Cochliobolus ade mutants, using a modified yeast procedure. No colonies were recovered on regeneration plates of five of the mutants but on plates of the sixth several dozen slow-growing colonies appeared, which achieved diameters of one-two cm. Next, the mutant that responded to transformation, Cochliobolus ade3, was transformed with each of the yeast ADE plasmids individually. Colonies appeared only on plates that received Cochliobolus ade3 cells treated with yeast ADE3 DNA. This experiment has been done five times using several variations in protocol. In four of the five trials, colonies appeared after treatment with DNA but not in the controls with no DNA treatment. In all experiments, colonies were transferred to media either containing or lacking adenine. However, none of the colonies tested so far have been prototrophic as expected; they have all retained the ade3$^-$ phenotype, and thus appear to be abortive transformants.

One possible explanation for the ephemeral nature of the apparent transformants is that the yeast ADE3 gene functions in Cochliobolus cells but the vehicle that carries the gene (YEp13) cannot be maintained there. It cannot replicate itself autonomously because it lacks a Cochliobolus origin of replication. Integration into chromosomal DNA is unlikely because the plasmid has no homology with Cochliobolus chromosomes. If this explanation is correct, it should be possible to obtain transformants that can be maintained by constructing recombinant plasmids containing the yeast ADE3 gene and a Cochliobolus ars sequence. Such a plasmid would have two possible mechanisms for maintenence in recipient cells. If the ars functions as an origin of replication in Cochliobolus, it may permit autonomous replication of the plasmid. If not, the ars sequence will provide homology with Cochliobolus chromosomal DNA so that integration by homologous recombination will be possible. At the time of this writing the recombinant plasmids are being prepared but have not yet been analyzed.

After a transformation system, including an efficient cloning vehicle, has been developed for Cochliobolus it will be possible to clone and characterize the genes that control pathogenicity and virulence in this fungus. The first gene we wish to clone is Tox1, (35) which controls production of T-toxin. The experience, expertise, and technology acquired during the cloning of this gene will then be turned to other genes and other fungi of economic importance. There are two long-range goals. One is to identify and characterize all of the genes that a fungus needs to persist as a pathogen in nature. The other is to develop technology that will aid the genetic engineering of fungi for use in agriculture and industry.

ACKNOWLEDGEMENTS

Work from the author's laboratory was supported by grants 75002 from the Rockefeller Foundation, 78-59-2361-0-1-004-1 from the US Dept. Agriculture Competitive Research Grants Office, and PCM-8104104 from the National Science Foundation. Karen Kindle, Gillian Turgeon, Jeff Stein, Sharon Powell, and Ann Whitney were involved in performing the recombinant DNA experiments. Dr. G.R. Fink provided technology and advice to initiate the molecular biological investigation of Cochliobolus.

REFERENCES

1. Beach, D., and P. Nurse. 1981. High-frequency transformation of the fission yeast Schizosaccharomyces pombe. Nature 290:140-142.

2. Beer, S.V., and H.S. Aldwinckle. 1976. Lack of correlation between susceptibility to Erwinia amylovora and sensitivity to amylovorin in apple cultivars. Proc. Amer. Phytopathol. Soc. 3:300.

2a. Behnke, M. 1980. General resistance to late blight of Solanum tuberosum plants regenerated from callus resistant to culture filtrates of Phytophthora infestans. Theor. Appl. Genet. 56:151-152.

3. Botstein, D., C.S. Falco, S.E. Stewart, M. Brennan, S. Scherer, D.T. Stinchcomb, K. Struhl, and R.W. Davis. 1979. Sterile host yeasts (SHY): A eukaryotic system of biological containment for recombinant DNA experimetns. Gene 8:17-24.

4. Brettell, R.I.S., and D.S. Ingram. 1979. Tissue culture in the production of novel disease-resistant crop plants. Biol. Rev. 54:329-345.
5. Carlson, P.S. 1973. Methionine sulfoximine-resistant mutants of tobacco. Science 180:1366-1368.
6. Case, M.E., M. Schweizer, S.R. Kushner, and N.H. Giles. 1979. Efficient transformation of Neurospora crassa by utilizing hybrid plasmid DNA. Proc. Natl. Acad. Sci. 76:5259-5263.
7. Chan, C.S.M., and B.K. Tye. 1980. Autonomously replicating sequences in Saccharomyces cerevisiae. Proc. Natl. Acad. Sci. 77:6329-6333.
8. Chilton, M.-D., D.A. Tepfer, A. Petit, C. David, F. Case-Delbart, and J. Tempe. 1982. Agrobacterium rhizogenes inserts T-DNA into the genomes of the host plant root cells. Nature 295:432-434.
9. Couch, T.L., and C.M. Ignoffo. 1981. Formulation of insect pathogens. In Microbial Control of Pests and Plant Diseases, 1970-1980, pp. 621-634. H.D. Burges, ed. London: Academic Press. 949 pp.
10. Dimond, A.E., and P.E. Waggoner. 1953. On the nature and role of vivotoxins in plant disease. Phytopathology 43:229-235.
11. Dulmage, H.T. 1981. Insecticidal activity of isolates of Bacillus thuringiensis and their potential for pest control. In Microbial Control of Pests and Plant Diseases, 1970-1980, pp. 193-222. H.D. Burges, ed. London: Academic Press. 949 pp.
12. Durbin, R.D., ed. 1981. Toxins in Plant Disease. New York: Academic Press. 515 pp.
13. Durbin, R.D., T.F. Uchytil, J.A. Steele, and R.L.D. Ribeiro. 1978. Tabtoxinine-β-lactam from Pseudomonas tabaci. Phytochemistry 17:147.
14. Earle, E.D. 1978. Phytotoxin studies with plant cells and protoplasts. In Frontiers of Plant Tissue Culture 1978, Proc. 4th Int. Cong. Plant Tiss. Cell Cult., Calgary, pp. 363-372. T.A. Thorpe, ed.
15. Earle, E.D. 1981. Application of plant tissue culture in the study of host-pathogen interactions. In Applications of Plant Cell and Tissue Culture to Agriculture and Industry, pp. 45-62. D.T. Tomes, B.E. Ellis, P.M. Harney, K.J. Kasha, and R.L. Peterson, eds. Univ. Guelph Press.
16. Fried, H.M. and J.R. Warner. 1981. Cloning of yeast gene for trichodermin resistance and ribosomal protein L3. Proc. Natl. Acad. Sci. 78:238-242.
17. Fried, H.M., and J.R. Warner. 1982. Molecular cloning and analysis of yeast gene for cycloheximide resistance and ribosomal protein L29. Nucl. Acid Res. 10:3133-3148.
18. Galston, A.W. 1974. Molding new plants. Natural History 83:94-96.
19. Gengenbach, B.G., C.E. Green, and C.M. Donovan. 1977. Inheritance of selected pathotoxin resistance in maize plants

regenerated from cell cultures. Proc. Natl. Acad. Sci. 74:5113-5117.
20. Gonzalez, C.F., and A.K. Vidaver. 1979. Syringomycin production and holcus spot disease of maize: plasmid-associated properties in Pseudomonas syringae. Current Microbiol. 2:75-80.
21. Goodman, R.N., P.R. Stoffl, and S.M. Ayers. 1978. The utility of the fire blight toxin, amylovorin, for the detection of resistance of apple, pear, and quince to Erwinia amylovora. Acta Hort. 86:51-56.
22. Gracen, V.E., M.J. Forster, K.D. Sayre, and C.O. Grogan. 1971. Rapid method for selecting resistant plants for control of southern corn leaf blight. Plant Dis. Rep. 55:469-470.
23. Graniti, A. 1972. The evolution of the toxin concept in plant pathology. In Phytotoxins in Plant Disease, pp. 1-18. R.K.S. Wood, A. Ballio, and A. Graniti, eds. New York: Academic Press. 530 pp.
24. Gross, D.C., and A.K. Vidaver. 1981. Transformation of Pseudomonas syringae with nonconjugative R plasmids. Can. J. Microbiol. 27:759-765.
25. Henikoff, S., K. Tatchell, B.D. Hall, and K.A. Nasmyth. 1981. Isolation of a gene from Drosophila by complementation in yeast. Nature 289:33-37.
26. Hinnen, A., J.B. Hicks, and G.R. Fink. 1978. Transformation of yeast. Proc. Natl. Acad. Sci. 75:1929-1933.
27. Holenstein, J.E. 1982. On the transmission of naphthazarine production and pathogenesis in Nectria haematococca Berk. et Br. Ph.D. Thesis, Confederate Technical University, Zurich (transl. from German by Leo Kanner Assoc., P.O. Box 5187, Redwood City, CA 94063). 48 pp.
28. Jamieson, A.F., R.L. Bielesky, and R.E. Mitchell. 1981. Plasmids and phaseolotoxin production in Pseudomonas syringae pv. phaseolicola. J. Gen. Microbiol. 122:161-165.
29. Jimenez, A., and J. Davies. 1980. Expression of a transposable antibiotic resistance element in Saccharomyces. Nature 287:869-871.
30. Kono, Y., H.W. Knoche, and J.M. Daly. 1981. Structure: fungal host-specific. In Toxins in Plant Disease, pp. 221-257. R.D. Durbin, ed. New York: Academic Press. 515 pp.
31. Kuo, M.S., and R.P. Scheffer. 1964. Evaluation of fusaric acid as a factor in development of Fusarium wilt. Phytopathology 54:1041-1044.
32. Lacy, G.H., and R.B. Sparks, Jr. 1979. Transformation of Erwinia herbicola with plasmid pBR322 deoxyribonucleic acid. Phytopathology 69:1293-1297.
33. Leach, J., and O.C. Yoder. 1982. Heterokaryosis in Cochliobolus heterostrophus. Exp. Mycol. (in press).
34. Leach, J., B.R. Lang, and O.C. Yoder. 1982. Methods for selection of mutants and in vitro culture of Cochliobolus heterostrophus. J. Gen. Microbiol. 128 (in press).

35. Leach, J., K.J. Tegtmeier, J.M. Daly, and O.C. Yoder. 1982. Dominance at the Tox1 locus controlling T-toxin production by Cochliobolus heterostrophus. Physiol. Plant Pathol. 21 (in press).
36. Luke, H.H., H.E. Wheeler, and A.T. Wallace. 1960. Victoria-type resistance to crown rust separated from susceptibility to Helminthosporium blight in oats. Phytopathology 50:205-209.
37. Moore, L.W., and G. Warren. 1979. Agrobacterium radiobacter strain 84 and biological control of crown gall. Ann. Rev. Phytopathol. 17:163-179.
38. Nester, E.W., and T. Kosuge. 1981. Plasmids specifying plant hyperplasias. Ann. Rev. Microbiol. 35:531-565.
39. Nishimura, S., R.P. Scheffer, and R.R. Nelson. 1966. Victoxinine production by Helminthosporium species. Phytopathology 56:53-57.
40. Piwowarski, J.M., and P.D. Shaw. 1982. Characterization of plasmids from plant pathogenic Pseudomonads. Plasmid 7:85-94.
41. Ratzkin, B., and J. Carbon. 1977. Functional expression of cloned yeast DNA in Escherichia coli. Proc. Natl. Acad. Sci. 74:487-491.
42. Roberts, D.W. 1981. Toxins of entomopathogenic fungi. In Microbial Control of Pests and Plant Diseases 1970-1980, pp. 441-464. H.D. Burges, ed. London: Academic Press. 949 pp.
43. Roberts, T.M., S.L. Swanberg, A. Poteete, G. Riedel, and K. Backman. 1980. A plasmid cloning vehicle allowing a positive selection for inserted fragments. Gene 12:123-127.
44. Rochaix, J.-D., and J. Van Dillewijn. 1982. Transformation of the green alga Chlamydomonas reinhardii with yeast DNA. Nature 296:70-72.
45. Rudolph, K. 1976. Non-specific toxins. In Encyclopedia Plant Physiology, New Ser., Vol. 4, Physiological Plant Pathology, pp. 270-315. R. Heitefuss and P.H. Williams, eds. New York: Springer-Verlag. 888 pp.
46. Scheffer, R.P. 1976. Host-specific toxins in relation to pathogenesis and disease resistance. In Encyclop. Plant Physiology, New Ser., Vol. 4, Physiological Plant Pathology, pp. 247-269. R. Heitefuss and P.H. Williams, eds. New York: Springer-Verlag. 888 pp.
47. Scheffer, R.P., and R.B. Pringle. 1967. Pathogen-produced determinants of disease and their effects on host plants. In The Dynamic Role of Molecular Constituents in Plant-Parasite Interaction, pp. 217-236. C.J. Mirocha and I. Uritani, eds. St. Paul, Minn.: Bruce. 372 pp.
48. Scheffer, R.P., and O.C. Yoder. 1972. Host-specific toxins and selective toxicity. In Phytotoxins in Plant Diseases, pp. 251-272. R.K.S. Wood, A. Ballio, and A. Graniti, eds. New York: Academic Press. 530 pp.

49. Schnepf, H.E., and H.R. Whiteley. 1981. Cloning and expression of the Bacillus thuringiensis crystal protein gene in Escherichia coli. Proc. Natl. Acad. Sci. 78:2893-2897.

50. Schweizer, M., M.E. Case, C.C. Dykstra, N.H. Giles, and S.R. Kushner. 1981. Identification and characterization of recombinant plasmids carrying the complete qa gene cluster from Neurospora crassa including the qa-1+ regulatory gene. Proc. Natl. Acad. Sci. 78:5086-5090.

51. Stahl, U., P. Tudzynski, U. Kuck, and K. Esser. 1982. Replication and expression of a bacterial-mitochondrial hybrid plasmid in the fungus Podospora anserina. Proc. Natl. Acad. Sci. 79:3641-3645.

52. Staskawicz, B.J., N.J. Panopoulos, and Hoogenraad. 1980. Phaseolotoxin-insensitive ornithine carbamoyltransferase of Pseudomonas syringae pv. phaseolicola: Basis for immunity to phaseolotoxin. J. Bacteriol. 142:720-723.

53. Stinchcomb, D.T., M. Thomas, J. Kelly, E. Selker, and R.W. Davis. 1980. Eukaryotic DNA segments capable of autonomous replication in yeast. Proc. Natl. Acad. Sci. 77:4559-4563.

54. Storms, R.K., E.W. Holowachuck, and J.D. Friesen. 1981. Genetic complementation of the Saccharomyces cerevisiae leu2 gene by the Escherichia coli leuB gene. Molec. Cell. Biol. 1:836-842.

55. Straley, C.S., M.L. Straley, and G.A. Strobel. 1974. Rapid screening for bacterial wilt resistance in alfalfa with a phytotoxic glycopeptide from Corynebacterium insidiosum. Phytopathology 64:194-196.

56. Templeton, G.E., D.O. TeBeest, and R.J. Smith, Jr. 1979. Biological weed control with mycoherbicides. Ann. Rev. Phytopathol. 17:301-310.

57. Uchytil, T.F., and R.D. Durbin. 1980. Hydrolysis of tabtoxins by plant and bacterial enzymes. Experientia 36:301-302.

58. VanAlfen, N.K., and B.D. McMillan. 1982. Macromolecular plant-wilting toxins: Artifacts of the bioassay method? Phytopathology 72:132-135.

59. Wheeler, H.E., and H.H. Luke. 1955. Mass screening for disease-resistant mutants in oats. Science 122:1229.

60. Wheeler, H.E., and H.H. Luke. 1963. Microbial toxins in plant disease. Ann. Rev. Microbiol. 17:223-242.

61. Wilson, C.L. 1969. Use of plant pathogens in weed control. Ann. Rev. Phytopathol 7:411-434.

62. Yoder, O.C. 1980. Toxins in pathogenesis. Ann. Rev. Phytopathol. 18:103-129.

63. Yoder, O.C. 1981. Assay. In Toxins in Plant Disease, pp. 45-78. R.D. Durbin, ed. New York: Academic Press. 515 pp.

ASPECTS OF SALT AND DROUGHT TOLERANCE IN HIGHER PLANTS

R.G. Wyn Jones and J. Gorham

Department of Biochemistry and Soil Science
University College of North Wales, Bangor
Gwynedd LL57 2UW, Wales

ABSTRACT

In relation to salt tolerance in higher plants it may be suggested that two fundamental criteria must be fulfilled: (i) the maintenance of a cytoplasmic, ionic composition suitable for metabolic activity, particularly in apical meristems; and (ii) the maintenance of adequate turgor pressure by at least partial osmotic adjustment. A wide range of morphological/anatomical and physiological/biochemical characteristics observed in halophytes can be interpreted as contributing to their fulfillment. Recent data showing the extent of solute compartmentation in *Suaeda maritima* is described supporting this analysis. Cellular adaptations are only part of the total mechanism by which tolerance is acquired although these may be the more accessible to genetic manipulation. At the cellular level the accumulation of compatible cytosolutes is accepted to be of major importance although membrane transport must be equally so. Experiments are described comparing the effects of exogenous and endogenous proline and exogenous glycinebetaine on salt-stressed barley embryos and showing an interaction between organic solute and ion transport. The relevance of the data to the introduction of tolerance into cereals is assessed. The possible use of glycinebetaine in screening procedures and the potential significance of two different biosynthetic pathways in barley and in members of the Chenopodiaceae will be discussed briefly.

INTRODUCTION

In this paper some physiological and biochemical aspects of the salt and, to a minimal extent, drought tolerance of higher plants will be considered in relation to the fundamental aim of introducing either by conventional breeding or by genetic engineering enhanced tolerance into commercially important crops.

At the outset it is important to recognize that the ability of a crop plant to give a good yield in a particular environment is the summation of many aspects of its metabolism. In an agricultural context the attainment of an adequate yield of the crop and, of course, an economic return is the major, perhaps even the only relevant criterion. The degree of salt tolerance of the vegetative growth of the plant in question may be therefore only one factor in the overall equation. A high degree of tolerance to a saline environment if coupled to a very slow growth rate or poor harvest index may be of little economic value. However, if initially tolerance is considered in a rather less demanding sense, specifically the ability to grow in highly saline conditions, then there are many features which have been found to be associated, albeit loosely, with tolerance. In Figure 1 an attempt is made to summarize some of these characters in a formalized diagram and to divide them somewhat arbitrarily into anatomical/morphological and physiological/biochemical features. Not all these characters are found in a single halophyte and some are indeed probably mutually exclusive. Nevertheless this formalized diagram serves to emphasize the variety and complexity of the processes at the cellular, tissue, and whole plant level that are involved.

Two important problems are posed immediately by this variety. In the face of such complexity can any useful generalizations be made about the mechanisms of tolerance and can these in turn direct us to rational breeding criteria or allow us to deduce which particular characters need to be transferred into commercially important plants in order to confer tolerance?

Previously we have suggested (1,2) that two basic requirements need to be fulfilled at the cellular level and that the features represented in Figure 1 may be interpreted as contributing to them. These are: (i) the maintenance of a cytoplasmic ionic composition compatible with metabolic activity especially in the immature growing cells of the apical regions. The cytoplasmic composition is suggested to consist of a relatively high K^+ level (*circa* 150 mM), a partial exclusion Na^+ and Cl^- and a total ionic strength which does not exceed about 200-250 mM (350-500 mOsmol kg^{-1}). The cells must also be supplied with adequate levels of carbon, nitrogen, phosphate, and other nutrients. (ii) The maintenance of an adequate turgor pressure in mature and immature cells alike by solute accumulation, although some indirect contribution by an adjustment in the threshold turgor pressure (P_{th}) or elongation modulus (E) cannot be excluded.

The reconciliation of these basic criteria requires that osmotic adjustment in the *cytoplasmic* phase be achieved by the

accumulation of non-toxic, compatible, organic solutes, although a wide range of solutes including NaCl can be used as vacuolar osmotica. This hypothesis can be expressed in an idealized model of solute compartmentation in plant cells shown in Figure 2.

Solute Compartmentation

This formalized model, expressed in Figs. 1 and 2, is crucial to our thinking about the problem of salt tolerance and to attempts to develop a rational approach to genetic manipulation. Since this model was originally based largely on indirect evidence and theoretical considerations, it is critically important to establish a more secure experimental base for it. At the recent International Botanical Congress, Storey, Pitman, and Stelzer (5) reported an X-ray microprobe analytical study of apical and mature cells of the halophyte, *Atriplex spongiosa* which strongly supported the model. Recently we have applied the same technique to the extreme halophyte *Suaeda maritima* and combined with it semi-microchemical analyses of small segments from the apical meristem which allowed the quantitative determination of the major cations K, Na, and the major compatible solute, glycinebetaine, in this region of increasing cell extension and vacuolation. Characteristic X-ray energy dispersive spectra from, respectively, an apical cell and the vacuole of a mature mesophyll cell are shown in Figure 3. Even taking into account the multiplication factor of 3 which must be applied to Na, the selective accumulation of K, P, and S in the dividing, cytoplasm-rich cells is very apparent and contrasts with the exclusive accumulation of Na and Cl in the vacuoles of mature cells. The percentage of the total counts found for the various elements, an approximate value for the total counts, and the comparative chemical analyses for the same plant are given in Table 1.

Several points emerge clearly. While the K:Na ratio in vacuoles of the cells of older leaves is slightly lower than that found in the whole tissues by chemical analysis, the K:Na ratio observed in the leaf primordia indicates a tenfold greater discrimination for K^+ in this region. The total X-ray signal also suggests that the total ionic strength may be lower in the primordium and, thus, it may be assumed that to maintain osmotic equilibrium this tissue accumulates an organic solute. The energy dispersive X-ray technique does not lend itself to precise quantitation as the absolute signal received is subject to various sources of error but, nevertheless, it provides excellent comparative data. These trends, revealed in Table 1, were confirmed and supplemented by semi-micro chemical analysis of the K, Na, and glycinebetaine contents of segments dissected from the apex of *S. maritima* (Table 2). Glycinebetaine was determined by a new more sensitive technique recently developed by Dr. Gorham (8). These data show also much higher K:Na discrimination in the apices than in mature tissue; the former also having very high glycinebetaine levels (in excess of 200 µmol g^{-1} fresh weight). Since photomicrographic examination of this tissue showed there to be substantial vacuolation even in the smallest samples we were able to analyze and since glycinebetaine has previously been shown to be partially excluded from vacuoles (9), it is probable that the free glycinebetaine concentration in the cytoplasms of apical cells exceeds 400 mol m^{-3}. These data will be reported in detail elsewhere (10). This evidence, together with all the previously assembled data, suggests that the cellular model of tolerance outlined earlier is fairly reliable and, although by no means supplying

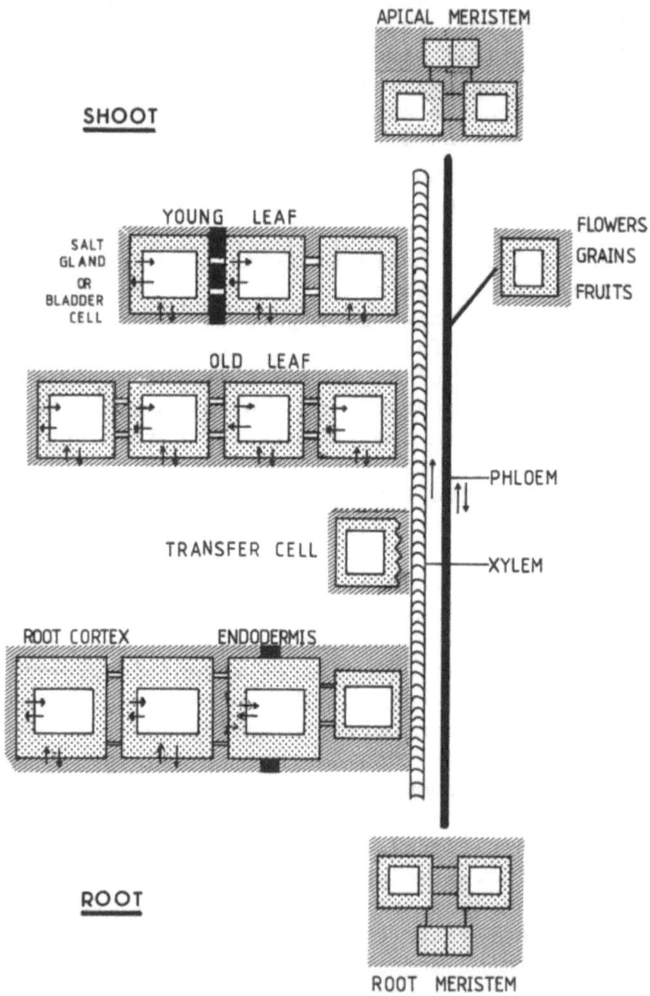

Fig. 1. Formalized schematic diagram of a plant showing some of the characters associated with salt tolerance (1,2).

LEGEND: ANATOMICAL/MORPHOLOGICAL FEATURES

ROOTS :	FEWER CORTICAL CELLS	(I)
	SECOND ENDODERMIS	(II)
ROOT STEM :	SALT EXTRACTING XYLEM PARENCHYMA	(III)
SHOOTS :	SALT EXCRETION : SALT GLANDS BLADDER CELLS	(IV)
	DIFFERENTIAL SALT RETENTION IN OLDER LEAVES - LEAF FALL C-4-LIKE ANATOMY.	

BIOCHEMICAL/PHYSIOLOGICAL CHARACTERISTICS

ROOT :	PLASMALEMMA K^+ SELECTIVE INFLUX) Na^+ SELECTIVE EFFLUX)	(A)
	TONOPLAST VACUOLAR INFLUX Na^+ K^+) EFFLUX K^+ Na^+)	(B)
	COMPATIBLE SOLUTE COMPARTMENTATION	
XYLEM LOADING :	GLYCOPHYTE K^+ Na^+; Cl^- EXCLUSION) HALOPHYTE Na^+ K^+) (High Na physiotype))	(C)
XYLEM TRANSPORT:	SELECTIVE SALT EXTRACTION BY XYLEM PARENCHYMA	(D)
LEAF CELLS :	PLASMALEMMA/TONOPLAST CYTOSOLIC SELECTIVITY FOR K^+, $H_2PO_4^-$ AND COMPATIBLE SOLUTE VACUOLAR SELECTIVITY FOR Na^+ Cl^-	(E)
APICES :	PREFERENTIAL ACCUMULATION OF CYTOSOLUTES	(F)
	RAPID VACUOLATION?	(F)
GAS EXCHANGE :	TIGHT CONTROL OF H_2O LOSS/CO_2 GAIN - C-4 AND CAM.	
TRANSPORT :	SOME RE-EXPORT OF NaCl?	(G)
	LOW SALT LOAD (NaCl) TO FLOWERS, GRAIN AND FRUITS	(H)

1. The list of characters is not complete and not all are shown in any one species, indeed some are mutually exclusive as they are characteristic of different adaptive mechanisms.

2. See Greenway and Munns (3), Wyn Jones (1), and Poljakoff-Mayber (4) for references.

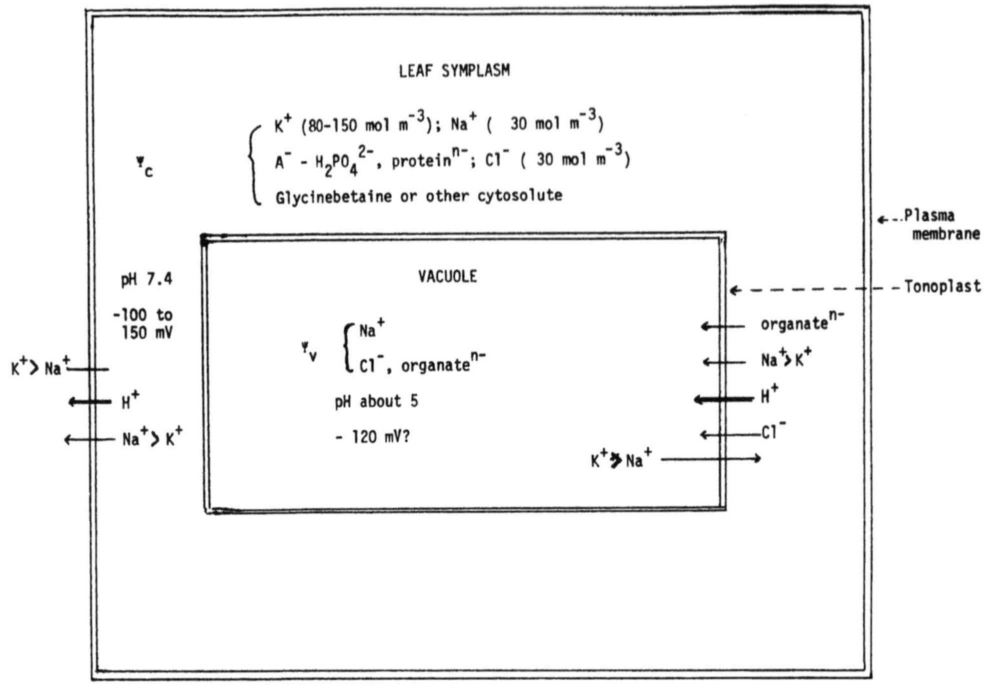

Fig. 2. Idealized model of solute compartmentation in higher plant cells. Values quoted are intended to indicate a hypothetical "norm." $\Psi_v = \Psi_c$; Ψ_v mainly due to Ψ^πNa salts; Ψ_c mainly due to Ψ^πK salts plus Ψ^π cytosolutes. Based on data from refs. 2,6, and 7.

all the answers, can be considered as a basis for developing rational criteria for breeding for enhanced tolerance by conventional or unconventional means.

It must however be emphasized that many critical questions remain unanswered, even unexplored: for example, whether cytoplasmic K^+:Na^+ discrimination may be relaxed as cells mature and decrease in their biosynthetic activity and whether the supply of large quantities of a nitrogen containing compatible solute to a growing apex may present a major problem in terms of C and N allocation. It is also important to establish whether the ability to utilize NaCl as the major vacuolar solute is a significant factor in facilitating rapid growth in saline habitats. The latter point will be discussed later.

Proline in Cultured Barley Embryos

It appears probable that at the cellular level, both specific ion transport characteristics and the ability to accumulate a

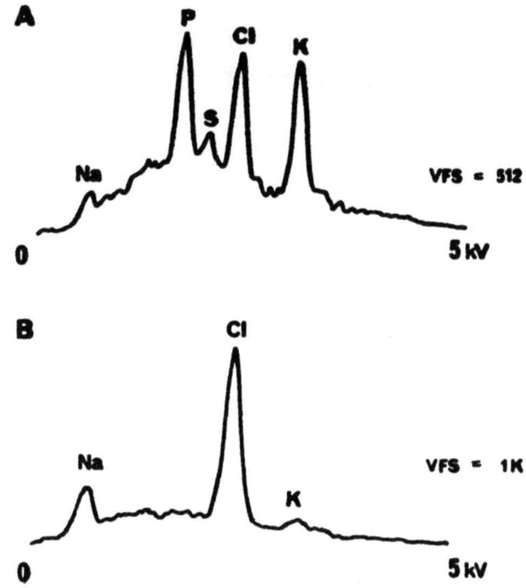

Fig. 3. Characteristic energy dispersive X-ray spectra of preparations of Suaeda maritima grown at 300 mol m^{-3} NaCl. (A) Leaf primordium cell (cytoplasm). (B) vacuole of mature mesophyll cell.

compatible organic cytosolute are important in tolerance. However since our biochemical knowledge of the former is so restricted attention has centered largely by default on the latter, particularly the possible manipulation of the biosynthetic capacity for proline (see Strøm et al., this volume). In order to gain additional knowledge on the value or otherwise of high proline accumulation, we in Bangor have studied the relationship of the accumulation of this solute to salt tolerance in barley embryos, both in wild type barley, cv. Maris Mink, and in high proline mutants isolated from this cultivar. In this work, being carried out by Iqbal Pervez, we have been collaborating with Dr. Simon Bright and Mr. Joseph Kueh in Dr. Miflin's Department at Rothamsted who isolated the high proline mutants (see Miflin and Bright, this volume).

Early studies on the exogenous application of choline and glycinebetaine to plants, while showing an enhanced salt tolerance, highlighted the major problems of microbial contamination associated with these experiments. These difficulties have been resolved by resorting to axenic embryo culture which is also a much more flexible experimental system. Dealing initially with work on proline, comparisons have been made between the effects of exogenous proline added to the agar medium on the tolerance of wild type Maris Mink embryos and the tolerance of embryos from the high proline strains selected for azetidine resistance by our colleagues in

Table 1. X-ray energy[1] dispersive and analytical chemical data for tissues of Suaeda maritima grown in 300 mol m^{-3} NaCl.

Tissue (no. replicates)	K	Na	Cl	P	S	Ratio K/Na	Total counts (K,Na,Cl,P)
	Element Analysed % of total counts						
Leaf Primordium (12)	25	21	24	17	4	1.19	4,250
Young leaf							
Epidermis (6)	14	31	43	3	0	0.45)	18,000
Mesophyll (11)	17	32	35	4	5	0.53)	
Old leaf							
Epidermis (12)	5	47	39	0	0	0.11)	
Palisade (13)	5	51	35	0	0	0.10)	22,000
Mesophyll (8)	3	46	42	0	0	0.06)	
Chemical analysis of same plants (whole shoots)							
Sap (mmol kg^{-1})	75±11	476±25	376±36	-	-	0.16	
Extract (mmol kg^{-1}FW)	62±7	460±24	404±7	-	-	0.15	

[1] Mean counts (60 sec) corrected to silicon standard.

Rothamsted. Difficulties were initially experienced in correlating the results obtained in Bangor and Rothamsted and these prompted us to study the effects of different containers on the responses of wild type barley embryos to proline. Summarizing these data, it was observed that the shoots of embryos grown in petri dishes were more affected by an equivalent salt stress than those grown in crystallizing dishes and that proline (10 mol m^{-3}) only ameliorated the effects of salt in the latter (Figure 4). An examination of ion contents of the shoots from these experiments suggests the following explanation (Fig. 4). In the crystallizing dishes the shoots were able to grow for 6-7 days before making contact with either the sides of the dishes or the agar medium, whereas in the shallow petri dishes this was not possible. It has been established for many years that in barley discrimination at the root symplasm-xylem boundary is an important, possibly critical, factor in determining the shoot salt load, and hence the salt tolerance of different cultivars (11,12). The contact between the leaves and the agar and/or salinized water on the sizes of the shallow vessels would then short-circuit that plant's own defense mechanisms. These data show that proline in the culture medium is principally promoting tolerance by decreasing the export of salt from the roots to the shoots and is apparently inactive if this mechanism is short-

Table 2. K, Na, and glycinebetaine contents of tissues of increasing age and size from Suaeda maritima grown at 150 mol m^{-3} NaCl.

Tissue	K	Na	K/Na	GLYCINEBETAINE[2]
	Solute concentration[1] (mmol kg^{-1} fresh weight)			
Shoot apex (<2 mm)	262	225	1.16	237
Axillary buds	211	379	0.56	168
Leaf (2-5 mm)	385	318	0.89	125
Leaf (5-10 mm)	174	310	0.58	86
Leaf (10-20 mm)	209	314	0.67	43
Leaf (>20 mm)	189	490	0.40	20
Stem	50	412	0.12	14

[1] Standard error <10%, values presented are the means of 3 replicates.

[2] determined as described in Gorham et al. 1982.

circuited. The magnitude of the growth response to proline in the salinized conditions increased up to 10 mol m^{-3} proline but not thereafter.

It is of interest to compare the response of wild type Maris Mink with that of the high proline mutants derived from it. Some comparative data are shown in Table 3. Two major points emerged. Although the constitutive proline content of the mutant is about 2.5 times that of the wild type shoots, this concentration was low compared with that found in the shoots of the embryos fed exogenous proline. Proline levels in the latter were similar to the concentrations found in the leaves of proline-accumulating halophytes, e.g., Puccinellia species, and Triglochin maritima (13,14).

A second significant observation is that the elevated proline content of the mutant is not fully expressed under the stressed conditions. These experiments indicate that the proline enhancement achieved in the mutants currently isolated is insufficient to induce any large improvement in tolerance (see also 15). Since even endogenous proline had little effect on the embryos grown in petri dishes and the solute appeared to have its effect on the embryos grown in crystallizing dishes by influencing xylem-loading, there is little to indicate that even a further 10- or 20-fold increase in shoot proline would of itself increase tolerance. It is worth

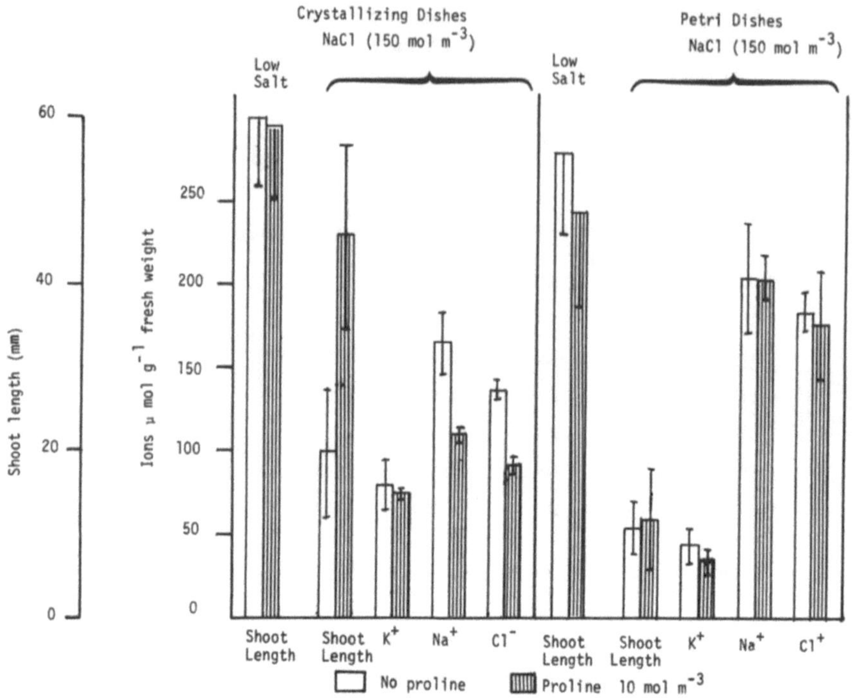

Fig. 4. Comparison of the effect of NaCl (150 mol m^{-3}) stress on the growth and ion contents of barley embryos grown in crystallizing and petri dishes in the presence and absence of proline (10 mol m^{-3}). (Unpublished data of Iqbal Pervez.)

noting that these experiments further underline the importance of plant anatomy, morphology, and tissue physiology in salt tolerance.

Glycinebetaine Accumulation

In view of evidence linking glycinebetane accumulation to salt tolerance in some higher plants (3,14), we have also examined the effects of this solute in the embryo culture system. Glycinebetaine at 10 mol m^{-3} consistently increased the salt tolerance of barley embryos although usually to a lesser extent than proline (Fig. 5). The glycinebetaine content of the shoots was also about 60 µmol g^{-1} fresh weight (cf. proline levels in Table 3) and this content is similar to that found in halophytic species (14). In an attempt at a solute control experiment, glycine (10 mol m^{-3}) was added to the agar medium but did not increase the salt tolerance of the cultured embryos.

Table 3. A comparison of the effects of exogenous (10 mol m^{-3}) and enhanced endogenous proline on wild type Maris mink barley and high proline mutant (5201) subjected to salt stress (2).

	Maris Mink		5201	
Growth condition	Shoot growth (mm)	Proline (μmol g^{-1}FW)	Shoot growth (mm)	Proline (μmol g^{-1}FW)
Low salt	60±12	0.82±0.06	54±8	2.1±0.4
High salt 150 mol m^{-3} NaCl [1]	38±4	6±1	39±3	6.2±1
plus proline at				
1 mol m^{-3}	41±3	13±1		
3 mol m^{-3}	37±5	26±5		
10 mol m^{-3}	52±5	61±12		

[1] Unpublished data of Iqbal Pervez

An examination of the ion contents of the shoots from the glycinebetaine experiments suggested a significant difference in the physiological effects of the betaine and proline. Glycinebetaine loading had little effect of K^+, Na^+, and Cl^- content of the shoots and it appears unlikely that the enhanced tolerance can be attributed to a decrease in the shoot NaCl burden. In this respect the effect of glycinebetaine, although smaller than that of proline, may be more promising as it resembles a more typically halophytic behavior, i.e., growth in the presence of a high salt load in the shoots. In other studies we have reported that glycinebetaine-loading of barley roots increased the preferential retention of Na^+ in the vacuoles (16). Perhaps the enhanced tolerance noted in the shoots of barley embryos may also be due to a similar phenomenon.

These data provide further support for the quite substantial prima facia case now established linking glycinebetaine accumulation to salt tolerance in certain families and tribes (3,14). It is equally clear as discussed in the first part of this paper that this is not the only factor involved. Nevertheless the definitive identification of one factor even in a limited range of species would be a significant advance and unambiguous proof will require genetic studies using high and low betaine lines. It is also worth emphasizing that the recent studies of Hanson and his colleagues (17,18) also suggest, albeit tentatively, that a positive relationship may exist between glycinebetaine accumulation and drought tolerance in barley. Both in Bangor and in Michigan it has been found that

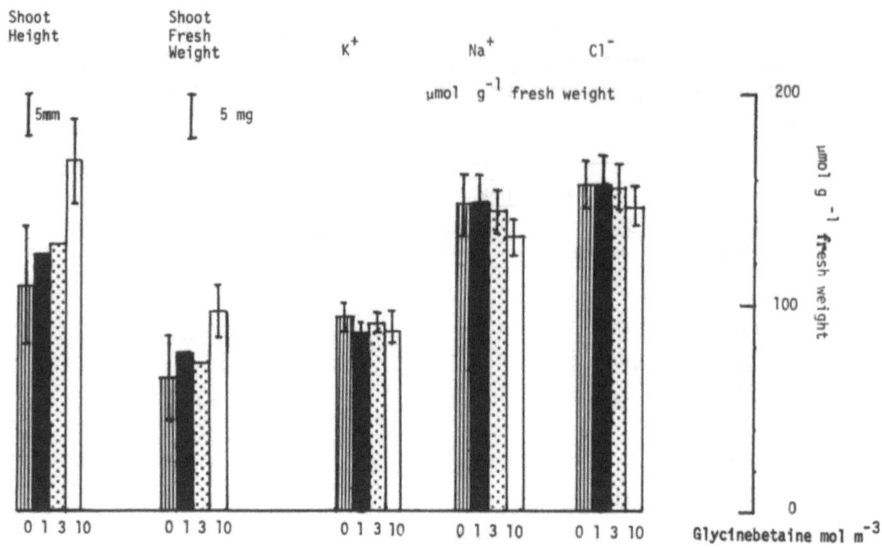

Fig. 5. Effect of different glycinebetaine concentration in the growth and ion contents of barley embryos grown in 150 mol m^{-3} NaCl. (Unpublished data of Iqbal Pervez.)

within this tribe Triticeae, which contains a number of the world's major crops, there are variations in the constitutive glycinebetaine levels and these could be exploited in the wide crossing programs now being instituted in a number of centers. Early work indicated that there was a strong relationship between constitutive and stress-induced glycinebetaine levels in a wide range of species (13) and more detailed work on Hordeum varieties by Hanson's group has shown that there is a strong correlation between betaine levels in young seedlings and mature plants in the field and between unstressed and stressed levels: these observations are important prerequisites to any useful selection program.

One of the most interesting recent developments in relation to glycinebetaine metabolism and one potentially capable of exploitation by genetic engineering has been the recognition of two biosynthetic pathways, the one in spinach and beet (Chenopodiaceae) differing for that in barley (Gramineae) (19,20). Discussion of the detailed biochemistry lies outside the scope of this paper but the crucial difference may be summarized as follows.

In the two chenopods studied, glycinebetaine is synthesized from ethanolamine through water soluble intermediates including choline. There is still some uncertainty whether the main carbon flux occurs through the phosphoryl intermediates from phosphoryl ethanolamine to phosphoryl choline or through the free bases directly to choline. This alcohol is subsequently reduced via the aldehyde to the acid, glycinebetaine. However, this pathway contributes little to betaine synthesis in barley in which phosphoryl choline must be incorporated into phosphatidyl choline before the release of free choline which is then capable of being oxidized to glycinebetaine. Thus in barley, but not in the two chenopods, synthesis is dependent on the turnover of membrane components. Several interesting questions are immediately posed. Do the more halophytic grasses such as *Spartina*, which accumulate much higher betaine levels than barley (see 14) use the 'chenopod' or the 'barley' pathway? What is the significance of the association of glycinebetaine synthesis with membrane turnover? Has the disassociation of betaine synthesis from membrane-turnover in beet and presumably the highly salt tolerant chenopods which accumulate much more of this solute, been a major factor in the evolution of salt tolerance in this family? If the answer to this last question is positive (and that is *not* proven), then we are approaching the point where the techniques of genetic engineering could find a role. One could envisage a condition in which the introduction of a gene for a choline phosphatase into barley could be a factor in enhancing the synthesis of this solute. However this speculation merely serves to emphasize our ignorance and to reinforce the comments of Dr. Cashmore (this volume) that in plants our knowledge of the biochemical and physiological base capable of being exploited by gene and chromosome manipulation is quite inadequate.

CONCLUDING REMARKS

In discussing salt tolerance in relation to genetic engineering and conventional breeding two facts clearly emerge. Despite the fairly rapid progress in this field since the mid 1970s, our detailed biochemical knowledge is inadequate even in relation to the compatible solutes and is essentially non-existent in relation to specific membrane changes which we assume to be associated with the changed ion transport activities in tolerant plants. Secondly, there is to date no reason to expect that the introduction of a single character, be it high proline or high glycinebetaine, will alone cause a marked increase in tolerance to either salt or probably to drought. Nevertheless in the case of glycinebetaine there can be a reasonable expectation that a high capacity for betaine synthesis would be one of several characters involved in tolerance in some agriculturally important species. In the introduction (see Fig. 1) the wide range of potentially adaptive traits, anatomical, morphological, physiological, and biochemical, were alluded to, but it was suggested that at the cellular levels a limited number of unifying requirements could be recognized - turgor maintenance by cytoplasmic and vacuolar osmotic regulation and partial cytosolic ionic homeostasis. However, it is also clear that these requirements at the cellular level are met within the totality of the plant's metabolism and that events at the organ or whole plant level, xylem loading for example, are as important as the characteristics of individual cells in determining whether these criteria can be met.

To return finally to a problem alluded to earlier, in the few cases where reliable comparative data on cultivars of agricultural crops are available, e.g. barley, it appears that enhanced tolerance is associated with greater efficiency of NaCl exclusion from the shoots (11,12). This contrasts with the situation in the halophyte Suaeda maritima outlined earlier when NaCl is used as a major vacuolar osmoticum (although fluxes must still be tightly controlled). The question which arises is whether or not selection for more efficient exclusion as has occurred in barley is compatible with rapid growth rates and high yields. An added complexity in this context is the difference between Graminaceous indeed most monocotyledenous species and many dicotyledenous species in their salt responses (21,22).

In the Gramineae high leaf NaCl levels are uncommon and a higher K:Na ratio is observed than in the Chenopodiaceae. If we are to achieve rational selection criteria, these physiotypic differences and their metabolic costs must be explored in greater physiological and biochemical detail. Until such work is undertaken, we can only agree with the point made by several contributors to this volume that the biochemical and physiological base on which to launch a rational program of genetic manipulation for salt tolerance is inadequate.

ACKNOWLEDGEMENTS

We would like to acknowledge the financial assistance of the Overseas Development Administration and the Royal Society in this work.

REFERENCES

1. Wyn Jones, R.G. 1981. Salt tolerance. In Physiology Processes Limiting Plant Productivity. C.B. Johnson, ed. Butterworth, London, pp. 271-292.
2. Wyn Jones, R.G., C.J. Brady, and J. Speirs. 1979. Ionic and osmotic relations in plant cells. In Recent Advances in the Biochemistry of Cereals. D.L. Laidman and R.G. Wyn Jones, eds. Academic Press, London, pp. 62-103.
3. Greenway, H., and R. Munns. 1980. Mechanisms of salt tolerance in non-halophytes, Ann. Rev. Plant Physiol. 31:149.
4. Poljakoff-Mayber, A. 1975. Morphological and anatomical changes in plants as a response to salinity stress. In Plants

in Saline Environments. A. Poljakoff-Mayber and J. Gale, eds. Springer Verlag, Berlin, pp. 97-117.

5. Storey, R., R. Stelzer, and M.G. Pitman. 1981. X-ray microprobe analysis of frozen hydrated cells of Atriolex spongiosa. In Abstracts of XIII International Botanical Congress. Sydney, p. 261.
6. Jeschke, W.D. 1979. Univalent cation selectivity and compartmentation. In Recent Advances in the Biochemistry of Cereals. D.L. Laidman and R.G. Wyn Jones, eds. Academic Press, London, pp. 37-62.
7. Raven, J.A., and F.A. Smith. 1977 Characteristics, functions and regulation of active proton extrusion. In Regulation of Cell Membrane Activities in Plants. E. Marre and O. Ciferri, eds. North Holland, Amsterdam, pp. 25-40.
8. Gorham, J., E. McDonnell, and R.G. Wyn Jones. 1982. Determination of betaines as ultraviolet-absorbing esters, Anal. Chim. Acta. 138:277.
9. Leigh, R.A., N. Ahmad, and R.G. Wyn Jones. 1981. Assessment of glycinebetaine and proline compartmentation by analysis of isolated beet vacuoles, Planta. 153:34.
10. Gorham, J., and R.G. Wyn Jones. Solute distribution in Suaeda maritima, Planta, (submitted for publication).
11. Greenway, H. 1973. Salinity, plant growth and metabolism. J. Aust. Inst. Agric. Sci. 39:24.
12. Wyn Jones, R.G., and R. Storey. 1978. Salt stress and comparative physiology in the Gramineae IV Comparison of salt stress. In Spartina townsendii and three barley cultivars. Aust. J. Plant Physiol. 63:839.
13. Stewart, G.R., and J.A. Lee. 1974. The role of proline accumulation in halophytes. Planta. 120:279.
14. Wyn Jones, R.G., and R. Storey. 1981. Betaines. In Physiology and Biochemistry of Drought Resistance in Plants. L.G. Paleg and D. Aspinall, eds. Academic Press, Sydney and London, pp. 171, 204.
15. Kueh, J.S.H., and S.W.J. Bright. Biochemical and genetical analysis of three proline-accumulating barley mutants. Plant Sci. Lett. (in press).
16. Ahmad, N., R.G. Jones, and W.D. Jeschke. Effect of exogenous glycinebetaine on Na^+ fluxes in barley roots. Physiol. Plant., (submitted for publication).
17. Ladyman, J.A.R., K.M. Ditz, R. Grumet, and A.D. Hanson. Genotypic variation for betaine accumulation by cultured and wild barley in relative to water stress. Crop Sci. (in press).
18. Hitz, W.D., J.A.R. Ladyman, and A.D. Hanson. Betaine synthesis and accumulation in barley during field water-stress. Crop Sci. 22:47-54.
19. Hitz, W.D., D. Rhodes, and A.D. Hanson. 1981. Radiotracer evidence implicating phosphoryl and phosphatidyl bases as intermediates in betaine synthesis by water stressed barley leaves. Plant Physiol. 68:814.

20. Coughlan, S.J., and R.G. Wyn Jones. 1982. Glycinebetaine biosynthesis and its control in detached secondary leaves of spinach. Planta. 154:6.
21. Albert, R., and M. Popp. 1977. Chemical composition of halophytes from the Neusiedler Lake region in Austria. Oecologia. 27:157.
22. Gorham, J., Ll. Hughes, and R.G. Wyn Jones. 1980. Chemical composition of salt marsh plants from Ynys Mon (Anglesey): the concept of physiotypes. Plant Cell Environ. 3:309.

RAISING THE YIELD POTENTIAL:

BY SELECTION OR DESIGN?

L.T. Evans

CSIRO Division of Plant Industry
Canberra, A.C.T. 2601
Australia

YIELD INCREASE

The most appropriate measure of yield has changed in the past, and may well continue to do so as agriculture itself evolves. The earlier criterion of number of grains harvested per grain sown has been largely displaced by yield per hectare per crop, which is the basis of the following discussion. But in multiple cropping systems, especially in the tropics, yield per hectare <u>per day</u> is becoming ever more important, while yields per unit of applied energy or water or phosphorus or labor are also important considerations.

Increased yield per hectare per crop has been the key to increased crop production in most developed countries over the last 30-40 years. In developing countries, especially Africa and Latin America, increase in the area under crop has been as important as increase in yield, but in Asia, where new arable land is in short supply, increase in yield per crop, and to some extent also in the number of crops per year, have been the key to the striking gains in crop production, whether of wheat in India or of rice in Indonesia.

It is important at the outset, therefore, to examine yield trends for any sign of slackening progress, and to see to what extent increases in yield have been due to improved genetic yield potential as against improved agronomy. Crop yields are profoundly influenced by the levels of water, fertilizer, pesticide and herbicide application, and these in turn depend not only on how responsive the crop is to such inputs, but also on whether pricing policies and economic constraints encourage such inputs. For example, although the yield of rice in Japan has increased more or less exponentially over the long term, it was relatively static from 1920 to 1940 due to the availability of cheaper rice from Taiwan and Korea, and again in recent years due to local overproduction (Figure 1). Similarly, the yield of sugar beet in the U.K. has fallen since 1960, following a period of rapid increase (Fig. 1) and coinciding with the adoption of a succession of agronomic innovations aimed at containing labor costs rather than increasing yields (39). Many other crops have been subject to similar constraints in recent years, which has led to statements (e.g., refs. 42, 52) that the rate of genetic improvement of economic plants has slowed down and to the conclusion that the emerging techniques of genetic engineering are in urgent need of application to crops.

However, where economic conditions have been favorable to increase in yield, as for wheat in the U.K., maize in the U.S.A., or rice in the Republic of Korea, steady increases in yield have occurred (Fig. 1), indeed quite striking gains when one considers how productive these crops already are, and the long periods over which their yields have already been rising. At many stages along the way, even knowledgeable scientists have been tempted to say that we have already reached the limit. For the British wheat crop, Austin (1) regards the current record yields as close to the possible limit, but so too did Vavilov in the 1930s, and Hallett in the mid-19th century. Of the latter's conclusion, Charles Darwin wrote to Hooker (August 7, 1869) that it would be rash to assume, on past experience, that further improvement could not be made, a comment which is appropriate today, and tomorrow, as it was then.

YIELD MAINTENANCE

Before we consider the components of increase in yield, the great effort required simply to sustain current high yield levels deserves notice. Such "maintenance research" may seem less glamorous than raising yield potential still further, but given the unrelenting evolution of new pest biotypes (e.g., of brown plant hopper on tropical rice) or new races of pathogens such as wheat rust, it is often more urgent, and absorbs much of the plant breeder's effort. Moreover, such "maintenance research" probably takes an increasing proportion of the plant breeder's time as yield levels rise and as successful varieties are grown over greater areas, offering a greater challenge to pest and pathogen. IR-36 rice is currently being grown on more than ten million hectares, and at least two Russian wheat cultivars and two wheat lines from CIMMYT covered even greater areas in their heyday. But successful cultivars may now have relatively short useful lives, and the breeding of a steady stream of adapted and resistant new cultivars is often more important than raising the yield potential still further.

For example, the remarkably successful breeding program at IRRI (the International Rice Research Institute) has produced a long

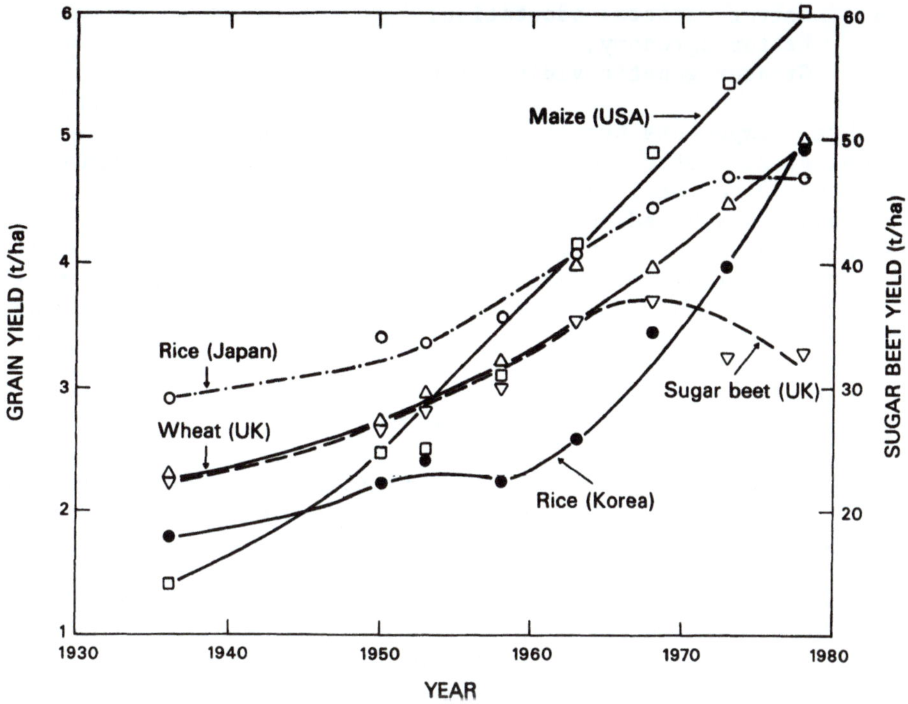

Fig. 1. Changes in the national yields of maize in the U.S.A. (□), wheat (△) and sugar beet (▽) in the United Kingdom, and brown rice (paddy x 0.8) in Japan (○) and the Republic of Korea (●), based on averages for successive 5 year periods (FAO Production Yearbooks).

series of cultivars with improved resistance to pests and diseases, wider adaptation to adverse soil and climatic conditions, better quality and shorter duration. But yield potential has not been increased beyond that of their first cultivar, IR-8 (17). While there may still be a substantial gap between record or experiment station yields and average farm yields, implying that there is ample spare yield capacity, this may not always be so (4). Greater yield potential must therefore remain as an important objective in most plant breeding programs.

THE COMPONENTS OF INCREASING YIELD

Crop yields in any one year depend greatly on weather and other local conditions, but longer term increases in yield can be ascribed to:

1. Better varietal adaptation,
2. Better agronomy,
3. Greater genetic yield potential.

These three components overlap and interact too much to be clearly differentiated. Better adaptation encompasses better adjustment of the life cycle to local conditions (often very important in the early stages of a breeding program), greater resistance to pests and diseases, greater competitiveness with weeds (in some situations), and greater tolerance of adverse environmental conditions. Better agronomy encompasses a great range of practices, including increased or more effective use of water, fertilizers, pesticides, herbicides, inoculants, regulants, etc.; closer spacing, for example in maize (14); better control of pests; improvement of soil fertility, and more effective and timely cultivation.

In this context, genetic yield potential may be defined as the yield of a cultivar grown in environments to which it is adapted, with nutrients and water non-limiting, and with pests, diseases, weeds, lodging and other stresses effectively controlled. These are rare, almost hypothetical conditions, but only with an approach to such conditions can we assess progress in the assembling of productivity genes as distinct from genes for adaptation to environment, adaptedness to modern agronomy, and resistance to pests and diseases (16).

One implication of this definition is that the genetic yield potential of a cultivar may be dependent on favorable conditions and good agronomy for its expression. Whether or not the greater yield potential of improved cultivars is also evident in poor conditions or with low inputs is an important and much debated question which cannot be answered here. However, four findings might be noted at this stage. Comparing old and new barley cultivars, Sandfaer and his colleagues at first found a pronounced interaction with the level of nitrogen fertilizer application, but when the old cultivars were freed of barley stripe mosaic virus, the yield advantage of the new varieties was greatly reduced, but was evident at all levels of nitrogen (47). The relative yields of 12 winter wheat cultivars grown in Britain at various times over the last 70 years were similar at both high and low levels of soil fertility, but some tendency for the modern cultivars to be better adapted to high fertility may be discerned (Figure 2). A differential response is even clearer in experiments with 40 pearl millet lines at ICRISAT, in which there was only a modest correlation between their rankings at high and low levels of soil fertility and a highly significant interaction between cultivar and fertility level. Some lines were among the best at both fertility levels, but as many were not (29). Interaction is also clear with different densities of planting. With maize hybrids grown in Iowa over the last 50 years, Duvick (14) found a yield gain of only 16 kg/hectare per year at a density of

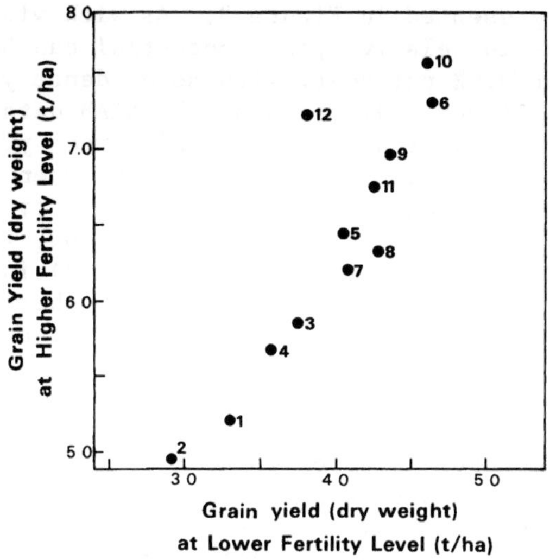

Fig. 2. Grain dry weight yields of 12 winter wheat cultivars grown at two levels of soil fertility at Cambridge (2).

32,000 plants per hectare, compared with 82 kg/hectare per year at a density of 66,000. Clearly, assessments of the increases in genetic yield potential depend to some extent on agronomic conditions.

INCREASE IN GENETIC YIELD POTENTIAL

Increases in genetic yield potential can be assessed from records of national yield trials conducted over a period of years, provided that the successive standard cultivars overlap long enough for their relative performance to be adequately compared. The results of Swedish trials with winter wheat and spring wheat, barley, and oats have been analyzed by MacKey, each series beginning with an unbred indigenous variety (37). His data for winter wheat can be fitted quite well by an exponential curve increasing at a rate of 0.42% per year, but the progress for oats has been less consistent and striking, and is better fitted by a straight line. In fact, plant breeders often quote a percent increase per year figure derived simply by dividing the overall increase by the number of years spanned, which is not only confusing but gives a higher figure than the true exponential rate.

While at Cambridge in 1978, I applied the approach developed by MacKey to the results of yield trials conducted by the National Institute of Agricultural Botany (NIAB) in the United Kingdom for winter and spring wheat and winter and spring barley. The results for the wheats have been published elsewhere (15), so those for spring barley are presented in Figure 3. As with winter and spring wheat, the increase in relative yield potential can be fitted to an exponential rate of 0.9% per year, with no evidence yet of approaching an asymptote. Silvey (49) has used the NIAB data to estimate what proportion of the increase in national cereal yields can be credited to improved genetic yield potential between 1947 and 1978, by taking account of the areas on which each variety is grown: for wheat, barley, and oats these proportions were 60%, 42%, and 29% respectively. Moreover, her earlier analysis (48) indicated that, for both wheat and barley, this proportion rose sharply in the last decade, genetic improvement being more important than agronomic improvement in recent years.

Such trials have the advantage that cultivars are compared under the agronomic conditions for which they were bred. However, they could over-estimate the increases in yield potential if the standard variety begins to fail in its resistance to pests and diseases before the new standard variety takes over. Many direct comparisons of yield potential have therefore been made, but these are valid only if the older cultivars are protected from lodging, pests, and diseases. When this was done using winter wheat cultivars released in the U.K. over the last 70 years (see Fig. 2), the increase in yield potential was found to be about 40% at both high and low levels of fertility (2), somewhat less than that estimated from the NIAB trials. A comparable experiment (43) with spring barley varieties grown in Britain between 1880 and 1980 also gave a somewhat lower rate of increase in genetic yield potential than that suggested by the NIAB trials.

Wheat yields in New York state have increased steadily since 1935 and from the results of his "living museum" of New York cultivars, Jensen (30) has concluded that about half of this increase has been due to the 47% increase in genetic yield potential. Comparable increases in other wheat breeding programs in the U.S.A., and in the U.S.S.R., have been reviewed by Frey (22), who concludes that the rise in genetic yield potential is often irregular. Such stepwise progress has been found in MacKey's analysis for oats in Sweden (37), and in experiments on wheat in several countries (16) and on soybeans in the U.S.A. (5). With maize in the U.S.A., on the other hand, the improvement in genetic yield potential has been steady, and accounts for about 60% of the increase in yield in Iowa (14,46).

Taken overall, the results with several major crops indicate substantial increases in genetic yield potential and no sign of diminishing progress in recent years, with the possible exception of soybeans. These increases have been a major component, at least half, of recent increases in crop yields, and in some cases this proportion appears to be rising. The nature of these increases in genetic yield potential will now be examined.

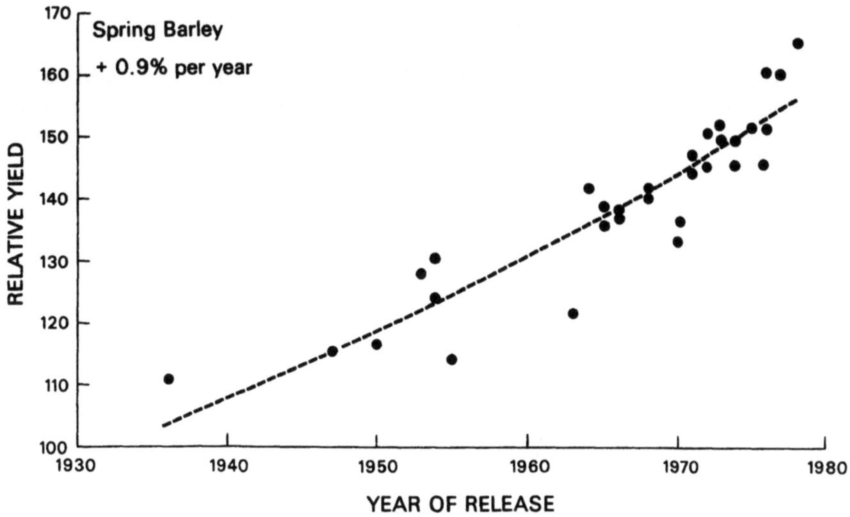

Fig. 3. Estimated increase in relative yield potential of English spring barley cultivars based on the results of annual trials conducted by the National Institute of Agricultural Botany. Yields are relative to Spratt Archer = 100, the standard cultivar until 1952. Other standard cultivars were Kenia (1950/4), Proctor (1954/75), Zephyr (1969/75), Julia (1976-), and Sundance (1977-).

COMPONENTS OF INCREASE IN YIELD POTENTIAL

Up to now, increase in the genetic yield potential of crops has not come from a rise in the rates of photosynthesis or growth, but rather from a progressive rise in the proportion of biomass which is partitioned into the harvested organs. This proportion is called the harvest index.

Van Dobben (11) first drew attention to this feature of increased yield potential in Dutch winter wheat cultivars, but it has since been found to apply to many other crops, not only in comparisons among cultivars of different vintage, but also in comparisons with wild progenitors of the crops (25).

For example, there has been no increase in crop biomass associated with the substantial increase in genetic yield potential in a succession of British wheat (2,3), barley (43), and oat (34) cultivars, or in soybean varietal comparisons (24). Even with pasture grasses, there has been little increase in production potential over 100 years in either the Netherlands or Britain (44).

Similarly, there has been no increase in the relative growth rates of young plants associated with greater yield potential in wheat, maize, tomato, or cowpea (25) or rice (8).

Increase in harvest index with increase in yield potential has been found in wheat (2,3,11), barley (43), peanuts (13), soybeans (24), and other crops. For wheat and its wild progenitors, grown in the field, the harvest index ranged from 5-17% in the wild diploids to 51% in Hobbit, a modern cultivar (3). Given the rather low correlation between the yield of spaced plants and of plots, Fischer and Kertesz (21) suggested that harvest index of the main shoot could be a useful criterion in selection for yield in early generations involving spaced plants.

Harvest index varies greatly between crops depending on their life cycle and growth habit, reaching its highest values among the root and tuber crops. How much further it can be raised in grain crops is a matter for speculation, but it could perhaps reach 63% in wheat (2), giving up to 25% further increase in yield potential. However, the harvest index is merely the outcome of many interacting component processes, each itself complex. Five factors influencing the harvest index will be considered, but others could be added to this list.

1. Reduced Investment in Other Organs

In crops harvested for their grain, a rise in the harvest index could involve reductions in stems, leaves, roots, and reserves. Current evidence is equivocal for roots, and clearest for stems and leaves, as illustrated in Figure 4. This indicates that for a series of winter wheat cultivars grown at both high and low fertility levels, the reductions achieved in straw weight have been matched, more or less, on a one to one basis, by gains in grain yield. The genetic controls, in this case, have operated primarily on stem length, and the assimilate thus freed during the stem elongation phase has been invested in greater grain growth. Just how this has been achieved is still not clear: some could be used directly to promote greater inflorescence development (more spikelets, more florets, and more advanced ovaries), some could be temporarily stored until grain growth begins several weeks later, and some could promote the development of more ear-bearing tillers. In tropical maize populations subjected to selection for reduced height, there has been no change in biomass at flowering, but the reduction in stem weight of 1-2 tons per hectare has been associated with an increase in ear weight of only about 0.1 t/ha at flowering, but of more than 3 t/ha at maturity (20). Clearly, although only a small part of the assimilate released by reducing stem growth was immediately invested in greater inflorescence growth, its ultimate impact on grain yield was greatly enhanced.

Further increase in the harvested organs at the expense of the remainder of the plant is likely to depend on the provision of

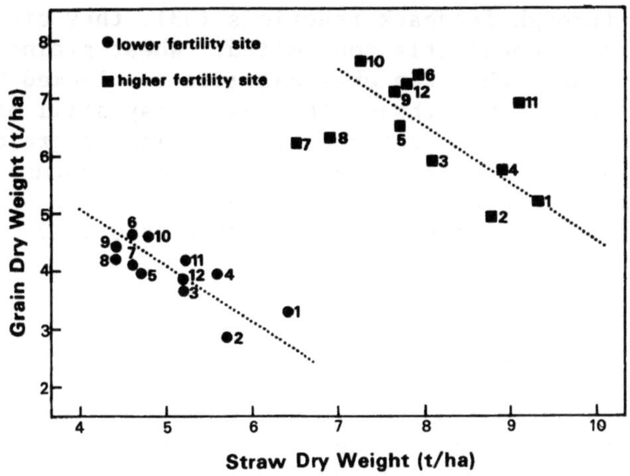

Fig. 4. Grain dry weight plotted against straw dry weight at maturity, at two levels of soil fertility, for the 12 winter wheat cultivars in Fig. 2 (2). The lines show a slope of -1.

greater agronomic support for the crops. The reduction of stem height and leaf size is practicable only if weeds are controlled. Root systems can be reduced only if irrigation and fertilizers reduce the need for them to grow in search of water and nutrients. Reserves can be reduced only when environmental stresses, and attacks by pests and diseases, are controlled.

2. Changed Phases of Development

Climatic and other constraints often place limits on when crop life cycles can begin and end, and also on when certain particularly susceptible stages are reached, such as the beginning of ear development and shoot elongation, meiosis, and grain set. Nevertheless, some adjustment of the relative lengths of the phases of crop life cycles has been possible, with effects on the harvest index. Artificial grain drying, for example, allows the grain growth period to be extended. Higher levels of fertilizer application not only allow early vegetative growth to be accelerated, thereby making it possible for the final storage phase to be to be extended, but also allow leaves to remain photosynthetically active for a longer time in the storage phase, by reducing the need to mobilize nitrogen and phosphorus out of the upper leaves and into the grain. Thus, they have made it feasible to select cultivars with both a longer duration of photosynthetic activity in the upper leaves and a longer duration of grain growth. Although more prolonged photosynthesis makes more prolonged grain growth possible, and vice versa through feedback reactions (33), they are not always coupled, and are independently controlled. Wheat grains, for example, cease growth when the chalazal zone is blocked by the deposition of lipids, even though the leaves may still be photosynthetically active (51). But greater duration of grain growth, often associated with larger kernels, has been a feature of the domestication and improvement of wheat, as well as of other crops, e.g., maize (9) and soybeans (24), and has undoubtedly contributed to the rise in harvest index.

3. Greater Rates of Storage

Grain growth often proceeds linearly in spite of substantial variations in crop irradiance, suggesting that it is often limited by factors other than the supply of photosynthetic assimilate, such as the vascular system and transport processes supplying the grains, or their enzymatic capacity for starch and protein storage. Certainly, in both wheat (19) and oats (28), the vascular system supplying the inflorescence has increased in parallel with grain growth rate, so that the rate of assimilate transfer per unit cross-sectional area of phloem has remained constant. The enzymatic capacity for storage has presumably also increased, and faster grain growth is evident in some higher yielding lines, e.g., in wheat (50) and maize (9).

4. Greater Remobilization of Reserves

The remobilization of reserves into the harvested organs varies among crops and cultivars, but also depends greatly on environmental and agronomic conditions, making generalization difficult. Fifty years ago, with much lower levels of fertilizer application, photosynthesis during grain growth was restricted and the "uplift" of remobilized reserves contributed substantially to grain growth. With more nitrogenous fertilizer, and faster and more prolonged leaf photosynthesis, current assimilates are more important than earlier reserves. On the other hand, the greater sink capacity of higher yielding cultivars may lead to greater remobilization of reserves. But such reserves may also play a valuable role in osmotic adaptation to water stress and in resistance to lodging and disease, e.g., in maize and sorghum. As a result of these cross-currents, there is no clear evidence that higher yielding cultivars consistently remobilize more of their reserves into the grain (2).

5. Enhanced Competitiveness by the Storage Organ

The mechanisms and rules which control the partitioning of assimilates among competing organs have been little studied. Genes which reduce the growth of competing organs such as the stem may allow more assimilate to be partitioned into the storage organs, but several characteristics of the latter may also augment their share. Experiments with wheat have shown that, for competing sinks of similar kind, the greater the relative size of a sink, or the closer it is to the source of assimilates, or the more direct and plentiful its vascular connections to the source, the greater is its share of current assimilates (7). Thus, the position, potential size, and geometry of the storage organ may all influence the partitioning of assimilates and therefore the harvest index.

Other factors also influence yield via the harvest index, but these few illustrate how complex, interacting, and dependent on environmental and agronomic conditions the yield determining process is. They also suggest that increase in genetic yield potential requires the coordination of changes in several processes rather than increase in any one limiting process.

YIELD POTENTIAL AND PHOTOSYNTHETIC RATE

Of the many processes having a major effect on yield potential, photosynthesis would seem to be a prime candidate for improvement. It generates most of the substrate for crop yield, and the enhancement of photosynthetic rate by higher irradiance or CO_2 levels frequently leads to substantial increases in yield. Moreover, there is substantial variation in the rate of photosynthesis per unit leaf area among cultivars of many crops. Although it has not been selected for directly by plant breeders in the past, one might have expected some indirect selection to have taken place, given the central role of the photosynthetic process.

However, there is no evidence of this having occurred, so far at least, in that modern high yielding cultivars have not been found to have maximum photosynthetic rates per unit leaf area higher than either older cultivars or wild progenitors in wheat, maize, sorghum, pearl millet, sugar cane, cotton, or cowpea (25). In fact, the highest rates for four of these crops are found among the wild relatives. Similarly in rice there appears to have been no increase in photosynthetic rate in the course of evolution and improvement (8). Average photosynthetic rates for 50 cultivars of rice, grouped by decade of their introduction in the Philippines (Table 1), indicate no change. The two crops in which higher photosynthetic rates may be associated with higher yield potential are beans (41) and soybeans (6), but for neither is the available evidence conclusive.

An important distinction needs to be made at this point, between cultivar rankings for photosynthetic rate at its maximum, and those at some later stage. Among the wheats, for example,

Table 1. Rates of photosynthesis in Philippine rice cultivars used in successive decades. (mg CO_2 dm^{-2} h^{-1} at 23°C, 1100 µE m^{-2} s^{-1}, 300 ppm CO_2). (Unpublished data of R.M. Visperas, L.T. Evans, and B.S. Vergara.)

	Decade of introduction					
	1916/26	1927/36	1937/46	1947/56	1957/66	1967/76
Leaf 5	33.2	32.9	32.5	32.2	31.9	32.5
Leaf 8 or 9	33.6	34.9	35.7	34.9	35.7	35.0
No. of cultivars	4	9	4	17	12	4

photosynthesis reaches its highest rates among the wild diploids, but these also fall most rapidly, with the result that modern wheats display the highest rates several weeks after anthesis (3,18). In cowpeas, although there were no differences in maximum photosynthetic rate that could be associated with yield, later measurements were related to yield rankings (36), just as in soybeans rankings for photosynthetic rates during flowering were not related to yield, whereas those during pod filling were (6). Thus, on present evidence, correlations between photosynthetic rate and yield probably reflect differences in the duration rather than the maximum rate of photosynthesis.

Selection for higher photosynthetic rate has been carried out with several crops. In alfalfa the selection was actually for high specific leaf weight (SLW), which was closely correlated with photosynthetic rate, but neither bore much relation to forage yield, which was more closely related to leaf area (27). Similarly with peas, lines selected for higher photosynthetic rates had smaller leaf areas and lower crop growth rates than lines selected for low photosynthetic rates (38). Selection for higher photosynthetic rate in maize is possible (10), but earlier results do not encourage expectations of yield increase from this quarter (40). With soybeans, on the other hand, selection for higher photosynthetic rate is difficult (54), but when lines were selected for higher net photosynthesis by crop canopies, seed yields tended to be higher. However, the leaf area index of these canopies was not measured, and the differences in canopy photosynthesis could have been due to differences in leaf area index rather than rate. We still lack evidence, therefore, for any increase in yield following selection for higher photosynthetic rate, not only in grain crops but even in forage crops.

The major reason for this situation would seem to be the frequently negative relation between the area and the maximum photosynthetic rate of leaves, resulting in a compensation between them. Such a negative relation has been found many times among both species and cultivars of wheat (3,18,23,32), and was compensatory in that all produced the same biomass in the field (3), as occurred also with peas (38).

Since the proportion of biomass and nitrogen which is invested in leaf tissue tends not to vary greatly among the cultivars of a crop, a particular cultivar can have either larger leaves with smaller SLW, or smaller leaves with greater SLW. The greater the SLW, and the higher the N content per unit leaf area, the higher is the photosynthetic rate. In both rice and wheat, for example, variation in photosynthetic rate among many species and cultivars is closely related to N content per unit leaf area (8,32).

Greater leaf area is most advantageous in the early stages of crop growth, when ground coverage and light interception is limiting, but once the crop canopy is closed, smaller leaves with faster photosynthesis would be advantageous. The optimum compromise between these requirements, apart from an ontogenetic shift from larger to smaller leaves, would be intermediate values, and it is notable that, in both wheat and rice, the most productive cultivars are found near the middle of the area/rate curve (8). However, as planting densities continue to increase, a shift towards the higher photosynthetic rate/smaller leaf area end of the curve should be advantageous, and this may also be so with higher rates of nitrogen fertilizer application.

Thus, selection for higher photosynthetic rate remains a valid objective, but it is likely to succeed only when counterproductive associations with smaller cells and leaves and shorter duration of photosynthetic activity are taken into account. Two further points should be made. One is that to achieve a significant increase in the rate of photosynthesis by the crop canopy will require a proportionately much greater increase in the maximum photosynthetic rate per unit leaf area, because only the uppermost leaves of a crop are not light-limited. The other is that photosynthesis consists of a large number of tightly linked processes within an anatomical setting of considerable complexity. Although classical selection procedures may result in some increase in photosynthetic rate through coordinated change in many components, the techniques of genetic engineering would seem less well suited to so complex a task, unless it can be resolved into a few key components under known or identifiable genetic control.

Some lessons to be learned from work on two related processes will now be considered.

PHOTORESPIRATION AND DARK RESPIRATION

At least until recently, the substantial loss of carbon dioxide by photorespiration seemed wasteful, with no apparent function, and was therefore an attractive target for elimination by plant breeding. By contrast, dark respiration was thought to be tightly coupled to function (being biochemically uncoupled only when local heat production was required to penetrate snow, attract nocturnal pollinators, etc.) and therefore unlikely to be amenable to selection. However, it now seems that the reverse may prove to be the case.

Despite a wide search among many crop plants with C_3 photosynthesis, non-photorespiring forms have not been found, but rather a close positive correlation between photorespiration and gross photosynthesis, e.g., in sunflowers (35). Given the substantial increase in net photosynthetic rate and the consequent competitive advantage that would accrue from the absence of photorespiration, and the enormous time span available for its elimination by natural selection, the fact that this has not happened suggests that photorespiration has adaptive value, at least in many environments. The large catalytic subunit of ribulose bisphosphate carboxylase-oxygenase, which is genetically encoded in the chloroplasts, has undergone some changes in amino acid composition in the course of evolution, but has, nevertheless, been highly conserved, even across the immense span from *Anabaena* to maize. The central catalytic region of the large subunit has been even more highly conserved, and the likelihood of greatly modifying the carboxylation/oxygenation ratio would appear to be slender.

On the other hand, selection for reduced dark respiration, which seemed theoretically undesirable, now appears to have provided an effective route to greater yield. Following earlier work with maize, Wilson and his colleagues selected lines of *Lolium perenne* S23 for slow dark respiration by mature leaves, and have now shown that swards of slow respiring lines are more productive in closed canopies during many cycles of regrowth in both growth rooms and field (45,55,56). Further advance in this direction may be limited, and disadvantages of selection for slow dark respiration may become apparent, but this brief consideration of photorespiration and dark respiration illustrates some of the problems of suggesting which physiological process should receive attention from plant breeders and genetic engineers.

CONCLUSIONS IN RELATION TO GENETIC ENGINEERING FOR YIELD

For those crops in which environmental and economic conditions make it worthwhile, genetic yield potential is still rising, and the power of conventional plant breeding is by no means exhausted. In this centennial year of Charles Darwin's death we should not forget that it was his recognition of the power of sustained artificial selection that led him to evolution by natural selection. The question is, therefore, not how genetic engineering will achieve greater yield potential, but what it can add to the armory of plant breeding.

Crop yield is the ultimate outcome of the whole life cycle of the crop, and of the rates, durations, and interlinkages of many processes at all stages of development, any one of which may be limiting to yield in a particular environment. No one process provides the key to greater yield potential, and the yield of different cultivars or hybrids at one site may be limited by quite different processes (9). Yield potential is an extremely complex character, which is a matter for recognition rather than for delight.

Moreover, the characteristics often associated with greater yield potential, such as more prolonged photosynthesis, faster or longer grain growth, the development of more florets, or the setting of more grains, are themselves quite complex. Even much simpler characteristics may have unexpected associations and feedback effects. For example, when an attempt was made to reduce the water use of barley crops by selecting plants with lower stomatal frequency, it turned out that this characteristic was associated with greater stomatal size, leaf area, and leaf longevity, and resulted in greater rather than less use of water (31). Counter-intuitive effects may also be involved in selecting individual plants for performance within the crop community. Donald (12) argues that we should be seeking the weakly competitive "communal" plant, not the aggressive competitor that commands attention in the F_2 row.

Thus, crop physiologists are not in a position to specify desirable single characters which are touchstones for greater yield potential and which have a simple genetic basis for the ready application of genetic engineering techniques. Nevertheless, there are several ways in which these techniques may be helpful to breeding for greater yield potential.

Wherever possible, plant breeders prefer to use adapted and high yielding lines as parents for crossing, in order to maintain yield potential while seeking to add other characteristics, such as improved disease resistance. When these can be obtained only from low yielding cultivars or wild species, a temporary penalty on yield potential may be unavoidable. But if the desired genes for resistance, quality, etc., could be specifically transferred by molecular techniques, the adverse effects of linked genes on yield potential are likely to be reduced, and the gain in yield potential could be indirectly accelerated.

Secondly, it appears that the introduction of new alien germplasm in advanced breeding programs often raises the yield potential (22). The mechanisms by which it has done so and the processes contributing to the increase usually remain unknown. In these circumstances, the introduction of small random segments of alien DNA into high yielding adapted cultivars may be an effective, although not intellectually satisfying, approach to the further raising of yield potential. However, the huge number of independent clones that would need to be grown on in order to examine the effect of each unique sequence would be daunting.

Consequently, every effort should be made to take advantage of the greatest strength of genetic engineering techniques, namely their specificity, by using them to examine the effects on yield potential of individual genes. Crop physiologists have been very limited in work of this kind by the small array of characters available for comparison from back-crossing programs. Molecular plant breeding could provide a valuable opportunity for crop physiologists and biochemists to analyze the components of yield potential at a level not hitherto possible.

Whether we could then design physiological masterpieces of greater yield potential remains to be seen. In the meantime, such is the power of selection that I am sure plant breeders will continue to raise yield levels. As von Karman once said: "The scientist describes what is; the engineer creates what never was." So, too, may the genetic engineer, helping both the plant breeder and the crop physiologist in their tasks, so that both chance and design can raise yield potential still further.

REFERENCES

1. Austin, R.B. 1978. Actual and potential yields of wheat and barley in the United Kingdom. ADAS Quart. Rev. 29:76.
2. Austin, R.B., J. Bingham, R.D. Blackwell, L.T. Evans, M.A. Ford, C.L. Morgan, and M. Taylor. 1980. Genetic improvements in winter wheat yields since 1900 and associated physiological changes. J. Agric. Sci. Camb. 94:675.
3. Austin, R.B., C.L. Morgan, M.A. Ford, and S.G. Bhagwat. 1982. Flag leaf photosynthesis of *Triticum aestivum* and related diploid and tetraploid species. Ann. Bot. 49:177.
4. Barker, R., S.K. de Datta, K.A. Gomez, and R.W. Herdt. Phillipines 1974, 1975, 1976. 1977. In Constraints to High Yields on Asian Farms, 121. Los Banos: IRRI.
5. Boerma, H.R. 1979. Comparison of past and recently developed soybean cultivars in maturity groups VI, VII and VIII. Crop Sci. 19:611.
6. Buttery, B.R., R.I. Buzzell, and W.I. Findlay. 1981. Relationships between photosynthetic rate, bean yield and other characters in field grown cultivars of soybean. Can. J. Plant Sci. 61:191.
7. Cook, M.G., and L.T. Evans. 1978. Effect of relative size and distance of competing sinks on the distribution of

phytosynthetic assimilates in wheat. Aust. J. Plant Physiol. 5:495.
8. Cook, M.G., and L.T. Evans. Some physiological aspects of the domestication and improvement of rice (Oryza spp.). Field Crops Research (in press).
9. Crosbie, T.M., and J.J. Mock. 1981. Changes in physiological traits associated with grain yield improvement in three maize breeding programs. Crop Sci. 21:255.
10. Crosbie, T.M., R.B. Pearce, and J.J. Mock. 1981. Recurrent phenotypic selection for high and low photosynthesis in two maize populations. Crop Sci. 21:736.
11. Van Dobben, W.H. 1962. Influence of temperature and light conditions on dry matter distribution, development rate, and yield in arable crops. Neth. J. Agric. Sci. 10:377.
12. Donald, C.M. 1981. Competitive plants, communal plants, and yield in wheat crops. In Wheat Science--Today and Tomorrow, 223. L.T. Evans and W.J. Peacock, eds. Cambridge: Cambridge University Press.
13. Duncan, C.M., D.E. McCloud, R.L. McGraw, and K.J. Boote. 1978. Physiological aspects of peanut yield improvement. Crop Sci. 18:1015.
14. Duvick, D.N. 1977. Genetic rates of gain in hybrid maize yields during the past 40 years. Maydica 22:187.
15. Evans, L.T. 1980. Response to challenge: William Farrer and the making of wheats. J. Aust. Inst. Agric. Sci. 46:3.
16. Evans, L.T. 1981. Yield improvement in wheat: empirical or analytical? In Wheat Science--Today and Tomorrow, 203. L.T. Evans and W.J. Peacock, eds. Cambridge: Cambridge University Press.
17. Evans, L.T., and S.K. de Datta. 1979. The relation between irradiance and grain yield of irrigated rice in the tropics, as influenced by cultivar, nitrogen fertilizer application and month of planting. Field Crops Res. 2:1.
18. Evans, L.T., and R.L. Dunstone. 1970. Some physiological aspects of evolution in wheat. Aust. J. Biol. Sci. 23:725.
19. Evans, L.T., R.L. Dunstone, H.M. Rawson, and R.F. Williams. 1970. The phloem of the wheat stem in relation to requirements for assimilate by the ear. Aust. J. Biol. Sci. 23:743.
20. Fischer, K.S., and A.F.E. Palmer. Maize. In Potential Productivity of Field Crops under Different Environments. Los Banos: IRRI (in press).
21. Fischer, R.A., and Z. Kertesz. 1976. Harvest index in spaced populations and grain weight in microplots as indicators of yielding ability in spring wheat. Crop Sci. 16:55.
22. Frey, K.J. 1981. Capabilities and limitations of conventional plant breeding. In Genetic Engineering for Crop Improvement, 15. K.O. Rachie and J.M. Lyman, eds. New York: Rockefeller Foundation.
23. Gale, M.D., J. Edrich, and F.G.H. Lupton. 1974. Photosynthetic rates and the effects of applied gibberellin in some

dwarf, semi-dwarf and tall wheat varieties (Triticum aestivum). J. Agric. Sci. Camb. 83:43.
24. Gay, S., D.B. Egli, and D.A. Reicosky. 1980. Physiological aspects of yield improvement in soybeans. Agron J. 72:387.
25. Gifford, R.M., and L.T. Evans. 1981. Photosynthesis, carbon partitioning, and yield. Ann. Rev. Plant Physiol. 32:485.
26. Harrison, S.A., H.R. Boerma, and D.A. Ashley. 1981. Heritability of canopy-apparent photosynthesis and its relationship to seed yield in soybeans. Crop Sci. 21:222.
27. Hart, R.H., R.B. Pearce, N.J. Chatterton, G.E. Carlson, D.K. Barnes, and C.H. Hanson. 1978. Alfalfa yield, specific leaf weight, CO_2 exchange rate and morphology. Crop Sci. 18:649.
28. Housley, T.L., and D.M. Peterson. 1982. Oat stem vascular size in relation to kernel number and weight I. Crop Sci. 22:259.
29. ICRISAT. Annual Report 1979/80, p. 53.
30. Jensen, N.F. 1978. Limits to growth in world food production. Science 201:317.
31. Jones, H.G. 1977. Transpiration in barley lines with differing stomatal frequencies. J. Exp. Bot. 28:162.
32. Khan, M.A., and S. Tsunoda. 1970. Evolutionary trends in leaf photosynthesis and related leaf characters among cultivated wheat species and its wild relatives. Jap. J. Breed. 20:133.
33. King, R.W., I.F. Wardlaw, and L.T. Evans. 1967. Effect of assimilate utilization on photosynthetic rate in wheat. Planta (Berl.) 77:261.
34. Lawes, D.A. 1977. Yield improvement in spring oats. J. Agric. Sci. Camb. 89:751.
35. Lloyd, N.D.H., and D.T. Canvin. 1977. Photosynthesis and photorespiration in sunflower selections. Can. J. Bot. 55:3006.
36. Lush, W.M., and H.M. Rawson. 1979. Effects of domestication and region of origin on leaf gas exchange in cowpea (Vigna unguiculata L.). Photosynthetica 13:419.
37. MacKey, J. 1979. Genetic potentials for improved yield. In Proc. Workshop on Agricultural Potentiality Directed by Nutritional Needs, 121. S. Rajki, ed. Budapest: Akad. Kiado.
38. Mahon, J.D. 1982. Field evaluation of growth and nitrogen fixation in peas selected for high and low photosynthetic CO_2 exchange. Can. J. Plant Sci. 62:5.
39. Milford, G.F.J., P.V. Biscoe, K.W. Jaggard, R.K. Scott, and A.P. Draycott. 1980. Physiological potential for increasing yields of sugar beet. In Opportunities for Increasing Crop Yields, 71. R.G. Hurd, P.V. Biscoe, and C. Dennis, eds. Boston: Pitman.
40. Moss, D.N., and R.B. Musgrave. 1971. Photosynthesis and crop production. Adv. Agron. 23:317.
41. Peet, M.M., A. Bravo, D.H. Wallace, and J.L. Ozbun. 1977. Photosynthesis, stomatal resistance, and enzyme activities in relation to yield of field grown dry bean varieties. Crop Sci. 17:287.

42. Rachie, K.O., and J.M. Lyman. 1981. Preface to Genetic Engineering for Crop Improvement. New York: Rockefeller Foundation.
43. Riggs, T.J., P.R. Hanson, N.D. Start, D.M. Miles, C.L. Morgan, and M.A. Ford. 1981. Comparison of spring barley varieties grown in England and Wales between 1880 and 1980. J. Agric. Sci. Camb. 97:599.
44. Robson, M.J. 1980. A physiologist's approach to raising the potential yield of the grass crop through breeding. In Opportunities for Increasing Crop Yields, 33. R.G. Hurd, P.V. Biscoe, and C. Dennis, eds. Boston: Pitman.
45. Robson, M.J. 1982. The growth and carbon economy of selection lines of Lolium perenne cv. S23 with differing ratés of dark respiration I. Ann. Bot. 49:321.
46. Russell, W.A. 1974. Comparative performance for maize hybrids representing different eras of maize breeding. Proc. 29 Ann. Corn and Sorghum Res. Conf., 84.
47. Sandfaer, J., and V. Haar. 1975. Barley stripe mosaic virus and the yield of old and new barley varieties. Z. Pflanzenzucht 74:211.
48. Silvey, V. 1978. The contribution of new varieties to increasing cereal yield in England and Wales. J. Natl. Inst. Agric. Bot. 14:367.
49. Silvey, V. 1981. The contribution of new wheat, barley and oat varieties to increasing yield in England and Wales 1947-1978. J. Natl. Inst. Agric. Bot. 15:399.
50. Sofield, I., L.T. Evans, M.G. Cook, and I.F. Wardlaw. 1977. Factors influencing the rate and duration of grain filling in wheat. Aust. J. Plant Physiol. 3:785.
51. Sofield, I., I.F. Wardlaw, L.T. Evans, and S.Y. Zee. 1977. Nitrogen, phosphorus and water contents during grain development and maturation in wheat. Aust. J. Plant Physiol. 4:799.
52. Strobel, G.A. 1980. Potentials for improving crop production. In Genetic Improvement of Crops: Emergent Techniques, 3. I. Rubenstein, B. Gengenbach, R.L. Phillips, and C.E. Green, eds. Minneapolis: Univ. Minnesota.
53. Thompson, L.M. 1975. Weather variability, climatic change, and grain production. Science 188:535.
54. Wiebold, W.J., R.M. Shibles, and D.E. Green. 1981. Selection for apparent photosynthesis and related leaf traits in early generations of soybeans. Crop Sci. 21:969.
55. Wilson, D. 1982. Response to selection for dark respiration rate of mature leaves of Lolium perenne and its effects on growth of young plants and simulated swards. Ann. Bot. 49:303.
56. Wilson, D., and J.G. Jones. 1982. Effect of selection for dark respiration rate of mature leaves on crop yields of Lolium perenne cv. S23. Ann. Bot. 49:313.

AMINO ACIDS, NUTRITION AND STRESS: THE ROLE OF BIOCHEMICAL MUTANTS IN SOLVING PROBLEMS OF CROP QUALITY

B.J. Miflin, S.W.J. Bright, S.E. Rognes, and J.S.H. Kueh

Biochemistry Department
Rothamsted Experimental Station
Harpenden, Herts. AL5 2JQ, United Kingdom

SUMMARY

Amino acids affect crop quality in a number of ways. We consider here the deficiencies of lysine and threonine which adversely affect the nutritional quality of cereal diets and the potential role of proline in stress tolerance. To probe the biosynthesis of these amino acids further, it was decided to select for mutants of barley that synthesize enhanced amounts of them. For several reasons we considered it preferable to carry out our selection using whole embryos. The genetical and biochemical characteristics of the mutants selected are described; mutants Rothamsted (R) 906 and R4402 contain the gene aec, which causes impaired uptake of lysine; R2501 and R3202 the gene Lt1, which leads to the desensitization of one isoenzyme of aspartate kinase (AK); R3004 the gene Lt2, which causes the desensitization of another isoenzyme of AK; and R5201, R6102, and R6902 the Hyp1 gene, which confers resistance to hydroxyproline. R3004 and R2501 accumulate threonine and R5201, R6102, and R6902 proline.

CROP QUALITY AND AMINO ACIDS

Amino acid metabolism of crop plants affects their growth and composition in a number of ways; in this article we will focus on two of these--nutritional quality and resistance to stress--particularly in barley, which is the major cereal crop in the U.K. and is chiefly used for animal feed.

Nutritional Quality

Non-ruminant animals require a diet containing a balanced content of amino acids. Barley, wheat, and maize grain provide an unbalanced diet which requires supplementation. However, if this is done, they can be a very effective food source; for example, supplementation of barley with lysine, threonine, and histidine provided a diet for pigs with protein of a biological value of 0.97 (28). Because Europe has to import large quantities of soybean meal to supplement cereals in diets for non-ruminant animals, considerable effort is being devoted to improving the nutritional quality of cereal seed. The reason for the poor quality is the amino acid composition of the cereal grain and in particular that of the prolamin storage protein which makes up about half of the total protein. Approaches to improvements include:

- altering the proportion of prolamin proteins; low prolamin mutants of barley exist (22,53); however, these generally have depressed yields (23);

- changing the composition of the prolamin fraction, either by breeding or genetic manipulation; although isolation of the hordein genes is progressing, the fact that the hordeins are a polymorphic series of proteins under the control of three families of multiple genes (71) limits the prospects of success (see refs. 55, 72);

- increasing the amount of high-lysine proteins in the seed; lysine-rich proteins have been identified in barley and the high-lysine line Hiproly and cultivars derived from it have enhanced amounts of these (37);

- increasing the amount of the desired amino acids in the soluble (free) amino acid fraction of the grain.

The approaches are not mutually exclusive and success in breeding terms may depend on combining more than one of them. This paper, however, is concerned only with the last topic. The synthesis of most nutritionally essential amino acids has been shown to be limited by feedback regulation in both bacteria and higher plants (54,79). Mutants of bacteria exist in which this regulation is relaxed either by changing the enzyme or removing the natural regulation due to an auxotrophic mutation, and certain amino acids consequently produced are accumulated in large amounts; currently, commercially available lysine is produced by fermentation using such mutant bacteria (21,25). It is thus practical to consider developing crop plants which also have modified regulatory mechanisms and which produce enhanced amounts of amino acid. Described below are the approaches taken to achieve this aim.

Stress

It is well established that barley and many other plants accumulate proline as a response to water or salt stress (see ref. 76). Analysis of various lines of barley has led Singh et al. (74) to suggest that the ability to accumulate proline may be of adaptive significance; those varieties with the greatest ability to do this should be the most resistant to water stress. Hanson et al. (35) and Hanson and Hitz (36) have argued, however, that this is unlikely to be the cause of present differences between cultivars and that the amount of free proline in the tissues reflects the degree of stress experienced by the plant. This question is unlikely to be resolved by studying the currently available genotypes, due to lack of sufficient variation in their genetic capacity to accumulate proline. One approach that has been used in bacteria is to select for proline accumulation, and a mutant of *Salmonella typhimurium* isolated with this characteristic was found to have enhanced salt tolerance (19). Concurrently, we have been following the same approach with barley (11,12,45,46).

Amino Acid Biosynthesis and Its Regulation

The pathways for the synthesis of the aspartate and glutamate families of amino acids are given in Figures 1A and 1B. Evidence in favor of these pathways has been reviewed recently (14,30,54,78); briefly, there is good evidence to support the scheme in Fig. 1A, but the amount of experimental work in plants on which Fig. 1B is based is small, and serious doubts exist as to the way in which proline is synthesized (24,51).

Regulation of the aspartate pathway is probably achieved by a complex series of feedback loops, as indicated in Fig. 1A. These operate chiefly at the enzyme level by changing the activity of a given amount of enzyme (allosteric feedback inhibition or stimulation), but there is also evidence of controls which change the amount of certain enzymes (repression or depression). Aspartate kinase, which is the first enzyme common to the whole pathway, is polymorphic, with forms sensitive to feedback control either by threonine or by lysine. The sensitivity to lysine is increased in the presence of S-adenosylmethionine (AdoMet) (65). Lysine further regulates its synthesis by feedback inhibition of dihydrodipicolinic acid synthase (82), and threonine inhibits some, but not all forms of homoserine dehydrogenase (14,67). Apart from its synergistic effect on lysine-sensitive aspartate kinase, AdoMet markedly stimulates the activity of threonine synthase (49), and the enzyme is virtually inactive in its absence. Methionine (or a product derived from it, such as AdoMet) also represses the amount of cystathionine synthase present in certain green plants (77).

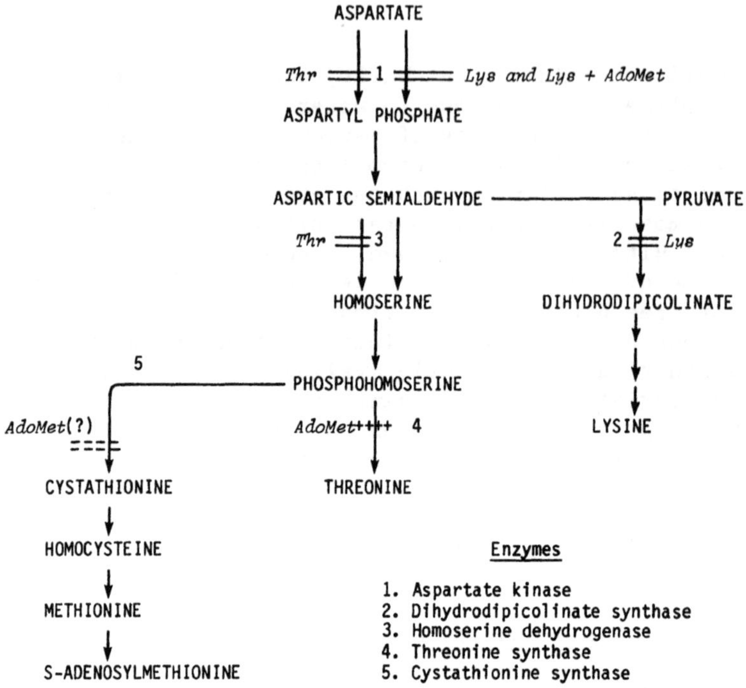

Fig. 1A. The pathway of lysine, threonine, and methionine synthesis in plants. Regulation of the various enzymes is indicated, ═ indicates feedback inhibition, +++ feedback stimulation, and --- repression. The actual repressor for cystathionine synthase is not known other than it is methionine or a closely related metabolic product.

Regulation of proline synthesis has not been well characterized, but there is evidence for feedback inhibition controls operating on proline synthesis in maize (60) and barley (2).

TECHNIQUES FOR THE SELECTION OF BIOCHEMICAL MUTANTS

Two prerequisites are required to select crop plants altered in their ability to accumulate certain amino acids: an adequate population of individuals, treated in such a way that there is a reasonable chance that the desired mutant exists within the population, and a suitable protocol for identifying that mutant plant in a form suitable for use by the plant breeder.

Mutant Plant Populations

Ideally the system used should enable a high number (10^5-10^6) of individuals to be screened using relatively limited space. Furthermore, it should ensure that the desired mutants can be

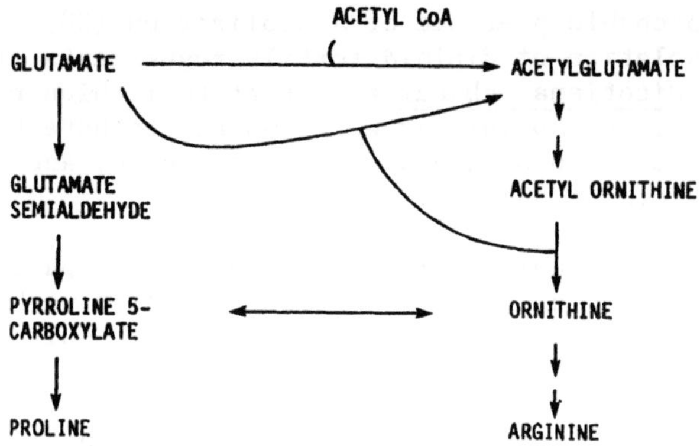

Fig. 1B. Possible pathways of proline synthesis in plants.

recovered as fully differentiated, fertile, cytologically normal plants which are changed as little as possible other than in the desired characteristic. Much effort has been expended on using tissue and protoplast cultures, particularly of Nicotiana and Daucus species, for selecting mutant or variant lines with altered amino acid metabolism. Undoubtedly the use of such systems satisfies the first criterion--that of having large populations in a small space. However, in many instances, if not all, the tissue and protoplast culture systems have failed to meet the other requirements. While useful information has been gained from variants selected in tissue cultures, and lines that overproduce proline, lysine, methionine, and tryptophan (85,86) have been found, the fact that they have proved incapable of regeneration has limited their usefulness. Where variant lines have been regenerated, problems have been encountered; first, expression in tissue culture is not always associated with expression in the plant, and secondly, regenerated plants are frequently abnormal cytologically and morphologically with reduced fertility. Bourgin (3) selected for valine resistance using protoplasts of Nicotiana tabacum. Of several resistant calli obtained, only two gave resistant plants--one of these was male sterile with an unstable number of chromosomes, and the other was a cross-fertile diploid; the paper discusses some of the reasons for the problems encountered in successfully regenerating resistant lines. Reisch et al. (64) regenerated plants from alfalfa cell lines resistant to ethionine. Of these, only 7 out of 26 of the plants tested produced calluses with measurable levels of ethionine resistance and an unusually high frequency of morphological and cytological (63) abnormalities were observed. Hibberd et al. (39) and Hibberd and Green (38) have recovered two maize plants from maize tissue cultures resistant to lysine plus threonine, one of these, D33, was infertile and no genetic analysis was possible (39). In contrast, LT19 was fertile and the resistance was ascribed to a single gene; however, the authors report that cytological analysis has shown the probable presence of a duplication (38). Carlson (15) reported the isolation of diploid fertile regenerants from haploid protoplasts of *Nicotiana tabacum* resistant to methionine sulphoximine. Unfortunately, no further work appears to have been done on them, either in Carlson's laboratory or elsewhere, and the nature of the resistance remains unknown.

Because of the problems with tissue cultures, particularly in cereals, we developed a system for barley in which we have carried out selection on mature embryos, grown in sterile culture (8). The advantages of this system are:

- each embryo gives rise to a normal fertile plant;

- the embryos germinate uniformly and they respond directly to medium constituents because they have few reserves in the scutellum;

- it is possible to use established seed mutagenesis techniques, such as sodium azide (43), to induce high numbers of mutations with little chromosomal damage;

- after growth and self-fertilization of plants from mutagen--treated seed, recessive mutations can be isolated in the diploid homozygous condition.

It appears to us that the system suffers from three disadvantages. First, relatively fewer genetic individuals can be screened and the system requires more space and effort per individual than protoplast systems. However, if the relative effort is assessed on the number of fertile mutants obtained, this technique appears much more successful, as on average one biochemically and genetically characterized single gene mutant has been obtained for every 10^4 embryos plated out. Secondly, the selection is tedious; however, the effort can be reduced, where it is appropriate to use whole seeds, for example, where nutrients supplied from the endosperm are unlikely to interfere with the screen (chlorate resistance) or where the seeds are very small (e.g., Arabidopsis). An example of how the presence of maize endosperm can interfere with amino acid analogue selection schemes has been documented by Phillips et al (62). Finally, the principle is simple and lacks the current glamor and high technology gloss encompassing protoplasts and tissue culture. This last disadvantage, like the others, is trivial, and to us none of them override the positive aspects of whole seed or embryo selection systems. A similar conclusion has been reached by Jacobs et al. (41) after a comparison of selection for lysine plus threonine and AEC-resistance in protoplasts of *Nicotiana sylvestris*, embryoids of carrot, embryos of barley, and seeds of *Arabidopsis thaliania*.

Selection Protocol

The aim is to exert a selection pressure on the above population such that any organism that produces enhanced amounts of the required amino acid will be able to grow, whereas the wild type will not.

The implication is occasionally made that biochemical selection screens only work for plants grown in tissue culture. Thus, Chaleff (18) writes, "Plant mutants were always available to the geneticist, but these variant forms were identified on a visual rather than a biochemical basis. Mutants...in general (they) could not be selected deliberately as modifications of specific biochemical function.... But now the methods of cell culture can be applied to isolate biochemically defined mutants of plants...." Such a generalized implication is not supportable by the facts; while it is true that certain changes, such as gradually acquired resistance (or ability to utilize a substrate) resulting from increasing sequential selection or changes in organelle genomes, are most easily selected in cell cultures, far more biochemical mutants have been selected by whole plant or embryo techniques than by cell culture.

The problems lie not in the level of the organism but in the design of the selection strategy and in matching it to the organism under test. In doing this, there were a number of criteria with which we tried to comply.

(1) To have a clear distinction between the inhibited wild type and potential mutants without setting the selection pressure to such a high degree that genuine mutants could not survive. We did this by testing a range of concentrations of inhibitors and choosing those that decreased growth by about 80% and yet could be relieved by relatively low concentrations of the relevant amino acid (e.g., see ref. 4). Examples of concentrations finally chosen are shown in Fig. 2.

(2) To use conditions and inhibitors which gave a uniform response and thus led to the minimum number of false positives. In the lysine plus threonine screen, the SE was small and we only selected those plants which had a shoot length greater than three standard deviations of the wild-type controls.

(3) To devise the procedure most likely to lead to the isolation of the desired mutants. For any given selection pressure there are likely to be a number of alternative mutations that will allow the plant to survive. We have therefore tried to ensure that we obtain the desired mutations by attempting to understand the nature of the inhibitory effect (not always possible) and choosing inhibitors that are specifically relieved by the desired amino acid.

We have chosen to use two approaches to select for mutants in the aspartate pathway, firstly to use the lysine analogue S-2-aminoethylcysteine (Aec), and secondly the inhibitory combination of lysine plus threonine. Aec probably inhibits growth by being incorporated into proteins (7); it is a relatively weak inhibitor of aspartate kinase (5,70) and thus unlikely to inhibit by false feedback inhibition. The inhibition is relieved by lysine (Figure 2) and to a lesser extent by other basic amino acids and also methionine (6).

Combinations of amino acids (particularly incomplete addition of members of an amino acid family) have long been known to be inhibitory to the growth of members of the plant kingdom from bacteria to barley embryos (13,31,42,52); such inhibition can often be overcome by supplying the missing member. Thus, Fig. 2 shows that lysine plus threonine inhibition of barley embryos grown on agar can be relieved by adding methionine, and labelling studies have shown that lysine plus threonine blocks methionine synthesis (5) (Figure 3). We now believe that the inhibitory effect of lysine plus threonine is due to the additive feedback inhibition of lysine- and threonine-sensitive aspartate kinases blocking the entry of aspartate into the pathway (66). However, other inhibitory effects of amino acids have been noted--under certain conditions lysine appears to interfere with arginine metabolism in barley (16,52) and threonine with nitrate uptake by tobacco cells (27).

Both trans-4-hydroxy-L-proline (Hyp) and azetidine-2-carboxylic acid (Azc) inhibit growth, an effect that can be relieved by proline (45); because the relief of Hyp inhibition was more specific (Fig. 2) this analogue was used in the selection studies.

MUTANTS OF AMINO ACID METABOLISM SELECTED IN BARLEY

From the results shown in Fig. 2, we set out to screen for resistance to S-2-aminoethylcysteine, lysine plus threonine and hydroxyproline using isolated barley embryos. The procedure used is shown in Figure 4. Table 1 summarizes the results of our selections to date and some characteristics of the mutants. The latter are described in more detail below.

Aec Resistance

We have selected and characterized two mutants containing recessive mutations conferring resistance to Aec (6,7). These have been given the Rothamsted (R) numbers R906 and R4402. Genetic analysis shows that these two independently isolated mutants contain nuclear, allelic, recessive genes (6, manuscript in preparation). The basis for resistance is the specifically decreased capability of

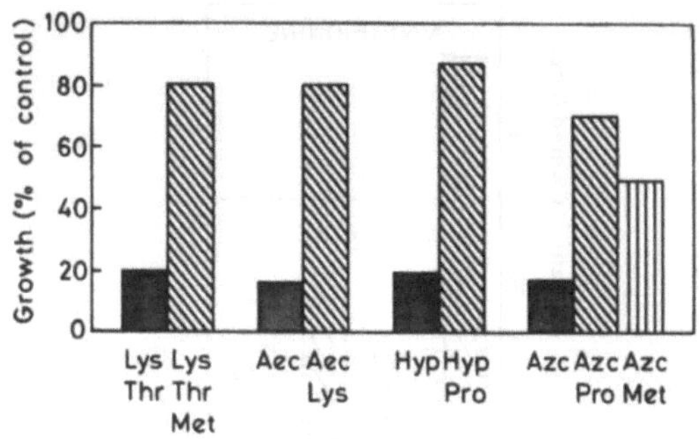

Fig. 2. The effects of various amino acids, singly and in combination on the growth of barley plants. Embryos were dissected out from mature seed, sterilized and plated out in petri dishes on agar containing inorganic salts, sucrose and various amounts of amino acids. Concentrations of amino acids used were: lysine (Lys) plus threonine (Thr) 2.5 mM each, with methionine (Met) 0.5 mM, aminoethyleysteine (Aec) 0.25 mM, lysine 1 mM, hydroxyproline (Hyp) 4 mM, azetidine carboxylic acid (Azc) 1 mM with proline (Pro) and methionine both 3 mM. Full information is given in Bright et al. 1978a (4), 1979a (6), and Kueh and Bright 1981 (45).

the roots of the mutant to take up the basic amino acids, lysine and its analogue Aec, and arginine; the capacity of R906 to take up neutral amino acids such as leucine is unimpaired (7). Although such mutants do not produce enhanced amounts of lysine, they may aid in the dissection of amino acid transport mechanisms.

Lysine Plus Threonine Resistance

Of four mutants selected, three have been studied in some detail. Genetic analysis has shown that resistance is due to dominant genes which we have designated Lt1a, Lt1b, and Lt2, and which are present respectively in R2501, R3202, and R3004. The gene Lt2 is not linked to either Lt1a or Lt1b, whereas the latter two behave, in crosses so far tested, as if they were allelic (10). Aspartate kinase, extracted from seedlings of normal barley or sensitive plants selected from the heterozygous mutant population, can be separated into three peaks of activity (AKI, AKII, and AKIII) on DEAE-cellulose (9,66). Investigations of the feedback sensitivity of these peaks shows that AKI is predominantly sensitive to threonine, whereas AKII and AKIII are inhibited by lysine or lysine plus AdoMet. In R2501 and R3202, AKII was found to be much less sensitive to lysine inhibition. The mutants differed in that, whereas AKII in R3202 had lost all sensitivity to lysine and lysine plus AdoMet, large concentrations of the two latter effectors

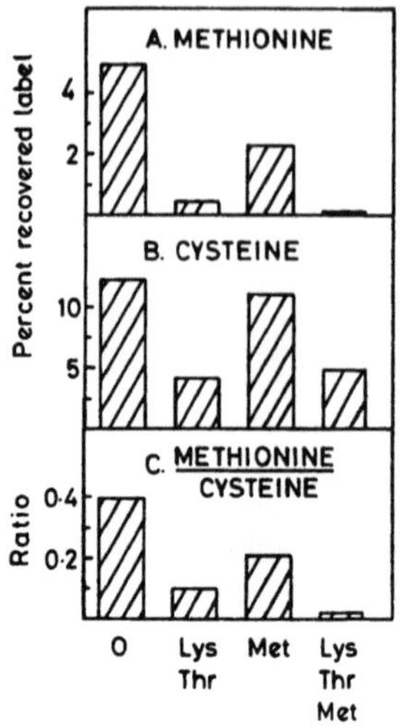

Fig. 3. Methionine and cysteine synthesis from ^{35}S-sulphate in barley roots and its inhibition by lysine plus threonine and methionine. Roots were taken from 20 d plants (cv Bomi) grown in aerated Hoaglands solution. Duplicate samples of 0.5 g material were incubated in 5 ml sterile Murashige and Skoog medium plus 50 mg/l gentamicin sulphate (0); or the same medium containing lysine and threonine each 2 mM (LT); methionine 0.5 mM (M), or lysine and threonine 2 mM and methionine 0.5 mM (LTM). After 3 hr the medium was replaced with the same medium plus 0.2 mCi $^{35}SO_4$ and incubated for a further 8.5 hr. Total labelled compounds were extracted from the roots after the addition of carrier methionine, performic acid oxidation and 6N HCl protein hydrolysis. Labelled compounds were separated by electrophoresis at pH 2.5. Results are expressed as the percentage of recovered label (1-2 x 10^4 dpm/20 µl extract) found in A: methionine sulphone; B: cysteic acid; and C: the ratio of A:B.

together could reduce activity by 80% in R2501. AKI and AKIII are unchanged in these mutants. In contrast, AKII (and AKI) were unaffected in R3004, but AKIII required ten times as much lysine for half-maximal inhibition. There were no changes in the properties of homoserine dehydrogenase in the mutants. We have interpreted these results as indicating that there are at least three isoenzymes (AKI, AKII, and AKIII) of aspartate kinase in barley. We suggest that the gene ltl is a structural locus for AKII and that we have selected two independent mutational events in this gene (Lt1a and Lt1b). Lt1a has had a less drastic effect on the enzyme in that the extractable activity is near normal and only the lysine binding site

Stage	Operation
1	Mutagenesis of seed with Na azide, γ rays or fast neutrons to give M_1 generation
2	Plant seed in field and harvest M_2 generation
3	Dissect out M_2 embryos and plate out on selective nutrient agar
4	Select M_2 plants growing better than defined standard (usually shoot length greater than 3 standard deviations from mean)
5	Grow on M_2 plants and harvest M_3 progeny
6	Retest M_3 progeny, determine segregation ratios of resistant M_3 families
7	Establish pure breeding lines from M_3 generation, bulk up seed for biochemical analysis, carry out crossing programmes to establish the genetics of the mutation

Fig. 4. Embryo selection scheme.

Table 1. Characteristics of selected barley mutants.

Numbers screened	Selection pressure	Mutants selected	Gene designation	Biochemical basis
20,000	0.25 mM Aec	R906	*aec1a*	decreased lysine uptake
		R4402	*aec1b*	
20,000	2.5 mM Lys + 2.5 mM Thr	R2501	*Lt1a*	desensitized AKII
		R2506	?	?
		R3004	*Lt2*	desensitized AKIII
		R3202	*Lt1b*	desensitized AKII
20,000	4 mM Hyp	R5201	*Hyp1a*	proline accumulation
		R6102	*Hyp1b*	
		R6901	?	
		R6902	*Hyp1c*	

appears to be affected. Lt1b causes a more marked change with much less extractable activity being present in R3202 and with complete loss of the lysine regulatory site and possibly also of the AdoMet site. We have not excluded the alternative possibility that the effectors are still bound, but that their inhibitory effect is not transmitted to the active site. Similarly, lt2 is considered to be a structural gene for AKIII, of which Lt2 is a mutant allele specifying an isoenzyme 10-fold less sensitive to feedback inhibition by lysine.

We have examined the consequences of these mutations on amino acid accumulation in shoots and in seeds (Table 2). The soluble pool of threonine is increased in both tissues of mutant R2501 and in seeds of R3004 (shoots not tested). However, no changes were found for R3202. No significant increases in free methionine were observed in any of the mutants. The changes in the soluble threonine pool of the seeds of R2501 and R3004 were sufficient to increase the total content of threonine by about 6-10%. Although this is sufficient to improve the nutritional quality with respect to threonine deficiency, it does not completely alleviate the problem.

Hydroxyproline Resistance

Four mutants have been obtained, R5201, R6102, R6901, and R6902 (11,45,46). We have concentrated our studies on R5201, R6102, and R6902 as these were obtained as homozygous, pure-breeding, resistant lines. Genetic analysis shows that all three of these mutants are allelic. The resistance seen in offspring between the mutants and normal barley can be categorized into three types: (a) resistant with good growth, green leaves and penetration of the agar by roots, (b) intermediate, with smaller, pale green leaves, some swelling and curling of roots, and (c) sensitive, poor growth, pale, swollen, and curled leaves. Although absolutely precise scoring is difficult, it has been possible to obtain sufficient results to show that the resistance behaves as if coded for by a single, semi-dominant nuclear gene. These genes termed *Hyp1a*, *Hyp1b*, and *Hyp1c* also confer resistance to L-thiazolidine-4-carboxylic acid, and partially to azetidine-2-carboxylic acid. The mutants are not resistant to Aec or to lysine plus threonine.

Amino acid analysis has shown that all the mutants contain about three times as much soluble proline in the leaves (Table 2), which is further increased by the imposition of water stress (40% w/v polyethylene glycol) (45). R5201 accumulates proline more rapidly and to a greater final amount than the parental, wild-type plants. However, attempts to show that these mutants have superior resistance to stress suggest that the presence of the mutant gene confers only a marginal advantage under NaCl stress (46).

ASSESSMENT OF PRESENT PROGRESS AND FUTURE PROBLEMS

The results summarized above show that it is possible to isolate mutants of barley that have the desired characteristics to investigate problems of crop quality. These results have been confirmed by other workers using our techniques (17,41). These selections can be achieved without the need to use tissue culture and protoplast systems. The mutations are stable, expressed in the

Table 2. Soluble amino acids in seeds and young plants of selected barley mutants.

A Soluble amino acids (nmol/mg N) in seeds[1]

Mutant	Thr	Lys	Ala	Glx	Total (17/20)
Wild type	12	12	47	140	740
R3004	116	15	41	122	745
R2501	147	19	29	112	743
R3202	16	18	68	199	905

B Soluble amino acid (nmol/g FW) in young plants

Mutant	Thr	Lys	Met	Pro
Wild type[2]	1300	350	140	n.d.
R2501[2]	5200	520	300	n.d.
Wild type[3]	n.d.	n.d.	n.d.	72
R5201[3]	n.d.	n.d.	n.d.	224
R6902[3]	n.d.	n.d.	n.d.	239
R6102[3]	n.d.	n.d.	n.d.	231

[1] Data from Bright et al. 1982b

[2] Data from Bright et al. 1982a

[3] Data from Kueh and Bright 1982

mature plant, and have been characterized with respect to genetics and in some cases biochemistry. Both recessive and dominant mutations have been identified. Besides providing potentially valuable variation in agronomic properties, the mutants also provide important tools for increasing our understanding of biochemical mechanisms; this latter aim is also furthered by certain tissue culture variants. The work described here is seen as only a beginning in terms of what may be achieved with respect to our understanding of amino acid biochemistry or in crop improvement. Some of the outstanding problems are discussed below.

Selection Protocols

Conventional opinion appears to hold that the way to obtain mutants of higher plants with defined alterations in their biochemistry and/or with improved agronomic traits is via tissue or protoplast culture techniques; little attention is paid to using whole plants. We are firmly of the opinion that the latter method is the one of choice. Even though people have been using tissue and protoplast techniques for the purpose of selecting mutants of amino acid metabolism since the early pioneering work on tissue culture variants by Widholm (85), and the first successful regenerants were achieved by Carlson (15) in 1973, Chaleff, in his recent book (18), listed only three examples of regenerated plants for which the genetics of a lesion in amino acid metabolism is known [the valine resistant line of tobacco (3), the lysine and threonine resistant mutant of maize (38), and Carlson's original methionine sulphoximine resistant mutants]. To these examples we can add the threonine deaminase minus mutant of *Nicotiana* (73) and the Aec resistant line of *Nicotiana* (41).

Over a somewhat shorter period of time, many mutants of amino acid metabolism have been selected using whole plant techniques. We have reported here on eight barley mutants with characterized changes in three pathways; Jacobs and colleagues (17,41) have selected some similar mutants using the same methods; and Somerville and Ogren (75) have identified and characterized mutants lacking serine-glyoxylate transaminase, ferredoxin-dependent glutamate synthase, and the ability to convert glycine to serine. If the discussion is widened somewhat more, the number of biochemical mutants becomes very large. Somerville and Ogren (75) actually selected 200 mutants; from these, about 50 have been analyzed as being defective in one of seven different genes, of which 6 have been characterized at the biochemical level. Lundqvist and Wettstein-Knowles (48) summarize work on 1355 induced and selected mutants of barley with altered synthesis and excretion of wax which include recessive mutations of some 70 loci and 18 mutants at one dominant locus; the biochemical changes associated with many of these mutants are known (e.g., see ref. 83). Again in barley, Møller, von Wettstein, and colleagues have devised screens for identifying biochemical mutants perturbed in their photochemistry and have characterized many of them (57,84).

Whereas there appear to be fundamental problems in regeneration of large numbers of fully fertile, cytologically normal plants in a wide range of species, a problem urgently in need of more study, there are no such restrictions on the use of whole plants. What is limiting the number of mutants selected is an insufficient effort being put into whole plant selection and into devising simple selection or screening techniques. With sufficient thought and ingenuity, apparent drawbacks can be overcome without recourse to tissue culture. For example, it might be thought that the lysine plus threonine selection system could not be used on legumes because of the large reserves of protein and amino acids in the embryo; however, Sainis and Rao (68) have shown that this can be overcome simply. They have found that lysine plus threonine inhibit greening of pieces of dark grown leaves, an effect presumed to be due to the lack of methionine for chlorophyll synthesis because methionine relieves the inhibition. Using this screening test, they have identified two presumptive mutant lines of Vigna in initial tests.

Lysine Accumulation

So far, none of the barley mutants selected accumulate enhanced levels of lysine. We ascribe this failure to the fact that barley contains only very low levels of threonine-sensitive aspartate kinase and has a dihydropicolinate synthase sensitive to lysine. This latter enzyme from wheat is very sensitive in that it is 50% inhibited by 11 μM lysine (50). The effect of these controls are probably that mutations in both enzymes would be required before lysine can accumulate; the chances of this occurring are small. We are therefore following a two-stage approach in which we have remutagenized seed of R3004 prior to selecting for Aec resistance. Hopefully there will be sufficient flux through the aspartate kinase reaction in this mutant to allow lysine to accumulate in plants with altered dihydrodipicolinate synthase and thus permit growth in the presence of Aec. Evidence to suggest that this strategy should prove successful is the isolation of Aec-resistant plants of N. sylvestris, which accumulate lysine (58). This species has a larger proportion of threonine-sensitive (i.e., lysine-insensitive) aspartate kinase activity than barley (41).

Even if double mutants are found, there may still be factors, particularly degradation, that mitigate against large amounts of lysine (or any other amino acid) building up in the tissue. In the case of lysine, an enzyme has recently been described which condenses lysine with 2-oxoglutarate to give saccharopine (1); this step is thought to be the first reaction in lysine breakdown in a wide range of organisms (40,56,81). The enzyme appears to be particularly active during endosperm development in maize (Arruda, personal communication). Thus, depending on the respective subcellular locations of it and lysine, it may modify the degree to which soluble lysine may be enhanced.

Complexity of Aspartate Kinase

The Lt mutants have provided possibly the only means of determining that there are at least two lysine-sensitive aspartate kinase isoenzymes in barley under the control of separate genes. However, this is unlikely to be the complete story. Studies with different tissues and different species suggest that the ratio and amount of lysine- and threonine-sensitive isoenzymes can vary as a function of developmental age and tissue (20,47,69). Recently Gonzales et al. (32) have reported that there is a 4-fold increase in the amount of lysine-sensitive aspartate kinase in ethionine-resistant tobacco cells. Consequently we may expect to uncover mutations which control the amount and ratios of the different forms of the isoenzymes, and these may have some value in understanding changes taking place in plants during their developmental cycle.

Proline and Stress

As yet, the increased proline levels present in R5201 and the other Hyp resistant lines do not appear to be sufficient to confer any significant resistance to water and salt stress. This, perhaps, is not too surprising, given that the amount of the increase is only about 3-fold (45,46), whereas proline can accumulate 10- to 50-fold under stress. However, studies in which exogenous proline has been fed to barley suggest that it may provide a significant relief of NaCl stress (Wyn Jones, this volume; Pervez et al., manuscript in preparation). Consequently, we consider that it may be necessary to carry out further selections on our mutants to find lines having higher concentrations of free proline in the absence of stress before we can really resolve the questions posed at the beginning of this paper.

As yet, the Hyp mutants have not allowed us any greater insight into the biochemistry of proline synthesis and its regulation in the way that Lt mutants have. However, we remain optimistic that progress will be made in this area.

Side Effects

The most important side effect of any mutational change is that it may decrease yield. Such decreases may be direct, due to pleiotropic effects, or may be secondary, due to increased susceptibility to pests and diseases. Work with the high-lysine mutants of maize (opaque 2) and barley (Risø 1508 and Hiproly) have shown that prolamin-deficient mutants have smaller seed, and that this results in a yield penalty. Although much effort over nearly 10 years has been put into breeding for higher yield in these mutants, it is by no means clear that high-lysine varieties with acceptable yields will become widely available (33,59,80). While it is relatively early to expect to be able to test the effect of the mutations reported here on yield, preliminary studies show that at least one of the original selections has reasonable productivity, seed size, and nitrogen content (Table 3). All the Lt mutants

Table 3. Agronomic characteristics of field-grown barley mutant R3004.

	Bomi	R3004	%
Yield Trial I	91	87	96
Yield Trial II	93	79	85
100 seed weight	3.91	3.75	96
Nitrogen content	1.42	1.48	105

Yields are expressed relative to a pooled variety control of 100% from triplicate small plots. L.s.d. (5%) trial I = 10%, trial II = 8.5%. Data from Dr. A. Hayter, SCRI Edinburgh. 100 seed weights (grams) are from triplicate air dried samples from yield trials. Nitrogen content (gN/100 g dw) are from whole meals of the same material. Data from Dr. M. Fuller, RRI Aberdeen.

possess the normal spectrum of storage proteins (Figure 5). Since the material used has not been backcrossed and reselected, it may still have other mutations than the ones which were selected.

Besides decreased seed size, the opaque-2 mutation can also cause increased susceptibility to diseases such as ear rots. It has also been suggested that enhanced proline levels may produce plants more susceptible to attack by insects (34). To test this latter possibility, R5201 was exposed to attack by locusts, slugs, aphids, and a mildew fungus, but there was no evidence that the mutant was more susceptible than the wild type (12). However, despite these encouraging results, it is obvious that agronomic use of these or any subsequent mutations will be dependent on their successful incorporation in varieties capable of passing the rigorous field tests used in commerce.

CONCLUSIONS

There are a number of reasons for seeking biochemical variants, including those in amino acid metabolism, among which are:

- to identify mutants of potential agronomic value and/or having a specific biochemical trait;
- to study the genetics of resistance mechanism;
- to provide genetic markers for classical genetic studies;

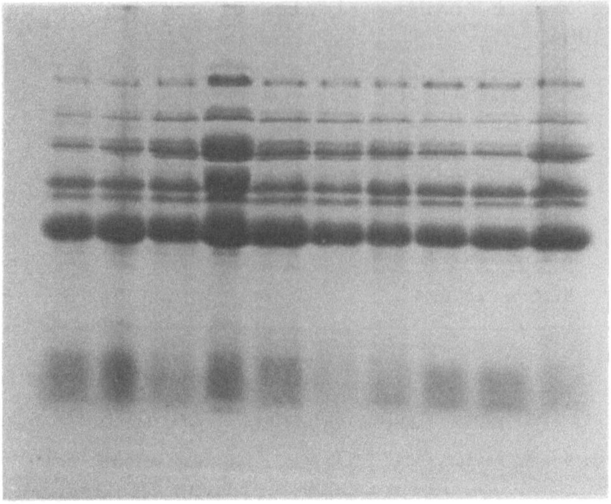

Fig. 5. Hordein storage proteins in single seeds of barley mutants resistant to lysine plus threonine. A,B: R2501 resistant lines; C: R2501 sensitive line; D: cv Bomi (parent line); E,F: R2506 resistant line; G: R3202 resistant line; H: R3202 sensitive line; I: R3004 resistant line; J: Bomi.

- to study the biochemistry of resistance mechanisms;
- to study variation induced or affected by different stages of plant de- and re-differentiation;
- to provide markers for somatic cell genetics and cell fusion studies.

While protoplasts and tissue cultures have provided useful material for the latter three aims, they have provided little for the first three. In contrast, selections using whole seed or embryo systems have provided a range of mutants in a variety of pathways; in many cases, both the biochemical and genetical basis of the lesion has been determined (this paper; 26,29,44,48,61,75,84). Many of these mutants, besides providing material for studies in the first three categories above, are useful for the latter three areas as well. We would therefore suggest that if people are interested in isolating variants or mutants of higher plants, even from Nicotiana species, they would be well advised to concentrate first on whole plant selection techniques and only to use protoplast and tissue cultures where there is a specific need or desire to study these systems as such. Providing that sufficiently good screening or selection procedures can be devised, the whole organism approach can yield dominant and recessive mutants capable of providing answers to important biochemical and agronomical questions; we believe that the barley mutants described here provide support for this statement.

ACKNOWLEDGEMENTS

We are grateful to Drs. M. Fuller and A. Hayter for the data in Table 3, and to Dr. P. Shewry for the separation shown in Fig. 5.

REFERENCES

1. Arruda, P., L. Sodek, and W.J. Da Silva. 1982. Lysine-ketoglutarate reductase activity in developing maize endosperm. Plant Physiol. 69:988.
2. Boggess, S.F., D. Aspinall, and L.G. Paleg. 1976. The significance of end-product inhibition of proline biosynthesis and of compartmentation in relation to stress induced proline accumulation. Aust. J. Plant Physiol. 3:513.
3. Bourgin, J-P. 1978. Valine resistant plants from in vitro selected tobacco cells. Mol. Gen. Genet. 161:225.
4. Bright, S.W.J., E.A. Wood, and B.J. Miflin. 1978a. The effect of aspartate-derived amino acids (lysine, threonine, methionine) on the growth of excised embryos of wheat and barley. Planta 139:113.
5. Bright, S.W.J., P.R. Shewry, and B.J. Miflin. 1978b. Aspartate kinase and the synthesis of aspartate-derived amino acids in wheat. Planta 139:119.
6. Bright, S.W.J., P.B. Norbury, and B.J. Miflin. 1979a. Isolation of a recessive barley mutant resistant to S-2-aminoethylcysteine. Theor. Appl. Genet. 55:1.
7. Bright, S.W.J., L. Featherstone, and B.J. Miflin. 1979b. Lysine metabolism in a barley mutant resistant to S-2-aminoethylcysteine. Planta 146:629.
8. Bright, S.W.J., P.B. Norbury, and B.J. Miflin. 1980. Isolation and characterisation of barley mutants resistant to aminoethyl cysteine and lysine plus threonine. In Plant Cell Cultures: Results and Perspectives, 179-182. F. Sala, B. Parisi, R. Cella, and O. Ciferri, eds. Elsevier: North Holland Biomedical Press.
9. Bright, S.W.J., B.J. Miflin, and S.E. Rognes. 1982a. Threonine accumulation in the seeds of barley mutant with an altered aspartate kinase. Biochem. Genet. 20:229.
10. Bright, S.W.J., J.S.H. Kueh, J. Franklin, S.E. Rognes, and B.J. Miflin. 1982b. Two genes for threonine accumulation in barley seeds. Nature 299:278.
11. Bright, S.W.J., J.S.H. Kueh, J. Franklin, and B.J. Miflin. 1982c. Proline-accumulating barley mutants. Barley Genetics IV, Proc. Fourth Int. Barley Genet. Symp. July 1981 (in press).

12. Bright, S.W.J., P.J. Lea, J.S.H. Kueh, C. Woodcock, D.W. Holloman, and G.C. Scott. 1982d. Proline content does not influence pest and disease susceptibility in barley. Nature 295:592.
13. Burlandt, L., P. Datta, and H. Gest. 1965. Control of enzyme activity in growing bacterial cells by concerted feedback inhibition. Science 148:1351.
14. Bryan, J.K. 1980. Synthesis of the aspartate family and branched-chain amino acids. In The Biochemistry of Plants, vol. V, Amino Acids and their Derivatives, 403. B.J. Miflin, ed. London and New York: Academic Press.
15. Carlson, P.S. 1973. Methionine sulphoximine-resistant mutants of tobacco. Science 180:1366.
16. Cattoir, A., E. DeGryse, M. Jacobs, and I Negrutiu. 1980. Inhibition of barley and Arabidopsis callus growth by lysine analogues. Plant. Sci. Lett. 17:327.
17. Cattoir-Reynaerts, A., E. DeGryse, and M. Jacobs. 1981. Selection and analysis of mutants overproducing amino acids of the aspartate family in barley, Arabidopsis and carrot. In Induced Mutations a Tool in Plant Breeding, 353-361. Vienna: IAEA.
18. Chaleff, R.S. 1981. Genetics of Higher Plants: Applications of Cell Culture. Cambridge: Cambridge University Press. 184 pp.
19. Csonka, L.H. 1981. Proline overproduction results in enhanced osmotolerance in Salmonella typhimurium. Mol. Gen. Genet. 182:82.
20. Davies, H.M., and B.J. Miflin. 1978. Regulatory isoenzymes of aspartate kinase and the control of lysine and threonine biosynthesis in carrot cell suspension culture. Plant Physiol. 62:536.
21. Demain, A.L. 1970. Overproduction of microbial metabolites and enzymes due to alteration of regulation. Adv. Biochem. Engin. 1:113.
22. Doll, H., B. Koie, and B.O. Eggum. 1974. Induced high-lysine mutants in barley. Radiation Bot. 14:73.
23. Doll, H., and B. Koie. 1978. Influence of the high lysine gene from barley mutant 1508 on grain carbohydrate and protein yield. In Seed Protein Improvement by Nuclear Techniques, 107. Vienna: IAEA.
24. Erickson, S.S., and W.V. Dashek. 1982. Foliar soluble proline accumulation does not originate by either synthesis from glutamate or proteolysis in SO_2-stressed Hordeum vulgare cv Proctor seedlings. Plant Physiol. 69:Suppl., 43.
25. Eveleigh, D.E. 1981. The microbiological production of industrial chemicals. Scientific American 245:120.
26. Feenstra, W.J. 1964. Isolation of nutritional mutants in Arabidopsis thaliana. Genetica 35:259.
27. Filner, P. 1969. Control of nutrient assimilation. A growth-regulatory mechanism in cultured plant cells. Dev. Biol. Suppl. 3:206.

28. Fuller, M.F., R.M. Livingstone, B.A. Baird, and T. Atkinson. 1979. The optimal amino acid supplementation of barley for the growing pig 1. Response of nitrogen metabolism to progressive supplementation. Br. J. Nutr. 41:321.
29. Gavazzi, G., M. Nava-Rachi, and C. Tonelli. 1975. A mutation causing proline requirement in maize. Theor. Appl. Genet. 46:339.
30. Giovanelli, J., S.H. Mudd, and A.H. Datko. 1980. Sulphur amino acids in plants. In The Biochemistry of Plants, vol. V. Amino Acids and their Derivatives. B.J. Miflin, ed. London and New York: Academic Press.
31. Gladstone, C.P. 1939. Inter-relationships between amino acids in the nutrition of B. anthracis. Brit. J. Exp. Path. 20:189.
32. Gonzales, R.A., P.K. Das, and J.M. Widholm. 1982. Characterization of cultured tobacco cells resistant to a methionine analog. Plant Physiol. 69, suppl.:33.
33. Hagberg, A., G. Persson, R. Ekman, K.E. Karlsson, A.M. Tallberg, V. Stoy, N.-O. Bertholdsson, M. Mounla, and H. Johansson. 1979. The Svalov protein quality breeding programme. In Seed Protein Improvement in Cereals and Grain Legumes, vol. II, 311-314. Vienna: IAEA.
34. Haglund, B.M. 1981. Proline and valine--cues to stimulate grasshopper herbivory? Nature 288:697-698.
35. Hanson, A.D., C.E. Nelsen, A.R. Pedersen, and E.H. Everson. 1979. Capacity for proline accumulation during water stress in barley and its implications for breeding for drought resistance. Crop. Sci. 19:489.
36. Hanson, A.D., and W.D. Hitz. 1982. Metabolic responses of mesophytes in plant water deficits. Annu. Rev. Pl. Physiol. 33:163.
37. Hejgaard, J., and S. Boisen. 1980. High lysine proteins in Hiproly barley breeding: Identification, nutritional significance and new screening methods. Hereditas 93:311.
38. Hibberd, K.A., and C.E. Green. 1982. Inheritance and expression of lysine plus threonine resistance selected in maize tissue culture. Proc. Natl. Acad. Sci. U.S.A. 79:559.
39. Hibberd, K.A., T. Walter, C.E. Green, and B.G. Gengenbach. 1980. Selection and characterisation of a feedback-insensitive tissue culture of maize. Planta 148:183.
40. Hutzler, J., and J. Dancis. 1968. Conversion of lysine to saccharopine by human tissues. Biochem. Biophys. Acta 158:62.
41. Jacobs, M., A. Cattoir-Reynaerts, I. Negrutiu, I. Verbruggen, and E. DeGryse. 1982. Abstracts of International Plant Tissue Conference, Tokyo (in press).
42. Joy, K.W., and B.F. Folkes. 1965. The uptake of amino acids and their incorporation into proteins of excised barley embryos. J. Exp. Bot. 16:646.
43. Kleinhofs, A., C. Sander, R.A. Nilan, and C.F. Konzak. 1974. Azide mutagenicity-mechanisms and nature of mutants produced. In Polyploidy and Induced Mutations in Plant Breeding STI/PUB/359. Vienna: IAEA.

44. Kleinhofs, A., R.L. Warner, F.J. Muehlbauer, and R.A. Nilan. 1978. Induction and selection of specific gene mutations in Hordeum and Pisum. Mutation Research 51:29.
45. Kueh, J.S.H., and S.W.J. Bright. 1981. Proline accumulation in a barley mutant resistant to trans-4-hydroxy-L-proline. Planta 153:166.
46. Kueh, J.S.H., and S.W.J. Bright. 1982. Biochemical and genetical analysis of three proline accumulating barley mutants. Plant Sci. Lett. 27:233.
47. Lea, P.J., W.R. Mills, and B.J. Miflin. 1979. The isolation of a lysine-sensitive aspartate kinase from pea leaves and its involvement in homoserine biosynthesis in isolated chloroplasts. FEBS Lett. 98:165.
48. Lundqvist, U., and P. von Wettstein-Knowles. 1982. Dominant mutations at Cer-yy change barley spike wax into leaf blade wax. Carlsberg Res. Commun. 47:29.
49. Madison, J.T., and J.F. Thompson. 1976. Threonine synthase from higher plants - stimulation by S-adenosymethionine and inhibition by cysteine. Biochem. Biophys. Res. Commun. 71:684.
50. Mazelis, M., F.R. Whatley, and J. Whatley. 1977. The enzymology of lysine biosynthesis in higher plants. FEBS Letts. 84:236.
51. Mestichelli, L.J.J., R.N. Gupton, and I.D. Spenser. 1979. The biosynthetic route from ornithine to proline. J. Biol. Chem. 256:640.
52. Miflin, B.J.. 1969. The inhibitory effects of various amino acids on the growth of barley seedlings. J. Exp. Bot. 20:810.
53. Miflin, B.J., and P.R. Shewry. 1979. The synthesis of proteins in normal and high lysine barley seed, In Recent Advances in the Biochemistry of Cereals, D.L. Laidman and R.G. Wyn Jones, eds., London: Academic Press.
54. Miflin, B.J., and P.J. Lea. 1982. Ammonia assimilation and amino acid metabolism. In Encyclopedia of Plant Physiology New Series, Vol 14A. D. Boulter and B. Partheir, eds. Berlin, Heidelberg, New York: Springer-Verlag, pp 1-64.
55. Miflin, B.J., S.W.J. Bright, and E. Thomas. 1982. Towards the genetic manipulation of barley. In Barley Genetics IV, R.N.H. Whitehouse and M.J. Allison, eds. (in press).
56. Møller, B.L. 1976. Lysine catabolism in barley (Hordeum vulgare L.). Plant Physiol. 57:687.
57. Møller, B.L., J.H.A. Nugent, and M.C.W. Evans. 1981. Electron paramagnetic resonance spectroscopy of photosystem I mutants in barley. Carlsberg Res. Commun. 46:373.
58. Negrutiu, I., A. Cattoir-Reynaerts, I. Verbruggen, and M. Jacobs. 1981. Lysine overproduction in an S-aminoethylcysteine-resistant mutant, isolated in protoplast culture of Nicotiana sylvestris, Spegg and Comes. Societe Belge de Biochemie B, 188.
59. Nelson, O.E. 1980. Genetic control of polysaccharide and storage protein synthesis in the endosperms of barley, maize,

and sorghum. In Advances in Cereal Science and Technology, vol. III, Y. Pomeranz, ed. St. Paul, MN: American Association of Cereal Chemists, 41.
60. Oaks, A., I.J. Mitchell, R.A. Barnard, and F.J. Johnson. 1970. The regulation of proline biosynthesis in maize roots. Can. J. Bot. 48:2249.
61. Oostindier-Braaksma, F.L. and W.J. Feenstra. 1973. Isolation and characterisation of chlorate-resistant mutants of Arabidopsis thaliana. Mutat. Res. 19:175.
62. Phillips, R.L., P.R. Morris, F. Wold, and B.G. Gengenbach. 1981. Seedling screening for lysine-plus-threonine resistant maize. Crop Sci. 21:601-607.
63. Reisch, B., and E.T. Bingham. 1981. Plants from ethionine-resistant alfalfa tissue cultures: Variation in growth and morphological characteristics. Crop Sci. 21:783.
64. Reisch, B., S.H. Duke, and E.T. Bingham. 1981. Selection and characterisation of ethionine-resistant alfalfa (Medicago sativa L.) cell lines. Theor. Appl. Genet. 59:89.
65. Rognes, S.E., P.J. Lea, and B.J. Miflin. 1980. S-Adenosylmethionine, a novel regulator of aspartate kinase. London: Nature 287:357.
66. Rognes, S.E., S.W.J. Bright, and B.J. Miflin. 1983. Feedback-insensitive aspartate kinase isoenzymes in barley mutants resistant to lysine plus threonine. Planta (in press).
67. Sainis, J.K., R.G. Mayne, R.M. Wallsgrove, P.J. Lea, and B.J. Miflin. 1981. Localisation and characterisation of homoserine dehydrogenase isolated from barley and pea leaves. Planta 152:491.
68. Sainis, J.K., and S.R. Rao 1982. A simple procedure for rapid preliminary screening for lysine plus threonine resistance in green gram (Vigna radiata) Plant Sci. Lett. 25:91.
69. Sakano, K., and A. Komamine. 1978. Changes in the proportion of two aspartokinases in carrot root tissue in response to in vitro culture. Plant Physiol. 61:115.
70. Shewry, P.R., and B.J. Miflin. 1977. Properties and regulation of aspartate kinase from barley seedlings (Hordeum vulgare L.). Plant Physiol 59:69.
71. Shewry, P.R., and B.J. Miflin. 1982. Genes for the storage proteins of barley. Qual. Plant.: Plant Foods Hum. Nutr. 31:251.
72. Shewry, P.R., and S.W.J. Bright. 1983. Improvement of protein quality in cereals. Crit. Rev. Plant Sci. (in press).
73. Sidorov, V.A., L. Menczel, and P. Maliga. 1981. Isoleucine requiring Nicotiana plant deficient in threonine deaminase. Nature 294:87.
74. Singh, T.N., C.G. Paleg, and D. Aspinall. 1973. Stress metabolism III. Variations in response to water deficit in the barley plant. J. Biol. Sci. 26:65.
75. Somerville, C.R., and W.L. Ogren. 1982. Genetic modification of photorespiration. Trends Biochem. Sci. 7:171.

76. Stewart, G.R., and F. Larher. 1980. Accumulation of amino acids and related compounds in relation to environmental stress. In Biochemistry of Plants, Vol V, B.J. Miflin, ed. London: Academic Press, pp 609-635.
77. Thompson, G.A., A.H. Datko, S.H. Mudd, and J. Giovanelli. 1982. Methionine biosynthesis in Lemna. Plant Physiol. 69:1077.
78. Thompson, J.F. 1980. Arginine synthesis, proline synthesis and related processes. In The Biochemistry of Plants, Vol V, B.J. Miflin, ed. Amino Acids and Their Derivatives. London, New York: Academic Press, pp 375-402.
79. Umbarger, H.E. 1978. Amino acid biosynthesis and its regulation. Annu. Rev. Biochem. 47:533.
80. Vasal, S.K., E. Villegas, and R. Bauer. 1979. Present status of breeding quality protein maize. In Seed Protein Improvement in Cereals and Grain Legumes, Vol. II, IAEA, Vienna, pp 127.
81. Wade, M., D.M. Thomson, and B.J. Miflin. 1980. Saccharopine an intermediate of L-lysine biosynthesis and degradation in P. oryzae. J. Gen. Microbiol. 120:11.
82. Wallsgrove, R.M., and M. Mazelis. 1981. The enzymology of lysine biosynthesis in higher plants: Partial purification and characterization of spinach leaf dihydrodipicolinate synthase. Phytochemistry 20:2651.
83. Wettstein-Knowles, P. von. 1972 Genetic control of β-diketone and hydroxy-β-diketone synthesis in epicuticular waxes of barley. Planta 106:113.
84. Wettstein, P. von, B.L. Møller, G. Hoyer-Hansen, and D. Simpson. Mutants in the analysis of the photosynthetic membrane polypeptides. In Origin of Chloroplasts. J.A. Schiff and R.Y. Stanier, eds. Amsterdam: Elsevier, North Holland Biomedical Press.
85. Widholm, J.M. 1972. Anthranilate synthetase from 5-methyl-tryptophan-susceptible and resistant cultured Daucus carota cells, Biochim. Biophys. Acta 279:48.
86. Widholm, J.M. 1976. Selection and characterization of cultured carrot and tobacco cells resistant to lysine, methionine, and proline analogs. Can. J. Bot. 54:1523.

CHALLENGES TO CROP IMPROVEMENT:

CHAIRMAN'S INTRODUCTION

Albert Ellingboe

Plant Pathology
International Plant Research Institute
853 Industrial Road
San Carlos, California 94070

In the preceding chapters there are many references to the need to improve yield, quality, and the stability of yield and quality. Particular references have been increasing the yield potential. Probably equally important is to find means to evaluate the yield potential. We know, for example, that present corn cultures have high potentials for yield. About a decade ago a world record for corn production on non-irrigated land was set in southeastern Michigan. The yield was 306 bushels per acre on a 20 acre field. Clearly a standard hybrid like Funks G4444, the one grown in this field, has a genetic potential for high yield. What was unique about the procedures by this one farmer who was able to realize this high yield? Can it be reproduced? One fact that is clear is that we as researchers are unable to set up demonstration plots to show farmers how to consistently get 300 bushels per acre. That one year and one field are not just a querk because 350 bushels per acre were produced in a commercial field in Michigan a few years later, again with a commercial hybrid, but this time with some irrigation.

The point that I'm trying to make in the above discussion is that we know the genetic potential for yield is very high in several commercial corn cultivars. Why can't we realize that potential on a more routine basis? Why can't we at least set up a demonstration to show the genetic potential for yield? What restricts our ability to demonstrate that potential?

When I have read about increasing yield there usually is an implication that it is necessary to accumulate genes that have a positive function. The implication is that if more active alleles can be introduced, more of the critical enzymes will be produced and the plant will be more productive. Is the accumulation of additive and/or dominant genes that have a positive function for growth the appropriate way to view the process of increasing the actual yield? Can it be that much of the gains made by breeders are due to a fortuitous selection for genes that control losses? We know about losses to some pathogens and insects. We know very little about others. And what other components are there that we cannot conjecture to be there?

We can debate whether the above argument is a question of semantics, but it can't be ignored. If the biological phenomena have not been properly described (or at best conceived) the possibility of explaining those phenomenon at a molecular level is almost nil. Furthermore, if the effort is to try to clone genes involved in these functions, the chance of cloning a gene for a null function is almost zero. So it becomes extremely important to get the concepts of the basic biology correct. It is important to critically analyze the data to see if our hypotheses are realistic, and to devise experiments to test the hypotheses rather than just to support them.

The discussions throughout this volume reinforce an observation made many years ago. Even though a large number of people consider themselves geneticists, there is a wide gulf in the communication between plant breeders and molecular geneticists that is hard to fathom. It is difficult to believe that they were instructed in the same scientific language. The thought processes seem to be entirely different. Breeders like to talk about statistical analyses. A trait is basically a meaningless term to a molecular geneticist. If a trait can be expressed in terms of a limited number of defined genes, the molecular geneticist has a handle to begin research. In the present state of knowledge, the molecular geneticist usually is confined to studying one gene (or one DNA fragment). What constitutes an agronomically useful gene seems to be outside the scope of many molecular geneticists. It is easy to conjecture what genes are useful in commercial production, and it is also easy to produce new cultivars. What is commercially useful and accepted for commercial production is another problem. What is perceived as being useful by the researcher may not be accepted by the commercial producers. There is a dire need for the breeders and molecular geneticists to learn enough of the same language, and each others thought processes, so that they can converse. The breeders will have to learn to state their problems in a framework that is meaningful to molecular geneticists. Molecular geneticists will have to describe to the breeders what they can and cannot do so that the breeders do not develop unreasonable expectations as to what can be done, and in what time frame.

Disease resistance is a complex trait. For example, wheat has more than 30 genes for resistance to leaf rust, more than 30 genes for resistance to stem rust, more than 8 genes for resistance to powdery mildew, and many more genes for resistance to other pathogens. The one fortunate characteristic of disease resistance is that it is possible to work with one gene at a time. There are methods for scoring for each of the Sr genes for stem rust essentially independent of the other. The commercially and biologically important problem of disease resistance can be stated in a manner that makes these genes amenable to analysis by molecular geneticists. Disease resistance, therefore, will probably be one of the first economically important traits to be studied by molecular genetics. Our challenge is to proceed with a system that we can deal with in a realistic way, where we have defined an agronomically important problem, stated the problem in a manner so that it can be examined experimentally and critically, and to see if our perception of the problem and what we could do about it was part of reality.

THE EVOLUTION OF HOST-PARASITE INTERACTION

P.R. Day, J.A. Barrett*, and M.S. Wolfe

Plant Breeding Institute, Cambridge, England
*Department of Genetics, University of Liverpool, England

INTRODUCTION

The extent of economic damage to crop plants due to pests and diseases is very large and well documented. Minimizing this loss through the use of pesticides is expensive. The high cost of pesticides and the threat they pose to non-target organisms make breeding for disease resistance a cost-effective and attractive alternative. Seed for most annual crops, except potatoes, is still under 10% of the total production cost to the farm gate. Clearly, the returns from improvements in yield, quality, and resistance due to plant breeding are so great that public and private investment in this activity will grow. Disease resistance is thus a primary objective in every plant breeding program. For example, at the Plant Breeding Institute, of some 50 research staff who work with wheat, 16 are engaged solely on aspects of resistance to fungus diseases and aphids. National and local trials all over the world emphasize resistance in evaluating new cultivars. However, breeding for resistance has had many failures; the appearance of new races of old parasites and sometimes of new parasites has frequently caused the withdrawal of new and promising cultivars in almost every crop. In this paper we briefly review the major constraints to the use of disease resistance by breeders and farmers and discuss their consequences in terms of parasite evolution. Clearly, most plant breeders would like eventually to use genetic transformation to confer resistance on breeding lines. This is not yet possible, and the problem is to establish where and how to start this activity.

MAN-GUIDED EVOLUTION

Plant breeders setting out to breed for disease resistance either collect material from the wild or make use of collections of early primitive cultivars and land races (13). Each accession represents a small sample of an original population, and if the breeder is to stand a chance of detecting resistance, then its frequency in that population probably has to approach one or two per- cent or higher. In order for a resistance gene to attain such a frequency, it must increase fitness and also be capable of being selected as a heterozygote. A tendency for increased expression will enlarge its selective advantage. If its effects on increasing fitness are important, there will be selection for modifiers which promote dominance of resistance. One would therefore expect that the genes for resistance to disease that are selected by breeders would tend to be dominant. At the same time the breeder will be concerned to identify genes with clear-cut, easily scored effects that do not complicate his breeding program. Recessive resistance will tend to be discarded in favor of dominant resistance which can be selected as a heterozygote. During successive generations of selection the breeder will accumulate a genetic background which enhances the expression of resistance. It is not surprising, therefore, to find, in analyses of disease resistance used by plant breeders, that the majority of examples involve dominant genes. However, genes that are recessive or which have intermediate dominance, and genes with small effects on the phenotype that are not clear-cut, nevertheless restrict the growth of the parasite and can therefore be classified as determining resistance. They tend not to be used unless they are all that is available. The resistance genes that breeders have preferred to use represent an extreme. That they are dominant has little bearing on the mechanism by which they bring about their effects. It merely reflects how they were detected and the way they have been used.

A crop plant protected by a dominant major gene is changed in such a way that it provides a substrate which the pathogen is unable to colonize because it is beyond the normal limits of variation of the pathogen population. Although some extremes of continuous variation could be selected, they may be so unbalanced that they would have strongly disruptive effects on pathogen fitness. Mutations at single loci appear to provide a suitable phenotypic effect without any wholesale disruption of the rest of the genome and consequent detrimental effects on other aspects of fitness. The population will likely include them at a low frequency (10^{-6} - 10^{-8}). The great majority will be poorly adapted but nevertheless able to survive on the resistant host because there is not competition from the non-mutant parent pathogen. As a consequence of intense selection, one form will emerge which will rapidly predominate. Clonally propagated, by vegetative spores for example, the combination of the primary mutation and the particular collection of adaptive background modifiers will persist and will eventually be recognized by the plant pathologist as a new race of the pathogen. By this time other, more poorly adapted forms will have been eliminated by natural selection. In a diploid or dikaryotic fungal pathogen, rare deleterious mutants are more likely to be recessive and, under the intense selection imposed by a cultivated crop, the newly favored variant will be selected as a recessive homozygote. There will be no opportunity for heterozygotes to be formed and dominance to evolve. This is presumably why virulence is recessive. Again, as in the case of dominant resistance, this is unlikely to have any biochemical or functional significance in terms of mechanism. The idea that different mutants may have the same phenotype could be tested by comparing the virulent forms of a pathogen that arise on geographically separated populations of the host protected by the same resistance gene. One might find different virulence genes and modifiers in the pathogen populations from different environments. A much simpler test is to select mutants of independent origin from an avirulent pathogen that are all virulent on a resistant host and to compare them to see if they include alterations at several loci. So far as we are aware, no such tests have been carried out.

The effect of releasing cultivars with different resistance genes is to direct the evolution of their parasites. Several authors have discussed the unstable equilibria that are set up in these interacting populations (see ref. 1). Various strategies for minimizing the damaging effects of parasites that have overcome major gene resistance have been proposed. One strategy is to deploy mixtures of cultivars carrying different resistance genes. Single genes that are ineffective in monocultures may be very useful in mixtures with three or more components (20) (Table 1). The mixtures are reformulated frequently to take advantage of the yield, quality, and disease resistance of new cultivars. The mixtures also exploit important effects introduced by the variation in genetic backgrounds among the components (Table 2). This adds a useful new dimension that is neglected when only dominant resistance genes are taken into account. The pathogen population is thus forced to adapt to a host which is heterogeneous in both space and time. Provided the supply of resistance genes is sufficient, the host population can be changed more rapidly than the pathogen can respond so that disease control is stable and maintained (19).

A CONSEQUENCE OF HOST-PARASITE EVOLUTION

One of the most important general principles to have emerged from genetic comparisons of a cultivated resistant host and one of its parasites is Flor's gene-for-gene hypothesis. Studies of the inheritance of resistance in flax to the rust *Melampsora lini*, coupled with parallel studies of the inheritance of virulence in the rust to the flax host, led Flor to the conclusion that "for each gene conditioning rust reaction in the host, there is a specific gene conditioning pathogenicity in the parasite" (9). Flor's reference to a specific gene may be misleading because, while there were no doubt specific genes segregating in the F_2 host and parasite

Table 1. Percentage of powdery mildew (*Erysiphe graminis hordei*) on the upper leaves of four mixtures of three barley cultivars, each compared with the mean infection on the same three cultivars grown separately.

Components	Mildew infection	
	Observed	Expected
Athos-Georgie-Mazurka	6.5	20.3
Hassan-Georgie-Wing	9.3	19.4
Athos-Ambre-Mazurka	2.8	9.7
Athos-Sundance-Mazurka	1.7	7.9
Mean	5.1 36.0%	14.3

From Wolfe and Barrett (20).

Table 2. Powdery mildew (whole plot scores relative to the means of the components grown alone) on three barley cultivar mixtures, in each of which the components have the same identified resistance genes but different unidentified resistances, and a fourth mixture in which each cultivar has different identified and unidentified resistance genes.

Identified genes	Cultivars	Dates 4/6	11/6	20/6	28/6
Mlg+Mlv	Abacus-Georgie-Sundance	100	83	82	47
Mlg+Mlas	Aramir-Athos-Porthos	116	83	74	34
Mla4/7	Ark Royal-Mazurka-Wing	102	72	68	64
(Mlg+Mlas)+ Mla4/7+(Mlg+Mlv)	Athos-Mazurka-Sundance	33	36	31	22

From Wolfe, Barrett, and Jenkins (21).

populations, he did not establish that the same specific rust genes were always responsible for the reactions he observed in rust races from geographically isolated populations. We should also remember that while the compelling simplicity of the gene-for-gene hypothesis led to its extension to a number of other host-parasite interactions (see ref. 4), very few other systems have been examined as rigorously as flax-flax rust. Neither has the hypothesis yet been shown to apply in a rigorous analysis of a natural host-parasite interaction involving a non-cultivated host plant. Probably the biggest influence that the hypothesis has had on plant pathology has been to encourage us to believe that elucidating the mechanisms of host-parasite interaction will most easily be accomplished by studying its apparently simple single gene differences. Many papers discussing gene-for-gene systems have speculated on the nature of specificity, and yet in no case has it been clearly explained. Certainly the gene-for-gene hypothesis is a useful statement of where plant breeding has led us, but it must not be allowed to limit our thinking.

The gene-for-gene hypothesis does not, of course, itself imply a general mechanism. But it has led to the expectation that the basis of our recognition of specificity is the interaction of the product of an allele for avirulence with the product of an allele for resistance in the host (6), and that the key to understanding host-parasite interaction is to identify these gene products and show how they produce their effects. Of course, it may be equally important to understand the reaction between the products of a host resistance gene and a pathogen virulence gene. This results in a susceptible or compatible phenotype which may, or may not, be qualitatively similar to compatible phenotypes in which a resistance gene apparently plays no part. In those pathogens that produce pathotoxins, the basis of specificity is also accommodated by the gene-for-gene hypothesis. Specificity depends on the interaction of a pathogen gene product for virulence -- the pathotoxin -- with the product of a host gene for susceptibility (see ref. 3).

No doubt the mechanism of specificity determination will soon be understood for several resistant hosts and avirulent strains of their pathogens, but whether or not important and useful generalizations will follow from this is by no means clear. Our hope is that it will lead to generalized methods for identifying the resistance genes we would like to clone and move into plants by the vector systems we are all presently engaged in developing.

IS THE GENE-FOR-GENE HYPOTHESIS AN ARTIFACT?

The gene-for-gene hypothesis is an elegant statement of the rules governing the interactions between one kind of host resistance and the pathogens forced to respond to it. It ignores most other forms of resistance. These include: host variation that affects the probability of infection, rate of its establishment, and the quantities of inoculum produced. The host's tolerance to infection, the durability or stability of resistance over time, genotype-environment interaction, and the effects that virulence may have on general pathogen fitness are other factors of epidemiological significance that the hypothesis does not take into account. Instead, it concentrates attention on evolved systems that have been pushed by the breeder toward an extreme which maximizes the differences between resistance and susceptibility. As a consequence, the role of background and modifiers in both host and pathogen is ignored. In a sense, gene-for-gene systems are the tips of icebergs. We cannot see, and we therefore neglect, what is underneath. When, by transformation, we can move the base sequence corresponding to a resistance gene into a susceptible plant, we might only be able to obtain the phenotype we expect if the susceptible plant is virtually isogenic with the donor, if expression of the resistance gene is dependent on its genetic background.

We must also ask if each gene-for-gene effect is a special case. Every day, potential hosts are bombarded with the spores of potential pathogens, some of which establish themselves and cause disease. Any variation in the host population which reduces the probability of infection will be favored by natural selection, and, conversely, any variant in the parasite which enables it to overcome the defenses of the host and reproduce successfully will also be favored. Thus there will be a constant flux of reciprocal adaptive responses in host and pathogen as each is influenced by the presence of the other. Evolution is an ad hoc process; it improvises on the variation available. It would therefore be very surprising if, under these circumstances, a single mechanism underlay all plant host-parasite interactions.

ACCELERATING EVOLUTION

The attractions of molecular biology include the prospect of manipulating plant disease resistance and making better use of it than we can at present. There will likely be a number of steps. First, we need to understand the mechanisms of the genetic controls of host-parasite specificity we wish to work with (see Day, ref. 5, in press). This will be important for isolating and cloning the host and parasite genes that control interaction. With cloned resistance genes, it should then be possible to mutate or construct sequences (artificial genes). At the same time, progress in plant transformation and the design of vector systems will allow us to test isolated genes. For the moment, such tests are limited to the prokaryotic plant parasites, bacteria, and viruses. An efficient method of plant transformation will also enable us to develop tests for evaluating non-host resistance. The remainder of this paper explores these ideas further.

THE DISTRIBUTION OF RESISTANCE GENES

The steady growth of linkage maps for the major crop plants provides an opportunity to establish whether or not there are patterns in the way resistance genes are distributed over chromosomes and chromosome arms that are relevant to cloning. For example, if there are general mechanisms for disease resistance, then one might expect the host genes controlling them to be grouped because they are functionally related. This is certainly borne out for mildew resistance in barley, where as many as 30 alleles, or genes, of the Ml-a series (10) plus five other loci, all controlling resistance to *Erysiphe graminis hordei*, are located on the short arm of chromosome 5 (12). Two other loci, *Ml-g* and *ml-o*, are on chromosome 4 (18). Shewry et al. (17) found that Ml-a is between the two structural genes *Hor-1* and *Hor-2* that control hordein polypeptides of the seed storage proteins. The sequence and map distances in centimorgans of this region of 5S are:

Hor-2 .082 ± .024 *Mla* .064 ± .020 *Hor-1* .....centromere

However, genes for resistance to other pathogens of barley are less obviously grouped. Two genes for resistance to barley yellow mosaic virus are on different chromosomes (1 and 2) and of four genes for resistance to brown rust (*Puccinia hordei*) two are on chromosome 3 and one each on chromosomes 1 and 5. The latter gene, *Rph 4*, is distal to *Hor-2* and a gene for resistance to yellow rust (*P. striiformis*) is proximal to *Hor-1* (12). Clearly the short arm of barley chromosome 5 is particularly rich in genes for disease resistance. The clustering of mildew resistance genes probably reflects their functional properties in directing resistance. The genes for rust resistance may be near by as a result of chance. On the other hand, they may have originated through duplication and subsequent divergence resulting in a chromosomal region with a broader function directed at a range of pathogens that may well extend beyond mildew, brown rust and yellow rust.

In wheat, Sears (16) and, in annual supplements since then, McIntosh (14), have summarized mapping information for 32 genes for resistance to stem rust (*P. graminis tritici*), 24 genes for resistance to leaf rust (*P. recondita*), and 8 genes for resistance to powdery mildew (*E. graminis tritici*). The major associations of resistance genes are chromosome 2B with 8 genes (6 stem rust loci, 1 leaf rust locus and 1 mildew locus), 4A with 6 genes (2 stem rust loci, 3 leaf rust loci and 1 mildew locus) and 2A, 2D and 7A each with 5 genes. The three homoelogues of group 2 (2A, 2B, 2D) together carry 18 resistance loci, group 7 has 11, group 4 has 9 and group 6 has 6. The distribution of the remaining resistance genes shows no obvious pattern except that only two chromosomes have none: 4D and 5A.

In maize there is not enough evidence to establish whether genes for resistance are clustered. Thus, two genes governing reaction to *Helminthosporium carbonum* *hm* and *hm* 2 are on chromosomes 1 and 9 respectively; *Ht* for resistance to *H. turcicum* is on chromosome 2; *Rp3*, *Rp4* and *rp7* governing reaction to *Puccinia sorghi* are located on chromosomes 3, 4 and 2 respectively (14).

Clearly barley 5S and the wheat group 2 chromosomes are of unusual interest in determining resistance and for this reason should repay further study using the newer methods of physical mapping by *in situ* hybridization with labelled probes (11).

CLONING RESISTANCE GENES

No resistance genes have yet been knowingly cloned so that what follows is an outline of possible approaches. If the host exhibits a mutator system that is based on controlling element sequences or transposons as in maize then these might be used to select mutants of loci controlling disease reactions. A general method of this kind has been established in *Drosophila* which is based on the *copia* sequence (2). The method depends on the identification of the base sequence of the *copia* controlling element in DNA digests from mutant flies. The sequences flanking the element are thus identified as belonging to the mutated gene. Although instability in disease expression has been observed, it is usually ignored by breeders. An example is the variable, or mesothetic reaction shown by some cereal cultivars to rust infection. Whether such variable disease resistance reactions are likely to be produced by insertions is unknown.

A second approach involves "genome walking". This requires one or more markers with known DNA sequences near the resistance gene. A series of overlapping genomic clones extending from the known marker would be expected to include the sequence coding for resistance. Its precise identification does, however, present difficulties which may be solved with the aid of the third or fourth methods. Genome walking may well prove to be worthwhile in those parts of the genome where resistance genes are clustered such as chromosome 5S of barley. The *ml-a* locus is especially interesting because of the *Hor-1* and *Hor-2* loci that bracket it and the fact that the sequences corresponding to these two endosperm protein genes should soon be available. Unfortunately the high frequency of interspersed DNA repeats in plant genomes (7,8) make this method unworkable unless the distances between markers are very small, say less than 100 kilobases.

A third method involves a comparison of the mRNAs that are formed when resistant host tissue is challenged with a virulent and with an avirulent race of the pathogen. The assumption is that mRNAs only formed in the incompatible reaction include the transcription products of the resistance gene and could then be used to form a cDNA probe or be used as a probe directly.

A fourth method assumes that genetic manipulation of the pathogen is presently easier than the host which is certainly the case for bacterial pathogens and viruses. An avirulent strain of the pathogen is mutated to produce a number of virulent forms. If avirulence is dependent on a gene product formed by an avirulence locus, then the mutants may be expected to include deletions covering this locus. By comparing restriction maps of the mutants and the wild type, one or more deletion sites should be identified which will correspond to the avirulence allele in the parent. If this fragment is cloned, its role in conferring avirulence can be confirmed by using it to transform virulent mutants which would then be expected to become avirulent. It should be possible to isolate the product of this gene and, by using a labelled product, to identify the corresponding host gene product with which it interacts, depending on the complexity of the host gene product and its nature. For example, if it is a fairly small polypeptide it may be possible to prepare an antiserum from which a polysome fraction rich in mRNA can be recovered. The mRNA, or a cDNA clone, can then be used as a probe for the resistance gene. The procedure is so laborious that, even if successful, it is very unlikely to become a general method for recovering resistance genes.

It is important to note the influence of the gene-for-gene hypothesis in formulating these methods. Each involves a comparison of resistant versus susceptible, or virulent versus avirulent, forms derived from each other by mutation. The underlying assumption in the gene-for-gene hypothesis is that incompatibility is due to a single locus with an overriding effect in either the host or the pathogen, or both. This assumption has never been rigorously tested. A second underlying assumption, which is perhaps more a pious hope than anything else, is that once a resistance gene has been identified by one or other of these methods, the detailed knowledge of its mechanism of action will lead to some more general methods. For example, many genes for resistance have multiple alleles. It is possible that further alleles could be induced in cloned resistance genes and that these would have different race specificities when reintroduced into the host plant by transformation. However, the resistance locus may have other functions in directing host metabolism that impose constraints on the range of variants that could be tolerated.

A major difficulty with most systems that obey the rules of the gene-for-gene hypothesis is that the genes for resistance are race specific. Indeed, methods three and four require the availability of both virulent and avirulent pathogen races. The breeder may well ask whether the technology investment needed to clone such a gene is worthwhile, because even if he is presented with a transformant, it is likely to be very shortlived in agriculture. This objection can be met for a number of diseases by the strategy of using multilines or mixtures of cultivars. Indeed, such strategies require a ready supply of new resistance genes in adapted backgrounds. However, whether such a supply would be at all readily available by any of the means outlined above is by no means clear.

AN ALTERNATIVE METHOD

There seems to be little prospect, apart from barley chromosome 5S, that systematic work on a small part of a host genome in which several resistance genes are located could generate the range of new forms that the breeders would like to have. An alternative approach, which may be rewarding in the short-term, would be to transform susceptible host cells with random DNA fragments either from a related resistant species or from a totally unrelated species which was not a normal host for the pathogen. The products of such a shotgun transformation experiment would of course be most easily tested if the pathogen produced a pathotoxin which could be used to select resistant from susceptible protoplasts or cultured cells. Only cells or calluses that had incorporated useful resistance would be grown on to recover plants for further tests. Shotgun transformation would test the value of non-host resistance and whether it can be transferred to and will function in a true host. The intergeneric hybrid, triticale, is an example of the benefits in transferring alien resistance genes carried on rye chromosomes into a wheat background. However, as with all the methods outlined so far, the test of their usefulness awaits the development of a reliable and efficient method of transformation.

CONCLUSION

Whatever we do to obtain new resistant cultivars we still have to be concerned about how to use them to the best advantage. No ideas that have yet emerged from molecular biology encourage us to believe in very long-term stability coupled with complete control in a monoculture.

ACKNOWLEDGEMENT

We thank Dr. Albert Ellingboe for discussions about ways of identifying genes for disease resistance during his visit to England in December 1982.

REFERENCES

1. Barrett, J.A. 1982. Plant-fungus symbioses. In Co-evolution. D.J. Futuyama, M. Slatkin, J. Roughgarden, and B.R. Levin, eds. Sinauer Press (in press).
2. Bingham, P.M., R. Levis, and G.M. Rubin. 1981. Cloning of DNA sequences from the white locus of Drosophila melanogaster by a novel and a general method. Cell 25:693-704.
3. Daly, J.M. 1981. Mechanisms of action. In Toxins in Plant Disease, pp. 331-394. R.D. Durbin, ed. New York: Academic Press.
4. Day, P.R. 1974. Genetics of Host-Parasite Interaction. San Francisco: Freeman.
5. Day, P.R. Genetics of recognition systems in host-parasite interactions. In Encyclopedia of Plant Physiology, New Series, Intercellular Interactions. H.F. Linskens and J. Heslop-Harrison, eds. Berlin: Springer (in press).
6. Ellingboe, A.H. 1982. Genetical aspects of active defense. In Active Defense Mechanisms in Plants, pp. 179-192. R.K.S. Wood, ed. London: Plenum.
7. Flavell, R.B. 1980. The molecular characterization and organization of plant chromosome DNA sequences. Ann. Rev. Plant Physiol. 31:569-596.
8. Flavell, R.B. 1982. Sequence amplification, deletion and rearrangement: Major source of variation during species divergence. In Genome Evolution, pp. 301-323. G.A. Dover and R.B. Flavell, eds. London: Academic Press.
9. Flor, H.H. 1956. The complementary genic systems in flax and flax rust. Adv. Genet. 8:29-54.
10. Giese, H. 1981. Powdery mildew resistance genes in the Ml-a and Ml-k regions on barley chromosome 5. Hereditas 95:51-62.
11. Hutchinson, J., R.B. Flavell, and J. Jones. 1981. Physical mapping of plant chromosomes by in situ hybridisation. In Genetic Engineering, pp. 207-222. J. Setlow and A. Hollaender, eds. New York: Plenum Press.
12. Jensen, J. 1981. Coordinator's report: Chromosome 5. Barley Genetics Newsletter 11:87-88.
13. Knott, D.R., and J. Dvorak. 1976. Alien germ plasm as a source of resistance to disease. Ann. Rev. Phytopath. 14:211.
14. McIntosh, R.A. 1981. Catalogue of gene symbols for wheat, 1981 supplement. Cereal Research Comm. 9:71-72.
15. Neuffer, M.G., and E.H. Coe, Jr. 1974. Corn (Maize). In Handbook of Genetics: 2, pp. 3-30. R.C. King, ed. New York: Plenum Press.
16. Sears, E.R. 1974. The wheats and their relatives. In Handbook of Genetics, pp. 59-91. R.C. King, ed. New York: Plenum Press.
17. Shewry, P.R., A.J. Faulks, R.A. Pickering, I.T. Jones, R.A. Finch, and B.J. Miflin. 1980. The genetic analysis of barley storage proteins. Heredity 44:383-389.

18. Tsuchiya, T. 1981. Revised linkage maps of barley, 1981. Barley Genet. Newsletter 11:96-98.
19. Wolfe, M.S. 1981. Pathogen population control in powdery mildew of barley. Proc. IX Int. Congr. Plant Protect., Washington 1:145-148.
20. Wolfe, M.S., and J.A. Barrett. 1980. Can we lead the pathogen astray? Plant Disease 64:148-155.
21. Wolfe, M.S., J.A. Barrett, and J.E.E. Jenkins. 1981. The use of cultivar mixtures for disease control. In Strategies for the Control of Cereal Disease, pp. 73-80. J.F. Jenkyn and R.T. Plumb, eds. Oxford: Blackwell.

DISEASE LESION MIMIC MUTATIONS

Virginia Walbot, David A. Hoisington*, and M.G. Neuffer*

Department of Biological Sciences
Stanford University
Stanford California 94305

ABSTRACT

Local lesion formation on plant leaves can result from physiological stress, wounding or disease. The lesions are often diagnostic for the factor which promoted lesion formation. For example, the lesion size, shape and pattern of spread on particular genotypes are used as identification criteria for many plant diseases. In this paper we will discuss a class of mutations of maize called disease lesion mimics. These mutations promote the production of discrete leaf lesions in the absence of obvious stress, wounding or disease on the plant. As found in authentic disease and stress responses, lesion mimic expression depends on the plant genotype and environmental conditions. We propose that the disease lesion mimic mutations of maize provide a simplified system for the study of plant response to disease and stress. Using the disease lesion mimics it is possible to study the plant response and lesion formation without the causative agent present.

INTRODUCTION

Figure 1 is a simple model of how plants may respond to pathogens and stress. The key features of this model are that the inciting agent produces physical disruption of the plant as well as inducing physiological responses. In response to the initial physical and chemical damage the plant produces local, discrete lesions of chlorotic and necrotic tissue. Physical disruption can be a

* Department of Agronomy, University of Missouri, Columbia, Missouri 65211.

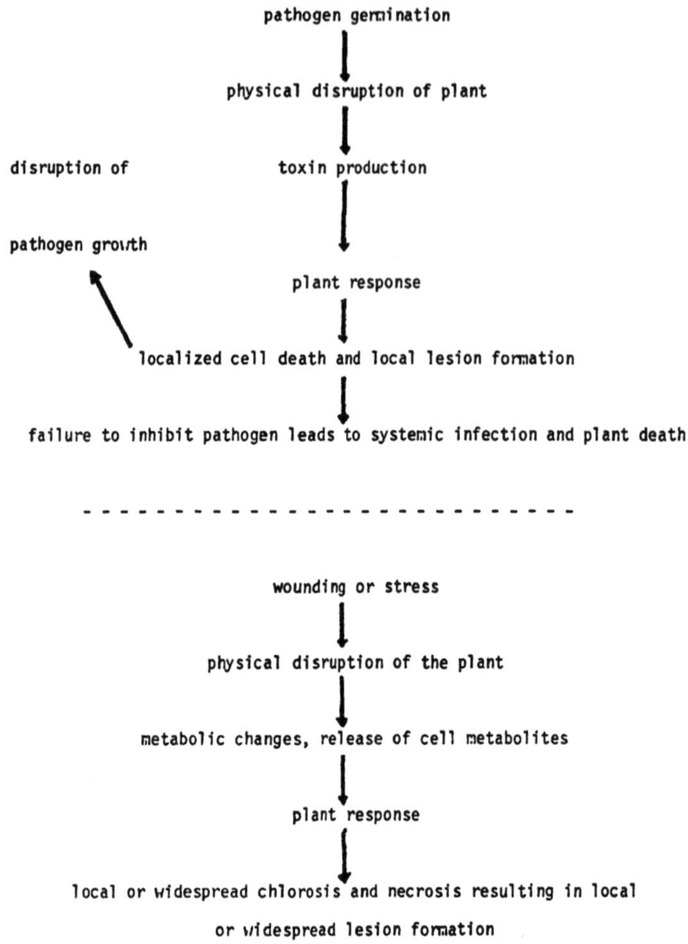

Fig. 1. General model of plant response to disease or stress resulting in lesion production.

pathogen deforming plant tissue by its growth or wounding resulting in cell death, loss of cuticle, etc. The chemical aspects of wounding and pathogen effects on plants are caused by toxins and/or enzymes produced by the pathogen, the actual chemical used to cause wounding (fertilizer, pesticide or herbicide damage, for example) or the release of cell metabolites from physically disrupted plant cells.

Plant cell death in the diseased or wounded area results from two processes: (1) direct killing by the pathogen or stress and (2) death resulting from the plant's response to the pathogen or stress. The propagation of the initial response to form a local lesion requires the participation of cells not initially affected by physical disruption or the chemical damage caused by pathogens or wounding.

The propagation of the plant response can take the form of metabolic changes that may include the synthesis and release of pathotoxic compounds (reviewed 1) to counteract the pathogen and the synthesis of stress-induced proteins which may serve to protect the plant against thermal, anaerobic or other stress (2,3,4).

The plant response provides protection but also results in plant cell death. It has been hypothesized that plant response-mediated plant cell death serves an important role in terminating the growth of some pathogens by creating a zone of dead tissue around the site of infection. Such a zone inhibits the spread of pathogens requiring or preferring living plant material as substrate or by production of pathotoxic compounds. In plant response to stress, the production of proteins such as alcohol dehydrogenase in response to anaerobiosis allows the plant to survive in the short term, but after 2 to 3 days of alcohol production catalyzed by this enzyme, the tissue becomes necrotic, poisoned by the stress-induced protection response (3 and personal communication, M. Freeling).

Local lesion production is the outcome of the complex interactions between host and pathogen or within the plant in response to stress or injury. That different discrete lesions of a specific morphology are produced in response to different pathogens and injuries provides strong evidence that these complex interactions and plant responses are very specific for different inciting agents.

Occurrence of Disease Lesion Mimics in Maize and Other Plants

Both spontaneous and mutagen-induced cases of dominant and recessive mutants causing discrete leaf lesion formation have been reported in maize and other plants (Table 1). Each gene identified results in different symptomology on affected plants. This class of mutants in maize has been designated disease lesion mimics, because each gene causes stereotypic symptoms of different corn diseases. The symbol Les is used for dominant genes and les for recessive mutations (5).

In addition to the cases in which individual genes have been identified as lesion mimic genes, there are numerous informal reports of "phantom pathogens" in major crop plants (J. Berry, personal communication). In these cases plants die of disease-like symptoms, but no pathogen can be recovered from the affected leaves. Typically whole fields of plants spontaneously demonstrate symptoms unlike the normal spread of pathogen from initial sites of infection through the field over the course of days or weeks. Consequently, the cases of phantom pathogen behave like additional cases of a genetic defect rather than authentic pathogen-induced disease.

Of the maize mutants so far detected, none are allelic nor identical in symptomology. Photographs of some of the maize mutants

Table 1. Genetic lesion mutants of higher plants.

MUTANT *	GENETIC DATA	PHENOTYPE	REFERENCE
Autogenous necrosis of tomato	Dominant	Severe necrosis	17
Lethal leaf spot	Recessive	Resembles *H. carbonum* infections	18
Blotched leaf	Recessive	*Helminthosporium* infections	19
Maize necrotic	Dominant	Severe necrosis	20
Necrotic leaf spot	Recessive	Leaf spots	21
"Target spot"	Recessive	Scattered lesions	22
Dominant necrotic spot	Dominant	Numerous small lesions	23
Les1	Dominant	Low-temperature expression of numerous medium-sized lesions	10
Les2	Dominant	Small white spots often with a dark border	10
Les3	Dominant	Late expression of large lesions	24
Les-1375	Dominant	Late expression of various sized necrotic and chlorotic lesions	25
Les-1378	Dominant	Late expression of numerous large necrotic and chlorotic lesions	25
Les-1438	Dominant	Like *Les1*	25
Les-1442	Dominant	Early expression of numerous small necrotic lesions	25
Les-1449	Dominant	Like *Les2*	25
Les-1450	Dominant	Late expression of numerous chlorotic spots	**
Les-1451	Dominant	Late expression of large necrotic lesions	25
Les-1453	Dominant	Early expression of numerous medium necrotic and chlorotic lesions	25
Les-1461	Dominant	Late expression of numerous small chlorotic spots	25
Les-A607	Dominant	Numerous uniform medium-sized lesions	**
les-1395	Recessive	Numerous chlorotic lesions	**
les-F26514	Recessive	Numerous irregular-shaped chlorotic lesions	**
les-A467	Recessive	Extremely large necrotic lesions with concentric dark rings	**
Spontaneous necrosis in maize	Probably Recessive	Several instances, mimics of various diseases	(J. Berry personal communication)

* Mutants in maize, unless otherwise noted
** Field observations, summer 1981

are presented in Figure 2. This evidence suggests that there are many as yet undiscovered lesion mimic mutants. Based on the frequency of mutant recovery thus far, it is likely that at least 60 - 70 different loci in maize exist which can cause lesion formation (6). Until duplicate recovery of alleles has been achieved it is difficult to predict the final number of genes.

Aspects of the host defense mechanisms can also occur in non-leaf tissues. Production of host defense chemicals has been reported in mutants and genetic tumors. For example, a dwarf pea mutant accumulated pisatin in the absence of an inciting agent (7). Tobacco tumors can accumulate high levels of polyphenolic compounds (8), and there is increased flavanol and terpenoid aldehyde synthesis in genetic incompatibility tumors of cotton (9). These mutant and tumor cases could also be considered mimics of aspects of the host defense mechanisms in the absence of the inciting pathogen.

Factors Affecting Expression of Leaf Lesion Mimic Genes

Environmental conditions, plant genotype, and the physiological and developmental state of the plant appear to be the major factors affecting the severity of leaf lesion expression. Many of the disease lesion mimic genes are conditional lethals; in the most extreme conditions the genes result in plant death. Neuffer and Calvert (10) reported that Les1 required low temperature for expression whereas high temperature induces Les2 expression. Nonpermissive conditions have been carefully defined for Les1, but much less is known about the other lesion mimic genes.

We have previously established that leaves must be fully expanded but not yet senescent for Les1 to cause lesions and that lesion expression occurs at low temperatures only (11; see Table 2). Les1 expression mimics symptoms caused by Helminthosporium maydis race T on susceptible T cytoplasm maize (10). Expression of Les1 in an unrestored T cytoplasm source of unknown nuclear background was mild. To adequately test the importance of cytoplasmic type on Les expression, we are currently constructing Les stocks in W23 background with normal and the T, C, and S sources of cytoplasmic male sterility.

The nuclear background of the plant does substantially influence the frequency of lesions and the non-permissive temperature conditions required to induce lesions. When Les1 is in maize variety MO20W, a nuclear background with high natural resistance to H. maydis, there are few lesions and cold (20°C) temperatures are required to induce lesion formation. When Les1 is in a susceptible background such as W23, lesion formation is more frequent and occurs at any temperature below 28°C (11).

Table 2. Factors affecting expression of Les1.

Factors promoting Les1 expression:
nuclear background of the plant: expression is enhanced in W23 compared to M020W
low temperature
high humidity
light
wounding of young tissue
Factors suppressing Les1 expression:
high temperature
dark
senescence of leaves

Model of Lesion Mimic Gene Action

It is interesting that both dominant and recessive lesion mimic genes exist. It is unusual that over half of all the lesion mimic mutations are dominant gene mutations (Table 1). The ratio of recessive to dominant mutations after ethyl methane sulfate treatment of maize pollen is 200:1 (12), but eleven of the lesion mimic mutants so far recovered after this mutagenic treatment have been dominant gene mutations (6).

In our model, we propose that lesion expression has two phases: initiation and propagation. We propose that leaf lesions are usually initiated by the physical and chemical damage caused by pathogens or wounding. The initiation phase involves the overproduction of a toxic cell metabolite in affected plant cells. This metabolite diffuses to nearby cells where it autocatalytically induces further synthesis of the compound and produces cell death. Discrete lesions are produced because the plant also contains a suppression system that finally reduces the concentration of the initiating-propagating molecule at the edges of the lesion. The suppression system is responsible for maintaining normal cell metabolism when toxic molecules are produced; enzymes such as catalase, esterases, and superoxide dismutase could be involved in reducing the concentration of toxic cell metabolites.

We propose that each dominant Les gene increases the production of or cell sensitivity to a toxic cell metabolite(s). Lesions form when the concentration of these toxic products exceeds the normal capacity of the suppression system in the initiating cells. The toxic molecules cause the initiation and propagation of a lesion

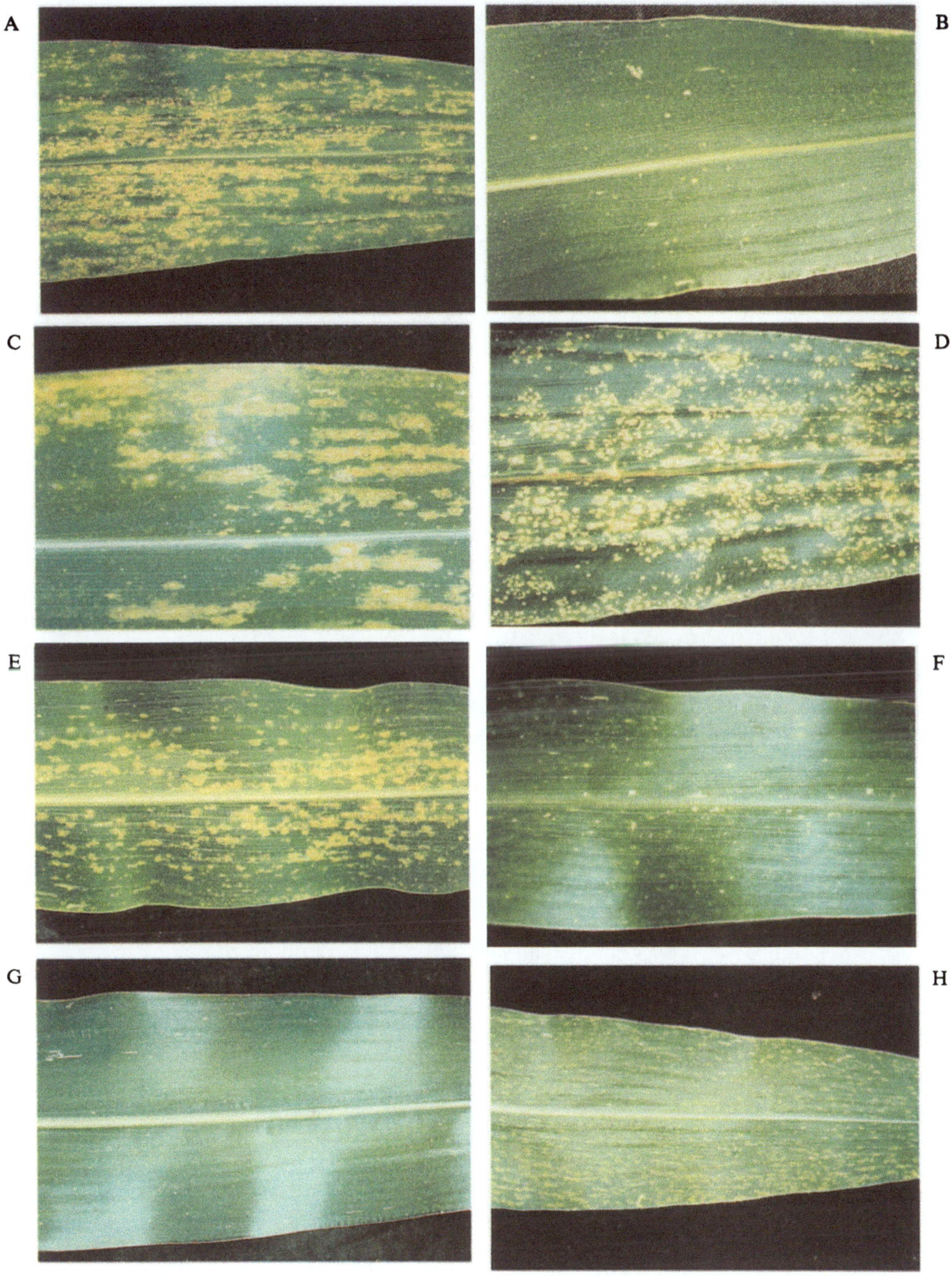

Figure 2. Disease lesion mimics of maize. Photographs of mature field-grown leaves showing typical lesions of various lesion mimic genes. (A) *Les1* in W23; (B) *Les2* in W23/N6J; (C) Les-1375 in W23/N6J; (D) Les-1378 in W23; (E) Les-1438 in W23; (F) Les-1442 in W23; (G) Les-1449 in Mo20W; (H) Les-1450 in M14/W23.

Continued overleaf

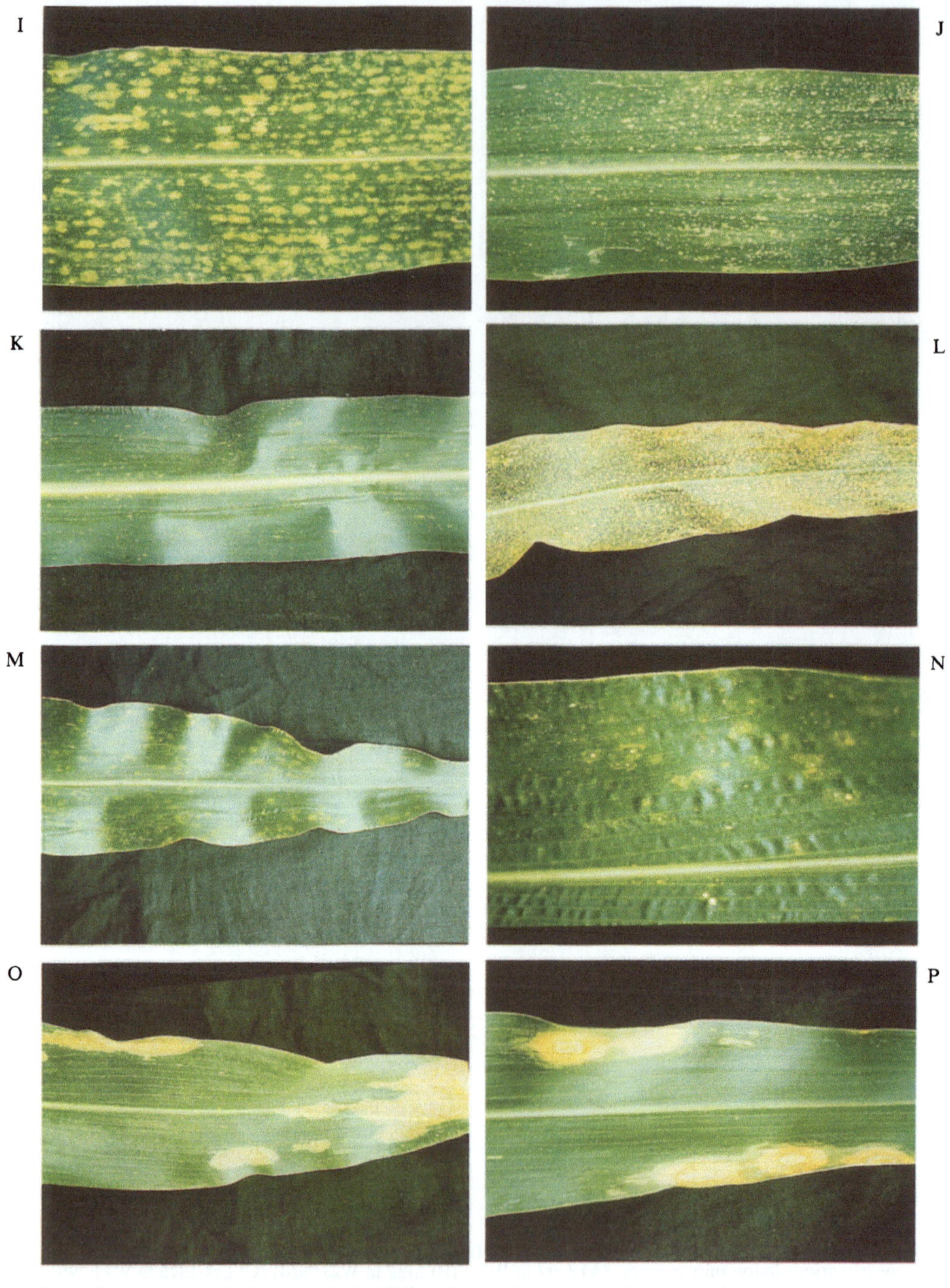

Figure 2 (continued). (I) Les-1451 in Mo20W; (J) Les-1453 in W23; (K) Les-1461 in Mo20W; (L) Les-A607 in W23; (M) les-1395 in Aho; (N) les-F26514 in W23/K55; (O) les-A467 in W23; (P) 11sl in Mo20W.

until the suppression system can reduce toxin concentration or cell sensitivity. In this way the dominant Les genes are a very close mimic of pathogen or stress-induced lesions. The Les gene causes a response to a toxic molecule usually produced by the plant in response to a specific pathogen or stress. Subsequent propagation of lesions and eventual suppression are identical for both the mutants and authentic diseases.

We propose that the lesions produced in the presence of a recessive les gene occur due to a failure of the suppression system. If toxic molecules are produced by plant cells as a result of normal metabolic activities, the suppression system prevents cell damage and lesion induction. The suppression system could act within the cell to prevent autocatalytic over-production of the toxic metabolite, could act outside the cell to prevent diffusion of the toxic molecules to neighboring cells, or could act to suppress the cell's response or perception of toxic molecules diffusing from other cells. Mutation in any aspect of the suppression system could result in more extensive or more frequent propagation of the lesion initiation signal. We hypothesize that the suppression system is responsible for the discrete nature of lesions of a typical size produced in authentic disease. The suppression system prevents the propagation of lesion formation throughout the leaf and minimizes the extent of leaf damage.

The model makes several interesting predictions about the lesion mimic mutants. First, both dominant and recessive mutants with the same phenotype can exist, but they will be non-allelic. The dominant mutants result from the production of substances that initiate and propagate lesions whereas the recessive mutants result from a lack of suppression of lesion spread. It is unlikely that a single locus could be involved in both lesion initiation and suppression. We are currently seeking recessive alleles at Les loci to determine the phenotype of such mutants.

Second, this model allows for considerable variability in lesion expression depending on the quality of the suppression system. Multiple suppressors may exist for each lesion phenotype as well as non-specific suppressors that may be involved in suppression of all types of damage. Inbred lines should differ considerably in the quality of different suppression systems, and indeed we find substantial variability in the extent of Les and les expression in different inbred lines. Some lines such as MO20W show less expression for almost all Les genes than does W23 background. These differences could reflect the quality of the non-specific suppression system as well as the presence of specific suppressors of individual Les genes.

We propose that genetic backgrounds with lower lesion expression should contain a more efficient suppression system(s) such that

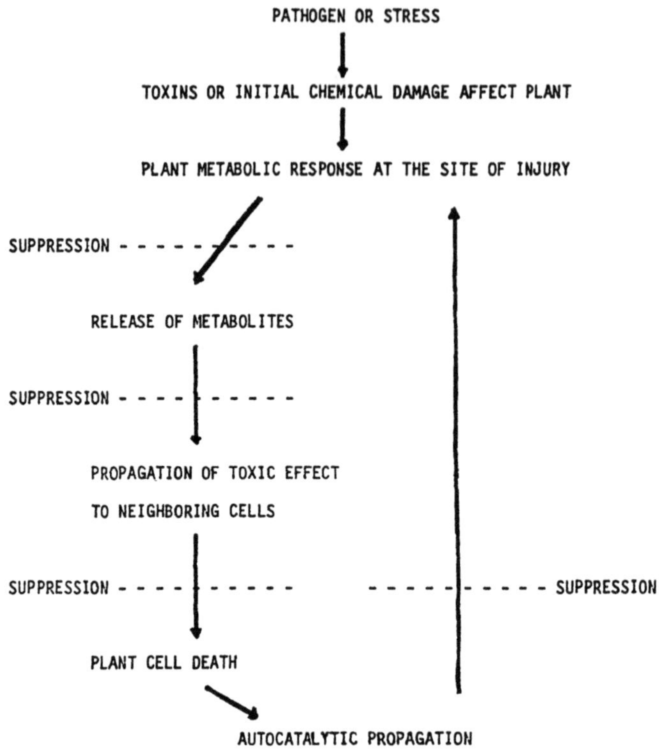

Fig. 3. Model of dominant and recessive disease lesion mimic gene action. Legend: Pathogens or stress stimulate plant metabolic response at the site of injury. This metabolic response is propagated into neighboring cells which also undergo metabolic changes resulting in cell death. The response and propagation process is autocatalytic resulting in an ever-widening ring of plant cell death. Because discrete leaf lesions are formed rather than whole-leaf necrosis, we postulate that additional genes suppress the plant response at any point in the autocatalytic cycle. Genes could suppress propagation, plant cell death, or plant cell response to dying cells.

We propose that the dominant Les mutants inappropriately produce a factor(s) that is usually involved in plant response to particular pathogens or stress conditions. The presence of this metabolic product induces the formation of the local leaf lesion diagnostic for particular disease or stress conditions.

Recessive les mutations are viewed as a failure of the plant to produce sufficient factor(s) to efficiently suppress the plant metabolic changes similar to responses

fewer or smaller lesions are produced. We have investigated the genetics of the suppression system in a preliminary experiment with Les1 expression. In crosses of MO20W (low expression) by W23 (high expression) we found that the F1 hybrid progeny had Les1 expression typical of MO20W suggesting that the suppression system is dominant. This result is predicted by our model as wild type is the presence of the [better] suppression system, and mutants are defective [lacking] in suppression. In backcrosses to each parental line we have found a relatively simple segregation pattern suggesting that the suppression system for Les1 is composed of a few genes.

A third prediction of this model is that Les genes should express no matter how many copies of the wild type allele are present. We are currently testing Les expression in trisomics (Les++). None of the ten wild type trisomic stocks of maize have a lesion phenotype. This result suggests that Les is more than just a slight overproduction of a metabolite and more likely is gross overproduction. For the recessive les mutants, a single wild type allele is sufficient to restore a normal phenotype.

A fourth prediction of the model is that certain combinations of recessive and dominant genes should result in greatly enhanced lesion expression or total leaf necrosis. Les and les genes involved in the same response pathway should be at least additive in the expression of lesions as there is both over-production of the initiation-propagation agent and a lack of suppression of lesion spread. As the actual phenotype of Les and les genes in the same pathway may not be precisely identical in terms of lesion size or frequency, we must combine each Les and les gene to discover synergistic interactions.

Other genetic data (6) on the Les mutants demonstrate that this class of mutations are recovered spontaneously or following ethyl methane sulfate mutation which is expected to give point mutations. They are not recovered following x-irradiation which is expected to cause deletions. Thus, the Les mutants require the action of the gene and are not due to the loss of a dominant inhibitor of lesion formation.

So far, all available data is consistent with our model of Les and les gene action, however specific experiments will have to be performed to adequately test this model.

---

to a particular pathogen or stress condition. The lack of this metabolic product allows the formation of the local leaf lesion diagnostic for particular disease of stress conditions when low or normal levels of cytotoxic molecules are produced.

Potential Utility of the Disease Lesion Mimic Genes in Plant Breeding Programs for Disease and Stress Resistance

The genetic basis for host-pathogen interaction is an exciting area of research. Recognition of the role of both host and pathogen functions in this interaction (13) and of specific chemical defenses of the host (1, for review) has influenced breeding programs for disease resistant crop plants. We believe that there are two obvious applications of the lesion mimic mutants in future breeding programs for resistance to stress and disease. The first application involves the use of Les2, a mimic of the hypersensitive response in maize (10). The hypersensitive response is a rapid response of plant cell metabolism and structure to produce small disease lesions in which the pathogen never successfully colonizes the host. Such successful host responses can also result in acquired resistance. Acquired resistance predisposes the host plant to more rapidly and successfully deter potential pathogens. Plants with acquired resistance produce fewer lesions and other symptoms (1,14,15,16).

Les2 is expressed at high temperature (over 30°C) and results in the infrequent production of very small, 0.1 to 2 mm^2, lesions on mature leaves. If production of these lesions does offer some protection against subsequent pathogen infection, then Les2 could be used as a "self immunization" gene. Les2 seedlings would gain acquired immunity against corn pathogens after the first 30°C temperature treatment; this temperature is routinely reached within the first month of the season of most corn growing regions. New alleles of Les2 with lower temperature sensitivity could also be sought. Maintenance of plants with a dominant gene such as Les2 is simple as only one parent must be homozygous for the gene. In a seed production program, one inbred parental line would carry Les2 and on crossing to another line would yield all hybrid Les2 progeny.

A second strategy is to use the individual Les genes to search for suppressors of the lesion phenotype. The principle of this strategy is that the symptoms of the disease or stress are deleterious to yield. Consequently, suppression of symptoms would allow the plant to produce a high yield even if a pathogen or stress were present. Each Les or les gene would be crossed into a wide variety of backgrounds to identify lines containing factors which successfully suppress lesion formation. Such homozygous Les or les lines would be used in a breeding program to maximize suppression of lesion formation in the presence of the Les genes(s). When maximal suppression has been achieved, the Les gene can be removed from the background leaving a line with the ability to suppress a particular type of lesion formation.

As we learn more about the genetic, environmental, developmental and physiological factors that influence leaf lesion formation

other applications of this remarkable set of genes will undoubtedly arise. The very existence of disease lesion mimic genes focuses our attention on the role of the plant response to pathogens and stress as the major determining factor in the extent of symptom production. As we cannot eliminate disease and stress from the environment, basic research on the nature of plant resistance to wounding and disease will continue to be of fundamental importance to improving plant productivity.

ACKNOWLEDGEMENTS

This project was supported by National Science Foundation grants PCM 80-26847 and PCM 81-15441 awarded to MGN and VW respectively, and by USDA grant SEA 5901-0410-8-0121-0 to MGN. Additional support was provided by a Herman Frasch Foundation grant to VW. Contribution from the Missouri Agricultural Experiment Station, Journal Series Number 9195.

REFERENCES

1. Bell, A.A. 1981. Biochemical mechanisms of disease resistance. Ann. Rev. Plant Physiol. 32:21-81.
2. Barnett, T., M. Altschuler, C.N. McDaniel, and J.P. Mascarenhas. 1980. Heat shock induced proteins in plant cells. Dev. Genet. 1:331-340.
3. Sachs, M.M., M. Freeling and R. Okomoto. 1980. The anaerobic proteins of maize. Cell 20:761-767.
4. Key, J.L., C.Y. Lin, and Y.M. Chen. 1981. Heat shock proteins of higher plants. Proc. Natl. Acad. Sci. USA 78:3526-3530.
5. Coe, Jr., E.H., and M.G. Neuffer. 1977. Genetics of corn. In Corn and Corn Improvement. G.F. Sprague, ed. Amer. Soc. Agronomy, Madison, Wisconsin.
6. Neuffer, M.G. 1982. Genetics (in preparation).
7. Hadwinger, L.A., C. Sander, J. Eddyvean, and J. Ralston. 1976. Sodium azide-induced mutants of peas that accumulate pisatin. Phytopathology. 66:229-230.
8. Sheen, S.F., and R.A. Anderson. 1974. Comparison of polyphenols and related enzymes in the capsule and nodal tumor of Nicotiana plants. Can. J. Bot. 52:1379-1385.
9. Mace, M.E., and A.A. Bell. 1981. Flavanol and terpenoid aldehyde synthesis in tumors associated with genetic incompatibility in a Gossypium hirsutum X G. gosspiodes hybrid. Can J. Bot. 59:951-955.
10. Neuffer, M.G., and O.H. Calvert. 1975. Dominant disease lesion mimics in maize. J. Hered. 66:265-270.
11. Hoisington, D.A., M.G. Neuffer, and V. Walbot. 1982. Disease lesion mimics in maize. I. Effect of genetic background, temperature, developmental age, and wounding on necrotic spot formation with Les1, in press. Dev. Biol.

12. Neuffer, M.G., and W.F. Sheridan. 1980. Defective kernel mutants of maize. I. Genetic and lethality studies. Genetics. 95:929-944.
13. Day, P.R. 1974. Genetics of Host-Parasite Interaction. W.H. Freeman and Co., San Francisco.
14. Lobenstein, G. 1972. Localization and induced resistance in virus-infected plants. Ann. Rev. Phytopathol. 10:177-206.
15. Johnson, R. 1978. Induced resistance to fungal diseases with special reference to yellow rust of wheat. Ann. Appl. Biol. 89:107-110.
16. Kuc, J., and R. Hammerschmidt. 1978. Acquired resistance to bacterial and fungal infection. Ann. Appl. Biol. 89:313-317.
17. Langford, A.N. 1948. Autogenous necrosis in tomatoes immune from Cladosporium fulvum Cooke. Can. J. Res. 26:35-64.
18. Ullstrup, A.J., and A.F. Troyer. 1967. A lethal leaf spot of maize. Phytopathology. 57:1282-1283.
19. Emerson, R.A. 1923. The inheritance of blotch leaf in maize. Cornell Univ. Memoir. 70:3-16.
20. Ghidoni, A. 1974. Un gene dominante che produce lesioni necrotiche nel mais, associato al chromosoma 9e con effetto di letalita allo stato omozigote. Genet. Agar. 28:162-169.
21. Gardner, C.O. 1971. Induced "necrotic leaf spot" mutation allelic to zebra necrosis (zn1). Maize Genet. Coop. News Letter. 45:150.
22. Wright, J.E., and D. Foley. 1954. Inheritance of target spot. Maize Genet. Coop. News Letter 28:29.
23. Mortimore, C.G., and L.F. Gates. 1979. A genetically controlled necrotic spotting of corn leaves. Can. J. Plant Sci. 59:147-152.
24. Neuffer, M.G., and S.E. Pawar. 1980. Dominant disease lesion mutants. Maize Genet. Coop. News Letter. 54:34-36.

DETERMINATION OF PLANT ORGANS AND CELLS

I.M. Sussex

Department of Biology
Yale University
New Haven, Connecticut 06511

INTRODUCTION

In the development of a multicellular organism there occurs a progressive loss of developmental potential in each part as it differentiates, and it is a requirement that this occur for development to proceed normally. There must be mechanisms to suppress the expression of inappropriate developmental programs in each part of the organism. Biologists studying this phenomenon in animals have come to the conclusion that the restriction of developmental potential seen at the organ level occurred also at the cellular level. The concepts that emerged from a long series of studies over many years, first in experimental embryology, then in regeneration, and most recently in studies of cultured cells, were those of competence and determination. It was demonstrated that early in development organ initials and cells had multiple developmental potential, but that at some particular time and site cells or groups of cells became "competent" to react to a specific signal that would cause them to become "determined". Their subsequent development and differentiation was then independent of further exogenous signals. The fact that competence was site specific could be demonstrated by grafting cells from another region into a developmental site and showing that the grafted cells acquired the ability to develop in accordance with their new position, and determination could be demonstrated by removing the organ or cells to a new environment, either a new graft site or into sterile culture, where it continued to develop as it would have in the original site (1).

Plant biologists have, in general, come to quite different conclusions with regard to the relationship between organ and cell determination. Although there have been numerous studies of organ determinations in plants (2), there is little agreement that determination occurs at the cellular level. The argument most commonly advanced against the concept of cell determination is that because plant cells, at least in many cases, can be demonstrated to be totipotent they cannot be determined. The failure to identify a cellularly determined state in plants might indicate a significant difference in the developmental mechanisms of animals and plants, or it is possible that plant cells do become determined but that determination in them is less stable than in animal cells, or it is possible that determination has not been studied appropriately in plant cells.

In the following sections some examples of determination in animal development will be reviewed, primarily to indicate the methods that have been used to study organ and cell determination. Then examples of determination at different levels of organization in plants will be examined. A definition of determination that will cover the cases described is that it is an event that occurs in an organ, tissue, or cell at high frequency, that it is directed, that it is stable in the subsequent absence of its inducer, and that it is transmitted clonally. Determination is an epigenetic event which, unlike a genetic event which alters the base sequence or the organization of the genome and is therefore not readily reversible, should be reversible at some reasonably high frequency.

DETERMINATION OF ANIMAL ORGANS AND CELLS

A classical example of organ determination occurs in the amphibian Ambystoma, the forelimb of which has three developmental axes: anterior-posterior, dorsal-ventral, and proximal-distal. Each of these axes is determined independently of the others and at a different time. By excising the developing forelimb bud and transplanting it to different sites it was shown that the order of determination occurs in the sequence anterior-posterior, then dorsal-ventral, then proximal-distal (3). Transplants done after a particular axis had been determined resulted in development continuing independently of the new site and graft orientation thus indicating the stability of the determined organ axis.

The stability of determination in animal cells has been demonstrated convincingly in cell fusion studies carried out by Fougère and Weiss (4). They fused mouse melanoma cells which expressed the melanin pigmented phenotype in culture with rat hepatoma cells which expressed the phenotype of albumin secretion into the medium. Cell lines grown from fusion products initially expressed neither phenotype of the parental cells, but after about twenty cell generations pigmented cells appeared in the cultures and the culture medium was found to contain albumin. Pigmented cells were cloned and were found not to produce albumin, but cloned unpigmented cells did. In subsequent experiments they observed the reappearance of unpigmented cells in the pigmented cultures. When recloned these nonpigmented cells again gave rise to pigmented cells, and this appearance and disappearance of the pigmented phenotype was reversed four times. The interpretation of these experiments was that the pigmented phenotype was stably determined in the cells of the mouse melanoma, but that its expression could be suppressed and reexpressed according to the state of the cultured cells.

The questions of whether cellular determination can be reversed has been addressed in studies of limb regeneration. After amputation of a part of the amphibian limb there is extensive dedifferentiation of the remaining cells in the distal region followed by redifferentiation as new parts are formed. The two major cell types, muscle and cartilage, dedifferentiate to an apparently common mesenchymal form. The question is whether mesenchymal cells form a common pool that redifferentiate into both muscle and cartilage, or whether the mesenchymal cells retain the determined state and redifferentiate to form only that cell type from which they were derived. The evidence now points to the latter situation: that is, they retain their determined state even though this is not expressed in the mesenchymal cells (5).

Although none of the above examples point to reversal of the determined state in animals, recent evidence of complete reversal has been obtained for at least one case. Mintz and Illmensee (6) showed that teratocarcinoma stem cells were capable of becoming redetermined during embryonic development of mice so as to form essentially all types of somatic cells and also germ line cells. Teratocarcinoma stem cells are pluripotent, forming a variety of differentiated cell types in chaotic array in solid tumors. However, they fail to form at least kidney and liver cells and some stem lines fail to differentiate completely. When stem cells of teratocarcinoma origin are introduced into the blastocoel of genetically marked mouse embryos which were then allowed to develop in foster mothers, tumor-free mice were produced which contained cells of teratocarcinoma origin in virtually all tissues including liver and kidney. The teratocarcinoma origin of these cells was proven by the presence of isozyme markers and by tissue-specific products.

From the above examples, and from the extensive literature on determination in animal organs and cells, it can be concluded that the determined state of animal somatic cells is very stable and is not readily reversed.

DETERMINATION OF PLANT ORGANS

At the whole plant or organ level there have been numerous studies that have described determination in terms similar to those applied to animal systems; that is, the development of a stable state which persists in the absence of the inducing stimulus. The following are some examples. The apical meristems of the shoot and the root are determined as shoot forming and root forming structures, respectively, and there are no reliable reports of their interconversion in flowering plants (7). Similarly, the floral meristem appears to be determined for flower formation and can be removed from the plant and grown to maturity in sterile culture from the time it is first recognizable as a nonvegetative meristem (8). There is some evidence that the floral meristem determination, in some cases at least, is not irreversible and that under appropriate treatment the meristem can revert to the vegetative state (9).

Leaf development shows the progressive determination of morphological axes which is remarkably similar to limb axis determination in amphibians. Leaves arise in succession around the shoot apical meristem, appearing first as radially symmetrical outgrowths, then becoming bilaterally flattened and dorsiventral. Steeves (10) excised leaves of the fern *Osmunda cinnamomea* and grew them in sterile culture on a medium that contained only sugar and inorganic salts. Those that were excised at the youngest stage of development all grew into shoots that were radially symmetrical, grew indefinitely, and produced leaves. Successively older leaf initials showed an increasing proportion that developed in culture as leaves, and not as shoots, until the tenth leaf primordium when one hundred percent developed as leaves. These results were interpreted as showing that determination as a bilateral, determinate leaf occurred after growth of the primordium had begun. Experiments on the timing of determination of axes in the leaf of potato, in which the prospective leaf site on the margin of the shoot apical meristem or the young emerging leaf primordium were surgically isolated from the surrounding tissues, indicated that determinate growth was imposed on the primordium before bilaterality (11): that is, there was a difference in the time of determination of the proximal-distal and the dorsal-ventral axes as occurs in amphibian limb axis determination.

Although the meristems of plants appear to be stably determined, they are able to undergo changes in the state of determination that is related to the normal development of the plant. These changes usually take place in a progressive manner that has been termed by Brink (12) "phase change", and they are similar to the transdeterminations that have been described in *Drosophila* (13). Examples are the transition from vegetative to flowering meristem in the vascular plants, from juvenile to adult meristem in *Hedera helix*, the English ivy (14), and similar juvenile to adult changes that occur during meristem development in woody plants (15). Whether the change in the state of organ determination is accompanied by a change in the state of cell determination has not been critically examined in any of these cases, and a critical evaluation would involve cloning cells from each state and regenerating plants from such cloned cells. However, there is circumstantial evidence that the determined state of the organ is either not transmitted through its constituent cells or that cell determination is not stable and can be reversed easily. Banks (16) regenerated plants from callus derived from adult meristems of Hedera helix and found that these were always juvenile, and Pierik (17) regenerated buds on vernalized leaf cuttings of the vernalization-requiring plant Cardamine pratensis in vitro and found that none of the buds developed into shoots that would flower unless they underwent a further round of vernalization. So, in both these cases the determined state of the plant was not transmitted to a second plant regenerated from the first. Furthermore, it is evident that in the life cycle of each photoperiodic or thermoperiodic plant the information that leads to the determined state giving rise to the reproductive system must be obliterated. Otherwise the next plant generation would not require reinduction in order to flower.

Consideration of the above examples does not lead to a clear understanding of the relationship between the state of determination of the organ and its constituent cells. This is, in part, because the examples are drawn from a variety of experimental systems, and because many of the experiments described were not carried out specifically to examine the question of cell determination. However, there is one case in which cell determination has been examined in detail in relation to plant development and this will be described in the next section.

DETERMINATION OF PLANT CELLS

Plant cells in culture typically require an exogenous source of auxins, cytokinins, and certain B vitamins. But cases have been described in which the cultured cells have become independent of the previously required substance, having developed the capacity to synthesize it. In some cases the change results from a mutation from auxotrophy to autotrophy, but in other cases the frequency of this change is much higher than that expected for mutations, and it has been shown that the change does not involve alteration in the base sequence or organization of the genome. These latter kinds of changes are termed "habituation", and the most extensively studied is the habituation of tobacco cells for cytokinin (18). Cytokinin (CK) habituation conforms to the definition of determination in that it is directed, stable in the absence of the inducing stimulus, transmitted clonally, and is reversible. Thus it is a cellular epigenetic change. Meins and Binns (18) determined that the rate of habituation for tobacco pith cells was greater than 4×10^{-3} conversions per cell generation, a rate 100-1,000 times greater than somatic mutation rates reported for tobacco. They found also that habituation is not an all or none phenomenon, but that the degree of habituation measured as growth on a CK minus medium over growth on a CK plus medium could change to higher or lower values. For habituation to occur, pith explants had to be larger than 20-30 mg. Smaller explants appeared to be incapable of responding to the inductive signal, which was exogenous CK (19). The model proposed by Meins and his co-workers for CK habituation, and for its stable maintenance in the absence of the inducer was a positive feedback system. They assume in this model that a cell division factor (CDF) can enter and leave the cell and be degraded in the cell. Further, they assume that CDF induces its own synthesis. The cell can then exist in either of two stable states: one where the concentration of CDF is zero, and the other where CDF is above a threshold value and the cell is triggered to synthesize CDF. According to the model when nonhabituated cells are exposed to an exogenous source of CDF (in this case, CK), it is taken up and triggers CDF synthesis; the cells become habituated and can continue to grow in the absence of exogenous CDF. If habituated cells are treated so as to block CDF synthesis, CDF leaks from the cell or is degraded, its concentration falls below the threshold value, and cells revert to the nonhabituated phenotype.

When cells from other parts of the tobacco plant were examined for habituation they were found to differ from pith cells (20). Primary explants of leaf cells, or subcultures, or cloned leaf cells remained nonhabituated, whereas cortex cells were already habituated when removed from the plant. These results indicate that habituation, or changes similar to habituation, are a normal component of development. That is, in tobacco, cell determination does occur during development.

The relationship between explant size and CK habituation in tobacco resembles the requirement for a minimal size of explant in regeneration of alfalfa cells (21). For alfalfa cells to regenerate they must be transferred from a growth medium to an induction medium for four days and then to a regeneration medium. Tissue pieces of less than 105 μm diameter consisting of twelve or fewer cells failed to respond to the induction medium, but on further growth would respond. That is, there seemed to be a requirement for a specific number of cells, or mass of cells for them to be competent to respond to the inductive signal.

The above examples point to cell determination and its persistence in plants. Unfortunately, there are very few cases in which determined states of cells in plants have been shown to be expressed by the cultured cells. Thus, the stability of cell determination is still open to question. An example of the approach required to examine this question has been described for Prunus (22). Here antigenic determinants that were expressed in tissues of the style were also expressed in tissue cultured cells derived from the style. Experiments of this type, in which tissue-specific markers can be identified and then examined in cells derived from that tissue for expression in culture, as indicators for the stable expression of tissue-specific determinants should permit a deeper insight into the question of cell determination.

DISCUSSION

The question of whether plant cells are determined, and if so the question of how stable is their determination, is of importance in considerations of genetic manipulation of plants. The failure of plant cells in culture to regenerate into new plants, that is, to exhibit totipotence, has frequently been attributed to the failure of the experimenter to discover the appropriate conditions of culture for the expression of this developmental pathway. However, it is by no means certain that all living plant cells are totipotent. The observed differences in the capability of closely related cells to regenerate suggests that there are at the very lest different cell states and that different states require different conditions for regeneration (23). Alternatively those cells that fail to regenerate might be determined along a pathway that precludes regeneration.

Secondly, there are situations where it is desirable to maintain a differentiated function in cultured cells (24) as, for example, for the synthesis of a secondary product. Here, knowledge about the stability of the determined state that precedes expression of the differentiated state is likely to be of importance in the establishment of culture conditions that will be optimal.

Finally, since genes appear to be regulated in sets during development, it will be important to understand conditions that specify which gene set is active in those studies where foreign genes are transferred into cultured cells if expression is expected.

REFERENCES

1. Graham, C.F., and P.F. Wareing. 1976. The Developmental Biology of Plants and Animals. W.B. Saunders and Co., Philadelphia.
2. Sawhney, V.K., and R.I. Greyson. 1979. Interpretations of determination and canalization of stamen development in a tomato mutant. Canadian J. Bot. 57:2471.
3. Swett, F.H. 1937. Determination of limb axes. Quart. Rev. Biol. 12:322.
4. Fougère, C., and M.C. Weiss. 1978. Phenotypic exclusion in mouse melanoma-rat hepatoma hybrid cells: Pigment and albumin production are not expressed simultaneously. Cell 15:843.
5. Grant, P. 1978. Biology of Developing Systems. Holt,

Rinehart and Winston, New York.
6. Mintz, B., and K. Illmensee. 1975. Normal genetically mosaic mice produced from malignant teratocarcinoma cells. Proc. Natl. Acad. Sci. USA 72:3585.
7. Steeves, T.A., and I.M. Sussex. 1972. Patterns in Plant Development. Prentice-Hall, NJ.
8. Hicks, G.S., and I.M. Sussex. 1970. Development in vitro of excised flower primordia of Nicotiana tabacum. Canadian J. Bot. 48:113.
9. Krishnamoorthy, H.N., and K.K. Nanda. 1968. Floral bud reversion in Impatiens balsamina under non-inductive photoperiods. Planta 80:43.
10. Steeves, T.A. 1966. On the determination of leaf primordia in ferns. In Trends in Plant Morphogenesis, E.G. Cutter, ed. Longmans, Green, London.
11. Sussex, I.M. 1955. Morphogenesis in Solanum tuberosum L.: Experimental investigation of leaf dorsiventrality and orientation. Phytomorphol. 5:286.
12. Brink, R.A. 1962. Phase change in higher plants and somatic cell heredity. Quart. Rev. Biol. 37:1.
13. Hadorn, E. 1966. Dynamics of determination. In Major Problems in Developmental Biology, M. Locke, ed. Academic Press, NY.
14. Rogler, C.E., and M.E. Dahmus. 1974. Gibberellic acid induced phase change in Hedera helix as studied by deoxyribonucleic acid-ribonucleic acid hybridization. Plant Physiol. 54:88.
15. Robinson, L.W., and P.F. Wareing. 1969. Experiments on the juvenile-adult phase change in some woody plants. New Phytol. 68:67.
16. Banks, M.S. 1979. Plant regeneration from callus from two growth phases of English ivy, Hedera helix L. Z. Pflanzenphysiol. 92:349.
17. Pierik, R.L.M. 1967. Effect of light and temperature on flowering in Cardamine pratensis L. Z. Pflanzenphysiol. 56:141.
18. Meins, F., and A.N. Binns. 1978. Epigenetic clonal variation in the requirements of plant cells for cytokinins. In The Clonal Basis of Development, S. Subtelny and I.M. Sussex, eds. Academic Press, NY.
19. Meins, F., J. Lutz, and R. Foster. 1980. Factors influencing the incidence of habituation for cytokinin of tobacco pith tissue in culture. Planta 150:264.
20. Meins, F., and J. Lutz. 1979. Tissue-specific variation in the cytokinin habituation of cultured tobacco cells. Differentiation 15:1.
21. Walker, K.A., M.L. Wendeln, and E.G. Jaworski. 1979. Organogenesis in callus tissue of Medicago sativa. The temporal separation of induction processes from differentiation processes. Plant Sci. Lett. 16:23.
22. Ruff, J.W., and A.E. Clarke. 1981. Tissue-specific antigens

secreted by suspension-cultured callus cells of Prunus avium L. Planta 153:115.

23. Stamp, J.A., and G.G. Henshaw. 1982. Somatic embryogenesis in Cassava. Z. Pflanzenphysiol. 105:183.

24. Dougall, D.K., J.M. Johnson, and G.H. Whitten. 1980. A clonal analysis of anthocyanin accumulation by cell cultures of wild carrot. Planta 149:292.

GENETIC ENGINEERING OF PLANTS: SOME PERSPECTIVES ON THE CONFERENCE, THE PRESENT, AND THE FUTURE*

R.L. Phillips

Department of Agronomy and Plant Genetics
University of Minnesota
St. Paul, MN 55108

The need for continued genetic improvement in crop varieties is clearly evident. Steady progress in crop yields is expected from current plant breeding procedures, but will the advances be sufficient to meet future food demands? Recent advances in plant molecular and cellular biology have led to a number of approaches for molecular genetic manipulation of crop species. The purpose of this conference was to assess these advances in relation to crop improvement. When considering the potential impact of genetic engineering of plants, we must start with certain premises:

First, plant genetic engineering is a long-term effort and expanded research in the basic plant sciences is needed. Advancements in the genetic engineering of plants will depend, in part, on additional knowledge of the biochemical, physiological, and developmental processes of plants. Not only is there a need for pioneering efforts on the technological aspects of plant genetic engineering, but also for basic plant science research on which this technology depends. By the very definition of basic research, 20 to 30 years may be required before we realize benefits from some of the information. But if we do not start now, the information will not be available when needed.

Second, a multidisciplinary effort is required. Dean Hess pointed out in his welcome to the conference participants that "We

---

*This chapter reports the wrap-up presentation given by R.L. Phillips at the conclusion of the conference. These remarks, prepared during the conference, were to provide a brief integration and synthesis of the various conference presentations.

need a continuum from basic research to the applied." A team effort of scientists knowledgeable in molecular, cellular, and organismal biology is required for maximizing research progress directed toward applications in agriculture.

Third, research in plant genetic engineering is worth the effort, even if never directly applied. The basic plant science information deriving from this research is extensive, valuable, and of high quality. Simmonds aptly stated during the conference that "Improved genetic understanding has never failed to have an impact on plant breeding."

Several questions pertinent to a conference on "Genetic Engineering of Plants" are asked herein. The questions are used to highlight certain aspects of the conference and to present some personal perspectives as a plant geneticist; no attempt is made to fully answer the questions. Documentation for several of the remarks can be found in the various chapters of these proceedings.

What is plant genetic engineering? There seems to be some confusion about the definition of plant genetic engineering. Some people deliberately exclude tissue culture and wish to consider only the isolation, introduction, and expression of foreign DNA in a plant cell. Others define it simply as introducing genes into plants. These appear to be rather confining definitions, although accurate. A broader definition would emphasize that the focus is at the cell level and involves the interfacing of all aspects of cell and tissue culture, molecular biology, and gene transfer. This definition reflects the enormous scope of the effort and the many research activities that have to be intertwined in order to apply such basic information within an agricultural context.

How can these new techniques most effectively be combined with existing plant breeding methods in an integrated approach to crop improvement? "Plant breeding is a highly effective technology which will be further refined but not revolutionized;" this is a quote from Simmonds but is the contention of many others. It is my current perception that if something can be accomplished by non-laboratory means, that is the way it will be done. For a procedure simply to be more rapid is not a sufficient reason for its adoption. The new technology must do something that cannot be done by conventional means, or at least have some unique feature. The argument that the transfer of a single gene will be done by sexual means if at all possible, and not by molecular genetic approaches, may imply that the probable impact of genetic engineering will be nil. The transfer of single genes by sexual means, however, usually also involves the transfer of a block of linked genes. This necessitates an evaluation of the converted line to determine if the linked genes have a deleterious effect. Perhaps the transfer of single genes by molecular means could be more exacting and reduce, or even eliminate, the need for testing the converted line. The transfer of single genes from wild species could be especially useful. Although wild or otherwise divergent species often possess single genes of interest, such as for disease resistance, these genes are difficult to incorporate into adapted cultivars by conventional means. The sorting out of beneficial from deleterious traits is a major undertaking and often not successful simply due to the required number of generations and population sizes. Thus, the molecular transfer of single genes from divergent species would be valuable even in cases where sexual crosses are possible and fertility is not a problem.

To achieve maximum impact in plant improvement, it is clear that ways must be found for transferring combinations of genetic factors, such as those responsible for polygenic traits; but we must learn to "walk" before we can "run". Designing molecular techniques first for the transfer of single genes probably will lead to the development of methods for transferring more complex genetic traits.

Does the plant breeder need to modify current evaluation procedures to better accommodate materials generated by genetic engineering procedures? Dean Hess stated that "Agricultural industry is ready and able to adopt new approaches." For them to be adopted, however, the usefulness of the new approaches must be demonstrated. I would present a challenge to the plant breeders. I do not believe anyone expects the plant breeder to do molecular or cell biology, but it seems molecular or cell biologists are often expected to carry out an evaluation of materials to determine the worthwhileness of their new types. The plant breeder has considerable expertise in the evaluation phase and knowledge of the intricacies involved. I would appeal to the plant breeder to plan ahead to assist in the evaluation phase. This applies to both public and private breeders. Many companies, for example, with the aid of a plot combine can harvest a corn yield trial in 30 seconds or less and have the data on yield weight and moisture instantaneously ready for the computer. The point is, a plant breeder's expertise in evaluation is impressive, based on theory and experience, and is valuable to the overall enterprise of genetic engineering of plants. The genotype by environment component so important in plant breeding will be no less important in genetic engineering of plants.

Are new or modified plant breeding procedures suggested by the emerging information on plant molecular biology? Frey stated that plant breeding has made little use of the last 30 years of DNA research. Recent information emerging on the molecular genetics of plants may have more direct implications for crop improvement.

The understanding of regulatory genes and repeated sequences is important to plant breeding. The powerful approach of plant breeding probably does not provide progress by the modification, recombination, and selection of structural genes but by the selection of regulatory sequences. The fact that the majority of the DNA in a cell is repeated surely means that plant breeding acts on repeated sequences to a large degree. How changes in gene multiplicity affect character expression and how breeding procedures could be modified to accommodate selecting repeated sequences need more attention by the breeder and quantitative geneticist. Experience with long-term selection experiments shows that complete genetic fixation is never achieved; one can practice reverse selection and make progress. This situation might be expected with multiply-repeated genes and the possibility of unequal crossing over or some form of control of gene multiplicity.

What does the new technology offer that can assist conventional plant breeding? The technology offers a certain amount of protection against genetic vulnerability. I would argue that research knowledge is our best protection against problems caused by genetic uniformity. Heterogeneity is not the solution in itself. The devastating diseases of the American chestnut and the elm showed that heterogeneity does not necessarily protect the species. Knowledge coming from genetic engineering research about genetic heterogeneity at the DNA level is substantive and will provide the basis for making wise decisions.

The technology may provide means of modifying the genetic behavior of the plant without going to new varieties. Gardner reminded us of the use of cross-protection in tomato greenhouses where plants infected with a symptomless mutant of TMV (tobacco mosaic virus) were cross-protected against virulent strains of TMV. He suggested that a virus vector might be used in crops like apples where one might make an orchard resistant by treating the individual trees. This approach may be better than planting a new resistant variety because of the number of years required for new trees to become productive. Another advantage of such an approach is that the breeder would not have to spend time breeding for disease resistance, for example, and could spend the time on improving yield components and potential. He pointed out that for the vector to be useful it cannot cause a reduction in yield. Actually, this stipulation probably applies to all plant genetic engineering work. Even an obviously desirable gene, like the maize opaque-2 gene which contributes higher lysine, will likely not be used if yield is reduced.

The ability to couple molecular cytogenetics with breeding was suggested by Appels as a powerful approach. The ability to detect repeated sequences of telomeric heterochromatin in rye x wheat hybrids may prove beneficial in selecting Triticale with better seed quality; this approach is under investigation today.

What limits progress in genetic improvement of crop productivity and quality today? Simmonds outlined several needs related to fundamental information. He pointed out that a finite gene pool can only yield a finite response; it was emphasized that we need to have access to a broad genetic base and practice germplasm enhancement. But it is also recognized that genetic engineering research will more thoroughly describe the variation inherent in our crop species at the DNA level as well as provide means for directed changes in base sequences for specific purposes. John Bedbrook noted that molecular biology provides a precise identification of the genetic material.

Recombination is another genetic phenomenon that Simmonds indicated often limits progress. He stated that 90% of the recombination that occurs in an inbreeding procedure following hybridization occurs in the first three generations. More information on recombination and ways to selectively enhance it are clearly needed. In many ways, recombination is the "name of the game" in breeding improved varieties.

In terms of selection procedures, Simmonds explained that we need more understanding of multiple character selection. Why associations exist between certain characters and how the breeder can make most efficient use of the associations are important questions. Molecular genetic studies on gene regulation and tissue specific expressions as discussed by Freeling should lead to some clues.

Mutants altered in amino acid balance, like those selected by embryo procedures as described by Miflin, may lead to useful improvements in crop quality. The identification and selection of plants with favorable mutations can be arduous in labor and numbers. Improved *in vitro* selection procedures and directed gene mutation or substitution could improve the efficiency of the process, but Miflin argued that the efficacy of such procedures was yet to be established.

What additional genetic traits are being derived from tissue culture that would improve a crop species? Larkin convincingly illustrated that there exists a wide array of genetic variations among regenerants from virtually all tissue culture methods. Some of the changes are for desirable attributes. But perhaps the most exciting aspect is that quantitatively inherited traits vary as well as qualitative traits. The tissue culture process may provide variation that cannot be obtained, or rarely obtained, by other methods. The genetic mechanism may be a rather unusual one involving substitutions of chromosomes or parts of chromosomes for one another--a cytogenetic behavior that does not normally occur in somatic tissues of the intact plant.

What new information on plant biochemistry, molecular biology, or physiology is emerging from genetic engineering research? While a great deal of new information is emerging from genetic engineering research, what is lacking perhaps is more important than what is emerging. Cashmore and many others suggested that the lack of knowledge on plant biochemistry represents a formidable constraint. One would assume that to engineer something, we would have to know what it is we are engineering. Defining important characteristics in biochemical terms is of utmost importance. We need to define various biochemical pathways in order to accomplish the molecular isolation, transfer, and expression of appropriate genes.

Miflin pointed out that having only one gene for each step in a plant biochemical pathway may be rare. Two, three or more isozymes, each encoded by a different gene, may be present for a single biochemical step. An additional complexity is that one enzyme of a pathway may be localized in an organelle and another outside the organelle. These complications make it even more important to analyze in detail biochemical pathways pertinent to crop quality and production.

What proportion of traits are under the control of a family of highly homologous gene sequences? Will transformation procedures need to be altered for traits displaying multigene family inheritance? Bedbrook speculated that most of the major products in a plant cell are encoded by a multigene family. He indicated that the level of expression of the various genes of a multigene family is not easy to assess; this presents a problem in knowing what to engineer. Cashmore, Larkins, and Messing talked about the existence of multigene families for RuBP carboxylase small subunits and corn and soybean storage proteins. Will modification of only one from a family of storage protein genes have a significant impact on grain quality? Perhaps not, but the research may provide probes to detect plants carrying particular combinations of genes that code for proteins superior in terms of specific amino acids.

What more has been learned about the relationship of gene structure and chromosome organization to expression? Larkins and Cashmore reviewed the existence of introns in various genes. Introns also were noted to be absent in certain genes, such as the zein genes. Bedbrook indicated that we do not understand the complex processing that occurs to reduce the primary transcript to the size of the mature message, or how important splicing will be in transformation procedures.

Several investigators suggested during the conference that proper expression of genes may be the greatest obstacle in plant genetic engineering. Hundreds or thousands of genes will be isolated in the next few years and the correlation of gene structure and function will become increasingly clear. Engineered genes will need to be designed to express at the proper time, frequency, level, and developmental stage in appropriate tissues. Interesting insights are being developed. For example, Cashmore showed that the small subunit of RuBP carboxylase has two intervening sequences. He found that one of these introns was totally conserved, which was a surprise since the region is not represented in the mature messenger RNA.

Freeling suggested that every gene is a differentially regulated gene. To understand the regulation of expression of a trait, one has to start with the gene as a unit and then work backwards into higher levels of coordination. Obtaining bona fide regulatory mutations may not be easy; for example, Freeling found that over 100 EMS-induced Adh (alcohol dehydrogenase) maize mutants could all be explained on the basis of structural gene alterations. Regulatory mutations are being produced by heavy neutrons and other means. These should be especially useful for studying gene regulation.

Levings described the relationship of cytoplasmic male sterility (cms) to the existence of plasmid-like DNAs in mitochondria. However, the simple presence of plasmid-like DNAs is not sufficient for male sterility since fully fertile South American races of maize as well as *Zea diploperennis* possess them. He suggested that, with greater understanding, we may be able to create better forms of cms. I believe the capability of converting corn lines to male sterility by the molecular transfer of plasmid-like DNAs would be highly desirable because private company breeders may spend up to half their program on introducing cms. If this process could be done by genetic engineering, the breeder would be free to expend more effort on increasing the yield potential of the crop. L. Evans emphasized that a lot of "maintenance effort" goes into plant breeding research, precluding full attention to raising yield potentials.

Larkins and others pointed out the need to understand the secondary structure of proteins. Changes one might make in a protein have to be in places that do not functionally alter the secondary structure. Messing summarized the zein protein structure and indicated which region was the most variable; this region may be the best candidate for making amino acid substitutions.

That repeated sequences in plants are under complex controls was emphasized by Appels. He indicated that there appears to be communication between non-allelic ribosomal DNA regions. The nature of such controls needs to be further investigated.

What is known about introns and signal sequences such as TATA and CAT boxes that will be important in transformation experiments? Holland mentioned that learning how to introduce a gene without having it interfere with the function of other genes will be a challenge. Information on signal sequences in the DNA will enable the construction of genes that will express properly in cells. The more we know about these sequences, the better our chances for success.

Several examples of consensus sequences were described. These included ones for binding of RNA polymerase and of ribosomes, for termination of transcription and translation, and for polyadenylation. Messing presented evidence for a new signal sequence called the AGGA box that might be a plant gene sequence for modulating transcription.

The isolation of sequences from tobacco that can function as promoters in yeast was described by Malmberg. Again, knowledge of the structure and function of such sequences is essential in order to modulate expression of plant genes introduced into recipient cells.

Depicker presented considerable evidence on sequences related to the junctions between plant DNA and T-DNA from the *Agrobacterium* Ti plasmid. Variation in the border DNA exists suggesting incorporation can occur at various genomic locations.

Are new gene transfer vectors being developed, especially for use with monocots? Levings suggested that plasmid-like DNAs found in corn may be potential vectors for monocots. These elements have the capacity to move and insert into DNA and can cause phenotypic changes. The other intriguing aspect, in my opinion, is the possibility that they may transpose to the nucleus from the mitochondrion. Several of the revertants from male sterility to fertility were found to be due to a nuclear factor with each occurrence leading to a nuclear gene for fertility restoration mapping in a different chromosomal location. Does the plasmid-like DNA have the capability of moving from the cytoplasm to the nucleus in this monocot?

Nester pointed out that *Agrobacterium tumefaciens* infects dicots but suggested that perhaps the use of protoplasts from monocots may allow infection.

Viruses represent genetic material that can be introduced and expressed in plant cells. Gardner described aspects of viruses that would make them useful vectors.

Controlling elements of maize were mentioned by Döring as candidates for vectors in monocots. The transposable element might be useful because it can integrate into the chromosome and be stable. Starlinger and Döring and others have been isolating and characterizing *Ds* elements inserted at the sucrose synthase locus. Each *Ds* seems to be different, although homologies exist.

What progress is being made on physical (non-biological) means of introducing DNA? Most biological vectors suggested for use in transferring DNA suffer from limitations in the amount of DNA that can be packaged or otherwise transferred. Gardner said that the packaging limitation for viruses might be so severe that one cannot insert a full length gene and get expression. Physical means of transfer would appear to have the potential of transferring larger blocks of DNA. The molecular and cell biologists, I believe, have been offered the challenge of finding means of transferring the genetic determinants for complexly inherited traits. Can this only be done by vectorless means of transformation?

Fraley mentioned that one only has to look at the developments with yeast and mammalian cells to predict what lies ahead for plants. He showed how co-cultivation of petunia protoplasts and *Agrobacterium tumefaciens* cells could result in 10% of the protoplast colonies being transformed. With this high frequency one may not need a selectable marker to identify transformants. He also described advances in the use of liposomes for free DNA delivery.

Malmberg reviewed the advantages of being able to isolate and transfer whole chromosomes. This method would introduce large amounts of genetic information and may be more likely to transfer complex traits interesting to plant breeders. He presented evidence on the introduction of chromosomes to protoplasts; about 1% of the protoplasts appeared to contain a transferred chromosome. Bedbrook pointed out the potential usefulness of repeated sequences in detecting whole chromosome transfers.

Are there additional documented cases of transformation with plants? Nester reminded us that antibiotic resistance and a variety of other genes have been stably incorporated into plant cells but expression of these genes has not been demonstrated. The exception is the opine synthase gene from *Agrobacterium*. Depicker noted that tobacco plants were regenerated from tumors in tissue culture and that one in 250 contained opine synthase. When self fertilized, this plant gave seed progeny 75% of which contained octopine, suggesting that transmission was as a Mendelian dominant trait.

Are methods being developed to identify gene sequences associated with environmental stresses? Although the entire genome of a species presumably can be cloned via recombinant DNA techniques, associating specific DNA sequences with specific traits remains difficult. Part of the difficulty is the lack of demonstrated associations of biochemical entities with plant performance. For example, Valentine pointed out that we need to develop proof, for plants, that certain molecules are important in such traits as drought tolerance as well as tolerance to other environmental or biological stresses. He reviewed examples showing that whole-plant physiological changes are not necessarily desirable.

Possibilities exist for identifying specific sequences with stress preventive characteristics. Plasmids with proline-generating genes that confer tolerance to high levels of osmotic stress were described by Valentine. Wyn Jones reported on a high proline mutant of barley. He indicated that proline was not elevated any higher in the mutant than in wild-type strains when under salt stress. Thus, there was no stress advantage of this high proline mutant. Miflin reported a similar finding. Wyn Jones said there are indications that betaine may be worth selecting, but it is not proven.

Have sequences controlling disease resistance been identified among the various cDNA or genomic libraries? Disease resistance would appear to be a logical trait of choice for transfer by genetic engineering methods. Single gene resistance is available, although polygenic resistance often may be desirable. It is surprising that more advances have not been made in this area. The work described by Yoder on the Southern corn leaf blight fungus and the attempts to set up a transformation system for this pathogen represent steps in the right direction. The work on cloning the altered mitochondrial DNA fragment in fertile, toxin-resistant regenerants of T-cms maize, if successful, may lead to the isolation of a cytoplasmic gene involved in a disease reaction.

Duvick mentioned that plant breeding priorities often are (1) high yield with good quality, (2) stability of yield, and (3) durable resistance to diseases and pests. With this high priority on disease resistance, why isn't more effort going into identifying and isolating disease resistance genes? Day described several of the difficulties as well as possible approaches. The study of disease lesion mimics as described by Walbot may lead to the identification of some important plant genes involved in the disease reaction.

What additional information is available on genetic versus non-genetic changes in cell and tissue cultures? Chaleff reminded us that cultured plant cells have unique features and are not just like microbes. Selection for a novel phenotype in plant cells is often conducted at a level of differentiation distinct from that of expression of the phenotype. He suggested that certain traits lie beyond the somatic cell geneticist's reach. Sussex pointed out that the process of cell determination as documented with animal systems may be manifested in plants to the extent that not all cells will be capable of regeneration into plants.

The value of performing sexual crosses to distinguish epigenetic from genetic changes was emphasized, but Chaleff cautioned that certain genetic changes such as ones due to gene amplification may appear to be of an epigenetic origin because the changes may be unstable and not transmitted via sexual crosses. He pointed out that other genetic changes might not be detected because of shifts in isozyme proportions in the callus culture versus the plant. However, some notable successes of selecting in culture for whole-plant traits do exist. These include Chaleff's selection for picloram resistance. Another good example is Hibberd and Green's work where cultures resistant to lysine-plus-threonine gave rise to plants whose kernels possessed elevated threonine levels.

Does anther or microspore culture offer unique opportunities in addition to the production of haploid plants? The use of anther culture to produce haploid plants is in question from an agronomic point of view. Extensive evaluations of tobacco homozygous lines derived by anther culture indicate that on the average they are inferior to inbreeding-derived lines, although superior segregants from crosses may still occur. Another potential use of anther or microspore culture is for the selection of mutants since large numbers of haploid cells can be screened. However, Sunderland indicated that the use of microspore culture in deriving mutants had limitations because of the vast effort that would be required. He suggested that the major contribution of microspores in the future probably will be on the single cell level for the introduction of genetic material.

What is the nature of interspecific-hybrid derivatives recovered from tissue cultures of sexual or somatic hybrids? I ask this question because of the concern that straight somatic hybrids of divergent species will be of little immediate value. Derivatives of these hybrids where chromosome segments of one species have been exchanged with segments of the other probably will be of the most value. We were reminded by D. Evans that a number of somatic hybrids of tobacco species have been made, three of which could not be made sexually, Several other species can be used in forming somatic hybrids, including potato, carrot, petunia, and Brassica. Most of the somatic hybrids produced to date are sterile and also aneuploid.

I would suggest that one ultimate use of somatic hybrids will be fusions within the species to produce plants with different nuclear-cytoplasmic combinations. The idea of using cytoplasts without nuclei, as presented in the poster by Archer, Landgren, and Bonnett, is particularly appealing. Perhaps this procedure could allow transfer of cytoplasmic male sterility without the need for extensive backcrossing. Another potential outcome of cell fusion would be new combinations of organelles from the two parents, although for now we cannot be certain that new combinations persist. Also, the evidence for organellar recombination is not definitive since we know that mitochondrial DNA can be altered simply by passage through tissue culture. This phenomenon was revealed by the work on Texas male-sterile cytoplasm maize.

Culture of sexually-formed interspecific hybrids may yield some valuable derivatives. Chromosomal translocations, somatic recombination, and other forms of genetic exchange between chromosomes of the two species may occur in culture. Larkin reviewed several interspecific hybrids that are being cultured and derivatives studied at the present time.

What is the importance of the plant genetic engineering effort to continued progress in crop production capabilities? A technology assessment study of U.S. corn production was recently completed by Sundquist, Menz, and Neumeyer (Agricultural Experiment Station Bulletin 546-1982, University of Minnesota). Projections concerning the impact of various technologies on corn yields were derived from a variety of sources including a survey sent to plant breeders and geneticists and plant molecular and cell biologists. The results of this survey are summarized in the following table.

Projected Impacts of Various Technologies on Corn Yields for the Years 1981 to 2000

	Yield Increases (Bushels/Acre/Year)
Conventional Plant Breeding[1]	1.0 (1981 to 2000)
Additional Nitrogen	0.4 (1981) to 0.1 (1994) No impact 1995 to 2000
Production Management Technologies[2]	0.2 to 0.3 (1981 to 2000)
Emerging Biotechnologies[3]	No impact through 1988 0.1 (1989) to 1.7 (2000)

[1] Includes highly correlated factors such as plant density and moisture control.

[2] Includes aspects of plant nutrition, water management, pest control, and grain moisture management.

[3] Includes photosynthetic enhancement, plant growth regulators, genetic modification at the cellular level (cell and tissue culture, gene transfer), biological nitrogen fixation.

The plant breeding component of increased corn yields is expected to continue to have a major impact. L. Evans also indicated in this conference that no leveling off of the plant breeding contribution is envisioned. But it appears that if world population doubles in the next 40 years, as predicted, then even this impressive rate of increase through plant breeding will be insufficient. According to the study by Sundquist and coworkers, use of additional nitrogen will have little effect on increasing corn yields after the mid 1990s. Production technologies are expected to have a continued impact on increasing yields throughout the next twenty-year period. Emerging biotechnologies, including plant genetic engineering, are expected to commence contributing to corn yields around 1990 and have a major impact by 2000. Regardless of the accuracy of these predictions, the results indicate a high degree of optimism on the part of a broad array of plant scientists.

The tenor of the conference suggests we might expect more of a resurgence of plant genetics than a revolution in plant breeding. There is no question research on plant genetic engineering is worth the effort and obvious implications exist for crop improvement. A multidisciplinary team approach is needed. The basic plant science information that is forthcoming is extensive and useful, but more research is needed on plant biochemistry and molecular biology. The prognosis for success in meeting the world's ever-expanding food needs depends, in part, on basic research in agriculture and enhanced genetic manipulation capabilities. We cannot afford not to pursue every available avenue to improve food production.

GENETIC ENGINEERING IN PLANTS
An Agricultural Perspective

Roundtable Discussion on Research Priorities

Panelists:

Co-Chairmen

Calvin O. Qualset - University of California, Davis

Alexander Hollaender - Council for Research Planning in Biological Sciences

Mary Clutter - National Science Foundation

Donald N. Duvick - Pioneer Hi-Bred International, Inc.

J. Eugene Fox - ARCO Plant Cell Research Institute

Ramon L. Garcia - Allied Corporation

Ernest G. Jaworski - Monsanto Company

Robert H. Lawrence, Jr. - Agrigenetics Corporation

Alexander Hollaender

I am Alex Hollaender, a physical chemist who became interested in Biology a long time ago and who is now involved in seeing that the results of basic studies are applied to practical problems.

Research priorities are an especially important consideration in light of recent budget cuts and difficulties in getting financial support. It is increasingly necessary for us to decide upon essential topics for research that are most important to be developed to thereby promote the field of genetic engineering of plants.

In discussing the problems of genetic engineering, we must remember that we are "cashing in" on 50 to 70 years of pioneering plant breeding work. For example, in the research of genetics of maize, I remember discussing early problems with E.M. East and later with R.A. Emerson, L.J. Stadler, and D.F. Jones. There were other, equally important contributors to the development of hybrid corn which made the plant breeding field more economically and practically significant and which helped our country's agriculture to prosper. I see their early work as a basis for the "green revolution" which has been particularly consequential for less developed nations to become independent food producers.

This informal presentation will highlight the respective research areas of our panel members who will offer their interests and priorities in the development of this field. The purpose of this Roundtable is to suggest directions for future research in the area of genetic engineering of plants.

Mary E. Clutter

I must say I am delighted to remark about genetic engineering of plants. And, I would add, isn't it wonderful to be a plant biologist alive and working today? I have been waiting for this revolution for more years than I care to admit, and I believe that plant biology in today's climate is more exciting and filled with more potential for making major advances than almost any other area of science.

Having said that I will identify myself as a representative from the National Science Foundation (NSF). For my statement about research priorities I will confine my remarks to plant biology supported by NSF because that is what I know best.

Some people may not know that NSF is an agency of the United States government. It has been confused with the National Academy of Sciences which is a private organization. NSF is a Federal agency that has as its charge to maintain the health and vitality of science across the broad spectrum of all the sciences. That is our mission, just as the Department of Agriculture's (USDA) mission is to support agricultural research and the National Institutes of Health's (NIH) mission is to support health-related research. Thus, NSF, which is a very small agency, relatively speaking, has an awesome responsibility. What we tend to do is to confine the research we support to very fundamental questions. Most of you who are familiar with NSF know that the research we support is investigator-initiated. That means we do not set research priorities. We are reflective of the priorities set by the scientific community itself through the peer review process.

I thought it might help if I showed you trends in NSF's support of plant biology over the last five years. Since we attempt to support what is sometimes called "the cutting edge" the trends should illustrate your own research priorities. In Figure 1 there are three pies for 1977, 1979, and 1981, respectively. For the purposes of this meeting the most relevant is the slice labeled GM for genetic manipulation. We define genetic manipulation as research using unconventional methods for transferring genetic information. Thus, studies on vectors, gene-splicing, or protoplast fusion would be included. Classical genetic studies are not included in the slice, but are represented in the "Other" category.

In 1977, 2% of the $21.1 million spent by NSF on plant biology went to genetic manipulation studies. Six percent was invested in stress research, both biological and physical; fourteen percent in nitrogen fixation and 16% in photosynthesis. The largest slice, 61%, was invested in all other plant research. By 1979, the genetic manipulation slice had increased to 7% of $26.1 million, and by fiscal year 1981, the most recent and complete data available shows that the slice had increased to 13% of NSF's expenditures for plant research, $33.0 million. This enormous increase is readily understandable to those who have witnessed and are witnessing the remarkable progress in this area of science. This is the point I want to make, that NSF's investments reflect your interests; that advances in science establish priorities for funding.

What I heard coming from this meeting gave me some idea of what our research priorities will be over the next few years. Coming through loud and clear was that we should be supporting more biochemistry, elucidating metabolic pathways, etc. If that message is correct, we should begin to see really excellent proposals in those areas and we will provide support for them. The other message I heard coming through very strongly was that we need to be putting more money into interdisciplinary research. Again, it is up to you to send us good proposals. Currently at NSF we are placing an emphasis on support for the plant sciences. Therefore, we intend to do our part and provide support for strong research.

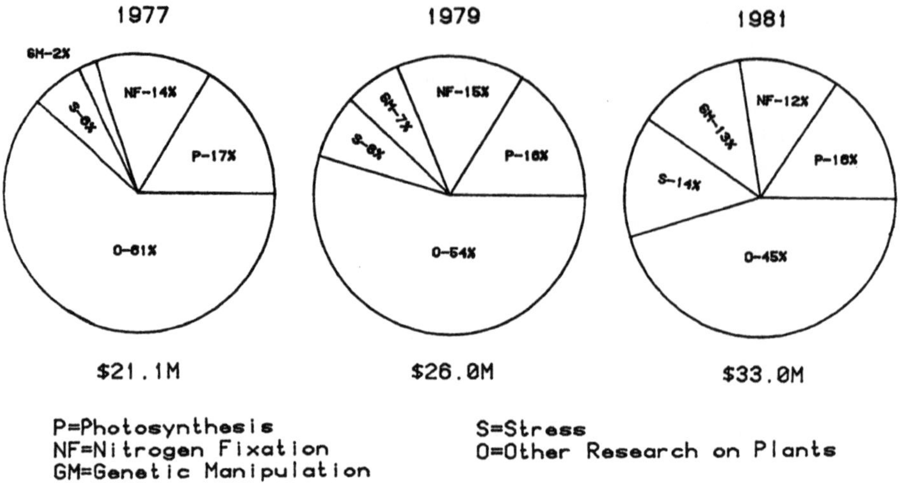

Fig. 1. National Science Foundation Research Awards in Plant Sciences (dollars in millions).

I'll make just two final observations. I won't call them priorities; rather I'll call them concerns. First, one of our concerns is about communication among scientists. I have not seen evidence yet, but people across the Federal agencies are concerned that the free flow of information within labs and among labs may be inhibited to some degree by current commercial interest in your research. Thus, one of our priorities or concerns is to encourage the free flow of information.

Secondly, we are very concerned about the training of the next generation of plant scientists. We see shortages in a number of areas that worry us. Therefore, to begin to address this problem, NSF has decided to invest a small amount of money in this next fiscal year, starting in October, in a post doctoral program for plant biology. There will be announcements in Science, in your Grants Office, and in other places about this program. The details are still being worked out. I'd be happy to answer questions about it.

In summary, your concerns are our concerns; your priorities are our priorities. We try to be responsive and to reflect your interests.

Robert H. Lawrence, Jr.

Rather than draw out a long list of research priorities as I see them from the perspective of a seed company, I would prefer to use this brief introductory presentation to make a few comments which might help guide our discussions later. In an attempt to define and rank research priorities for the application of genetic engineering of plants, it is important to recognize that the most obvious commercial application is in plant breeding and thus directly related to the development and release of new crop varieties and hybrids. From this perspective I would like to suggest that research priorities be considered: (a) within the context of the rapidly evolving structure of private and public involvement in plant breeding and genetics research (especially with respect to technical capabilities as well as amounts and sources of funding), and (b) within the framework of the various specific crop breeding strategies that are either currently in use or which have potential.

EVOLVING STRUCTURE OF PUBLIC AND PRIVATE RESEARCH

Figure 1 shows the flow of activities from fundamental research to the sale of seeds in a fully integrated seed company. This diagramatically represents the major components involved in the commercial development and utilization of plant breeding and may include many of the genetic technologies discussed in this symposium which are contributing to the foundation from which modern plant breeding will grow.

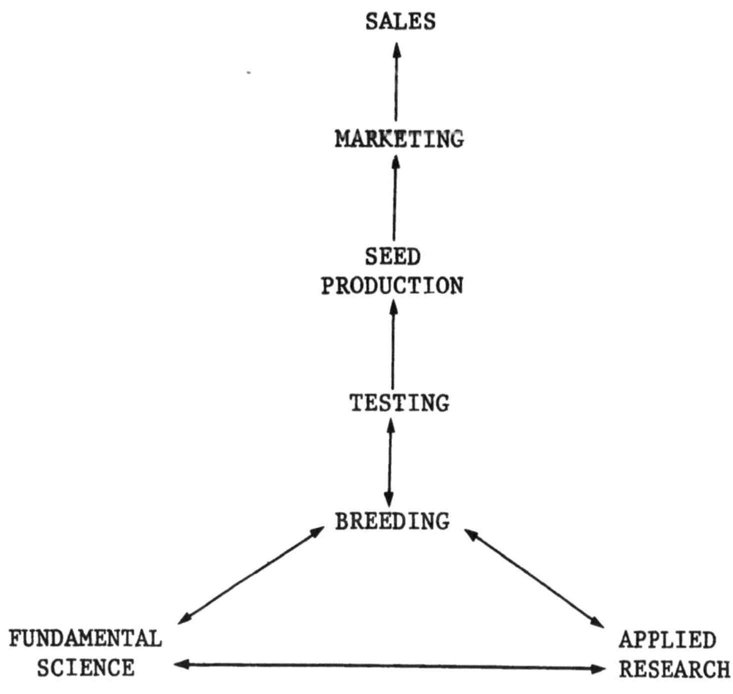

Fig. 1. Activities of a fully integrated seed company.

Traditionally, a large portion of seed companies relied on public development and release of new varieties and hybrids which are then further tested and improved before entering into seed production, marketing, and sales. A smaller number of seed companies developed in-depth plant breeding programs. Very few of these, however, were involved in basic genetics research which was generally assumed to be the domain of academic science and government research. Recently, however, this structure has begun to change dramatically primarily as a result of: (a) decreased availability (on a real dollar basis) of State and Federal funds for agricultural research, and (b) the increased availability of private funds for investment in the development and application of the so called "new genetic technologies". The latter spawned the biotechnology boom in plant genetics which resulted in an "alphabet soup" of technology-driven companies from Advanced Genetic Sciences (AGS) to Zoecon. This boom also stimulated a number of established seed companies to become more completely integrated from the top down (see Fig. 1) by establishing firm positions in both fundamental and applied genetics research (e.g., Agrigenetics, Asgrow/Upjohn, DeKalb/Pfizer, Dessert/ARCO, Pioneer Hi-Bred, etc.).

The subsequent mobilization of talented research scientists into private laboratories, coupled with the excellent facilities that a strong capital position furnished, has resulted in a shift in the balance of power in plant science research capabilities. Perhaps academia has experienced less of a decrease in research power than industry efforts and their depth positioned many of these private labs to make significant contributions in cellular and molecular genetics and, by nature of their commercial interest, removed some of the research partitioning (basic vs. applied) common to some academic efforts (although this, too, is rapidly changing). I feel that this evolving structure of research in the plant sciences can be most beneficial for all involved if we can successfully develop well balanced collaborative efforts between industry and academia. At the same time, we must do our best to structure these interactions in a manner that ensures the free interchange and development of the science and the scientist! As many of us are finding out, this is easier said than done.

One sure method of establishing collaborations is to fund these joint research efforts from the private sector. Investment capital in the last two years has become increasingly available to plant biotechnology areas through various mechanisms, such as private placements and public offerings. In several cases this resulted in substantial funding of university research. Although some of the problems inherent in funding from these sources have yet to be completely resolved (patents, ownership of germplasm, conflicts of interest, publication rights, etc.) it is encouraging that positive steps are being taken. Indeed, very powerful research collaborations are currently being established that, in my opinion, will be highly productive. Recently, numerous articles have directly addressed these issues (1-7).

RESEARCH PRIORITIES WITHIN CONTEXT OF BREEDING STRATEGIES

Another point that I think warrants consideration is that of research priorities in the application of genetic engineering of plants. To be realistically evaluated, these must be considered within the context of specific breeding strategies for crops of interest. Before I go further, I would like to recommend to my fellow cellular and molecular biologists a very informative and clearly written text entitled "Principles of Crop Improvement" by Professor N.W. Simmonds (this volume and ref. 8). This and similar texts on plant breeding (9-11) reveal the sometimes complex nature of breeding strategies for various crops and should help develop an appreciation for the assorted problems facing the breeder striving to make those small steps forward in genetic improvement. As Professor Simmonds expressed in his book, "Successful varieties are merely less imperfect than their predecessors." Crop improvement is a slow, tedious process involving operations which not only consume great amounts of time, such as, backcrossing and testing, but some which are simply very difficult to achieve, for example, tapping into wild gene pools or inbreeding naturally outcrossing crops. This is exactly where many of the cellular and molecular genetic technologies will have their greatest impact -- embedded within a breeding strategy, at various specific steps, assisting the breeder in making the difficult or "impossible" more readily achievable and compressing the time frame for achievement to a more manageable level.

Numerous considerations enter into the normal development of a well integrated breeding strategy, e.g., an understanding of the mating and population systems for the crop, a thorough assessment of the germplasm available and how to efficiently develop access to it, and the establishment of well targeted breeding objectives. In addition there is a wide assortment of breeding or genetic tools, which I believe consists of many of the technologies discussed in the symposium, that can be applied in various combinations to develop highly successful breeding programs. It is not essential that we get the big genetic "hit or home run" to be successful. As Agrigenetics Chairman, David Padwa, is fond of saying, there will be a few "bunts" and "singles", and perhaps a "double" or "triple" -- there are many ways to get on the score board.

For the sake of simplification, let us go back to the first diagram of the symposium shown in Professor Simmonds' talk. This generalized flow diagram consisted of the common elements of most breeding strategies (8): a) means of achieving flow of new and diverse germplasm into the program, b) generation of new gene combinations, c) identification of best combinations, d) exploitation for variety development, and e) trials and seed production. By reviewing the papers presented in this volume, the following list of technical capabilities can be generated (albeit an incomplete list) which have application at several of the steps:

- Somatic hybridization to generate new hybrids or cybrids
- Direct nuclear substitution into alien cytoplasms
- Somatic cell genetics, e.g., mutation/selection and cell level screening
- Generation of somaclonal variation
- Genotype cloning
- Germplasm preservation
- Gamete (haploid based) breeding schemes
- Gene isolation, modification, transfer, and expression
- Molecular - site directed mutagenesis, _in situ_
- Numerous others as reflected in various chapters

In addition, an area not covered in this volume which may prove valuable in developing highly specific genetic selection methods would be diagnostic biochemistry techniques involving monoclonal antibodies, molecular probes, and sophisticated methods ^f isozyme and protein analysis. Is there little wonder that the plant breeders of the future, in their role as the "central strategist" for crop improvement, need to become as intimately familiar with these technologies as with the market demands they must satisfy?

The point of all this is that by considering research within the context of breeding strategies for a particular crop it becomes possible to estimate the value of the potential applications, to assess the overall technical feasibility of these applications, and to evaluate the research requirements to bring the technology into actual use. All of these factors are important in the development and understanding of research priorities in this particular field.

REFERENCES

1. David, E. 1979. Science futures: The industrial connection. _Science_ 203(4383):837-840.

2. Edsall, J. 1981. Two aspects of scientific responsibility. Science 212(4490):11-14.
3. Heylin, M. 1981. Are we doing it right? Chem. & Engr. News, April 27:3.
4. Kiefer, D. 1980. Forging new and stronger links between university and industrial scientists. Chem. & Engr. News, Dec. 8:38-51.
5. Lepkowski, W. 1982. Balance in innovation involvement being sought. Chem. & Engr. News, May 24:9-14.
6. Meyer, J. 1979. Who should support agricultural research? BioSci. 29(3):143.
7. Ruttan, V. 1982. Changing role of public and private sectors in agricultural research. Science 216:23-29.
8. Simmonds, N.W. 1979. Principles of Crop Improvement. Longman, London and New York.
9. Allard, R.W. 1961. Principles of Plant Breeding. Wiley, New York and London.
10. Poehlman, J.M. 1978. Breeding Field Crops. 2nd printing, AVI, Westport.
11. Williams, W. 1964. Genetical Principles and Plant Breeding. Blackwell, Oxford.

J. Eugene Fox

Let me give you my "take-home" message straight away. Clausewitz is supposed to have said that, "War is too important to be left to the generals," and the message relevant to this discussion is "Research priorities are too important to be left in the hands of research directors and other kinds of management types."

The ARCO Plant Cell Research Institute (PCRI) is a collection of more than 50 scientists representing a variety of disciplines. We are engaged in a combination of some long-term fundamental research and some short-term, more applied projects. About 80% of what we do is the former type: basic and long-term. We will publish and talk about what we do to a large extent; we also have a post-doctoral program. So, rather than adopting a global view of research priorities, I would like to describe my perception of the way we have established research priorities within our own group.

I have written down four factors which influence the establishment of our priorities and they are: (i) the state of the art; (ii) the needs of the company; (iii) the needs of society at large; and (iv) the expertise of the research staff. Now, the first one, the state of the art, is rather obvious. The state of development in certain areas dictates priorities which, I think, are probably obvious to every reader. If we all generated a list, I suspect each wouldn't differ very much from the other. Thus, at our laboratory we have an intensive on-going effort in vector development based on the Ti plasmid of Agrobacterium as does nearly every other group in this field. We have considerable work of a long-term nature, dealing with regulation of gene expression, an area of concern so obvious to anyone who is the least bit acquainted with genetic systems that it really doesn't need any further justification here.

The establishment of research priorities obviously involves some judgment and some of ours might not be quite so obvious to everyone. For example, we have established as an institute-wide theme the molecular biology of stress in its broadest sense (that is, response to environment as well as response to disease-causing organisms) because, in our judgment, the state of the art in this area is just now getting to the point where it is "ripe" for exploitation, both scientifically and commercially. At the same time, we have decided against putting any effort into the field of nitrogen fixation. And that may be controversial, but at least in our opinion, that field has limited theoretical implications and almost no commercial potential in the coming decade. For many of the same reasons, we are not jumping onto the mycorrhizal "band-wagon" because we don't believe that organisms introduced into the soil have much chance against the "tough guys" already there which have survived selection against pretty formidable odds. At the same time, should scientific development substantially change the picture in those fields or other fields, in the future we could well alter our priorities. I think what I am saying is that we need to keep abreast of the state of the art.

Obviously another factor that influences all of us is the needs of the company which we represent. The seed companies associated with Atlantic Richfield have a shopping list of problems or areas requiring some scientific attention, just as do all seed companies and, clearly, the needs of the seed companies influence us heavily in the formulation of our research priorities. Indeed the most appealing problems which we undertake at PCRI are those which address both a short-term need of the seed companies and have at the same time a potential for a long-term, broader scientific impact or perhaps which can serve as a model system for other research studies. To those who know about the activities of the Dessert Seed Company, for example, it should come as no surprise that we have greatly interested ourselves in research problems involving the onion plant, triticale, and such.

The third factor is the needs of society at large. There are those of us at ARCO who believe strongly that scientific groups such as ours ought to be responsive to the needs of society. Our perception of such needs is an important factor in setting our research priorities, even though the research projects themselves may have no obvious commercial potential, or even any real scientific interest apart from the particular system being studied. For this reason, we've tackled a few research problems which are designed to bring about an improvement in the nutritional quality of seed proteins in important crop plants and we anticipate that we'll have an ongoing effort; some fraction of our effort will continue to be dictated by that particular research priority.

The last factor is equally as obvious: the expertise of the research staff. Clearly, the sorts of research problems that any organization undertakes are dictated to some very large extent by the particular talents of the people who are assembled. Although the reasoning here may perhaps be a little circular since we hire scientific staff with particular research goals in mind, I suspect that most research directors conclude that morale and productivity are related in a direct way to the extent to which people are doing the kinds of things they have been trained to do and in which they have some experience. And, as I pointed out earlier, it seems obvious to me, but apparently it is not obvious to everyone, that research priorities are too important and too complex really to be left entirely in the hands of research directors or management. Priorities, to some very large extent, ought to be generated from within rather than from the top down.

Ramon L. Garcia

There is no question that modern genetic engineering offers a very powerful tool for the improvement of crop productivity. It is evident to me that, if we are to utilize this new technology, it is imperative to identify and understand those genetic traits that will permit us to reach that goal.

As indicated in the justification and philosophy for these proceedings, to accomplish this objective will require the integration of many disciplines within the plant and biological sciences. In several occasions history tells us that scientists from different disciplines do not collaborate closely, but if we are successful in gaining this integration, and this volume is an indication of it, we are well on our way to being successful in reaching our objectives.

But, what are the research priorities for the application of Plant Genetic Engineering? I emphasize application because to me that indicates not only the successful genetic engineering of a plant, but the engineering of the ultimate end product - a plant with a gene or genes, or the derepression of a gene that is already present, that results in enhanced crop productivity. As a matter of fact this is the bottom line: we must enhance crop productivity, which can be measured as yield per hectare, better utilization of agricultural inputs, and improved quality.

This volume addresses many of the topics that would increase crop productivity, such as, disease resistance, photosynthetic efficiency, stress tolerance, increased nutritional value, and increased yield per acre. This list, of course, goes on and on. But what is a gene of choice to apply to genetic engineering of plants? We can spend a great deal of time speculating on which gene or characteristic to concentrate our efforts. Let me mention a couple of examples to illustrate my point and that also will allow me to lead into my basic recommendation on research priorities. Many researchers discuss plant resistance to certain pathogens which in some cases is conferred by a single gene. But has this approach been successful by conventional breeding? Not always. The pathogen or microorganism has a high capacity to mutate and adapt, thus in many cases overcoming the new resistance gene. Will it be any different for a new plant with an engineered resistance gene, even if this gene is from a source that could not, by conventional breeding, be placed into the population? We obviously don't know, but I feel that, given the versatility of fungi and bacteria a potential pathogen may be highly adaptable to the conditions of the engineered plant.

Other scientists talk of herbicide resistance. The potential is there, especially if resistance is confirmed by a single gene or a tightly linked gene complex. But for herbicide resistance, how many times will resistance be conferred by only a single gene?

The list of potential genes, of course, goes on. But for each one there seems to be a question. We cannot a priori guarantee that the gene or genes that are selected will result in increased crop productivity.

Thus, we feel, and I hope for the majority of you as well, that the first priority is to emphasize research in fundamental aspects of plant physiology and biochemistry as they relate to plant productivity. There must not only be research in gene location, delivery and transformation systems, and tissue culture, but an intensive integrated effort for the understanding of the processes that will result in increased plant productivity. I think that we are seeing a gap developing here between the molecular biology techniques and the understanding of the fundamental aspects of crop productivity. This might have an impact in the future in that there is a risk that the advances in these technologies might not totally interface with each other.

Basic studies are very much needed in areas such as biochemistry of nutrient assimilation by crops, remobilization of nutrients within the plant, developmental physiology, carbon metabolism and mobilization, senescence, host parasite interaction, environmental stress, and protein synthesis. A better understanding of these processes would lead, at a molecular level, to the identification and manipulation of genes that control them. This type of scientific endeavor offers, in my opinion, several unique advantages even though it might take longer for a closer collaboration among government, university, and industry.

Ernest G. Jaworski

This volume is composed of an impressive series of chapters many of which have underscored a variety of applications of genetic engineering to plants. Some are especially provocative.

My colleagues from Monsanto, Robert Fraley and Stephen Rogers, have made presentations here which illustrate the kinds of research we are interested in doing. I would like to provide an overview of issues that may not all be scientific but which I consider germane to the issue of research priorities. They can be broken into three segments:

Science - Basic plant science (generic needs)

Conservation - Germplasm and geography

Funding - Private sector and government

SCIENCE

It is apparent that the successful genetic engineering of plants must ultimately lead to the production of novel genotypes with new or improved agronomic characteristics - both quantitative (yield) and qualitative (nutrition). So much has been written on this subject in this volume and elsewhere that it seems redundant to repeat here what is all too obvious. Let me deal instead with a few of the broader issues that strike me as being important.

Whatever "YFG" we select ("YFG" being Mary-Dell Chilton's acronym for "Your Favorite Gene") that we feel will be beneficial either for insertion or amplification in plants will require considerable understanding as to its spacial and temporal regulation and expression. Thus, a high priority should be given to developing an understanding of the gene regulatory sequences at both the 5' upstream and 3' downstream regions from the structural gene that control when a gene is turned on, where (tissue/cell type) a gene is turned on, how much it is turned on, and for how long. If we are to rationally design or engineer plants, this type of knowledge would seem to be pivotal.

A second issue relates to gene organization. Once again we need much more information on how genes are organized, transported, and integrated in eucaryotic systems. Rational engineering, if we are to solve real agricultural productivity problems, demands a deeper knowledge base for the genetic engineer to build-in those properties around an important gene to ensure its integration into appropriate chromosomal regions in a stable and expressible form (cassette concept).

Finally, a third scientific issue recognized by those attempting to identify their favorite genes or those plant breeders charged with today's responsibility for developing improved crops is the paucity of plant biochemical data available.

CONSERVATION

Turning to resources, I would like to highlight the need for germplasm preservation and exploration -- especially in the tropical areas of the world. This, in my view, is important because such germplasm will serve us well as a reservoir of genetic sequences unlikely to ever be mimicked by man, even with his most sophisticated of technologies.

It is estimated that some three out of every five people in the world will be living in the tropics by the end of this century. Thus anyone interested in the future of agriculture and the feeding of a growing human population should have an interest in the biology of the tropics. While historically it is apparent that many plants, animals, and microbes have had great economic value, there is a very real possibility that these organisms and their germplasms may well disappear from the world before we have an opportunity to either study or take advantage of them.

With the large unexplored biological diversity that exists in the tropics, there may well be germplasms of interest to scientists working toward the applications of genetic engineering to improvements in plant and animal productivity. As an example, wild varieties of domesticated plants represent a valuable pool of genetic diversity from which resistance to diseases, insects, and stress may be recovered both for traditional genetic applications as well as to provide novel allelic genes for genetic engineering.

It is possible that these gene pools may well be more useful due to their evolution than any variants that can be created by modern man.

FUNDING

I would like to conclude with a final priority that deals with research support. Research priorities can certainly help focus upon relevant areas that have clear recognizable economic and societal impact. These are the criteria espoused by the Office of Technology Assessment and certainly the plant sciences fit into this category.

Mary Clutter (National Science Foundation) and I had the opportunity this year of speaking at a common forum at the University of Missouri regarding the funding of science, especially plant science, in the United States. In this context I reported that a variety of industries seemed supportive of attempting to provide resources for basic plant science and the development of scientists. As examples, I highlighted some of the programs that companies like Monsanto have been supporting including the Rockefeller University program, the Washington University program, and numerous post--doctoral fellowships including those that are now being sponsored in-house at Monsanto. These are important and perhaps essential contributions to the area but they are limited in their impact. It is still going to be incumbent upon the Federal government to supply the "lion's share" of support through granting agencies such as the National Institutes of Health, the National Science Foundation, and the U.S. Department of Agriculture Competitive Grants Program. There is no way that the private sector can adequately support the needed efforts for society in the scientific areas of plant biology. It is, therefore, necessary for all of us to take stronger, bolder positions and deal more effectively with our congressional leaders in Washington who, in fact, are the main route for effecting useful legislation to help augment the needs of society. I can only urge you to work through your professional societies and your State and Federal representatives to educate, to effect change, and to insure adequate support for the urgently needed development of basic plant science.

SUMMARY

There is a need for supporting basic plant science at a much higher level in this country and there is a need for doing something about the preservation and study of germplasm in the tropics. There is also a critical need for generating more knowledge in plant biochemisty. It is painfully clear to anyone attempting to conduct advanced plant breeding and to apply genetic engineering technology to genetic problems that we are desperately lacking fundamental knowledge about the biochemical mechanisms which exist in plant systems. It is almost impossible to approach the problems we are interested in solving without developing a better and more knowledgeable undergirding of plant biochemistry. Therefore, I believe it is necessary for all of us to do as much as we can in an organized and focused manner to convince our government agencies that the future of the United States and that of the world, in general, will be influenced by the ability of our country to contribute to the development of agriculture on a world-wide basis.

Donald N. Duvick

Some research priorities for plant breeders include: 1) higher yielding varieties with desired levels of quality, 2) varieties with more stability of yield, and 3) varieties with higher levels of durable resistance to disease, insect, and nematode pests. Other priorities, further down in the list, are varieties with greater salt tolerance to nutrient imbalances, and new kinds of genetic variability. All of these desirable traits are governed by large numbers of genes of unknown location with unknown primary products and with unknown interactions. Empirical techniques have long been in hand, nevertheless, that efficiently manipulate these unknown genes and effectively provide improvements in almost all desired traits. During the past 50 years these empirical techniques have been put on a sound scientific footing and have become so easy in application and productive in results that continually increasing numbers of plant breeding organizations, both public and private, have appeared. Yield gains due to genetics in all major U.S. crops have been substantial and continuous during the past 50 years (about 1% yr). They show no sign of reaching a plateau.

Two major new technologies have been added to the plant breeder's tool kit in the past 20 years: 1) mechanization of field plot trials has greatly increased output per worker and at the same time has greatly increased uniformity and precision of the trials, and 2) computer-assisted summarization and analysis of the data have allowed breeders to use and understand much larger volumes of data than previously was possible. Thus, the volume of effort has increased greatly, allowing breeders to increase selection differentials. Plant breeding today depends on probabilities, rather than on powerful genetic analyses and predictions, to give the desired results.

In general, detailed genetic physiological investigations have not been useful to the practical plant breeder because of the large gap between basic scientific knowledge and plant variety performance. Perhaps for this reason basic genetic and physiological research on plants has not been funded to the extent that it has been in medical and animal research and, in turn, this may be why fewer brilliant advances have been recorded in basic plant science than in other branches of biology. Another research priority - maybe the most important one for plant breeders - is to close the gap between basic and applied knowledge about our crop plants.

Genetic engineering now comes on the scene promising, according to some of its practitioners, to give us nitrogen-fixing maize and wheat, salt, and pest-tolerant crops, instant homozygous lines, and plants with built-in herbicides. It also says it can give us greatly improved nitrogen-fixing bacteria, bacteria capable of clearing the environment of unwanted left-over herbicides or insecticides, and microbial insecticides, and finally it promises to increase our knowledge of gene structure, function and regulation, and the molecular mechanisms important in plant productivity. Will this new group of interacting technologies - genetic engineering - become the next important addition to the plant breeder's tool kit?

Sometimes I think I am hearing a new version of the story about the simpleton whose technique for counting sheep was to count the legs and then divide by four. Many of the promises of genetic engineering are for results that already are achieved easily and economically with conventional plant breeding or with conventional crop ecology. Others are so visionary that no one who understands both plant breeding and genetic engineering expects them to be usefully applied within the next 50 years. Sometimes it seems that practitioners of the new technology describe the plant breeder's greatest needs as those that coincide with their particular specialty in genetic engineering. And sometimes I feel like the man, tarred and feathered and riding out of town on a rail, who was heard to say, "If it wasn't for the honor of the thing, I'd just as soon it was somebody else!"

But a large residue of potentially highly useful applications remains. I will list a few examples.

First, some of the technologies developed to do genetic engineering can be usefully applied to help in conventional plant breeding. We will have valuable new tools. For example, cloned DNA copies (cDNA), made from RNA viruses, can be used to identify these viruses in plant tissues, perhaps more precisely than with antisera and certainly more quickly than with grow-outs or cross-infections.

A second example is anther culture to develop homozygous lines. This could be very useful if it can be made to work easily in important crop plants and without introducing undesirable, aberrant genotypes.

A third example could be improvement in nitrogen-fixation systems of the rhizobia of alfalfa and soybeans. In soybeans such improvement could help push up the apparent yield "ceiling" for that crop. In alfalfa, nitrogen fixation and therefore the soil building attributes of the crop could be improved.

Fourth, and most important, the new technologies will increase our basic knowledge of the biology and genetics of our crop plants to a degree that was never before possible and force increases in biochemical and physiological knowledge. Such knowledge will help us to make better decisions as we use conventional plant breeding procedures. It should also help us think of ways to fill in the great gaps of knowledge that now exist between molecular and agronomic data. For we must admit that, today, very little of the impressive body of basic biological knowledge is or can be used in conventional plant breeding.

Calvin O. Qualset

It has been the goal of these proceedings to provide a forum for communication among scientists from the many fields that contribute to the stability and improvement of agricultural productivity. It is absolutely clear that research is needed at all levels of biological organization if the "new plant breeding" is to result ultimately in new plant products or improved methods in crop improvement. This volume obviously shows how crop improvement methods of the future intimately depend upon advances in biochemistry, plant physiology, soil science, plant pathology, and entomology as well as genetics, the disciplinary parent of plant breeding. I would like to make three comments which are intimately related to three of my roles in a public university.

First, as a researcher in plant genetics and a breeder of crop cultivars, I would like to emphasize the importance of integrating gene manipulation (insertion) techniques with the whole plant breeding process. The case for speed and efficiency in methods involving protoplast, cell, and tissue culture selection schemes is not well founded. It takes months or years to produce isolated cells or cultured tissues, to apply the mutation induction or gene insertion agent, to select desired progeny, to regenerate plants, to prove stability over time and during ontogenetic development in clonally propagated species or Mendelian inheritance of characters in sexually propagated species, to show that the character is agriculturally valuable, and to integrate it with all other necessary characters in a cultivar that will be usable in agriculture. This is especially true for perennial crops. Gene insertion schemes need to be developed that do not require manipulation of protoplasts. Thus, I would hope that whole-plant systems of plant molecular genetic engineering will be given more attention.

Second, as an administrator concerned with effective use of funds, the importance of individual creativity by scientists, and relevance of research to the public, I strongly urge that one or several model crop improvement systems that can be addressed by molecular plant genetics be identified and conducted to completion by an appropriate integrated research team. Individual contributions by scientists can be encouraged to meet the needs for genetic materials, biochemical and physiological characterizations, selection methodology, and finally for evaluation under user conditions. Such programs will require a rather large research fund investment, but could benefit from more efficient use of existing resources within most university research programs. I would hope that public and private funding organizations would take the opportunity to select a few such truly integrated research and development programs for support. The combined direct participation of both public and private research organizations should be explored. In my view, it is only when cellular-based plant improvement schemes are directly compared with conventional approaches that the problems and efficiencies of the various methods will be properly evaluated.

Third, as an educator involved in the training of plant geneticists and breeders for the future, I am greatly concerned about the ability of public institutions to provide the quality of training needed for a profession that is becoming even more complex. The recent recruitment of senior faculty members away from universities is one concern; a second is that the age structure of faculty in the rather specialized field of plant genetics and breeding shows a high proportion of faculty members approaching retirement and who will not make major programmatic changes in their research and teaching. We should make every effort to retain faculty positions in this area. In addition, greater consideration should be given to joint university and industry training programs.

Some years ago Dr. Paul Knowles in the Department of Agronomy and Range Science at U.C. Davis; Dr. Iver J. Johnson, then with Cal/West Seeds, Woodland, California; and Dr. David B. Ferguson, then with Northrup-King Seed Company, Woodland, California developed a program of industry-funded assistantships for graduate students (see Journal of Agronomic Education 2:89-90, 1973). In this program the students conduct thesis or dissertation research programs at the facilities of a seed company on a topic approved by the university professor and mutually interesting to the seed company. The assistantship funds are provided by the seed company to the university and the student receives support funds (stipend and fees) from the university. There are plenty of research topics that are of general interest and importance that do not impinge upon proprietary considerations. Even projects that do have such considerations can be accommodated with prior mutual agreement by all parties, including, of course, consideration for the best interests of the student involved. This program has been successfully implemented at Davis. Such a joint program is not unique to training in crop improvement research. I do believe, however, that because of the great expansion in private plant genetic and breeding activities that it is time to consider some formal arrangements for such programs. I suggest that industry-university training programs be considered at the Federal level by agencies concerned about strengthening education, research, and development in the United States. These programs could be structured similarly to National Institutes of Health training grants to universities and could be even more effective because of expanded training contributions by both university and industry scientists and could have greater impact because of cost-sharing.

INDEX

MIX
Papier aus verantwortungsvollen Quellen
Paper from responsible sources
FSC® C105338

If you have any concerns about our products,
you can contact us on
ProductSafety@springernature.com

In case Publisher is established outside the EU,
the EU authorized representative is:
Springer Nature Customer Service Center GmbH
Europaplatz 3, 69115 Heidelberg, Germany

Printed by Libri Plureos GmbH
in Hamburg, Germany